AF845228

Handbook of Peridynamic Modeling

Advances in Applied Mathematics

Series Editor: Daniel Zwillinger

Published Titles

Advanced Engineering Mathematics with MATLAB, Fourth Edition
Dean G. Duffy

CRC Standard Curves and Surfaces with Mathematica®, Third Edition
David H. von Seggern

Dynamical Systems for Biological Modeling: An Introduction
Fred Brauer and Christopher Kribs

Fast Solvers for Mesh-Based Computations *Maciej Paszyński*

Green's Functions with Applications, Second Edition *Dean G. Duffy*

Handbook of Peridynamic Modeling *Floriin Bobaru, John T. Foster, Philippe H. Geubelle, and Stewart A. Silling*

Introduction to Financial Mathematics *Kevin J. Hastings*

Linear and Integer Optimization: Theory and Practice, Third Edition
Gerard Sierksma and Yori Zwols

Markov Processes *James R. Kirkwood*

Pocket Book of Integrals and Mathematical Formulas, 5th Edition
Ronald J. Tallarida

Stochastic Partial Differential Equations, Second Edition *Pao-Liu Chow*

Advances in Applied Mathematics

Handbook of Peridynamic Modeling

Edited by

Florin Bobaru
John T. Foster
Philippe H. Geubelle
Stewart A. Silling

CRC Press
Taylor & Francis Group
Boca Raton London New York

CRC Press is an imprint of the
Taylor & Francis Group, an **informa** business
A CHAPMAN & HALL BOOK

First published in paperback 2024

First published 2017 by Chapman and Hall

Published 2019 by CRC Press
2385 NW Executive Center Drive, Suite 320, Boca Raton FL 33431

and by CRC Press
4 Park Square, Milton Park, Abingdon, Oxon, OX14 4RN

CRC Press is an imprint of Taylor & Francis Group, LLC

© 2017, 2019, 2024 Taylor & Francis Group, LLC

Reasonable efforts have been made to publish reliable data and information, but the author and publisher cannot assume responsibility for the validity of all materials or the consequences of their use. The authors and publishers have attempted to trace the copyright holders of all material reproduced in this publication and apologize to copyright holders if permission to publish in this form has not been obtained. If any copyright material has not been acknowledged please write and let us know so we may rectify in any future reprint.

Except as permitted under U.S. Copyright Law, no part of this book may be reprinted, reproduced, transmitted, or utilized in any form by any electronic, mechanical, or other means, now known or hereafter invented, including photocopying, microfilming, and recording, or in any information storage or retrieval system, without written permission from the publishers.

For permission to photocopy or use material electronically from this work, access www.copyright.com or contact the Copyright Clearance Center, Inc. (CCC), 222 Rosewood Drive, Danvers, MA 01923, 978-750-8400. For works that are not available on CCC please contact mpkbookspermissions@tandf.co.uk

Trademark notice: Product or corporate names may be trademarks or registered trademarks and are used only for identification and explanation without intent to infringe.

Publisher's Note
The publisher has gone to great lengths to ensure the quality of this reprint but points out that some imperfections in the original copies may be apparent.

ISBN: 978-1-4822-3043-7 (hbk)
ISBN: 978-1-03-291879-2 (pbk)
ISBN: 978-1-315-37333-1 (ebk)

DOI: 10.1201/9781315373331

Visit the Taylor & Francis Web site at
http://www.taylorandfrancis.com

and the CRC Press Web site at
http://www.crcpress.com

Contents

Foreword **xv**

Preface **xvii**

List of Figures **xix**

List of Tables **xxxv**

Contributors **xxxvii**

I The Need for Nonlocal Modeling and Introduction to Peridynamics **1**

1 Why Peridynamics? **3**

Stewart A. Silling

1.1 The mixed blessing of locality 3
1.2 Origins of nonlocality in a model 6
1.2.1 Long-range forces 7
1.2.2 Coarsening a fine-scale material system 10
1.2.3 Smoothing of a heterogeneous material system 12
1.3 Nonlocality at the macroscale 18
1.4 The mixed blessing of nonlocality 20
References 21

2 Introduction to Peridynamics **25**

Stewart A. Silling

2.1 Equilibrium in terms of integral equations 26
2.2 Material modeling 28
2.2.1 Bond-based materials 28
2.2.2 Relation between bond densities and flux 30
2.2.3 Peridynamic states 32
2.2.4 Ordinary state-based materials 35
2.2.5 Correspondence materials 37
2.2.6 Discrete particles as peridynamic bodies 39
2.2.7 Setting the horizon 40
2.2.8 Linearized peridynamics 41
2.3 Plasticity 42

2.3.1 Bond-based microplastic material 42
2.3.2 LPS material with plasticity 43
2.4 Damage and fracture 45
2.4.1 Damage in bond-based models 45
2.4.2 Damage in ordinary state-based material models 47
2.4.3 Damage in correspondence material models 49
2.4.4 Nucleation strain 50
2.5 Treatment of boundaries and interfaces 51
2.5.1 Bond-based materials 51
2.5.2 State-based materials 53
2.6 Emu numerical method 54
2.7 Conclusions 56
References 57

II Mathematics, Numerics, and Software Tools of Peridynamics 61

3 Nonlocal Calculus of Variations and Well-Posedness of Peridynamics 63

Qiang Du

3.1 Introduction 63
3.2 A brief review of well-posedness results 66
3.3 Nonlocal balance laws and nonlocal vector calculus 67
3.4 Nonlocal calculus of variations — an illustration 72
3.5 Nonlocal calculus of variations — further discussions 79
3.6 Summary 81
References 82

4 Local Limits and Asymptotically Compatible Discretizations 87

Qiang Du

4.1 Introduction 87
4.2 Local PDE limits of linear peridynamic models 90
4.3 Discretization schemes and discrete local limits 97
4.4 Asymptotically compatible schemes for peridynamics 100
4.5 Summary 103
References 104

5 Roadmap for Software Implementation 109

David Littlewood

5.1 Introduction 109
5.2 Evaluating the internal force density 112
5.3 Bond damage and failure 115
5.4 The tangent stiffness matrix 115
5.5 Modeling contact 119
5.6 Meshfree discretizations for peridynamics 122
5.7 Proximity search for identification of pairwise interactions 124

5.8 Time integration . . . 125
5.8.1 Explicit time integration for transient dynamics . . . 126
5.8.2 Estimating the maximum stable time step . . . 128
5.8.3 Implicit time integration for quasi-statics . . . 129
5.9 Example simulations . . . 131
5.9.1 Fragmentation of a brittle disk resulting from impact . . . 131
5.9.2 Quasi-static simulation of a tensile test . . . 133
5.10 Summary . . . 134
References . . . 136

III Material Models and Links to Atomistic Models 141

6 Constitutive Modeling in Peridynamics 143

John T. Foster

6.1 Introduction . . . 144
6.2 Kinematics, momentum conservation, and terminology . . . 146
6.3 Linear peridynamic isotropic solid . . . 150
6.3.1 Plane elasticity . . . 152
6.3.1.1 Plane stress . . . 153
6.3.1.2 Plane strain . . . 156
6.3.2 "Bond-based" theories as a special case . . . 156
6.3.3 On the role of the influence function . . . 158
6.3.4 Other elasticity theories . . . 159
6.4 Finite Deformations . . . 159
6.4.1 Invariants of peridynamic scalar-states . . . 162
6.5 Correspondence models . . . 163
6.5.1 Non-ordinary correspondence models for solid mechanics 164
6.5.2 Ordinary correspondence models for solid mechanics . . 165
6.6 Plasticity . . . 166
6.6.1 Yield surface and flow rule . . . 167
6.6.2 Loading/unloading and consistency . . . 167
6.6.3 Discussion . . . 168
6.7 Non-ordinary models . . . 169
6.7.1 A non-ordinary beam model . . . 169
6.7.2 A non-ordinary plate/shell model . . . 171
6.7.3 Other non-ordinary models . . . 171
6.8 Final Comments . . . 172
References . . . 173

7 Links between Peridynamic and Atomistic Models 179

Pablo Seleson and Michael L. Parks

7.1 Introduction . . . 179
7.2 Molecular dynamics . . . 181
7.3 A meshfree discretization of peridynamic models . . . 182
7.4 Upscaling molecular dynamics to peridynamics . . . 184

7.4.1 A one-dimensional nonlocal linear springs model * 184
7.4.2 A three-dimensional embedded-atom model † 189
7.5 Computational speedup through upscaling 194
7.6 Concluding remarks 196
References 197

8 Absorbing Boundary Conditions with Verification 201

Raymond A. Wildman and George A. Gazonas

8.1 Introduction 201
8.2 A PML for state-based peridynamics 202
8.2.1 Two-dimensional (2D), state-based peridynamics review . 202
8.2.2 Auxiliary field formulation and PML application 204
8.2.3 Numerical examples 210
8.3 Verification of cone and center crack problems 213
8.3.1 Dimensional analysis of Hertzian cone crack development in brittle elastic solids 213
8.3.2 State-based verification of a cone crack 216
8.3.3 Bond-based verification of a center crack 218
8.4 Verification of an axisymmetric indentation problem 219
8.4.1 Formulation 220
8.4.2 Analytical verification 221
References 222

IV Modeling Material Failure and Damage 227

9 Dynamic Brittle Fracture as an Upscaling of Unstable Mesoscopic Dynamics 229

Robert P. Lipton

9.1 Introduction 229
9.2 The macroscopic evolution of brittle fracture as a small horizon limit of mesoscopic dynamics 234
9.3 Dynamic instability and fracture initiation 238
9.4 Localization of dynamic instability in the small horizon-macroscopic limit 240
9.5 Free crack propagation in the small horizon-macroscopic limit . . 241
9.6 Summary 242
References 243

*First published in P. Seleson, M. L. Parks, M. Gunzburger, and R. B. Lehoucq. Peridynamics as an upscaling of molecular dynamics, *Multiscale Modeling and Simulation*, 8(1): 204–227, 2009. Copyright ©2009 Society for Industrial and Applied Mathematics. Reprinted with permission. All rights reserved.

†First published in P. Seleson, M. L. Parks, and M. Gunzburger. Peridynamic state-based models and the embedded-atom model, *Communications in Computational Physics*, 15(1): 179–205, 2014, published by Global Science Press.

10 Crack Branching in Dynamic Brittle Fracture **245**
Florin Bobaru and Guanfeng Zhang
10.1 Introduction . . . 246
10.2 A brief review of literature on crack branching . . . 249
10.2.1 Theoretical models and experimental results on dynamic brittle fracture and crack branching . . . 249
10.2.2 Computations of dynamic brittle fracture based on FEM . 251
10.2.3 Dynamic brittle fracture results based on atomistic modeling . . . 253
10.2.4 Dynamic brittle fracture based on particle and lattice-based methods . . . 253
10.2.5 Phase-field models in dynamic fracture . . . 254
10.2.6 Results on dynamic brittle fracture from peridynamic models . . . 255
10.3 Brief review of the bond-based peridynamic model . . . 256
10.4 An accurate and efficient quadrature scheme . . . 260
10.5 Peridynamic results for dynamic fracture and crack branching . . 262
10.5.1 Crack branching in soda-lime glass . . . 265
10.5.1.1 Load case 1: stress on boundaries . . . 265
10.5.1.2 Load case 2: stress on pre-crack surfaces . . . 269
10.5.1.3 Load case 3: velocity boundary conditions . . . 272
10.5.2 Crack branching in homalite . . . 275
10.5.2.1 Load case 1: stress on boundaries . . . 276
10.5.2.2 Load case 2: stress on pre-crack surfaces . . . 278
10.5.2.3 Load case 3: velocity boundary conditions . . . 282
10.5.3 Influence of sample geometry . . . 285
10.5.3.1 Load case 1: stress on boundaries . . . 285
10.5.3.2 Load case 2: stress on pre-crack surfaces . . . 287
10.5.3.3 Load case 3: velocity boundary conditions . . . 289
10.6 Discussion of crack branching results . . . 291
10.7 Why do cracks branch? . . . 296
10.8 The importance of nonlocal modeling in crack branching . . . 304
10.9 Conclusions . . . 307
References . . . 309

11 Relations between Peridynamic and Classical Cohesive Models **317**
Scot M. Breitenfeld, Philippe H. Geubelle, Olaf Weckner, and Stewart A. Silling
11.1 Introduction . . . 317
11.2 Analytical PD-based normal cohesive law . . . 319
11.2.1 Case 1 — No bonds have reached critical stretch . . . 322
11.2.2 Case 2 — Bonds have exceeded the critical stretch . . . 323
11.2.3 Numerical approximation of PD-based cohesive law . . . 325
11.3 PD-based tangential cohesive law . . . 329
11.3.1 Case 1 — No bonds have reached critical stretch . . . 331
11.3.2 Case 2 — Bonds have exceeded the critical stretch . . . 331

11.4 PD-based mixed-mode cohesive law 333
11.5 Conclusions 336
References 337

12 Peridynamic Modeling of Fiber-reinforced Composites 339
Erdogan Madenci and Erkan Oterkus
12.1 Introduction 339
12.2 Peridynamic analysis of a lamina 341
12.3 Peridynamic analysis of a laminate 347
12.4 Numerical results 349
12.5 Conclusions 354
12.6 Appendix A: PD material constants of a lamina 354
12.6.1 Simple shear 355
12.6.2 Uniaxial stretch in the fiber direction 356
12.6.3 Uniaxial stretch in the transverse direction 358
12.6.4 Biaxial stretch 359
12.7 Appendix B: Surface correction factors for a composite lamina . . 363
12.8 Appendix C: PD interlayer and shear bond constants of a laminate 368
12.9 Appendix D: Critical Stretch Values for Bond Constants 374
References 376

13 Peridynamic Modeling of Impact and Fragmentation 379
Florin Bobaru, Zhanping Xu, and Yenan Wang
13.1 Introduction 380
13.2 Convergence studies and damage models that influence the damage behavior 382
13.2.1 Damage-dependent critical bond strain 382
13.2.2 Critical bond strain dependence on compressive strains along other directions 383
13.2.3 Surface effect in impact problems 383
13.2.4 Convergence study for impact on a glass plate 384
13.3 Impact on a multilayered glass system 388
13.3.1 Model description 389
13.3.2 A comparison between FEM and peridynamics for the elastic response of a multilayered system to impact 391
13.4 Computational results for damage progression in the seven-layer glass system 393
13.4.1 Damage evolution for the cross section 394
13.4.2 Damage evolution in the first layer 396
13.4.3 Damage evolution in the second layer 398
13.4.4 Damage evolution in the fourth layer 399
13.4.5 Damage evolution in the seventh layer 400
13.5 Conclusions 402
References 403

V Multiphysics and Multiscale Modeling 405

14 Coupling Local and Nonlocal Models 407

Yan Azdoud, Fei Han, David J. Littlewood, Gilles Lubineau, and Pablo Seleson

14.1 Introduction 408
14.2 Energy-based blending schemes 410
14.2.1 The Arlequin method 410
14.2.1.1 Description of the coupling model 410
14.2.1.2 A numerical example 413
14.2.2 The morphing method 415
14.2.2.1 Overview 415
14.2.2.2 Description of the morphing method 417
14.2.2.3 One-dimensional analysis of ghost forces . . . 419
14.2.2.4 Numerical examples 419
14.3 Force-based blending schemes * 423
14.3.1 Convergence of peridynamic models to classical models . 424
14.3.2 Derivation of force-based blending schemes 425
14.3.3 A numerical example 427
14.4 Summary 431
References 431

15 A Peridynamic Model for Corrosion Damage 437

Ziguang Chen and Florin Bobaru

15.1 Introduction 438
15.2 Electrochemical kinetics 443
15.3 Problem formulation of 1D pitting corrosion 446
15.4 The peridynamic formulation for 1D pitting corrosion 449
15.5 Results and discussion of 1D pitting corrosion 453
15.5.1 Pit corrosion depth proportional to $\sqrt{t}$ 453
15.5.2 Activation-controlled, diffusion-controlled, and IR-controlled corrosion 456
15.6 Corrosion damage and the Concentration-Dependent Damage (CDD) model 459
15.6.1 Damage evolution 462
15.6.2 Saturated concentration 464
15.7 Formulation and results of 2D and 3D pitting corrosion 465
15.7.1 PD formulation of 2D and 3D pitting corrosion 466
15.7.2 The Concentration-Dependent Damage (CDD) model for pitting corrosion: example in 2D 469
15.7.3 A coupled corrosion/damage model for pitting corrosion: 2D example 472

*First published in P. Seleson, Y. D. Ha, and S. Beneddine, Concurrent coupling of bond-based peridynamics and the Navier equation of classical elasticity by blending, *International Journal for Multiscale Computational Engineering*, 13(2): 91113 (2015). Copyright 2015 by Begell House, Inc.

15.7.4 Diffusivity affects the corrosion rate 474
15.7.5 Pitting corrosion with the CDD+DDC model in 3D 475
15.8 Pitting corrosion in heterogeneous materials: examples in 2D . . . 476
15.8.1 Pitting corrosion in layer structures 476
15.8.2 Pitting corrosion in a material with inclusions: a 2D example 479
15.9 Conclusions 480
15.10 Appendix 481
15.10.1 Convergence study for 1D diffusion-controlled corrosion . 481
15.10.2 Convergence study for 2D activation-controlled corrosion with Concentration-Dependent Damage model 482
References 484

16 Peridynamics for Coupled Field Equations 489

Erdogan Madenci and Selda Oterkus

16.1 Introduction 490
16.2 Diffusion equation 492
16.2.1 Thermal diffusion 492
16.2.2 Moisture diffusion 493
16.2.3 Electrical conduction 493
16.3 Coupled field equations 494
16.3.1 Thermomechanics 494
16.3.1.1 Thermal diffusion with a structural coupling term 494
16.3.1.2 Equation of motion with a thermal coupling term 495
16.3.2 Porelasticity 495
16.3.2.1 Mechanical deformation due to fluid pressure . 495
16.3.2.2 Fluid flow in porous medium 496
16.3.3 Electromigration 497
16.3.4 Hygrothermomechanics 499
16.4 Numerical solution to peridynamic field equations 500
16.4.1 Correction of PD material parameters 500
16.4.2 Boundary conditions 502
16.4.2.1 Essential boundary conditions 502
16.4.2.2 Natural boundary conditions 503
16.4.2.3 Example 1 506
16.4.2.4 Example 2 507
16.4.2.5 Example 3 509
16.5 Applications 511
16.5.1 Coupled nonuniform heating and deformation 511
16.5.2 Coupled nonuniform moisture and deformation in a square plate 514
16.5.3 Coupled fluid pore pressure and deformation 519

16.5.4 Coupled electrical, temperature, deformation, and vacancy diffusion 524
16.6 Remarks 527
References 528

Index **531**

Foreword

The *Handbook of Peridynamic Modeling* offers a quick introduction to peridynamics and presents several advanced topics and applications. A balanced mixture of mathematical, numerical, and software implementation issues are discussed. The applications cover: dynamic brittle fracture; crack branching; connections to cohesive-zone models; failure in fiber-reinforced composites; impact and fragmentation; coupling local to nonlocal models; corrosion-induced damage; and multiphysics modeling. The Handbook provides engineers, computational scientists, applied mathematicians, and graduate students in all areas of Engineering, Scientific Computing, and Applied Mathematics, an easy-to-access window into peridynamic modeling and its relevance in solving some challenging problems.

Preface

The idea for this Handbook was born out of collaborative discussions at the "Workshop on Nonlocal Damage and Failure: *Peridynamics and Other Nonlocal Methods*" held on March 11-12, 2013 in San Antonio, Texas and sponsored by the US Association for Computational Mechanics. The attendees and presenters at the workshop included members of academia, industry, and national laboratories who specialize or with interest in nonlocal theories of solid mechanics, damage modeling, and particularly in research utilizing the peridynamic theory. The workshop featured a keynote presentation by the originator of peridynamic theory and co-editor of this volume, Dr. Stewart Silling. In his talk, Dr. Silling made a compelling "case for nonlocality", which became motivation for the first chapter of this volume. The Handbook presents an up-to-date account of research in peridynamics. The chapters are written by leading researchers from academia and national research laboratories, and cover the mathematics and numerics behind nonlocal models, a full description of implementing a peridynamic computer code, formulations of different material models, links between peridynamics and atomistic models. The last two parts of the Handbook are dedicated to modeling of material failure and damage, and to multiscale and multiphysics modeling.

The list of applications of peridynamic modeling expands continuously, and the present volume cannot offer but a glimpse into the state-of-the-art of the field. The number of papers published on peridynamic topics is rapidly growing. At the time of writing this preface, Stewart Silling's original paper published in the Journal of the Mechanics and Physics of Solids in 2000 passed 700 citations in Google Scholar. With a rapidly expanding body of work, we felt that a Handbook on Peridynamic Modeling would be appropriate at this point in time. Our goals with this Handbook were both to offer a quick and practical introduction into peridynamics and to present a brief look into more advanced topics and applications of peridynamics, and into some open questions. We have included aspects related to modeling of materials, the mathematics of nonlocal calculus, and clear descriptions of algorithms and their implementation for those interested in using peridynamics for modeling of materials' damage and failure.

The Handbook is relevant to engineers searching for novel modeling techniques that allow them to understand how materials fail under dynamic loadings, or under the influence of multiple physical phenomena (e.g. thermal, mechanical, environmental, etc.). The Handbook should also offer inspiration to computational scientists who can contribute faster algorithms for more efficient simulations of 3D systems on GPU-enhanced systems, CPU clusters, or supercomputers. The applied mathematicians will be able to find new problems of interest, generated by nonlocal models and their numerical implementations. The computational engineer working on physics-based computer graphics and animation of fracture and shattering of brittle objects, for example, will find new avenues waiting to be explored. The design engineer, interested in designing fracture-tolerant or damage-resistant systems, can now develop and utilize novel peridynamic-based tools for such optimization problems. In short, our hope is that the Handbook will be a useful reference for a broad readership. We would like to express our gratitude to each and every contributor to this Handbook for their efforts in writing chapters that are informative, clear, and practical.

The editors would like to take this opportunity to thank the team at CRC Press, Taylor and Francis Group: Robert Ross, Delroy Lowe, and Marcus Fontaine. Their patience is deeply appreciated and their continued support and encouragements have helped us greatly. Special thanks are addressed to Prof. Martin Ostoja-Starzewski who invited us to publish the Handbook in the CRC Press book-series he is an editor for: "Modern Mechanics and Mathematics [*]."

Much of the work contained in this volume would not have been possible without the support from funding agencies. It is our pleasure to have this opportunity to express our thanks to the following individuals who have been supportive of non-local models/peridynamic research: Larry Russell (ARO), Nam Phan (NAVAIR), William Nickerson (ONR), Chian-Fong Yen (ARL), George Gazonas (ARL), David Stargel (AFOSR/AFRL), Ali Sayer and James Fillerup (AFOSR), Fariba Fahroo (AFOSR/DARPA), Bill Fincham (DOE-NETL), Joe Meyers (ARO), John Aidun, Randy Summers, and Veena Tikare (Sandia National Laboratories).

[*] https://www.crcpress.com/Modern-Mechanics-and-Mathematics/book-series/CHMODMECMTH

List of Figures

1.1 Damage growth ahead of a dynamic fracture in a brittle plate. . . . 8
1.2 Force on a sphere interacting with a half-space. 9
1.3 Self-assembly of peridynamic rods interacting through long-range forces. 10
1.4 Coarse graining of a one-dimensional lattice. 11
1.5 Composite composed of alternating stiff and compliant layers. . . . 13
1.6 Axial displacement fields in the composite. 14
1.7 Strain fields in the layers. 15
1.8 Stress contours near a hole in a composite plate. 19
1.9 Stress distribution near a hole in a composite plate. 20

2.1 Forces between volume elements. 27
2.2 Computation of fluxes. 32
2.3 Meaning of the deformation state. 34
2.4 Zero energy modes in a correspondence material model. 39
2.5 Microplastic material model. 44
2.6 Quasi-static crack growth in a PMB plate. 47
2.7 Dynamic crack growth in a PMB plate. 48
2.8 Modified PMB model to specify nucleation strain. 51
2.9 Model of interface crack growth. 53

3.1 Four of the possible configurations for the domain Ω and interaction set Ω_I. 74

4.1 A diagram of possible paths between u_δ, u_δ^h, u_0^h, and u_0 via various limits. 88
4.2 Dispersion relation for nonlocal model with a constant interaction kernel. 94

5.1 Conversion of a standard finite element mesh to a meshfree representation. 123
5.2 Explicit dynamic simulation of brittle fracture resulting from impact. The different shades (colors for the electronic version) indicate percentage of broken bonds, or damage, at a particular material point. 133
5.3 Quasi-static simulation of a tensile test. 135

6.1 Schematic description of peridynamic kinematics. 147
6.2 Schematic of peridynamic body with a horizon. 149
6.3 Ordinary vs. non-ordinary materials. 150
6.4 A bond-pair model. 169
6.5 Deformation and force vector-states. 170

7.1 Illustration of a system of interacting atoms in molecular dynamics. An atom interacts with other surrounding atoms, closer to it than a given cutoff radius r_{cut}. The force that a neighbor atom j exerts on a given atom i is denoted by $\mathbf{f}_{ij}$. By Newton's third law, $\mathbf{f}_{ji} = -\mathbf{f}_{ij}$. 182
7.2 Illustration of a one-dimensional nonlocal linear springs molecular dynamics system. The spring constant of the interaction between an atom and its jth nearest neighbors is denoted by K_j. 185
7.3 Density evolution for (a) the molecular dynamics model, (b) the peridynamic model, and (c) the classical wave equation. 189
7.4 Radial displacement profile assigned to the initial configuration for the two-dimensional simulations in Figure 7.5. 193
7.5 Molecular dynamics (MD) potential energy (a) and upscaled peridynamic (PD) strain energy (b,c,d) profiles for two-dimensional wave propagation on a plate. The magnitude of the energy depends on the choices of $\bar{k}$ and $\bar{c}$ (cf. Table 7.1). 194
7.6 (a) Radial displacements relative to the center of the plate, in the simulations presented in Figure 7.5 at the final time, for atoms/-points with reference positions along the x-axis. We compare the results of a molecular dynamics (MD) simulation with lattice constant a and coarse upscaled peridynamic (PD) simulations with mesh sizes of $h = 2a, 4a$. (b) Comparison of the energy evolution between the MD and the coarse upscaled PD simulations. We plot the kinetic energy (black dotted lines), the potential/strain energy (blue dashed lines), and the total energy (red solid lines) for each of the simulations. 195

8.1 Gaussian ramp PML distribution 207
8.2 Total strain energy 211
8.3 PML-terminated simulation 212
8.4 PML-terminated simulation vs. fixed boundary 212
8.5 Exact solution comparison 214
8.6 Dimensional analysis schematic 215
8.7 Crack propagation schematic 217
8.8 Damage close up 218
8.9 Crack length scaling vs. load 219
8.10 Crack length scaling for a center crack 220
8.11 Axisymmetric verification, vertical displacement 223
8.12 Axisymmetric verification, horizontal displacement 224

9.1 Convex-concave potential. 231
9.2 Unstable constitutive relation between force and antiplane shear strain. 232
9.3 Domains D and D_α. 233
9.4 Jump discontinuity. 238

10.1 The conical micromodulus function 258
10.2 (a) One-point Gauss quadrature for the spatial integration of the peridynamic equations; (b) exact areas of various nodes that participate in the quadrature or node **p**. 261
10.3 (a) Sample geometry; three loading conditions: stresses suddenly applied on top and bottom boundaries (b), stress applied on crack surfaces (c), and velocity boundary conditions (d). In (e): the initial conditions for the case shown in (d). 263
10.4 δ-convergence study in glass for the crack propagation speed using δ = 3, 1.5, and 0.8 mm, for a suddenly applied tensile stress on the top and bottom boundaries of magnitude σ = 2 MPa. Load is maintained constant in time for the duration of the simulation. . . . 266
10.5 Damage maps in glass under suddenly applied tensile stress on the boundaries with different amplitudes. 267
10.6 Crack propagation speed profiles for glass under stress boundary conditions applied on the boundaries of the sample at different amplitudes. 269
10.7 δ-convergence study in glass for the crack propagation speed using δ = 1.5 mm and 0.8 mm, with a suddenly applied tensile stress on crack surfaces of magnitude σ = 3 MPa. The load is maintained constant in time for the duration of the simulation. 270
10.8 Damage maps for glass under stress on pre-crack surface with different stress amplitudes applied. 271
10.9 Crack propagation speed profiles for glass under stress boundary conditions applied on the pre-crack surfaces: (a) σ = 0.5 MPa, (b) σ = 3 MPa, (c) σ = 6 MPa. 272
10.10 δ-convergence in terms of crack propagation speed in glass for velocity boundary conditions with applied velocity on top and bottom boundaries of V_{max} = 0.02 m/s, using δ = 3, 1.5, and 0.8 mm. . . . 273
10.11 Damage maps in glass under velocity boundary conditions with different velocity amplitudes applied. 274
10.12 Crack propagation speed profiles for velocity boundary conditions: (a) V_{max} = 0.02 m/s, (b) V_{max} = 0.06 m/s 275
10.13 δ-convergence study in homalite for crack propagation speed using δ = 6.0, 4.0, and 3.0 mm, for a suddenly applied tensile stress (σ = 1 MPa) on the top and bottom boundaries. Load is maintained constant in time for the duration of the simulation. 276

10.14 Damage maps for homalite under tensile stresses of various amplitudes applied on the top and bottom boundaries. Load is maintained constant in time. 277
10.15 Crack propagation speed profiles in homalite with tensile stresses of various amplitudes applied on the top and bottom boundaries: (a) $\sigma = 0.2$ MPa, (b) $\sigma = 1$ MPa 278
10.16 δ-convergence study for crack propagation speed in homalite, using $\delta = 6$, 4, and 3 mm, and a loading magnitude of $\sigma = 2$ MPa applied on the pre-crack surfaces. 279
10.17 Damage maps for homalite under stress on pre-crack surface with different amplitudes. 281
10.18 Crack propagation speed profiles for homalite with stress applied on pre-crack surfaces at various amplitudes. 282
10.19 δ-convergence study for velocity boundary conditions using $\delta =$ 4.0, and 3.0 mm for a loading magnitude of $V_{max} = 0.02$ m/s. . . . 283
10.20 Damage maps in homalite under velocity boundary conditions with different velocity amplitudes applied. 284
10.21 Crack propagation speed profiles for homalite under velocity boundary conditions for different applied amplitudes. 285
10.22 Damage maps for narrow sample of glass under suddenly applied tensile stresses, of various magnitudes, on the top and bottom boundaries. 286
10.23 Crack propagation speed profiles for glass under sudden tensile stress applied on the boundaries of the narrower sample at different amplitudes. 287
10.24 Damage maps for narrow sample of glass under stresses of different magnitudes applied on pre-crack surfaces. 288
10.25 Crack propagation speed in narrow sample of glass under tensile stresses of different magnitudes applied on the pre-crack surfaces. . 289
10.26 Damage maps for narrow glass sample under velocity boundary conditions with different applied velocity amplitudes. 290
10.27 Crack propagation speed in narrow glass sample under velocity boundary conditions applied with different magnitudes for applied velocity: (a) $V_{max} = 0.02$ m/s, (b) $V_{max} = 0.06$ m/s 291
10.28 Velocity vector plot at: (a) 5.5 μs, (b) 15.0 μs, (c) 24.5 μs, (d) 31.5 μs, (e) 36.0 μs for the loading on the crack surfaces' case with 8 MPa loading amplitude. The color (for color figures please see the electronic version of the handbook, or the online version of reference [13]). of the nodal velocity vectors represents the damage index at that node. 297
10.29 Vertical component of nodal velocity in homalite sample under 0.3 MPa stress applied on pre-crack surfaces: (a) waves created by the applied load approach the boundaries, (b) after several reflections and wave reinforcements, the crack starts propagating, (c) crack continues to propagate in a straight line without branching. 301

10.30 Vertical component of nodal velocity in homalite sample under 2 MPa stress applied on pre-crack surfaces: (a) waves propagate towards the boundaries, (b) crack propagation starts before reflected waves interact with the tip, (c) interaction of waves reflected from the boundaries with those produced by the crack propagation process leads to crack branching. 302
10.31 Vertical component of nodal velocity in homalite sample under 6 MPa stress applied on pre-crack surfaces (a) waves propagate from pre-crack surface, crack propagation starts almost as soon as the loads are applied, (b) branching takes place before reflected waves meet the propagating crack, (c) waves reflected from the sample's boundaries meet with the two propagating branches and influence their future path growth. 303
10.32 Two peridynamic models with two different horizons and the same discretization grid: one with horizon factor $m = \frac{\delta}{\Delta x} = 4.01$, shown by a solid circle, and the other with $m = 1.45$, shown by a dashed-line circle. 305
10.33 Damage maps for load case 1 (sudden loading on the top and bottom boundaries) with 3 MPa for a homalite sample: (a) results from the peridynamic model that only uses nearest-neighbor bonds (see dashed-line circle in Fig. 10.32), and (b) results from the cracking particle method of Rabczuk and Belytschko [80]. 306

11.1 Geometrical representation of the integration bounds (11.4) for a horizon containing a bisecting fracture plane used in the "bond stretch" and "energy density" criteria. The total work to break all the bonds connecting point A to all points B in the spherical cap is the integration over all these connections for each point A along the line z. 319
11.2 Normal opening, Δ_n, of the fracture surface. The dark shaded area corresponds to the portion of the horizon of point A that contains the end points of the bonds crossing the fracture surface. 320
11.3 Case 2A integration bounds when $\xi_c < \delta$. 323
11.4 Case 2B integration bounds when $\xi_c > \delta$. 324
11.5 Analytical traction-separation law for normal opening. The area under the curve is the fracture toughness G_c (11.5). 325
11.6 Lattice arrangements for the study of normal opening cohesive traction laws using a discretized domain. 327
11.7 Cohesive traction-separation law obtained numerically for different particle spacings in a cubic arrangement compared to the exact solution (11.20). 328
11.8 The cohesive traction versus the crack opening displacement using different particle spacings in a non-uniform arrangement compared to the exact solution (11.20). 328
11.9 Tangential opening, Δ_t, of the fracture surface. 330

11.10 Shear opening stretch between points on plane $z/\delta \simeq 0$ and a point located at $z/\delta = 0.75$ for an opening displacement $\Delta_t/\delta = 0.5$ showing both tensile and compressive stretches. 331
11.11 Tangential traction-separation law obtained analytically (11.35) and by using a series of refined cubic lattices in the absence of bond failure. A very good agreement is observed for $dq/\delta \leq 0.125$. . . . 332
11.12 Traction-separation law shear when bonds break only in tension. The insets show the bond intact regions (shaded in gray) as the shear opening increases. 334
11.13 Traction-separation law obtained numerically with dq/δ=0.0625 when bonds are allowed to fail in compression. η is the ratio of critical stretch in compression to that in tension. 335
11.14 Variation of the fracture toughness G_c with respect to the ratio of critical stretch in compression to critical stretch in tension. 335
11.15 Normal T_n and tangential T_t cohesive traction versus the normal Δ_n and tangential Δ_t displacement jumps for the case $\eta = \infty$. 336

12.1 PD horizon (interaction domain) of material point i for a lamina with a fiber orientation of θ and PD bonds between material point i and other material points within its horizon. 342
12.2 Components of the initial bond length between material points i and p. 344
12.3 Force-stretch relation for PD bonds in a lamina. 345
12.4 Determination of critical stretch values s_{Ft} and s_{At}. 346
12.5 Surface effects in the domain of interest. 347
12.6 Four different bond constants for a fiber-reinforced composite laminate. 347
12.7 Loading and geometry of the unidirectional lamina under uniaxial tension and uniform temperature change. 349
12.8 Horizontal displacement along the central axis at the end of 8000 time steps. 350
12.9 Vertical displacement along the central axis at the end of 8000 time steps. 351
12.10 Variation of horizontal displacement along the central axis at the end of 8000 time steps when no failure is allowed. 352
12.11 Variation of vertical displacement along the central axis at the end of 8000 time steps when no failure is allowed. 352
12.12 Loading and geometry of a composite laminate under uniaxial tension. 352
12.13 Horizontal displacement along the central axis in the 90° ply of the $[0°/90°/0°]$ lay-up at the end of 8000 time steps. 353
12.14 Vertical displacement along the central axis in the 90° ply of the $[0°/90°/0°]$ lay-up at the end of 8000 time steps. 354
12.15 Horizontal displacement along the central axis in the 45° ply of the $[0°/45°/0°]$ lay-up at the end of 8000 time steps. 355

12.16 Vertical displacement along the central axis in the 45° ply of the [0°/45°/0°] lay-up at the end of 8000 time steps. 356
12.17 Loading and geometry of the unidirectional lamina with a crack under tension loading. 356
12.18 Damage plots for laminae having a central crack with fiber orientations of (a) $\theta = 0°$, (b) $\theta = 90°$, and (c) $\theta = 45°$. 357
12.19 Loading and geometry of a composite laminate with a crack under tension loading. 358
12.20 Damage plots for a laminate of [0°/90°/0°] with a pre-existing crack: (a) bottom ply, 0°; (b) center ply, 90°; and (c) top ply, 0°. . 359
12.21 Damage plots for a laminate of [0°/45°/0°] with a pre-existing crack: (a) bottom ply, 0°; (b) center ply, 45°; and (c) top ply, 0°. . 360
12.22 Shear bond damage plots for a laminate of [0°/90°/0°] with a pre-existing crack: (a) bottom ply, 0°; (b) center ply, 90°; and (c) top ply, 0°. 361
12.23 Shear bond damage plots for a laminate of [0°/45°/0°] with a pre-existing crack: (a) bottom ply, 0°; (b) center ply, 45°; and (c) top ply, 0°. 362
12.24 (a) PD horizon for a lamina with a fiber orientation of θ and PD bonds between material point i and other material points within its horizon and (b) a PD bond between material points i and p with an orientation of ϕ. 363
12.25 Relative displacement between material points i and p. 364
12.26 Relative displacement between material points i and p. 365
12.27 Relative displacement between material points i and p. 366
12.28 Relative displacement between material points i and p. 367
12.29 Material point $\mathbf{x}$ with a truncated horizon in a lamina. 367
12.30 Material point $\mathbf{x}$ far away from external surfaces of a lamina. . . . 368
12.31 Construction of an ellipse for surface correction factors. 368
12.32 (a) PD bond between material points at $\mathbf{x}_{(i)}$ and $\mathbf{x}_{(j)}$and (b) the ellipse for the surface correction factor. 369
12.33 Interlayer and shear bonds between neighboring plies (only some of the interactions are depicted explicitly for clarity). 369
12.34 Shear bonds between material points b and a and between material points d and c in both undeformed and deformed configurations. . 370
12.35 (a) A composite laminate subjected to isotropic expansion loading, and (b) deformation of interlayer and shear bonds between neighboring plies (only some of the interactions are depicted explicitly for clarity). 371
12.36 (a) A composite laminate subjected to simple shear loading, and (b) deformation of interlayer and shear bonds between neighboring plies (only some of the interactions are depicted explicitly for clarity). 372
12.37 Shear bonds between material points b and a and between material points d and c in both undeformed and deformed configurations. . 373

12.38 Interlayer and shear bonds between material point **x** and other material points located at the $(k+1)$th ply. 375

13.1 A schematic of damage observed after impact on a seven-layer glass system (from [1]). 381
13.2 Damage maps on the impact surface at 15 μs from the time of impact. Horizon size is fixed, δ=1 mm (m-convergence study) and m=3 in A, m=4 in B, and m=5 in C. 385
13.3 Damage maps on the impact surface at 15 μs from the time of impact. Horizon factor $m = \frac{\delta}{\Delta x}$ is kept fixed (equal to 4), and the horizon varies: in (A) δ = 1.5 mm; in (B) δ = 1.0 mm; in (C) δ = 0.5 mm. 386
13.4 Damage map on the strike face at 15 μs for different combinations of critical bond-strain corrections applied: in (A) CSD and DD, but no SC; in (B) SC and DD, but no CSD; in (C) SC and CSD, but no DD; (D) all corrections applied; in (E) all corrections disabled. . . 387
13.5 Strike face damage map for different perturbations of the grid or the impact location: (A) no perturbations used; (B) random perturbations of nodal coordinates; (C) off-center impact. 388
13.6 Sample geometry of the multi-layer glass target (with a polycarbonate backing plate) and boundary conditions used in the peridynamic model (see [1] and [3]). 389
13.7 Elastic impact of a rigid sphere onto a two-layer glass with a polycarbonate backing. Top: vertical displacement of the glass layer centroid obtained with the FEM or peridynamics. Bottom: evolution of the projectile velocity over 60 μs time span (from [3]). . . . 393
13.8 Time evolution of damage for the through-thickness cross section through the impact point for the system in Fig. 13.6. From left to right, and top to bottom, snapshots taken at 12, 27, 43, 79, 110, and 161 μs from the time of contact between the projectile and the glass laminate (from [3]). 395
13.9 Time evolution of damage for the first glass layer in the laminate: top row shows the impact face, bottom row shows the back of this first layer. From left to right, snapshots taken at 9, 27, 43, 60, and 127 μs from the time of contact between the projectile and the glass laminate (from [3]). 396
13.10 Damage in the impact layer first layer). The small figures show: (a) bundled radial cracks, (b) fan crack, (c) coarse radial cracks, (d) bow-tie crack, and (e) dicing cracks (from [1]). 397
13.11 Time evolution of damage for the second glass layer in the laminate: top row shows the impact face, bottom row shows the back of this second layer. From left to right, snapshots taken at 12, 18, 24, 43, and 127 μs from the time of contact between the projectile and the glass laminate (from [3]). 398

13.12 Time evolution of damage for the fourth glass layer in the laminate: top row shows the impact face, bottom row shows the back of this fourth layer. From left to right, snapshots taken at 27, 43, 79, 127, and 161 μs from the time of contact between the projectile and the glass laminate (from [3]). 399
13.13 Time evolution of damage for the last layer of glass (7th) in the multilayered system: top row shows the impact face, bottom row shows the back of the this seventh layer. From left to right, snapshots taken at 36, 50, 79, 110, and 161 μs from the time of contact between the projectile and the glass laminate (from [3]). 401
13.14 Fracture regions in the last (7th) glass layer (from [1]). 401

14.1 Continuum body $\mathcal{B}$ composed of two overlapping subdomains $\mathcal{B}_1$ and $\mathcal{B}_2$, such that $\mathcal{B}_1 \cup \mathcal{B}_2 = \mathcal{B}$ and $\mathcal{B}_1 \cap \mathcal{B}_2 = \mathcal{B}_c$. 411
14.2 Dimensions and boundary conditions (left plot) and meshes (right plot) for a plate with a crack (unit of length: μm). 414
14.3 Structure deformation and strain component ε_{xx} of (a) the Arlequin solution using the H^1 coupling operator, (b) the Arlequin solution using the L^2 coupling operator, and (c) the PD solution. 414
14.4 The x-component of the Lagrange multiplier field, $\varphi_{\mathbf{x}}$, using the H^1 and L^2 coupling methods with constant weight functions in $\mathcal{B}_c$ (unit of length: μm). 415
14.5 A nonlocal domain 416
14.6 Dimensions and boundary conditions on the cube benchmark . . . 420
14.7 Distribution of the error in strain of the cube in traction benchmark 421
14.8 Distribution of the error in strain of the cube in shear benchmark . 421
14.9 Dimensions and boundary conditions of a cracked bulk 422
14.12 (a) Geometry of a square plate with a horizontal rectangular slit in its center. (b) Decomposition of the domain in the PD problem into the domain of interest $\mathcal{B}^{0,\delta}_{\text{plate}} \setminus \mathcal{B}_{\text{slit}}$ and the boundary layer $\mathcal{B} \setminus \mathcal{B}^{0,\delta}_{\text{plate}}$; nonlocal Dirichlet-like boundary conditions, consistent with the blended problem, are imposed in the boundary layer. (c) Domain decomposition and boundary conditions for the blended problem. . 428
14.13 Profiles for the strain component ε_{xx}, in the deformed configuration, for the PD model (a) and the blended model (b). 430

15.1 Snapshots from a peridynamic simulation of corrosion damage. The initial size of the defect in the top layer has a length of 60 μm. The legend represents the mechanical degradation or damage caused by the corrosion process. Red corresponds to completely corroded material, while blue shows pristine metal without any corrosion damage. The colors between blue and red describe degrees of partially damaged material. 439

15.2 Left: SEM of the cross section of a corrosion pit on a Mg-alloy sample exposed to the immersion corrosion environment after 60 h; the wetted layer is hundreds of microns thick. Right: X-ray mapping of the distribution of Mg after corrosion shows diffuse damage near the pit surface (reprinted from [48]). 441
15.3 During anodic reaction in a corrosion process, metal atoms lose electrons and diffuse into the solution. The blue color * depicts a passive layer on the metal surface (redraw from [30]). 444
15.4 Corrosion modeled by diffusion with a moving interface. The evolution of the interface ($h(t)$) is controlled by the fluxes through both liquid/solid interface and the pitting hole surface. In the classical model (A), the diffusion in the solid is not considered and a Stefan condition is imposed at the interface. When the diffusion in the solid is considered, a flux continuity condition at the interface is required in a classical model (B). In our PD model (C), the motion of the interface is part of the solution. 447
15.5 Peridynamic description of the diffusion model. The region denoted by H_x is the horizon (nonlocal region around the point) of node x, which has concentration $C(x,t)$. Bonds connecting points have micro-diffusivity $k(\hat{x},x)$. 449
15.6 Peridynamic nodes distribution. An extra boundary with length δ is added onto the left end of the 1D bar for taking care of the surface effect (see [36]). 452
15.7 (A) Concentration and interface evolution within 20 s. (B) Zoom-in near the saturated concentration location. The location of the jump discontinuity is the location of the interface (the nodes with concentration lower than 5100 mol/m^3 are liquid nodes, while the others are solid nodes). Concentration gap between solid boundary node and liquid boundary node decreases to near zero in time ($K_1 = 2.0\ \mu$m^2/s, $\delta = 0.05\ \mu$m, $m = 4$). 454
15.8 (A) The m-convergence results for the location of the interface at different times. (B) Corrosion depth as a function of $\sqrt{t}$. Except for the early time (when corrosion is activation controlled), the dependence is linear (in $\sqrt{t}$), matching perfectly the experimental results (see [21]) 455
15.9 Corrosion depth versus time at different effective diffusivities in the solid. When the diffusivity is small (small applied voltage), the corrosion process over the time interval shown is activation controlled, and the interface moving speed is linearly dependent on the diffusivity. When the diffusivity is large (larger than 0.4 μm^2/s), the corrosion process over this time interval is dominated by the diffusion-controlled corrosion with a constant ratio s (see Eq. 15.20) 456

*For colors please see the electronic version of the handbook or [12].

15.10 Experimental polarization data from [23] (squares) and prediction for the polarization computed with our PD model (triangles). The experimental data Tafel slope and exchange current are determined by the Tafel's law according to the experimental results in the activation-controlled corrosion [23]: $b_a = 45.1$ mV, $i_0 = 1.422$ mA/cm^2. Using just one experimental data-point in the activation-controlled region for calibrating the model, we predict the entire profile of the polarization curve. 458
15.11 Concentration-dependent damage (CDD) model. The damage index for the mechanical bonds is linked to the ratio between the total concentration drop and the difference of C_{solid} and C_{sat}. When the concentration drops to the saturated concentration, the damage index is 1, i.e. mechanical bonds (lower picture) are totally damaged. The diffusivity bonds, upper picture, are not affected by damage, but they can suffer changes if their ends goes through a phase change, from solid to liquid, when concentration equals C_{sat} (the PD model for corrosion). When the concentration is C_{solid}, no bonds are damaged, i.e., the damage index is 0. 460
15.12 The evolution of damage profiles at different diffusivity: (A) $K_l = 2.0\ \mu\text{m}^2/\text{s}$; (B) $K_l = 4.0\ \mu\text{m}^2/\text{s}$. In this map, dark regions correspond to 0% damage in metal, while lighter regions to 100% damaged metal. The size of the damage layer does not change much during the diffusion-controlled pitting corrosion. Results computed with: $\delta = 0.05\ \mu\text{m}$, $m = 4$. 463
15.13 Damage maps at 20 s showing that the size of the damage layer is linearly dependent on the diffusivity. Although the interface moving speed is weakly related to the diffusivity in diffusion-controlled corrosion stage, the profile of the damaged layer is highly dependent on the diffusivity value. Since the diffusivity is an exponential function of overpotential, increasing the overpotential changes the damage profile. 464
15.14 The dependence of s and damage profile on the saturated concentration. Larger saturated concentration leads to: larger speed of corrosion progression but smaller size (thickness) of the damaged region. 465
15.15 Peridynamic description of the diffusion model in 2D. The region denoted by H_x is the horizon of node x. The concentration at x is $C(x,t)$. Bonds connecting points have micro-diffusivity $k(x,\hat{x})$. . . 467
15.16 Concentration-dependent damage (CDD) model in 2D. Diffusion bonds (are affected by phase change, but cannot break), and mechanical bonds that break due to the concentration drop. 469
15.17 A 2D example of bi-material diffusion: boundary conditions and initial conditions. There is no flux around the boundary of the metal, except over the solid/liquid interface. In the PD model, no conditions have to be imposed at the solid/liquid interface. 470

15.18 The concentration distribution around solid/liquid interface within activation-controlled corrosion. The concentration inside the liquid is always zero, while the one inside the solid is always larger than the saturated concentration inside the liquid. Thus, there is a concentration jump at the interface. 470
15.19 The damage profiles around a half-circular pit after corrosion in NaCl solution for 20 s. The horizon factor is 4.02 and the horizon size is 4.02 μm. The effective diffusivities are 2.0 $\mu m^2/s$, which is corresponding to an overpotential around 0.2 V (units on both axes are μm). 472
15.20 The horizon size effect for the CDD+DDC model. The horizon factors in three panels are 4.02. The horizon sizes are different: (A) 2.01 μm; (B) 4.02 μm; (C) 8.04 μm. The effective diffusivities are 2.0 $\mu m^2/s$, which is corresponding to an overpotential around 0.2 V. 473
15.21 m-convergence for the CDD+DDC model. The horizon sizes are kept as 4.02 μm, while the horizon factors are doubled each time: (A) 2.01; (B) 4.02; (C) 8.04. 474
15.22 The damage profile after 20 s of corrosion process for different diffusivities: (A) 0.5 $\mu m^2/s$; (B) 1.0 $\mu m^2/s$; (C) 2.0 $\mu m^2/s$. The corrosion depths are 16 μm, 21 μm, and 35 μm. 474
15.23 Geometry, initial, and boundary conditions for a 3D sample. There is zero-flux on the boundary of the metal cubic sample except over the solid/liquid interface disk region on the top-center. 475
15.24 The damage evolution of 3D pitting corrosion. Sample is cut through the center to allow better viewing of the progression of corrosion damage. Only the solid nodes are shown. The horizon size and horizon factor are 8.04 μm and 4.02, respectively. 476
15.25 Illustrations of two cases of double-phase heterogeneous materials used to study pitting corrosion. The geometry of the metal sample and the initial pit are the same as given in Section 15.7.2. The corrosion resistances are different for phase 1 from phase 2. 477
15.26 The damage profiles (top panels) and the corresponding concentration distributions (bottom panels) after 20 s corrosion for Case A. The material heterogeneity contrast α from left to right: 1 (A and D; this is the homogeneous case), 8 (B and E), and 64 (C and F). $K_{①}$ fixed at 2.0 $\mu m^2/s$. In the damage maps only the solid nodes are shown. 478
15.27 The damage profiles (top panels) and the corresponding concentration distributions (bottom panels) after 20 s corrosion for Case B. The material heterogeneity contrast α from left to right: 1 (A and D; this is the homogeneous case), 8 (B and E), and 64 (C and F). $K_{①}$ is fixed at 2.0 $\mu m^2/s$. In the damage maps only the solid nodes are shown. 478

15.28 A two-phase material with randomly distributed inclusions. Within a 200 μm $\times$ 200 μm area, 100 circular inclusions with diameter 10 μm are randomly distributed. The diffusivity of the metal is $K_{①}$, and of the inclusions is $K_{②}$. Only the top-center part (100 μm $\times$ 100 μm) is shown in Figure 15.29. $K_{①}$ is fixed at 2.0 μm^2/s . . . 479

15.29 Damage profiles after 20 s of corrosion for the case shown in Figure 15.28 (only the observation area is shown and only solid nodes are shown). The heterogeneity contrast α are: (A) 1 (homogeneous case), (B) 8, and (C) 64. 480

15.30 Convergence study for the slope of depth vs. $\sqrt{t}$ (s, see Eq. 15.20). With the horizon size decreasing, the ratio between the interface position and the square root of the time (s) converges to a value near 8 μm/s$^{0.5}$. In the calculation of s, we fit the data pointsfrom the diffusion-controlled region with $t \in [15, 20]$ seconds ($\sqrt{t}$ between $\sqrt{15s}$ and $\sqrt{20s}$ in Figure 15.8B). 482

15.31 The damage profiles from the half-disk original pit after corrosion in NaCl solution for 20 seconds. The horizon factor is the same in all three cases, $m = 4.02$. The horizon sizes are: (A) 8.04 μm; (B) 4.02 μm; (C) 2.01 μm. The effective diffusivities are 2.0 μm^2/s, corresponding to an overpotential around 0.2 V. Units on both axes are μm, and the color represents the damage index (see the legend). 483

15.32 The effect of horizon factor m for a constant horizon size ($\delta =$ 4.02 μm) on the corrosion damage profile at 20 seconds. The horizon factors are: (A) 2.01; (B) 4.02; (C) 8.04. The effective diffusivity is 2.0 μm^2/s, corresponding to an overpotential around 0.2 V. Units on both axes are μm. 483

16.1 Surface effects in the domain of interest. 501

16.2 Interaction of material points across the interface. 501

16.3 Imposing essential boundary conditions on potentials: (a) nonzero potential $\psi(x^*,y^*,t) = \psi^*$ and (b) zero potential $\psi(x^*,y^*,t) = \psi^* = 0$. 503

16.4 Imposing boundary conditions on displacement components: (a) constant displacement components $u(x^*,y^*,t) = U^*$ and $v(x^*,y^*,t) = V^*$ and (b) zero displacement components $u(x^*,y^*,t) = U^* = 0$ and $v(x^*,y^*,t) = V^* = 0$. 504

16.5 Imposing boundary conditions on displacement components: (a) fixed in the normal direction, $U^* = 0$ and (b) free to move in the tangent direction. 504

16.6 Material point and its image in the fictitious region for imposing (a) $q^* \neq 0$ and (b) $q^* = 0$. 505

16.7 Fictitious region along the boundary with a unit normal (a) $\mathbf{n} = \mathbf{e}_x$ and (b) $\mathbf{n} = \mathbf{e}_y$. 506

16.8 PD temperature variation along the bar at different times and its comparison with FEA solutions at the final time step. 507

16.9 PD temperature variation close to the left end of the bar at different times and its comparison with FEA at the final time step. 508
16.10 PD temperature evolution at the center of the bar as a function of time and its comparison with FEA solution. 508
16.11 Rectangular plate under uniform stretch: (a) geometry and loading and (b) PD model with fictitious regions. 509
16.12 Horizontal displacement variation: (a) PD solution and (b) FEA solution. 510
16.13 Vertical displacement variation: (a) PD solution and (b) FEA solution. 510
16.14 Fixed rectangular plate under tension: (a) geometry and loading and (b) PD model with fictitious regions. 511
16.15 Horizontal displacement variation: (a) PD solution and (b) FEA solution. 512
16.16 Vertical displacement variation: (a) PD solution and (b) FEA solution. 512
16.17 Description of (a) plate geometry, (b) fictitious regions for the thermal field, and (c) fictitious regions for the displacement field. . . . 513
16.18 Temperature predictions after 1000 sec: (a) PD and (b) FEA. . . . 514
16.19 Horizontal displacement variation and deformed configuration (magnified by 10): (a) PD and (b) FEA. 515
16.20 Vertical displacement variation and deformed configuration (magnified by 10): (a) PD and (b) FEA. 516
16.21 Description of (a) plate geometry, (b) fictitious regions for the concentration field, and (c) no fictitious region for the displacement field. 517
16.22 PD prediction of moisture concentration after 10 h of absorbtion. . 518
16.23 Horizontal displacement variation and deformed configuration (magnified by 1000): (a) PD and (b) FEA. 518
16.24 Vertical displacement variation and deformed configuration (magnified by 1000): (a) PD and (b) FEA. 519
16.25 Description of (a) plate geometry, (b) fictitious regions for the fluid pressure, and (c) fictitious regions for the displacement field. . . . 520
16.26 Fluid pore pressure predictions (Pa): (a) PD and (b) FEA. 521
16.27 Vertical displacement variation (m): (a) PD and (b) FEA. 521
16.28 Description of (a) copper strip geometry, (b) fictitious regions for the displacement field, (c) fictitious regions for the electrical potential, and (d) fictitious regions for the temperature field. 522
16.29 Comparison of voltage distribution: (a) ANSYS solution and (b) PD solution at $t = 5 \times 10^{-5}$s. 522
16.30 Comparison of temperature distribution(°C): (a)ANSYS solution and (b) PD solution at $t = 5 \times 10^{-5}$s. 523
16.31 Comparison of axial displacement (m): (a) ANSYS solution and (b) PD solution at $t = 5 \times 10^{-5}$s. 523

16.32 Comparison of transverse displacement (m): (a) ANSYS solution and (b) PD solution at $t = 5 \times 10^{-5}$s. 524
16.33 Variation of cross-sectional area (μm^2) due to electromigration at $t = 1\,h$. 524
16.34 Comparison of area change due to electromigration at $t = 1$ h. 525

List of Tables

1.1 Stress concentration and decay distance near an open hole in a plate. 18

5.1 Constitutive parameters for the disk, modeled using the *linear peridynamic solid* constitutive model and the *critical stretch* bond failure law. 132
5.2 Constitutive parameters for the projectile, modeled using the *linear peridynamic solid* constitutive model. No bond failure law was assigned to the projectile. 132
5.3 Constitutive parameters for the tensile bar. The projectile was modeled using a linear elastic correspondence constitutive law. 133

7.1 Parameters for the MD EAM and upscaled PD simulations. 192
7.2 Comparison between the computational cost of the molecular dynamics (MD) simulation and the coarse upscaled peridynamic (PD) simulations presented in Figure 7.5. 193

8.1 Constants appearing in dimensional analysis. 215

10.1 Crack propagation speed at branching, computed by peridynamics, as a fraction of the Rayleigh wave speed (V/c_R). 293
10.2 Crack branching angles (between the symmetry line and a branch) computed by peridynamics, estimated at the branching location from damage maps. 295

11.1 The relative error of the critical strain energy release rate, $(G_c/G_c^{exact} - 1)\,100\%$, as a function of grid spacing for a cubic arrangements of particles. 326
11.2 The relative error of the critical strain energy release rate, as a function of the degrees of freedom, for a non-uniform arrangement of particles. 329
11.3 The relative error of the area under the curve for $0 \leq \Delta_t/\delta s_c \leq 25$ (Fig. 11.11) as a function of the cubic lattice grid spacing, dq, for shear opening Δ_t and assuming no bonds have broken. 332
11.4 Relative error of the fracture toughness, with respect to the fracture toughness G_c presented in Section 11.1, Eq. (11.5), as a function of the cubic lattice grid spacing, dq, for shear opening Δ_t and assuming no bonds break in compression ($\eta = \infty$). 334

14.1 Parameters for the rectangular slit simulations. 430

Contributors

Yan Azdoud
King Abdullah University of Science and Technology (KAUST)
Thuwal, Saudi Arabia

Florin Bobaru
University of Nebraska-Lincoln
Lincoln, Nebraska

Scot M. Breitenfeld
University of Illinois
Urbana-Champaign, Illinois

Ziguang Chen
University of Nebraska-Lincoln
Lincoln, Nebraska

Qiang Du
Columbia University
New York, New York

John T. Foster
University of Texas at Austin
Austin, Texas

George A. Gazonas
US Army Research Laboratory
Aberdeen, Maryland

Philippe H. Geubelle
University of Illinois
Urbana-Champaign, Illinois

Fei Han
King Abdullah University of Science and Technology (KAUST)
Thuwal, Saudi Arabia

Robert P. Lipton
Louisiana State University
Baton-Rouge, Louisiana

David J. Littlewood
Sandia National Laboratories
Albuquerque, New Mexico

Gilles Lubineau
King Abdullah University of Science and Technology (KAUST)
Thuwal, Saudi Arabia

Erdogan Madenci
University of Arizona
Tucson, Arizona

Erkan Oterkus
University of Strathclyde
Glasgow, United Kingdom

Selda Oterkus
University of Strathclyde
Glasgow, United Kingdom

Michael L. Parks
Sandia National Laboratories
Albuquerque, New Mexico

Pablo Seleson
Oak Ridge National Laboratory
Oak Ridge, Tennessee

Stewart A. Silling
Sandia National Laboratories
Albuquerque, New Mexico

Yenan Wang
University of Nebraska-Lincoln
Lincoln, Nebraska

Olaf Weckner
Boeing Research and Technology
Seattle, Washington

Raymond A. Wildman
US Army Research Laboratory
Aberdeen, Maryland

Zhanping Xu
University of Nebraska-Lincoln
Lincoln, Nebraska

Guanfeng Zhang
University of Nebraska-Lincoln
Lincoln, Nebraska

Part I

The Need for Nonlocal Modeling and Introduction to Peridynamics

Chapter 1

Why Peridynamics?

Stewart A. Silling

Sandia National Laboratories

CONTENTS

1.1	The mixed blessing of locality	3
1.2	Origins of nonlocality in a model	6
	1.2.1 Long-range forces	7
	1.2.2 Coarsening a fine-scale material system	10
	1.2.3 Smoothing of a heterogeneous material system	12
1.3	Nonlocality at the macroscale	18
1.4	The mixed blessing of nonlocality	20
	References	21

1.1 The mixed blessing of locality

One of the things I like about continuum mechanics is its peculiar combination of abstraction and practicality. It endows many of the things we encounter in ordinary life with a certain dignity and even mystery. For example, most people go through life thinking that dirt is just, well, dirt. In continuum mechanics, we know better: dirt is really a nonlinear viscoplastic solid with phase transitions — or better yet, a micropolar Cosserat continuum — whose response is worthy of deep contemplation and sophisticated analysis. There are mathematical objects, such as tensor fields and Airy stress potentials, that cannot be observed directly, yet reside within it and by secret means control its strength and stability.

Therefore, to design a building or a bridge that stands up and is safe, one might assume that engineers must need to know a lot about these tensor fields and stress potentials, in all their mathematical glory. If not, then surely they must depend on specialists in continuum mechanics for guidance. Right?

Actually, it turns out that most of the time, engineers can design a perfectly good building — or tunnel, or coal mine, or computer, or automobile, or airplane, or refrigerator — without ever solving a partial differential equation. A glance at a civil engineering textbook, or a typical Professional Engineer exam or a design code, shows that while "strength of materials" concepts are the workhorses, partial differential equations (PDEs) are noticeably absent. To some extent, this absence is the result of the *process* of engineering, in which a complex system is broken down into component parts which are individually analyzed to ensure they meet standard, verifiable requirements. Furthermore, any trained engineer must be able to perform this verification and get the same answer, making the artistic and creative aspects of applying PDEs more of a detriment to the process than an advantage.

The non-essential status of the standard (PDE-based) theory of continuum mechanics in much of engineering is even more true of peridynamics. Most engineers have never even heard the word. However, as will become more clear in the rest of this book, peridynamics is in some ways more compatible than the standard theory with the concept that engineers are comfortable with. Peridynamics uses relatively primitive concepts to characterize the forces within a body, such as spring-like interactions between pairs of points. Tensors are mostly absent from peridynamics. Anyone who can draw a free-body diagram and has a freshman engineering student's knowledge of integral calculus can understand the equation of motion for bond-based peridynamics.

The standard theory of continuum mechanics was developed around 1822 by the great genius Augustin-Louis Cauchy, who also made many other immortal contributions to mathematics. There are four fundamental assumptions in the standard theory (in peridynamics we drop the first three of these):

- The medium is *continuous* (a continuous mass density field exists).
- Internal forces are *contact* forces (material points interact only if they are separated by zero distance).
- The deformation is twice continuously differentiable (this assumption is relaxed in the weak form of the equations).
- The conservation laws of mechanics apply (conservation of mass, linear momentum, and angular momentum).

It can be argued that the first three of these assumptions are introduced for convenience rather than a close correspondence with physical reality. They lead to a mathematical description of continuous media with appealing properties. The internal forces in the standard theory are completely characterized by a single tensor field, the stress field. A straightforward application of the divergence theorem leads to an intuitive and easy to apply concept of surface traction and its coupling to the stress field. Displacement boundary conditions on a surface are similarly compatible with Cauchy's theory. With proper restrictions on the material model, the system of PDEs is elliptic in equilibrium problems, establishing that the boundary value problems are generally well-posed for typical materials.

Returning to the first three assumptions in the above list, there is some ambiguity about how closely these correspond with the reality of materials. All materials are discontinuous. All materials are heterogeneous. All materials have internal forces that occur across nonzero distances. The usual assumption is that at macroscopic length scales, a material can be well approximated as continuous. However, as the example discussed below demonstrates, nonlocal (non-contact) forces arise due to the heterogeneity of materials, even when the underlying physical interactions are assumed to be contact forces. Many problems of great importance — including fracture — fail to satisfy the smoothness requirement.

One may speculate that the concept of contact forces in a continuum would have caused discomfort to Newton, whose *Principia* largely concerned the mechanics of discrete bodies interacting over finite distances. In fact, one of the controversies that arose in response to Newton's view of the mechanics of planets was whether this "action at a distance" was possible; the prevailing view at the time was that there was some continuum in the universe that guided the motion of planets. Newton, wisely, declined to conjecture on how gravity really worked. Indeed, this is a question that the greatest minds in physics have still not solved. Newton was satisfied with a nonlocal mathematical model that explained observed phenomena.

It is interesting to survey various topics of current technological interest in mechanics and ask which of these can really and truly be modeled by the standard theory of continuum mechanics without violating the assumptions listed above. A glance at the technical program for any large, international mechanics conference is replete with topics that challenge the relevance of the standard theory of continuum mechanics: defects, phase transformations, composites, fracture, dislocations, micromechanics, nanostructures, biological materials, colloids, large molecules, complex fluids, and many others. Many of these topics have some things in common: discontinuities and long-range forces. Due to the basic characteristics of the standard theory, notably its reliance on PDEs, many of the papers at such a conference are about *repairing* or *adding onto* the theory. In many cases, such modifications are appropriate only for certain problems and materials. For example, the standard theory does not admit discontinuities, so researchers devise a special technique that is to be applied wherever a crack grows. This technique may contradict the basic PDEs, but this fact is easy to overlook.

Peridynamics does not magically give good answers in these problems involving discontinuities and long-range forces. It does, however, provide a consistent framework for modeling them. Since peridynamics is compatible with discontinuities and long-range forces, modeling these features is a matter of finding the right material model and applying the same basic field equations everywhere. Peridynamics avoids the need for special techniques that are incompatible with the basic equations.

The objective of peridynamics is to unify the mechanics of discrete particles, continuous media, and continuous media with evolving discontinuities (see [31]). The fundamental viewpoint of peridynamics is that the traditional distinctions between continuous, discrete, and discontinuous media are artificial and cause practical problems. After all, the physical principles that are operative in a material, that is, the conservation laws, do not distinguish between particles and continua, or between "continuous continua" and "discontinuous continua" (that is, fractured media). The spirit of peridynamics, then, is that a mathematical model should not distinguish between them either.

A peridynamic body may or may not be continuous in the traditional sense: in a neighborhood of a given material particle, there may or may not be an infinite number of points representing a continuous distribution of mass. Even if the body is continuous, its deformation may or may not be. If it is continuous, it may or may not be differentiable.

Internal forces in a peridynamic body occur due to interactions between material particles separated by a finite distance. In this sense, the peridynamic theory is *strongly nonlocal*. This is in contrast to a *weakly nonlocal* description, in which higher order derivatives imply the presence of a length scale, yet the interactions across finite distances are not included explicitly. In the following sections, I discuss some origins of nonlocality in modeling and argue that it should be considered as a fundamental aspect of mechanical systems.

1.2 Origins of nonlocality in a model

One motivation for peridynamics is that the partial differential equations of the classical theory of continuum mechanics fail to make sense on a crack tip or growing crack surface, because the required partial derivatives do not exist there. Therefore, we seek to reformulate the continuum model to avoid this problem. The approach to is to use integral equations rather than partial differential equations.

Crack nucleation and growth must be the result of some process: some progression of changes within a material that weaken it to the point that it no longer can hold together as a continuum. To understand this process is surely the key to understanding how to design materials that are more resistant to fracture, or which undergo fracture in ways that are favorable to applications. Traditional linear elastic fracture mechanics (LEFM) is silent about this process, subsuming the details of what happens near a crack tip into parameters such as the stress intensity factor (SIF) and kinetic relations that attempt to model crack evolution in terms of these parameters. The LEFM approach, which has achieved important successes in engineering, attempts to provide quantitative relationships between the SIF and the evolution of a crack. The crack itself is treated not as a discontinuity in a deformation field defined on the original body, but instead as a redefinition of the body itself. In this redefined body, the crack is treated like a boundary, and PDEs of the local theory of solid mechanics are applied everywhere else. These assumptions in LEFM intentionally ignore the details of what causes a crack to grow. Even the term "crack tip" is loaded with a key assumption of LEFM: that there is a precise location where a crack discontinuity is present, and ahead of it there is a classical continuum.

In LEFM, parameters such as the SIF characterize the asymptotic stress field near a crack tip. The kinetic relations that associate these parameters with the motion of a crack tip have been determined by experiment only for a handful of cases, mostly in highly idealized settings. Since this kinetic relation must be known in advance of solving a problem involving crack growth, an analyst using LEFM must assume that the relation applies to his or her particular application. But it is difficult to look at an actual microstructure of a material such as an aerospace composite and say with certainty that all the required assumptions to apply a kinetic relation are satisfied, especially if this kinetic relation was obtained for a much simpler system. In spite of the considerable successes that LEFM has achieved, I believe that these relations are sometimes applied beyond where they have been shown to be valid, including everything from carbon nanotubes to composites to cell membranes to earthquakes.

I also observe, as many others have, that LEFM is an "add-on" to the fundamental theory of continuum mechanics. As such, it represents a sort of admission of defeat for the continuum theory, whose PDEs cannot be applied on a discontinuity such as a crack. From a mathematical viewpoint, LEFM is also *inconsistent* with the continuum theory that it is appended to, because the PDEs are making one prediction for the evolution of the deformation (a stationary crack), while the LEFM equations are saying something else (a growing crack).

In contrast, peridynamics provides a framework that is more attuned than LEFM to the modeling of processes taking place ahead of a crack, and especially how these processes ultimately lead to a discontinuity. The usual way of modeling such processes in peridynamics is with a time-dependent damage variable applied to individual bonds. (A *bond* is an interaction between a pair of material points.) Damage modeling in peridynamics will be discussed further in the next chapter. However, to help explain the difference between peridynamics and LEFM, consider the dynamic crack shown in Figure 1.1. The material is elastic and brittle. The shaded areas indicate where damage is actively taking place in the continuum. (In this example, the assumed damage model is simple bond breakage at a prescribed critical value of bond strain, as discussed more fully in Chapter 2.) From the figure, evidently the *process* of damage occurs far ahead of where the discontinuities finally emerge. While this particular damage model is perhaps too simplistic for real materials, the computation illustrates how the process of damage in a continuum leads to the evolution of discontinuities. To provide this level of detail, a model must incorporate both continuous and discontinuous fields in its fundamental equations, a capability that peridynamics provides.

At the atomic and molecular levels, interactions within a material are always nonlocal because, by definition, atoms are separated from each other by a finite distance. At larger length scales, it is often assumed that because interatomic distances are smaller than the geometrical features of the problem, locality must be an excellent approximation. However, the local approximation has limitations, even at the mesoscale and larger length scales. These limitations arise because of at least three factors, discussed in the next three sections.

1.2.1 Long-range forces

Short-distance forces combine into longer-range forces when many atoms are taken together and modeled as a continuum or as some other structure. In biological systems, mechanical forces are important in the interaction and transport of molecules and larger structures. The underlying physical interactions that create these forces,

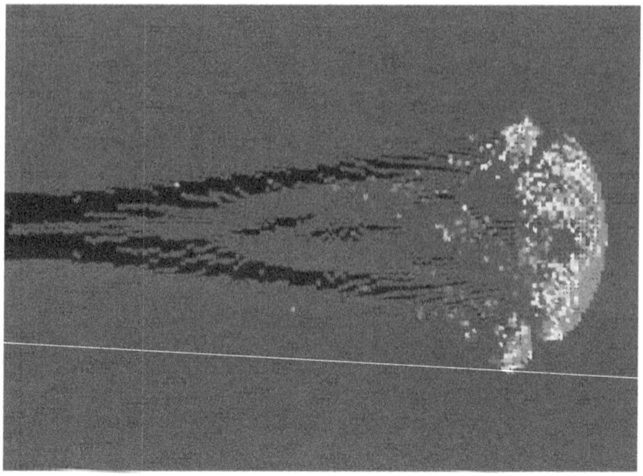

Figure 1.1: Damage growth ahead of a dynamic fracture in a brittle plate. The area of active damage growth is well ahead of the discontinuity. (Displacements shown are exaggerated 5 times.)

such as electrostatic charge and hydrogen bonding, are quantum mechanical and therefore short-range. However, the combination of many such molecules together have the effect of creating longer-range forces. If these interacting molecules are in solution, the solvent itself can further alter the interactions. This alteration must be included in the force potentials between the solute, unless we wish to explicitly model every molecule of the solvent.

A simple example is the interaction between a sphere and a half-space composed of the same material. The underlying physical interactions are quantum mechanical forces, such as van der Waals forces. The net force F_{ij} between two atoms i and j follows

$$F_{ij} = cr_{ij}^{-7}$$

where r_{ij} is the separation distance and c is a constant. Nonlocality arises from the combination of many such short-range interactions. Israelachvili [12] analyzes the net force on the sphere as a function of its distance from the surface of a half-space (Figure 1.2). He concludes that the net force on the sphere F_s follows

$$F_s = c_s D^{-2}$$

where D is the distance from the sphere to the surface. Evidently, when all the short-range interactions are taken into account, they combine to result in a longer-range force, because of the change in exponent between the last two equations. In modeling such systems, we could, in principle, model each atom in both bodies. This approach is usually impractical due to the number of atoms involved. Thus, a continuum approach is appropriate, and the relevant continuum forces are long-range, as well as longer range than the underlying physical forces at the interatomic level. Note that nonlocality arose when we made the *modeling decision* to no longer model individual atoms, but to model a continuum and a sphere instead.

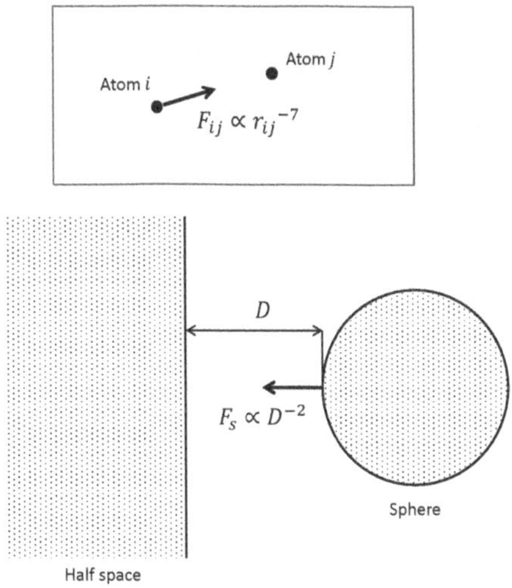

Figure 1.2: The net force on a sphere due to interatomic forces occurs over a much larger length scale than the force between individual pairs of atoms.

Long-range forces are of fundamental importance in the biology of cells at all length scales. Long molecules in solution, including DNA, interact with each other through complex potentials. These interactions control many of the processes within a cell. Many molecules can combine to form larger structures that interact with each other through other forces. Once again, due to the impracticality of modeling large numbers of atoms explicitly, we turn to a continuum-like model with appropriate force potentials. When we change models, we must also change the interactions between whatever mathematical entities are used to represent the molecules.

Self-assembly of structures at the nanoscale and mesoscale is another class of problem in which long-range interactions are of fundamental importance. In this type of process, smaller structural units interact with each other through some combination of long-range potentials to build up a larger structure. Whitesides and Grzybowski [24] discuss self-assembly as a general phenomenon in nature and nanotechnology over a wide spectrum of length scales involving many different types of physical interactions. All of these interactions have one thing in common: they involve long-range forces. Brinker, Lu, Sellinger, and Fan [5] discuss how the morphology of self-assembled nanostructures depends on the precise content of the solvent in which the process takes place. At all but the smallest length scales for self-assembled systems, some choice of modeling variables less burdensome than full molecular dynamics is needed. A continuum model using suitable long-range potentials is one option for modeling such systems.

As an example of how peridynamics might fit into a strategy for modeling self-assembly of nanoscale structures, Figure 1.3 illustrates the self-assembly of flexible peridynamic rods into larger structures. The rods themselves are continuous bodies,

Figure 1.3: Self-assembly of peridynamic rods interacting through long-range forces.

but they interact with each other through long-range forces. In this example, the ends of the rods attract each other strongly, while all other points along the rods repel each other weakly.

In all these classes of problems — surface physics, the mechanics of cell structures, and self-assembly of nanostructures — the *selection of the variables used to model the system* determines the length scale of the interactions. These length scales may be quite different from the forces at the interatomic level. The local theory of continuum mechanics does not lend itself to the modeling of these processes, because it does not incorporate long-range forces.

1.2.2 Coarsening a fine-scale material system

Nonlocality increases in importance through *coarsening* a body with short-range interactions. This process is, in a sense, the opposite of the smoothing process considered in the previous section. Suppose we want to model the mechanics of a long molecule, and suppose the positions of each atom are known, along with all the physical interactions between these atoms. These interactions are not restricted to pair potentials and could be complex, possibly affected by the presence of a solvent, as opposed to being in a vacuum.

It may be infeasible computationally to model each atom, in view of the large number of atoms and their high frequencies of vibration. The question is then how to reduce the number of degrees of freedom in a model of the system while retaining the key elements of the mechanics of the molecule, a process called *coarse graining*. Many approaches to coarse graining have been considered. Here, to illustrate the connection between this process and nonlocality, we consider the coarse graining of a string of identical point masses with mass m connected by springs of length a to nearest neighbors.

We consider only equilibrium configurations of the lattice. One option for modeling the system is to associate with each mass i a degree of freedom (DOF), namely a value of displacement u_i. In this case, the balance of forces in equibrium for mass i is

$$f_i = f_{i-1} \tag{1.1}$$

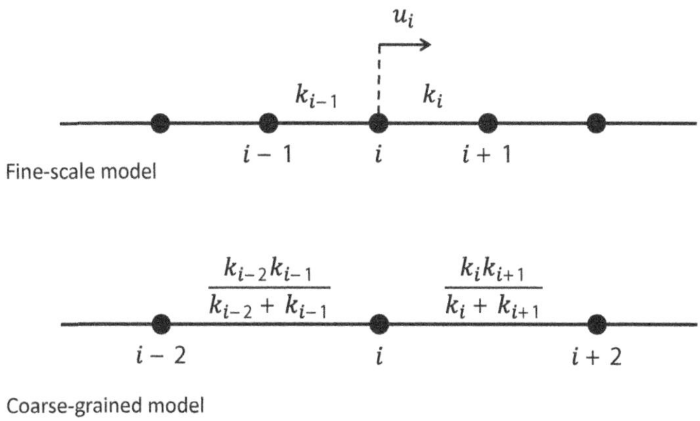

Figure 1.4: Coarse graining of a one-dimensional lattice increases the nonlocality of the model.

where

$$f_i = k_i(u_{i+1} - u_i), \qquad f_{i-1} = k_{i-1}(u_i - u_{i-1}), \tag{1.2}$$

where k_i is the spring constant of the spring connecting i to $i+1$ (Figure 1.4). Suppose we would like to save computational effort by omitting alternate DOFs, say, the odd numbered ones. How can we do this while reproducing the results of the full, fine-scale system? Under the assumption of equilibrium, (1.1) and (1.2) are easily combined to yield

$$u_i = \frac{k_i u_{u+1} - k_{i-1} u_{i-1}}{k_{i-1} + k_i}.$$

Combining this with the (1.1) and (1.2) leads to

$$f_{i-1} = f_i = \frac{k_{i-1} k_i}{k_{i-1} + k_i}(u_{i+1} - u_{i-1}).$$

Replacing i by $i+1$ or $i-1$ in this expression yields

$$f_i = \frac{k_i k_{i+1}}{k_i + k_{i+1}}(u_{i+2} - u_i), \qquad f_{i-2} = \frac{k_{i-2} k_{i-1}}{k_{i-2} + k_{i-1}}(u_i - u_{i-2}). \tag{1.3}$$

Using these expressions with (1.1) results in a coarse-grained model for the system:

$$f_i = f_{i-2}, \qquad i \text{ even} \tag{1.4}$$

where f_i and f_{i-2} are given by (1.3). What has been accomplished here is that we have eliminated the odd DOFs without changing the results that would be obtained for the remaining even DOFs. This procedure could be repeated any number of times to obtain a great reduction in the number of DOFs in the problem. An analogous process for continua is outlined in [19]. In this process, the volume that is modeled in a continuous body is successively reduced by a procedure similar to the one described above for a one-dimensional lattice.

The elimination of DOFs increases the distance across which interactions occur, because instead of springs with length a, we are left with springs of length $2a$. This increase in nonlocality is the by-product of reducing the number of DOFs in our model. The key point is that this increase in nonlocality arises from a decision about how to model the system, not from the underlying physical interactions.

The coarsening method described above, which is similar to static condensation, would apply to a dynamic model under the assumption that short-range interactions can be treated as being in equilibrium. Alternatively, and at the expense of much greater complexity, a fully dynamic version of the coarsening method can be obtained by considering the effect if one of the coarsened DOFs (i even) undergoes a displacement that follows a step function of time. By analyzing the resulting motion of this particle's neighbors, the time-dependent force on particle i can be derived. The resulting coarsened model is *nonlocal in time* and has a structure reminiscent of the Mori–Zwanzig formalism in statistical mechanics.

1.2.3 Smoothing of a heterogeneous material system

In the early 1980s, Bazant pioneered the application of nonlocal modeling to materials with damage. He demonstrated by a simple example that nonlocality is a necessary property of the elastic response in a material containing distributed defects [2]. In the same general spirit, the following discussion presents a derivation of how nonlocality is implied by the use of a smoothed displacement field to model a heterogeneous microstructure. Unlike Bazant's analysis, the discussion here does not include cracks or other damage.

In typical engineering analysis, a heterogeneous material is treated by applying the classical solid mechanics equations with smoothed (or "homogenized") material properties. These properties can be thought of as those that would be measured using macroscale laboratory test equipment. For example, we perform an unconfined compression test on a sample of concrete about 15 cm in diameter and measure the total force as a function of displacement at the ends. Concrete is a heterogeneous material. It consists of small pebbles (aggregate) of length scale about 2 cm held together by a weaker material (cement). We divide this displacement by the specimen length and call the result the "strain." We divide the force by the cross-sectional area and call the result the "stress." The ratio of stress to strain is defined to be the Young's modulus of the material, which is then treated as homogeneous, using the Cauchy local theory. This approximation is adequate for many purposes. However, it ignores the reality of how a load applied on the surface of a concrete body is transmitted internally. Since the aggregate inclusions may be in contact with each other, or nearly so, the actual force distribution within the material follows a tortuous path through the aggregate particles and their points of contact. This results in a quantitative and qualitative disparity between the local equations using smoothed material properties and how the material really behaves. Nonlocality in a random medium such as concrete has been treated in a number of references, for example [26, 8].

It has been recognized that in a heterogeneous material with a periodic microstructure, nonlocal interactions may arise as a result of homogenization [4, 11, 6].

In the following, we develop a derivation of nonlocality, and its implementation in the form of peridynamics, based on a simple model of the mechanics of a composite. This example also permits us to compute how mesoscale geometrical dimensions interact with material properties to produce internal force distributions within a heterogeneous material. It is demonstrated that nonlocality arises from the decision to model the composite in terms of a smoothed displacement field, rather than through direct physical interactions across a finite distance. Further details about this analysis are given in [20].

Consider a composite material composed of alternating layers of stiff (s) and compliant (c) materials (Figure 1.5). The two constituent materials have the same density. The layers have thickness $2h_s$ and $2h_c$. Let E denote Young's modulus and μ shear modulus. It is assumed that

$$E_s \gg E_c, \qquad \mu_s \gg \mu_c.$$

Only the displacements u_s and u_c in the x-direction, that is, parallel to the layers, appear in the following approximate analysis.

Figure 1.5: Composite composed of alternating stiff and compliant layers.

It is assumed that because of the disparity in elastic constants, u_s is independent of y. To model this system, it is convenient to introduce an *ansatz* for the general form of the x-displacement field based on intuition about how the compliant material deforms. In the compliant material, a displacement field of the form

$$u_c(x,y) = u_s(x) + (h_c^2 - y^2)w(x) \tag{1.5}$$

is assumed, where w is a continuous function, and $y = 0$ is located at the midplane of a typical compliant layer (Figure 1.6). The layers are connected rigidly at their interfaces, and a balance of shear tractions is also satisfied at these interfaces. It is assumed that any transverse displacements and body forces that are required to render the axial displacements in (1.5) equilibrated are small. In the absence of body forces, a force balance can be obtained by drawing a small control volume of length dx and height $2h_s$ that occupies a thin cross-sectional slice of a stiff layer. The force balance on this control volume yields

$$E_s h_s u_s''(x) + 2\mu_c h_c w(x) = 0. \tag{1.6}$$

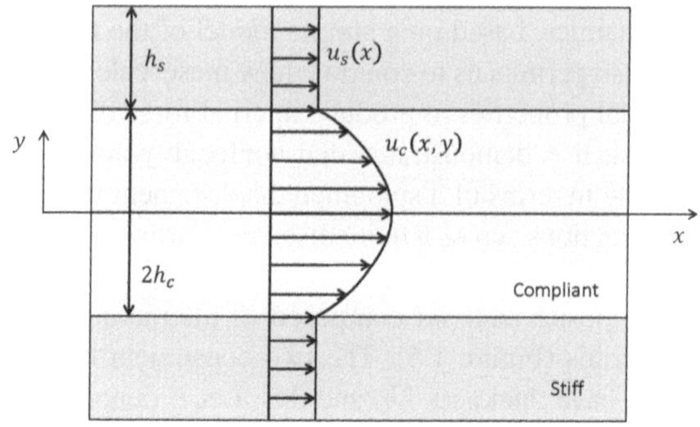

Figure 1.6: Axial displacement fields in the composite.

Now define the smoothed (homogenized) displacement by

$$\bar{u}(x) = \frac{1}{h_s + h_c}\left[h_s u_s(x) + \int_0^{h_c} u_c(x,y)\,dy\right]$$

which, using (1.5), evaluates to

$$\bar{u}(x) = u_s(x) + \alpha w(x), \tag{1.7}$$

where α is a constant. (See [20] for details.) Also define the smoothed normal stress by

$$\bar{\sigma}(x) = \frac{1}{h_s + h_c}\left[h_s \sigma_s(x) + \int_0^{h_c} \sigma_c(x,y)\,dy\right]. \tag{1.8}$$

Carrying out the integration in (1.8) using (1.5) leads to

$$\bar{\sigma}(x) = \bar{E}\bar{u}'(x) - \gamma w'(x) \tag{1.9}$$

where

$$\bar{E} = \frac{E_s h_s + E_c h_c}{h_s + h_c}, \qquad \gamma = \alpha \bar{E} - \frac{2E_c h_c^3}{3(h_s + h_c)}. \tag{1.10}$$

Observe that $\bar{E}$ is the effective Young's modulus that would be measured in a uniaxial tension test (in which the strain is constant and $w \equiv 0$ throughout). We will next investigate the relation between the smoothed displacement and stress fields from our micromechanical model in which the strain is allowed to vary with position.

Using (1.7), if $\bar{u}$ is prescribed, the force balance (1.6) may be written as the following second-order ordinary differential equation for w:

$$\bar{u}''(x) = \alpha w''(x) - \frac{2\mu_c h_c}{E_s h_s} w(x). \tag{1.11}$$

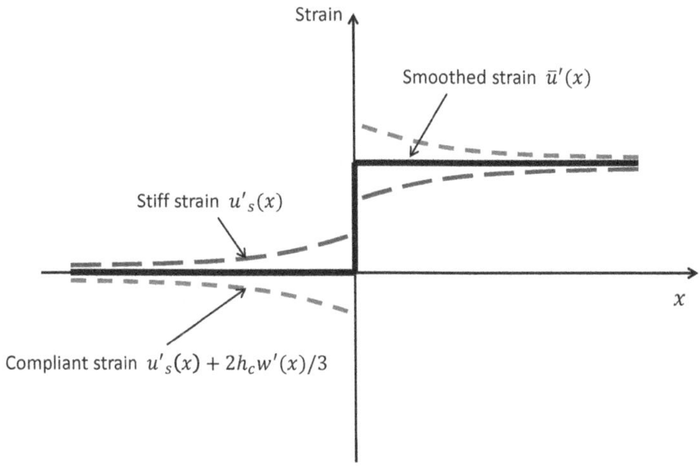

Figure 1.7: Strain fields in the stiff and compliant layers resulting from a prescribed smoothed displacement field in which there is a jump in strain at $x = 0$.

Of particular interest is the choice of prescribed $\bar{u}''$ given by

$$\bar{u}''(x) = \Delta(x)$$

where Δ is the Dirac delta function. Kinematically, this choice corresponds to a jump of 1 in the smoothed strain $\bar{u}'$ at $x = 0$ (Figure 1.7). For this choice, the solution, denoted G, to the ODE (1.11) is

$$w(x) = G(x) := -ke^{-\lambda|x|}, \tag{1.12}$$

where

$$\lambda = \sqrt{\frac{2\mu_c h_c}{\alpha E_s h_s}} = \sqrt{\frac{3\mu_c(h_s + h_c)}{E_s h_s h_c^2}}, \qquad k = \frac{1}{2\alpha\lambda}. \tag{1.13}$$

Continue to assume that $\bar{u}''$ is prescribed. Because the ODE (1.11) is linear, the properties of the Dirac delta function imply

$$w(x) = \int_{-\infty}^{\infty} \bar{u}''(p)G(x-p)\,dp. \tag{1.14}$$

Rewriting (1.9) using (1.14) and (1.10) and integrating by parts leads to

$$\bar{\sigma}(x) = E_c\bar{u}'(x) - \gamma\lambda^2 \int_{-\infty}^{\infty} \bar{u}'(p)G(x-p)\,dp. \tag{1.15}$$

Note that w no longer appears explicitly in this expression for the net stress. The nonlocality inherent in modeling heterogeneous materials using a smoothed displacement field is now evident: values of $\bar{u}'$ remote from x contribute to the stress at x. The expression (1.15) is similar to the nonlocal stress used in Eringen's nonlocal theory [10], because it involves taking a weighted average of strain.

In peridynamics, we seek to eliminate any reference to strain, because one of the goals is to avoid using spatial derivatives of the deformation. With this goal in mind, we next evaluate the acceleration at x at a given time t. From the linear momentum balance in the absence of body forces, using the assumption that $\rho_s = \rho_c$, our expression for the net normal stress (1.15) implies

$$\rho\ddot{\bar{u}}(x) = \bar{\sigma}'(x) = E_c\bar{u}''(x) - \gamma\lambda^2 \int_{-\infty}^{\infty} \bar{u}(p)G''(x-p)\,dp.$$

Simplifying the last term, and introducing a body force field b according to d'Alembert's principle, yields

$$\rho\ddot{\bar{u}}(x) = E_c\bar{u}''(x) + \gamma k\lambda^4 \int_{-\infty}^{\infty} (\bar{u}(p) - \bar{u}(x))e^{-\lambda|x-p|}\,dp + b(x). \tag{1.16}$$

In this form, the model is similar to the nonlocal theory proposed by DiPaola, Failla, and Zingales in Ref. [7]. This form, like Kröner's [14], retains both local and nonlocal terms.

Peridynamics, as noted above, seeks to eliminate the local term $\bar{u}''$, as well as $\bar{u}'$, in order to be compatible with the physical nature of cracks as discontinuities. In linearized peridynamics, as in Kunin's nonlocal theory [15], the equation of motion has the following general form:

$$\rho\ddot{u}(x) = \int_{\mathcal{H}_{\mathbf{x}}} C(p,x)(u(p) - u(x))\,dp + b(x) \tag{1.17}$$

where C is a function called the *micromodulus* and $\mathcal{H}_{\mathbf{x}}$ is a neighborhood of x called the *family* of x. The local term in (1.16), $E_c\bar{u}''$, does not appear in the peridynamic equation (1.17). However, it can be approximated by short-range interactions by using the same sort of tricks used above. The following identity holds as a result of integrating by parts twice:

$$u''(x) = \int_{-\infty}^{\infty} u''(p)\Delta(x-p)\,dp = \int_{-\infty}^{\infty} u(p)\Delta''(x-p)\,dp.$$

Suppose we approximate the Dirac delta function by a narrow tent-shaped function:

$$\Delta(x) \approx \phi(x) := \frac{\tau e^{-\tau|x|}}{2},$$

where τ is a large constant. Then

$$u''(x) \approx \frac{\tau^3}{2} \int_{-\infty}^{\infty} (u(p) - u(x))e^{-\tau|x-p|}\,dp.$$

Thus, our model for the smoothed composite displacement field, (1.16), can be approximated to any level of accuracy by choosing sufficiently large τ in the following peridynamic model:

$$\rho\ddot{\bar{u}}(x) = \int_{\mathcal{H}_{\mathbf{x}}} \left(\frac{E_c\tau^3}{2} e^{-\tau|x-p|} + \gamma k\lambda^4 e^{-\lambda|x-p|} \right) (\bar{u}(p) - \bar{u}(x))\, dp + b(x). \quad (1.18)$$

The foregoing example illustrates many of the key properties in understanding the role of nonlocality in the modeling of continua:

- Nonlocality is introduced by making the *modeling decision* to define the problem using a smoothed displacement field rather than the detailed microstructural fields.
- Although the underlying microstructural model (in terms of the u_s and u_c fields) is local, the resulting smoothed model (in terms of $\bar{u}$) is nonlocal.
- Expressions (called *bond force densities*) in the peridynamic model of the form

  $$C(p,x)(u(p) - u(x))$$

  do not necessarily direct nonlocal *physical* interaction (such as electrostatic forces) between p and x.
- The applicable *length scale* in the nonlocal term, $1/\lambda$, depends not only on the geometrical length scales (h_s and h_c), but also on the constituent material properties E_s and μ_c. Indeed, for the choices $h_s = h_c = 0.4$ mm, $E_s = 150$ GPa, $\mu_c = 4$ GPa, which are reasonably representative of typical composite material plies, one finds from (1.13) that

  $$1/\lambda = 1.41\text{mm}.$$

- A peridynamic model can contain multiple length scales; in the case of the composite model in (1.18) these are $1/\lambda$ and $1/\tau$.

The previous section demonstrated how nonlocality increases when we reduce the number of degrees of freedom in a model of discrete springs and masses. By reducing the number of masses by half, we increased the interaction distance by a factor of two. In the present section, we considered a similar effect in a continuous system. We started with two coupled displacement fields, $u_s(x)$ and $u_c(x,y)$. The equilibrium of these two fields was described by local interactions. When the decision was made to trade in these two fields for a single smoothed field $\bar{u}$, the price we paid was that the interactions then became nonlocal.

Alali and Lipton [1] also prove, by more abstract reasoning than the composite example presented here, that nonlocality is an essential feature of homogenization of heterogeneous media. In fact, the significance of nonlocality in homogenization has quite a long history; for an early example see [3].

Table 1.1: Stress concentration and decay distance near an open hole in a plate.

Material	n	$K_{\pi/2}$	r_0/r
Isotropic	2	3	1.65
Fabric ply	4.15	5.15	1.13
Unidirectional ply	10	11	1.07

1.3 Nonlocality at the macroscale

As noted in the previous section, the operative length scale in a heterogeneous material system depends on the constituent material properties as well as the geometrical length scale. However, the macroscale geometry of a body also combines with material properties to provide additional length scales. For example, consider the classic problem of an anisotropic plate under tension containing an open hole of radius r (Figure 1.8). The stress σ_{yy} as a function of position x along the midplane was derived analytically by Leknitskii [16]:

$$\sigma_{yy} = \frac{\sigma_\infty}{2}\left\{2+\left(\frac{r}{x}\right)^2+3\left(\frac{r}{x}\right)^4+(2-n)\left[5\left(\frac{r}{x}\right)^6-7\left(\frac{r}{x}\right)^8\right]\right\}$$

where

$$n=\sqrt{2\left(\frac{E_{11}}{E_{22}}-2\nu_{12}\right)+\frac{E_{11}}{G_{12}}}.$$

From this, the stress concentration at the edge of the hole is found to be

$$K_{\pi/2}=1+n.$$

The parameter n tends to increase in anisotropic materials, particularly those that have greater stiffness in the loading (y) direction. In addition to changing the stress concentration at the edge, anisotropy also changes the rate at which the stress decays at points moving away from the edge. Denote by r_0 the radius at which the stress drops off to half of its value at the edge, that is,

$$\frac{\sigma_{yy}(r_0)}{\sigma_{yy}(r)}=\frac{1}{2}.$$

Typical values of r_0 are given in Table 1.1. (This parameter plays an important role in certain laminate failure models that implicitly recognize nonlocality [25, 13].)

Thus, in a unidirectional ply, the stress decays to half its value at the edge over a distance of only 7% of the hole radius. If the hole radius is $r = 2.5$ mm, then this distance is $r_0 - r = 0.175$ mm, which is on the order of a typical ply thickness.

To put this result in the context of our previous discussion of length scales and nonlocality in heterogeneous media, compute the length scale implied by (1.13) with

$h_c = h_s = 0.1$ mm, $\mu_c = 4$ GPa, and $E_s = 150$ GPa. The resulting length scale for nonlocal interactions due to transfer of shear load between the constituent materials is

$$\frac{1}{\lambda} = \sqrt{\frac{E_s h_s h_c^2}{3\mu_c(h_s + h_c)}} = 0.25 \text{ mm}.$$

Comparing this with our stress concentration decay distance (0.175 mm), evidently nonlocality could play an important role in modeling the problem. Indeed, Toubal, Karama, and Lorrain [23] measured the stress in a fabric composite near an open hole as a function of position. Their measurements showed that the actual stress concentration is much lower than what is predicted by the analytical results from the local theory [16]. Does nonlocality explain this difference?

As shown in Figure 1.8, the contours of σ_{yy} (normal stress in the loading direction) show strong gradients in the vicinity of the hole. In fact, there are large gradients above and below the hole as well, due primarily to the relatively small shear modulus μ_{xy}, which is characteristic of fiber and fabric reinforced composites. In an isotropic material such as a typical metal, the contours of stress would be more diffuse. Figure 1.9 shows a comparison between the optically measured stress σ_{yy} and the local theory of anisotropic media [23]. The figure also shows results from a peridynamic computational model with length scale 2 mm. The results in this figure suggest that nonlocality helps improve the agreement between a continuum model and measured data for stress concentrations in composites. This also helps explain why failure criteria in composites that rely solely on the predicted stress in the local theory tend to underpredict the failure load in open-hole tension and compression tests: these criteria are based on an overprediction of the stress concentration [22].

Figure 1.8: Stress contours in a simulated open hole tension test in a composite. Note the large stress gradients.

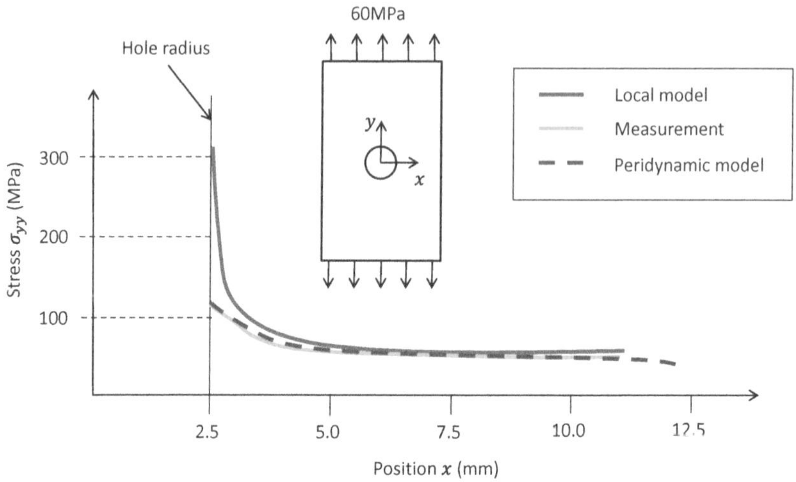

Figure 1.9: Comparison of stress along the midplane on an open hole tension test with a composite. The local theory overpredicts the stress concentration compared with optically measured data [23]. The peridynamic model offers improved agreement due to nonlocality.

1.4 The mixed blessing of nonlocality

In this chapter, we discussed how the assumptions in the standard theory of solid mechanics result in a mathematical system with desirable properties. Yet these assumptions lead to some limitations in applying the theory to many problems of current and future technological interest. We argued that the peridynamic theory (which includes the standard theory as a limiting case [9, 21]) may provide a suitable framework for extending the standard theory. We also showed that nonlocality, which is an essential aspect of any theory that sets out to do what peridynamics attempts to do, has a variety of origins. These origins include long-range physical interactions, such as van der Waals forces, that are significant in the mechanics of many small-scale physical systems. Nonlocality may also arise in a model due to the choice of variables, or degrees of freedom, that is made. For example, it was demonstrated in Section 1.2.3 that smoothing (homogenization) of a heterogeneous medium results in a nonlocal model that can be cast in the form of peridynamics.

In spite of all this, practical difficulties arise in modeling problems using a nonlocal method such as peridynamics. These include the following:

- A peridynamic computational model is usually slower than a finite element code that solves the standard (local) equations, due to the larger number of interactions that need to be computed.

- A peridynamic model of wave propagation in a macroscale system may exhibit excessive wave dispersion, due to the finite interaction distance between material points.
- The properties of a material are affected by proximity to boundaries and interfaces. (This leads to errors in applications in which the natural length scale of the material is much smaller than the horizon; see [17].)
- Certain material models are susceptible to unwanted zero-energy mode oscillations.

All of these practical difficulties are the subjects of research to make peridynamics a more useful tool in solving problems. For example, several groups are investigating coupling methods that will allow analysts to apply peridynamics in the vicinity of a subregion that is undergoing damage, while applying the standard theory everywhere else. Multiscale methods under development for peridynamics will adaptively allow greater resolution, and a smaller length scale, to be used near such features as shock waves and crack tips. New material models show promise for reduced effects from boundaries and interfaces, as well as resistance to zero-energy modes. Some of these advances are discussed in the remainder of this book.

References

[1] B. Alali and R. Lipton. Multiscale dynamics of heterogeneous media in the peridynamic formulation. *Journal of Elasticity*, 106(1):71–103, 2012.

[2] Z. P. Bazant. Why continuum damage is nonlocal: Micromechanics arguments. *Journal of Engineering Mechanics*, 117(5):1070–1087, 1991.

[3] M. Beran and J. McCoy. Mean field variations in a statistical sample of heterogeneous linearly elastic solids. *International Journal of Solids and Structures*, 6(8):1035–1054, 1970.

[4] C. Boutin. Microstructural effects in elastic composites. *International Journal of Solids and Structures*, 33(7):1023–1051, 1996.

[5] C. J. Brinker, Y. Lu, A. Sellinger, and H. Fan. Evaporation-induced self-assembly: Nanostructures made easy. *Advanced materials*, 11(7):579–585, 1999.

[6] A. Chakraborty. Wave propagation in anisotropic media with non-local elasticity. *International journal of solids and structures*, 44(17):5723–5741, 2007.

[7] M. Di Paola, G. Failla, and M. Zingales. Physically-based approach to the mechanics of strong non-local linear elasticity theory. *Journal of Elasticity*, 97(2):103–130, 2009.

[8] W. Drugan and J. Willis. A micromechanics-based nonlocal constitutive equation and estimates of representative volume element size for elastic composites. *Journal of the Mechanics and Physics of Solids*, 44(4):497–524, 1996.

[9] E. Emmrich, O. Weckner, et al. On the well-posedness of the linear peridynamic model and its convergence towards the navier equation of linear elasticity. *Communications in Mathematical Sciences*, 5(4):851–864, 2007.

[10] A. C. Eringen and D. Edelen. On nonlocal elasticity. *International Journal of Engineering Science*, 10(3):233–248, 1972.

[11] J. Fish, W. Chen, and G. Nagai. Non-local dispersive model for wave propagation in heterogeneous media: One-dimensional case. *International Journal for Numerical Methods in Engineering*, 54(3):331–346, 2002.

[12] J. N. Israelachvili. *Intermolecular and surface forces*: 2nd ed. Academic Press, 1992.

[13] W. L. Ko. Stress concentration around a small circular hole in the HIMAT composite plate. Technical report, NASA Technical Memorandum 86038, 1985.

[14] E. Kröner. Elasticity theory of materials with long range cohesive forces. *International Journal of Solids and Structures*, 3(5):731–742, 1967.

[15] I. A. Kunin. *Elastic media with microstructure I: One-dimensional models*. Springer, 1982.

[16] G. Lekhnitskii, W. S. Tsai, and T. Cheron. *Anisotropic plates*: 2nd ed. Gordon and Breach Science, 1968.

[17] E. Madenci and E. Oterkus. *Peridynamic theory and its applications*. Springer, 2014.

[18] S. A. Silling, Reformulation of elasticity theory for discontinuities and long-range forces, *Journal of the Mechanics and Physics of Solids*, **48**(1): 175–209, 2000.

[19] S. A. Silling. A coarsening method for linear peridynamics. *International Journal for Multiscale Computational Engineering*, 9(6):609–622, 2011.

[20] S. A. Silling. Origin and effect of nonlocality in a composite. *Journal of Mechanics of Materials and Structures*, 9(2):245–258, 2014.

[21] S. A. Silling and R. B. Lehoucq. Convergence of peridynamics to classical elasticity theory. *Journal of Elasticity*, 93(1):13–37, 2008.

[22] C. Soutis, N. Fleck, and P. Smith. Failure prediction technique for compression loaded carbon fibre-epoxy laminate with open holes. *Journal of Composite Materials*, 25(11):1476–1498, 1991.

[23] L. Toubal, M. Karama, and B. Lorrain. Stress concentration in a circular hole in composite plate. *Composite Structures*, 68(1):31–36, 2005.

[24] G. M. Whitesides and B. Grzybowski. Self-assembly at all scales. *Science*, 295(5564):2418–2421, 2002.

[25] J. M. Whitney and R. Nuismer. Stress fracture criteria for laminated composites containing stress concentrations. *Journal of Composite Materials*, 8(3):253–265, 1974.

[26] J. Willis. The nonlocal influence of density variations in a composite. *International Journal of Solids and Structures*, 21(7):805–817, 1985.

Chapter 2

Introduction to Peridynamics

Stewart A. Silling

Sandia National Laboratories

CONTENTS

2.1 Equilibrium in terms of integral equations 26
2.2 Material modeling 28
  2.2.1 Bond-based materials 28
  2.2.2 Relation between bond densities and flux 30
  2.2.3 Peridynamic states 32
  2.2.4 Ordinary state-based materials 35
  2.2.5 Correspondence materials 37
  2.2.6 Discrete particles as peridynamic bodies 39
  2.2.7 Setting the horizon 40
  2.2.8 Linearized peridynamics 41
2.3 Plasticity 42
  2.3.1 Bond-based microplastic material 42
  2.3.2 LPS material with plasticity 43
2.4 Damage and fracture 45
  2.4.1 Damage in bond-based models 45
  2.4.2 Damage in ordinary state-based material models 47
  2.4.3 Damage in correspondence material models 49
  2.4.4 Nucleation strain 50
2.5 Treatment of boundaries and interfaces 51
  2.5.1 Bond-based materials 51
  2.5.2 State-based materials 53
2.6 Emu numerical method 54
2.7 Conclusions 56
References 57

2.1 Equilibrium in terms of integral equations

Most of the literature on peridynamics treats the subject in general, mathematical terms. The original paper [31], the review article [36], and the book [23] contain derivations and details of the basic theory. In the spirit of the present volume as a handbook, rather than repeat this type of presentation, here we emphasize some techniques and material models that have proven to be useful in modeling applications. The following is a mechanically intuitive but less rigorous way of obtaining the same equations.

As discussed in the previous chapter, the purpose of peridynamics is to unify the mechanics of continuous media, continuous media with evolving discontinuities, and discrete particles. To accomplish this, peridynamics avoids the use of partial derivatives of the deformation with respect to spatial coordinates. Instead, it uses integral equations that remain valid on discontinuities. Discrete particles, as will be discussed later in this chapter, are treated using Dirac delta functions.

Imagine a continuous body $\mathcal{B}$ in a reference configuration. Let $\mathbf{x}$ and $\mathbf{q}$ be distinct points in the body, and let $dV_\mathbf{x}$ and $dV_\mathbf{q}$ be small volumes containing these points. The force that the material in $dV_\mathbf{q}$ exerts on $dV_\mathbf{x}$ is denoted $\mathbf{f}(\mathbf{q},\mathbf{x})\, dV_\mathbf{x}\, dV_\mathbf{q}$ where $\mathbf{f}$ is the *pairwise bond force density* function. The value of this function is determined by the deformation and material properties through the constitutive model, which will be discussed later in this chapter. In dynamic problems, the function $\mathbf{f}$ also depends on time, although in some cases we will shorten the notation by omitting the time variable.

The pairwise bond force density is required to obey the following antisymmetry relation:

$$\mathbf{f}(\mathbf{x},\mathbf{q}) = -\mathbf{f}(\mathbf{q},\mathbf{x})$$

for any $\mathbf{x}$ and $\mathbf{q}$. This property ensures that linear momentum is conserved in the model. It also can be regarded as a form of Newton's third law. The dimensions of $\mathbf{f}$ are force/volume^2. The vector in the reference configuration that connects $\mathbf{x}$ to $\mathbf{q}$ is called a *bond*.

Now consider the balance of forces on $dV_\mathbf{x}$ in equilibrium. Suppose there is a prescribed *body force density* field $\mathbf{b}(\mathbf{x})$ such that the net external force on $dV_\mathbf{x}$ is $\mathbf{b}(\mathbf{x})\, dV_\mathbf{x}$. Including interactions with other material particles, the total force on the small volume surrounding $\mathbf{x}$, which is required to vanish in equilibrium, is then

$$\int_{\mathcal{B}} [\mathbf{f}(\mathbf{q},\mathbf{x})\, dV_\mathbf{x}]\, dV_\mathbf{q} + \mathbf{b}(\mathbf{x})\, dV_\mathbf{x} = \mathbf{0}$$

which must hold for every $\mathbf{x}$ in the body (Figure 2.1). From this, the 3D peridynamic equilibrium equation follows immediately:

$$\int_{\mathcal{B}} \mathbf{f}(\mathbf{q},\mathbf{x})\, dV_\mathbf{q} + \mathbf{b}(\mathbf{x}) = \mathbf{0} \tag{2.1}$$

which holds for every point $\mathbf{x}$ in the body. The dynamic version of (2.1) is obtained from d'Alembert's principle and is called the peridynamic equation of motion, which has the following form in 3D:

$$\rho(\mathbf{x})\ddot{\mathbf{y}}(\mathbf{x},t) = \int_{\mathcal{B}} \mathbf{f}(\mathbf{q},\mathbf{x},t)\, dV_{\mathbf{q}} + \mathbf{b}(\mathbf{x},t)$$

for all $\mathbf{x}$ in the body and all time t, where $\mathbf{y}$ is the deformation map. Here, ρ is the density field and $\ddot{\mathbf{y}}$ is the acceleration field.

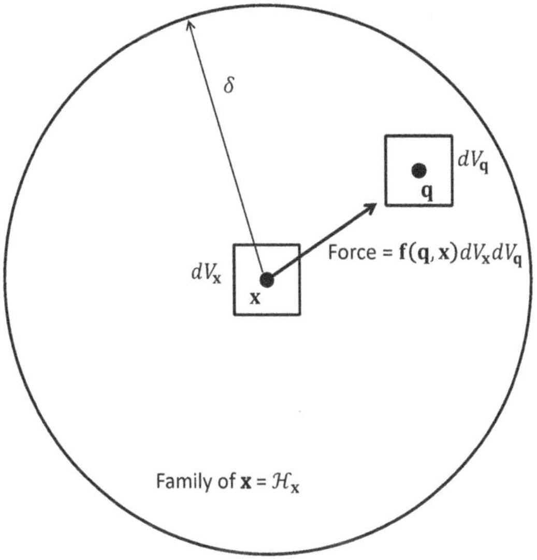

Figure 2.1: Volume element $dV_{\mathbf{q}}$ exerts a force $\mathbf{f}(\mathbf{q},\mathbf{x})\, dV_{\mathbf{x}}\, dV_{\mathbf{q}}$ on volume element $dV_{\mathbf{x}}$.

In peridynamics, it is assumed that for a given material, bonds longer than a certain distance δ do not interact directly through the material model. The parameter δ is called the *horizon* because the point $\mathbf{x}$ cannot "see" its neighbors $\mathbf{q}$ past this distance. Points $\mathbf{x}$ and $\mathbf{q}$ that are separated by a distance greater than the horizon in the reference configuration do not interact in the material model applied at either point. Material within the horizon of $\mathbf{x}$ is called the *family* of $\mathbf{x}$, denoted $\mathcal{H}_{\mathbf{x}}$. The equation of motion therefore may be written as follows:

$$\rho(\mathbf{x})\ddot{\mathbf{y}}(\mathbf{x},t) = \int_{\mathcal{H}_{\mathbf{x}}} \mathbf{f}(\mathbf{q},\mathbf{x},t)\, dV_{\mathbf{q}} + \mathbf{b}(\mathbf{x},t). \tag{2.2}$$

When working in 2D or 1D, rather than rewrite all the equations, it is convenient to change only the differential dV that appears in the integral according to the number of dimensions D. Define

$$dV = h_D \begin{cases} \text{differential volume} & \text{if } D = 3 \\ \text{differential area} & \text{if } D = 2 \\ \text{differential length} & \text{if } D = 1 \end{cases}$$

where

$$h_D = \begin{cases} 1 & \text{if } D = 3 \\ \text{plate thickness} & \text{if } D = 2 \\ \text{rod cross-sectional area} & \text{if } D = 1. \end{cases}$$

With this convention, the dimensions of $\mathbf{f}$ are force/volume^2 and the dimensions of ρ are mass/volume, regardless of D. Geometrically, $\mathcal{H}_{\mathbf{x}}$ is a sphere, disk, or line segment if $D = 3, 2, 1$, respectively. $\mathcal{H}_{\mathbf{x}}$ could also be a subset of any of these shapes if $\mathbf{x}$ is near a boundary. Differentials such as $dV_{\mathbf{q}}$ signify differential volume, area, or length, located at the point $\mathbf{q}$.

2.2 Material modeling

2.2.1 Bond-based materials

In peridynamics, a constitutive model provides values for $\mathbf{f}(\mathbf{q}, \mathbf{x})$ according to the deformation of the body and its material properties. The simplest peridynamic material model is *bond-based*, in which an $\mathbf{x}$ and $\mathbf{q}$ exchange forces through spring-like interactions:

$$\mathbf{f}(\mathbf{q}, \mathbf{x}) = f(|\mathbf{y}(\mathbf{q}) - \mathbf{y}(\mathbf{x})|, \mathbf{q}, \mathbf{x})\mathbf{M}, \qquad \mathbf{M} = \frac{\mathbf{y}(\mathbf{q}) - \mathbf{y}(\mathbf{x})}{|\mathbf{y}(\mathbf{q}) - \mathbf{y}(\mathbf{x})|} \tag{2.3}$$

where f is a function, $\mathbf{y}(\mathbf{x})$ is the deformed position of $\mathbf{x}$, and $\mathbf{M}$ is a unit vector in the direction of the deformed bond from $\mathbf{x}$ to $\mathbf{q}$. An example of a bond-based material is given by

$$\mathbf{f}(\mathbf{q}, \mathbf{x}) = \begin{cases} Ce\mathbf{M}, & \text{if } |\mathbf{q} - \mathbf{x}| \leq \delta, \\ \mathbf{0} & \text{otherwise,} \end{cases} \tag{2.4}$$

where C is a constant called the *micromodulus* and e is the *bond extension*, defined by

$$e = |\mathbf{y}(\mathbf{q}) - \mathbf{y}(\mathbf{x})| - |\mathbf{q} - \mathbf{x}|.$$

More generally, the micromodulus can depend on both endpoints of the bond, and is written $C(\mathbf{x}, \mathbf{q})$.

The distinguishing feature of bond based materials is that each bond force density $\mathbf{f}(\mathbf{q}, \mathbf{x})$ is independent of whatever happens in any other bond connected to either point. The material model (2.4) treats each bond like a linear elastic spring. Therefore, the bulk material properties are also elastic. This provides a way to calibrate C

in terms of measurable bulk elastic properties. The energy density (energy/volume^2) in each bond is called the *micropotential* and is denoted by w. In the case of the material in (2.4), this is given by

$$w = \frac{Ce^2}{2}. \tag{2.5}$$

The micropotential is related to the bond force density by

$$f = \frac{\partial w}{\partial e} \qquad \text{or} \qquad \mathbf{f} = \frac{\partial w}{\partial \mathbf{Y}}$$

where

$$\mathbf{Y} = \mathbf{y}(\mathbf{q}) - \mathbf{y}(\mathbf{x}).$$

To calibrate the peridynamic model in terms of the bulk elastic properties of a material, take the simplest case, in which C is a constant. Suppose the bulk modulus k is known. Consider an arbitrary point $\mathbf{x}$ in the interior of the body, far from any boundaries, and prescribe a deformation of the form

$$\mathbf{y}(\mathbf{x}) = (1+\varepsilon)\mathbf{x}$$

where ε is a small constant. The strain energy density W at $\mathbf{x}$ (energy/volume) is related to the micropotential by

$$W = \frac{1}{2}\int_{\mathcal{H}_\mathbf{x}} w \, dV_\mathbf{q}. \tag{2.6}$$

The factor of $1/2$ appears in (2.6) because, for bookkeeping purposes, each endpoint of a bond "owns" only one half of the energy stored in the bond. Combining (2.5) with (2.6), and setting

$$e = \varepsilon\xi, \qquad \xi = |\mathbf{q}-\mathbf{x}|$$

leads to

$$W = \frac{1}{2}\int_{\mathcal{H}_\mathbf{x}} \frac{Ce^2}{2}\, dV_\xi = \begin{cases} \pi C\varepsilon^2\delta^5/5 & \text{if } D=3, \\ \pi h_2 C\varepsilon^2\delta^4/8 & \text{if } D=2, \\ h_1 C\varepsilon^2\delta^3/6 & \text{if } D=1. \end{cases} \tag{2.7}$$

To complete the calibration process, we require that the peridynamic value of strain energy density W match the value from the standard theory of elasticity:

$$W = \begin{cases} 9k\varepsilon^2/2 & \text{if } D=3, \\ 2k'\varepsilon^2 & \text{if } D=2, \\ E\varepsilon^2/2 & \text{if } D=1, \end{cases} \tag{2.8}$$

where k is the bulk modulus, E is the Young's modulus, and k' is the two-dimensional bulk modulus, whose meaning is as follows. Under biaxial strain ($\varepsilon_{11} = \varepsilon_{22} = \varepsilon$), using the 2D dilatation and bulk modulus θ' and k',

$$\sigma_{11} = \sigma_{22} = k'\theta', \qquad \theta' = \varepsilon_{11} + \varepsilon_{22} = 2\varepsilon,$$

$$k' = \begin{cases} E/2(1-\nu) & \text{plane stress,} \\ E/2(1-\nu-2\nu^2) & \text{plane strain,} \end{cases}$$

where ν is the Poisson ratio. From (2.7) and (2.8), it follows that

$$C = \begin{cases} 45k/2\pi\delta^5 & \text{if } D = 3, \\ 16k'/\pi h_2 \delta^4 & \text{if } D = 2, \\ 3E/h_1 \delta^3 & \text{if } D = 1. \end{cases} \tag{2.9}$$

For more general bond-based material models, in which C depends on the initial bond vector, the same calibration process can be repeated, provided there is only one unknown parameter. For example, the choice

$$C(\mathbf{q}, \mathbf{x}) = \frac{c}{\xi}, \qquad \xi = |\mathbf{q} - \mathbf{x}|, \tag{2.10}$$

where c is a constant, corresponds to bond force that is proportional to *bond strain* (change in bond length divided by initial bond length), which is denoted by s. The force density and bond strain are given by

$$f = cs, \qquad s = \frac{e}{\xi}, \tag{2.11}$$

and the calibration for c results in

$$c = \begin{cases} 18k/\pi\delta^4 & \text{if } D = 3, \\ 12k'/\pi h_2 \delta^3 & \text{if } D = 2, \\ 2E/h_1 \delta^2 & \text{if } D = 1. \end{cases} \tag{2.12}$$

This material specified by (2.11) and (2.12) is called the *prototype microelastic* (PM) material. With various enhancements such as plasticity and damage, the PM material perhaps accounts in practice for about 90% of the material modeling done in applications of peridynamics. The constant c in the PM model is called the *spring constant*. Note that PM material is a special case of a microelastic material for the particular choice of micromodulus function given in (2.10). Anisotropy can be incorporated into the general framework of the PM model by allowing the spring constant to depend on bond direction [41, 16, 14, 15].

2.2.2 Relation between bond densities and flux

In relating peridynamic quantities to variables that are familiar from the standard theory of solid mechanics, it is often useful to derive a density of some quantity per unit area, which we will call a *flux*, from peridynamic densities.

An example is the derivation of a stress component, say σ_{11}, from a pairwise bond force density field $\mathbf{f}$. To make this connection, assume that the body is homogeneous and the deformation is uniform. Consider a plane normal to the x_1-axis. Also consider a thin cylinder to the left of the plane (Figure 2.2) with cross-sectional area dA and length δ, where δ is the horizon of the material. Force through the plane on the cylinder is carried by the bonds that have one end in the cylinder and the other end somewhere on the right side of the plane. A typical point $\mathbf{x}$ in the cylinder is located a distance z to the left of the plane, $0 < z \leq \delta$. The force density (per unit volume[2]) in a typical bond connecting this point to the right side of the plane is given by $\mathbf{f}(\mathbf{q}, \mathbf{x})$, where $q_1 > 0$. Using a spherical coordinate system in which ϕ is the angle from the x_1-axis, and ξ is the bond length, the total force on the cylinder is then (in 3D)

$$d\mathbf{F} = dA \int_0^{2\pi} \int_0^{\delta} \int_0^{\xi} \int_0^{\cos^{-1}(z/\xi)} \mathbf{f}(\xi, \phi, \theta) \xi^2 \sin\phi \, d\phi \, dz \, d\xi \, d\theta.$$

Since, under the present assumptions, the normal stress component is given by $\sigma_{11} = dF_1/dA$, it follows that

$$\sigma_{11} = \int_0^{2\pi} \int_0^{\delta} \int_0^{\xi} \int_0^{\cos^{-1}(z/\xi)} f_1(\xi, \phi, \theta) \xi^2 \sin\phi \, d\phi \, dz \, d\xi \, d\theta. \qquad (2.13)$$

The bond-based microelastic material whose pairwise force density is given by (2.4), under isotropic extension ($e = \varepsilon\xi$, where ε is a small constant) has a normal stress found from

$$f_1 = CeM_1 = C\varepsilon\xi \cos\phi;$$

hence, using (2.9),

$$\begin{aligned} \sigma_{11} &= 2\pi \int_0^{\delta} \int_0^{\xi} \int_0^{\cos^{-1}(z/\xi)} (C\varepsilon\xi \cos\phi) \xi^2 \sin\phi \, d\phi \, dz \, d\xi \\ &= 3k\varepsilon, \end{aligned}$$

as expected from our calibration process. (The 3 appears because the dilatation is 3ε.)

The computation of σ_{11} is an example of a flux determined from peridynamic bond quantities. More generally, a flux dP/dA_1 (per unit area in the plane normal to x_1) can be computed from the following expressions:

- Three dimensions:

$$\frac{dP}{dA_1} = \int_0^{2\pi} \int_0^{\delta} \int_0^{\xi} \int_0^{\cos^{-1}(z/\xi)} p(\xi, \phi, \theta) \xi^2 \sin\phi \, d\phi \, dz \, d\xi \, d\theta, \qquad (2.14)$$

- Two dimensions:

$$\frac{dP}{dA_1} = h_2 \int_0^{\delta} \int_0^{\xi} \int_0^{\cos^{-1}(z/\xi)} p(\xi, \phi) \xi \, d\phi \, dz \, d\xi, \qquad (2.15)$$

- One dimension:

$$\frac{dP}{dA_1} = h_1 \int_0^{\delta} \int_0^{\xi} p(\xi) \, dz \, d\xi, \qquad (2.16)$$

where p is some peridynamic bond density, either a scalar or a vector, defined per unit volume[2]. We emphasize that this expression applies only if the body is homogeneous and the deformation is uniform.

As an example, the interested reader may wish to confirm that for a 1D bar made of PM material under constant strain ε, the stress predicted by (2.16) comes out to $\sigma = E\varepsilon$. The flux in (2.13) is a special case (uniform deformation of a homogeneous body) of a more general, but more complex expression derived in [20]. In this work, a general nonlocal stress tensor is derived that permits the peridynamic integral to be written as the divergence of the stress tensor field. In other words, if this nonlocal stress is used, the peridynamic equation of motion is formally the same as that of the standard (local) theory. Another alternative formulation of the peridynamic equations uses a newly developed *nonlocal calculus* and is described in Chapter 3 of this book.

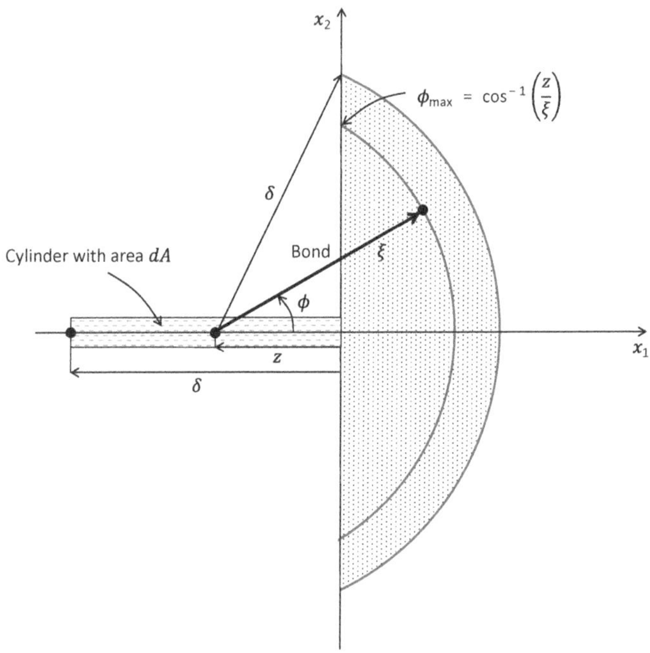

Figure 2.2: Computation of force per unit area from bond force densities. θ is the angle in the x_2-x_3 plane.

2.2.3 Peridynamic states

Material models of the form (2.3) carry with them the assumption that each bond connected to $\mathbf{x}$ responds independently of all the others. This assumption limits the types of material response that can be reproduced with this type of material model, even when the function $\mathbf{f}$ includes nonlinearities. In particular, the small-deformation elastic response of a bond-based material model is always 1/4.

This restriction to $\nu = 1/4$ also applies to a molecular dynamic system in which the atoms interact only through pair potentials. More general response in molecular dynamics must be modeled using multibody potentials. Similarly, in peridynamics, to obtain more general material response, it is necessary to introduce interdependencies between bonds beyond pair interactions. This generalization is provided using the idea of peridynamic states.

A peridynamic *state* is a function defined on the bonds in a family of some point in the body. The concept we wish to express is that the force density in a given bond $\mathbf{q}-\mathbf{x}$ depends not only on the deformation of that particular bond, but also the deformations of other bonds connected to $\mathbf{x}$ and $\mathbf{q}$. To express this, we write

$$\mathbf{f}(\mathbf{q},\mathbf{x}) = \mathbf{t}(\mathbf{q},\mathbf{x}) - \mathbf{t}(\mathbf{x},\mathbf{q})$$

where the $\mathbf{t}$ are bond force density vectors given by

$$\mathbf{t}(\mathbf{q},\mathbf{x}) = \underline{\mathbf{T}}[\mathbf{x}]\langle \mathbf{q}-\mathbf{x}\rangle, \qquad \mathbf{t}(\mathbf{x},\mathbf{q}) = \underline{\mathbf{T}}[\mathbf{q}]\langle \mathbf{x}-\mathbf{q}\rangle.$$

Here, $\underline{\mathbf{T}}[\mathbf{x}]$ and $\underline{\mathbf{T}}[\mathbf{q}]$ are the *force states* at $\mathbf{x}$ and $\mathbf{q}$, respectively. The state $\underline{\mathbf{T}}[\mathbf{x}]$ associates with each bond $\mathbf{q}-\mathbf{x}$ in the family of $\mathbf{x}$ some force density vector. By convention, the bond that a state is operating on is written in angle brackets, $\langle \mathbf{q}-\mathbf{x}\rangle$. The force state at a point is analogous to the stress tensor at a point in the standard theory.

A material model in state-based peridynamics associates a force state at $\mathbf{x}$ with the collective deformation of the entire family of $\mathbf{x}$. To express the deformation of the entire family, define the *deformation state* by

$$\underline{\mathbf{Y}}[\mathbf{x}]\langle \mathbf{q}-\mathbf{x}\rangle = \mathbf{y}(\mathbf{q}) - \mathbf{y}(\mathbf{x}).$$

Thus, the state $\underline{\mathbf{Y}}$ maps each bond onto its deformed image. By introducing state notation, it is easy to express a material model:

$$\underline{\mathbf{T}} = \hat{\underline{\mathbf{T}}}(\underline{\mathbf{Y}}).$$

Here, the material model is the function $\hat{\underline{\mathbf{T}}}$, which is a state-valued function of a state (Figure 2.3). The key concept in state-based material modeling is that the bond force density depends on the deformation *state*, which is itself a function, rather than on the deformation of just that particular bond.

To model an elastic material, we retain the basic notion of strain energy density $W(\mathbf{x})$ that is used in the standard theory of continuum mechanics. However, in peridynamics, the strain energy density depends on the deformation of the family of $\mathbf{x}$:

$$W(\mathbf{x}) = \hat{W}(\underline{\mathbf{Y}}[\mathbf{x}])$$

where $\hat{W}$ is a scalar-valued function of a state. To relate this function to the force state, we consider the change in strain energy density $d\hat{W}$ resulting from a small change

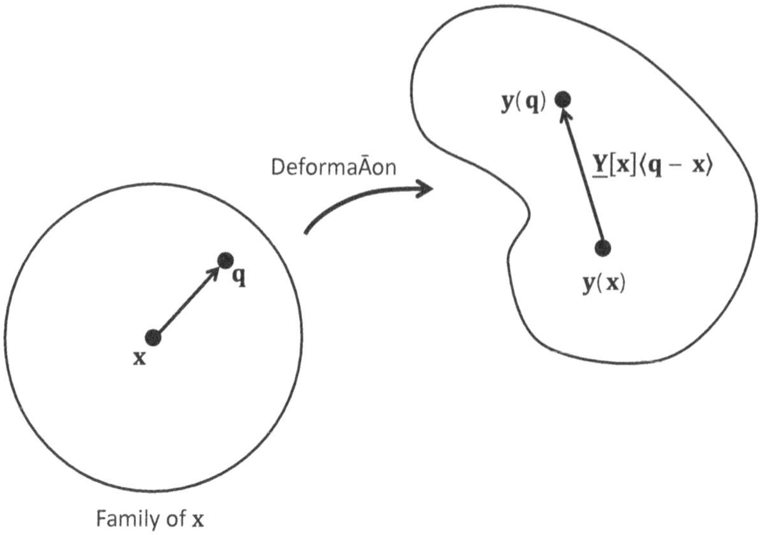

Figure 2.3: The deformation state $\underline{\mathbf{Y}}$ maps each bond to its deformed image.

in the deformation state $d\underline{\mathbf{Y}}$. The incremental change in strain energy density must equal the net work done by the bond forces on the incremental bond deformations, thus

$$d\hat{W} = \underline{\mathbf{T}} \bullet d\underline{\mathbf{Y}} \tag{2.17}$$

where the inner product of the two states is defined by

$$\underline{\mathbf{A}} \bullet \underline{\mathbf{B}} = \int_{\mathcal{H}_{\mathbf{x}}} \underline{\mathbf{A}}\langle\xi\rangle \cdot \underline{\mathbf{B}}\langle\xi\rangle \, dV_{\xi}$$

and where, for any two vectors $\mathbf{a}$ and $\mathbf{b}$,

$$\mathbf{a} \cdot \mathbf{b} = a_1 b_1 + a_2 b_2 + a_3 b_3.$$

The dimensions of both sides of (2.17) are energy/volume. The mathematical tool for defining the relation between $\hat{W}$ and $\hat{\underline{\mathbf{T}}}$ is provided by the Fréchet derivative. If $\Psi(\underline{\mathbf{A}})$ is a function of a state, then it has a Fréchet derivative $\Psi_{\underline{\mathbf{A}}}(\underline{\mathbf{A}})$ if, for every infinitesimal change $d\underline{\mathbf{A}}$,

$$\Psi(\underline{\mathbf{A}} + d\underline{\mathbf{A}}) = \Psi(\underline{\mathbf{A}}) + \Psi_{\underline{\mathbf{A}}}(\underline{\mathbf{A}}) \bullet d\underline{\mathbf{A}}.$$

Note that $\Psi_{\underline{\mathbf{A}}}$, if it exists, is a state-valued function. See [34] for further discussion and examples of Fréchet derivatives. Therefore, in an elastic material, the force state is the Fréchet derivative of strain energy density:

$$\hat{\underline{\mathbf{T}}} = \hat{W}_{\underline{\mathbf{Y}}}.$$

The total strain energy in a bounded body $\mathcal{B}$ is given by

$$E = \int_{\mathcal{B}} W(\mathbf{x})\, dV_{\mathbf{x}} = \int_{\mathcal{B}} \hat{W}(\underline{\mathbf{Y}}[\mathbf{x}])\, dV_{\mathbf{x}}.$$

This equation contains a fundamental and important statement: the total stored energy is distributed among all the points in the body according to the deformation of the family of each point. To see the significance of this statement, compare it with an alternative assumption that the total energy in the body is "a function of everything." This alternative statement can be expressed in the form

$$E = \hat{E}(\mathbf{y})$$

where $\hat{E}$ is a function of the deformation map $\mathbf{y}$, which is itself a function defined on the entire body. In this alternative view, there is no concept of a material model that is "portable" from point to point, yielding the same strain energy density, for a given deformation of the point's family, independent of the initial position of the point. The existence of such a material model is a fundamental assumption in both the standard theory of solid mechanics and in peridynamics.

It is worth emphasizing that in the peridynamic treatment of an elastic body, the two parts of the pairwise bond force $\mathbf{f}(\mathbf{q},\mathbf{x})$ due to the material models at $\mathbf{x}$ and $\mathbf{q}$ contribute separately to changes in strain energy density at these two points, even though the net force density is equal in magnitude between them. This is very different from the bond-based theory, in which the energy in the bond from $\mathbf{x}$ to $\mathbf{q}$ can be thought of as being stored in an elastic spring connecting these two points. Actually, a bond-based material model can be expressed as a state-based model by defining

$$\hat{W}(\underline{\mathbf{Y}}) = \frac{1}{2}\int_{\mathcal{H}_{\mathbf{x}}} w(|\underline{\mathbf{Y}}\langle \mathbf{q}-\mathbf{x}\rangle|, \mathbf{q}, \mathbf{x})\, dV_{\mathbf{q}}$$

where w is the bond-based micropotential. This demonstrates that the bond-based theory is a special case, indeed a very restricted special case, of the state-based theory.

In any force state, it is required that the net moment on $\mathbf{x}$ due to the bond forces is zero. This requirement is called *non-polarity* and is written

$$\int_{\mathcal{H}} \underline{\mathbf{T}}\langle \xi \rangle \times \underline{\mathbf{Y}}\langle \xi \rangle\, dV_{\xi} = \mathbf{0}. \tag{2.18}$$

Non-polarity assures that total angular momentum is constant in a bounded body, in the absence of external loads.

2.2.4 Ordinary state-based materials

State-based materials are categorized as *ordinary* or *non-ordinary*. In an ordinary material, the bond force density vector $\underline{\mathbf{T}}\langle \xi \rangle$ is always parallel to the deformed bond $\underline{\mathbf{Y}}\langle \xi \rangle$, where ξ is any bond in the family. In a non-ordinary material, these vectors are not necessarily parallel. Ordinary materials automatically satisfy the non-polarity requirement (2.18), because the vector cross-product in the integrand always vanishes. In a non-ordinary material model, non-polarity must be built into the material model and is not automatically satisfied.

If $\underline{\mathbf{T}}$ is the force state in an ordinary material, because the bond forces are always parallel to the deformed bonds, there must be a scalar-valued state $\underline{g}$ such that

$$\underline{\mathbf{T}} = \underline{g}\underline{\mathbf{Y}}$$

which is shorthand for

$$\underline{\mathbf{T}}\langle\xi\rangle = \underline{g}\langle\xi\rangle\underline{\mathbf{Y}}\langle\xi\rangle \qquad \text{for all } \xi \in \mathcal{H}.$$

If the material is also elastic, we also have $\underline{\mathbf{T}} = \hat{W}_{\underline{\mathbf{Y}}}$. So, from the definition of the Fréchet derivative, for any incremental change in the deformation,

$$d\hat{W} = \int_{\mathcal{H}} \underline{g}\langle\xi\rangle\underline{\mathbf{Y}}\langle\xi\rangle \cdot d\underline{\mathbf{Y}}\langle\xi\rangle \, dV_\xi = \frac{1}{2}\int_{\mathcal{H}} \underline{g}\langle\xi\rangle \, d|\underline{\mathbf{Y}}\langle\xi\rangle|^2 \, dV_\xi.$$

From this, it follows that $\hat{W}$ depends only on the deformed length of bonds, and

$$\underline{g} = 2\hat{W}_{|\underline{\mathbf{Y}}|^2}.$$

The significance of this result is that in formulating an elastic ordinary material model, only the length changes of bonds need to be considered, rather than the bond rotations.

An example of an elastic ordinary material model is obtained by assuming that the strain energy density depends on a weighted average of the bond extensions. Defining the *extension state* by

$$\underline{e}\langle\xi\rangle = |\underline{\mathbf{Y}}\langle\xi\rangle| - |\xi|$$

for any ξ in the family, we can define a nonlocal dilatation by

$$\vartheta = \frac{3}{m}\int_{\mathcal{H}} \underline{\omega}\langle\xi\rangle|\xi|\underline{e}\langle\xi\rangle \, dV_\xi, \qquad m = \int_{\mathcal{H}} \underline{\omega}\langle\xi\rangle|\xi|^2 \, dV_\xi \tag{2.19}$$

where $\underline{\omega}$ is a radially symmetric weighting function called the *influence function* and m is a normalization constant. The factor of 3 appears so that for a small isotropic expansion, that is, $|\underline{e}\langle\xi\rangle| = \varepsilon|\xi|$, the nonlocal dilatation agrees with the local dilatation, $\vartheta = 3\varepsilon$. If the strain energy density depends only on the nonlocal dilatation, after evaluating the necessary Fréchet derivatives, the force state is found to be

$$\underline{\mathbf{T}}\langle\xi\rangle = \frac{-3p\underline{\omega}\langle\xi\rangle|\xi|}{m}\mathbf{M}, \qquad \mathbf{M} = \frac{\underline{\mathbf{Y}}\langle\xi\rangle}{|\underline{\mathbf{Y}}\langle\xi\rangle|} \tag{2.20}$$

where the nonlocal pressure is given by

$$p = -\frac{\partial W}{\partial \vartheta}.$$

The force state given by (2.20) with the special choice

$$p = -k\vartheta$$

where k is the bulk modulus results in the material model for the *linear peridynamic fluid* (LPF). This model uses the nonlocal dilatation, that is, the mean length change of the bonds, to compute the nonlocal pressure. This nonlocal pressure is mapped to the bond forces. The LPF material model is an example of an ordinary state material model because the bond force density vectors are always parallel to the deformed bond vectors [34].

The LPF model can be extended to model a solid. In this extension, the peridynamic version of shear strain is found by subtracting the dilatational part of the deformation from the extension state, resulting in the *deviatoric extension state*:

$$\underline{e}^{\mathrm{d}}\langle\xi\rangle = \underline{e}\langle\xi\rangle - \frac{\vartheta|\xi|}{3}. \tag{2.21}$$

In an isotropic expansion, the deviatoric extension state would vanish. In the *linear peridynamic solid* (LPS) material model, the fluid response given in (2.20) is augmented by a deviatoric term to yield

$$\underline{\mathbf{T}}\langle\xi\rangle = \left(\frac{3k\vartheta\underline{\omega}\langle\xi\rangle|\xi|}{m} + \alpha\underline{\omega}\langle\xi\rangle\underline{e}^{\mathrm{d}}\langle\xi\rangle\right)\mathbf{M}, \tag{2.22}$$

where α is a constant that can be calibrated according to the measured shear modulus of the material. In the interior of a body (that is, no closer to any boundary or interface than δ) the calibration is

$$\alpha = \frac{15\mu}{m}.$$

However, near a boundary or interface, this relation does not hold, and in fact the response predicted by the LPS model is anisotropic. See [34] for more details about the LPS model.

The choice of influence function, although essentially arbitrary, affects the nonlocal response of a peridynamic state-based material model. In the numerical method described below, the smoothness of $\underline{\omega}$ can affect the accuracy of a discretized model. Seleson and Parks [30] have studied these and other aspects of the influence function in ordinary state models.

2.2.5 Correspondence materials

So far we have given examples of bond based and ordinary state material models. An example of a class of non-ordinary material model is the *correspondence* materials. In these materials, a classical (local) material model is adapted to the peridynamic setting. To do this, consider a classical elastic material model of the form $\sigma(\mathbf{F})$, where σ is the Piola stress tensor and $\mathbf{F}$ is the deformation gradient tensor, $\mathbf{F} = \partial\mathbf{y}/\partial\mathbf{x}$. In a correspondence model, the approximate deformation gradient tensor $\bar{\mathbf{F}}$ is computed from the bond deformations in a family:

$$\bar{\mathbf{F}} = \left(\int_{\mathcal{H}} \underline{\omega}\langle\xi\rangle \underline{\mathbf{Y}}\langle\xi\rangle \otimes \xi \, dV_\xi \right) \mathbf{K}^{-1}, \qquad \mathbf{K} = \int_{\mathcal{H}} \underline{\omega}\langle\xi\rangle \xi \otimes \xi \, dV_\xi \tag{2.23}$$

where the symbol $\otimes$ denotes the tensor (dyadic) product of two vectors and $\mathbf{K}$ is the *shape tensor*, which is always symmetric. It can be shown that if $\underline{\omega}$ is positive on $\mathcal{H}$, then $\mathbf{K}$ is also positive definite, implying that there is no doubt about the existence of its inverse, which is required in the expression for $\bar{\mathbf{F}}$. Once the approximate deformation gradient is known, the corresponding stress tensor is mapped into the bond forces:

$$\underline{\mathbf{T}}\langle\xi\rangle = \underline{\omega}\langle\xi\rangle \sigma(\bar{\mathbf{F}}) \mathbf{K}^{-1} \xi$$

for any bond ξ in the family. The correspondence material models defined in this way have the property that if the deformation is uniform throughout the body, that is, if $\mathbf{F}$ is independent of position, then $\bar{\mathbf{F}} = \mathbf{F}$ everywhere, including material points near boundaries (where the LPS breaks down, as discussed in the previous section). Otherwise, for non-uniform deformations, $\bar{\mathbf{F}} \neq \mathbf{F}$.

The correspondence materials also have the property of being nonpolar provided that the underlying classical material model satisfies the balance of angular momentum, $\tau = \tau^T$, where τ is the Cauchy stress tensor, $\tau = \sigma \mathbf{F}^T / \det \mathbf{F}$. Recall that nonpolarity is the peridynamic version of balance of angular momentum in a material model. Another property of correspondence materials is that they are elastic (in the peridynamic sense) if $\sigma(\mathbf{F})$ is elastic (in the sense of the standard theory). In other words, if there is a classical strain energy density function W' then

$$\sigma = \frac{\partial W'}{\partial \mathbf{F}} \qquad \Longrightarrow \qquad \underline{\mathbf{T}} = W_{\underline{\mathbf{Y}}}$$

where

$$W(\underline{\mathbf{Y}}) = W'(\bar{\mathbf{F}}).$$

These and other properties of correspondence materials are derived in [34]. This class of material models is non-ordinary because, in general, the bond force density vectors are not parallel to the deformed bond vectors.

The correspondence materials are useful in applying a known classical material model to peridynamics but suffer from some practical difficulties. These include non-invertibility in the following sense: there are many possible deformations of the family that result in the same force state. As a consequence there could be many possible deformations of the entire body for a given loading history. This has the practical effect in computations of resulting in zero-energy modes of deformation in the numerical mesh that need to be suppressed. Various remedies for the zero-energy modes are available. The simplest is to include *background bonds*. This means that the force state is determined partially by bond based forces (spring-like interactions), in addition to the intended correspondence model. The net force state then has the form

$$\underline{\mathbf{T}}\langle\xi\rangle = (1 - \gamma_z) \underline{\omega}\langle\xi\rangle \sigma(\bar{\mathbf{F}}) \mathbf{K}^{-1} \xi + \frac{\gamma_z c \underline{e}\langle\xi\rangle}{|\xi|}$$

where γ_z is a small parameter (typically on the order of 0.01) and c is the spring constant in the PM model given by (2.12). The small spring-like interactions have the effect of avoiding the nonuniqueness in the correspondence material models and therefore reducing the zero energy modes. Alternative methods for controlling zero energy modes have been proposed by Littlewood, Mish, and Pierson [22] and by Breitenfeld *et al.* [5]. Figure 2.4 illustrates velocity fields in an elastic plate containing a hole using a correspondence material, with and without control of zero energy modes. Without control, random oscillations are evident.

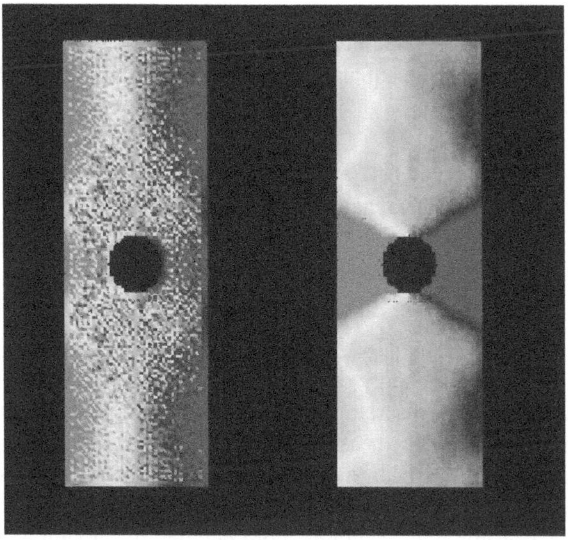

Figure 2.4: Effect of zero energy modes on the velocity field in a plate using a correspondence material model. Left: without zero energy mode control. Right: with control.

2.2.6 Discrete particles as peridynamic bodies

Peridynamics claims to unify the mechanics of continuous media and discrete particles, but so far we have only talked about continua. Consider a system of discrete particles that interact through an N-body potential of the form

$$U_i = \Phi(\mathbf{y}_1 - \mathbf{y}_i, \mathbf{y}_2 - \mathbf{y}_i, \ldots, \mathbf{y}_N - \mathbf{y}_i) \tag{2.24}$$

where U_i is the potential energy of a typical particle i with current position $\mathbf{y}_i$ that interacts with N of its neighbors. Such a system may be treated as a peridynamic body whose density distribution is of the form

$$\rho_i(\mathbf{x}) = \sum_i M\Delta(\mathbf{x} - \mathbf{x}_i)$$

where $\mathbf{x}$ is position in the reference configuration, M is the particle mass, $\mathbf{x}_i$ is the reference position of particle i, and Δ is the Dirac delta function. The material model is state-based and elastic with strain energy density function given by

$$\hat{W}_d(\underline{\mathbf{Y}}, \mathbf{x}) = \sum_i \Delta(\mathbf{x} - \mathbf{x}_i)\Phi(\underline{\mathbf{Y}}\langle\mathbf{x}_1 - \mathbf{x}\rangle, \underline{\mathbf{Y}}\langle\mathbf{x}_2 - \mathbf{x}\rangle, \ldots, \underline{\mathbf{Y}}\langle\mathbf{x}_N - \mathbf{x}\rangle).$$

As shown in [36], after evaluating the Fréchet derivative of this W to get the force state, and after applying the equation of motion (2.2), the expected particle equation of motion is recovered:

$$M\ddot{\mathbf{y}}_i = -\frac{\partial U_i}{\partial \mathbf{y}_i}.$$

The main feature that peridynamics offers in modeling discrete particles is the prospect of coupling them with continua without coupling dissimilar models. For example, an elastic material model could include both discrete and continuum interactions:

$$\hat{W} = \hat{W}_d + \hat{W}_{\text{LPS}}$$

where $\hat{W}_{\text{LPS}}$ is the strain energy density in the LPS model. Such a model might be useful in cases in which discrete particles float around and interact with a continuum half-space. A method for averaging a discrete peridynamic system to obtain a continuous material has been discussed in [36]. Seleson, Parks, Gunzburger, and Lehoucq describe a method for upscaling molecular dynamics within peridynamics [29]. Chapter 7 of this book contains more detailed discussion of the link between peridynamics and atomistics.

2.2.7 Setting the horizon

The horizon δ is similar to a material property. However, as we have seen in the various calibration exercises discussed in this chapter, for a given value of δ, the other parameters in a peridynamic material model can usually be chosen to match a given set of measured bulk elastic properties. So, how should one choose the horizon?

In some applications, the physics of the interactions between material points determines the horizon. For example, in modeling discrete particles, the properties of the interatomic potential Φ in (2.24) would dictate the maximum interaction distance and therefore δ.

As in the example of a layered composite in Section 1.2.3, nonlocal interactions are a fundamental aspect of working with homogenized deformations and material properties in heterogeneous media. The distance that characterizes these interactions depends not only on the microstructural features of the material, but also on differences in the properties of the component materials. These distances determine the peridynamic horizon in the smoothed model.

What about a material that is ideally continuous and homogeneous, in which the standard (local) theory applies perfectly? In such a situation, we might still wish to use peridynamics as a way to model damage and fracture. In this case, the horizon would be chosen "as small as possible." The discretization spacing that is used in a numerical model would usually determine the value of the horizon. Experience in working with such problems has suggested that $\delta = 3\Delta x$, where Δx is the grid spacing, usually gives good results. For δ much smaller than this value, the direction of crack growth can be strongly influenced by the orientation of the grid. δ much larger than this value not only causes excessive wave dispersion, but also increases the computational effort with no meaningful gain in accuracy. See [2] for further insight into the meaning and choice of horizon.

2.2.8 Linearized peridynamics

Under the assumption of small deformation (not necessarily small deformation *gradient*) the peridynamic equation of motion (2.2) is accurately approximated by the following *linear* integro-differential equation:

$$\rho(\mathbf{x})\ddot{\mathbf{u}}(\mathbf{x},t) = \int_{\mathcal{H}_\mathbf{x}} \mathbf{C}(\mathbf{q},\mathbf{x})(\mathbf{u}(\mathbf{q},t) - \mathbf{u}(\mathbf{x},t))\, dV_\mathbf{q} + \mathbf{b}(\mathbf{x},t) \tag{2.25}$$

where $\mathbf{u}$ is the displacement field and $\mathbf{C}$ is a tensor valued function. The linearized equation of motion (2.25) is essentially the same as in Kunin's linear nonlocal model [19]. Specific values of $\mathbf{C}$ are inherited from the full nonlinear material model $\hat{\mathbf{T}}$ by retaining the first-order terms. Details of this process may be found in [32], which also discusses how to obtain the fourth-order classical elasticity tensor components from a linearized material model. Linearization may also be applied to small displacement fields superposed on a larger field, which may include nonlinearities in material response. This is one approach to analyzing material stability, since the character of the $\mathbf{C}$ field determines how small perturbations evolve in the presence of nonlinear material response. Damage and fracture may be included in a linearized model which, as noted above, does not require small displacement gradients.

In practical numerical computations, the linearized version of peridynamics is not as much of a simplification as one might expect over the full nonlinear theory. The need in a numerical model, except in special cases, to evaluate and carry around the linearized $\mathbf{C}$ field creates an unwelcome burden that is generally not needed in the full theory. The assumption of linearized kinematics does not fit in well with simulations of fracture, in which we expect large bond deformations in the vicinity of a crack tip. The linearized theory is, however, more compatible with implicit solvers than the full theory. It is also much easier to obtain theoretical results about the properties of the linear equations, such as well-posedness, than those of the nonlinear model. The properties of waves are also much easier to analyze within the linearized theory.

2.3 Plasticity

2.3.1 Bond-based microplastic material

So far we have only talked about elastic materials. What about plasticity? Plasticity can be added to a microelastic bond-based model such as the PM material by treating each bond as though it has its own yield stress:

$$\mathbf{f} = f\mathbf{M}, \qquad \dot{f} = \begin{cases} c\dot{s}, & \text{if } |f| < f_y \text{ or } f\dot{s} < 0, \\ 0 & \text{otherwise} \end{cases}$$

where $\mathbf{M}$ is the deformed bond direction vector, and f_y is a prescribed force density at yield (Figure 2.5). This bond-based material model is called *microplastic*. The bond initially loads elastically and always unloads elastically. The yield force density f_y can be calibrated according to the measured material yield stress. To do this, assume conditions of uniaxial strain in the x_1 direction, and use (2.14) setting $p = f_1 = f_y \cos\phi$. Then in 3D,

$$\begin{aligned} \sigma_{11} &= 2\pi \int_0^{\delta} \int_0^{\xi} \int_0^{\cos^{-1}(z/\xi)} (f_y \cos\phi)\xi^2 \sin\phi \, d\phi \, dz \, d\xi \\ &= \frac{\pi\delta^4 f_y}{6}. \end{aligned}$$

After repeating this calculation for 2D and 1D, if σ_y is the normal stress in uniaxial strain after all the bonds have yielded, the bond parameter f_y can be calibrated according to

$$f_y = \begin{cases} 6\sigma_y/\pi\delta^4 & \text{if } D = 3, \\ 12\sigma_y/\pi h_2 \delta^3 & \text{if } D = 2, \\ 2\sigma_y/h_1\delta^2 & \text{if } D = 1. \end{cases} \tag{2.26}$$

For the bond-based material whose elastic response is given by the PM material (2.11) and (2.12), a similar computation for uniaxial strain *prior* to initial yielding gives (in 3D)

$$\sigma_{11} = \frac{\pi c \delta^4 \varepsilon_{11}}{10}$$

where ε_{11} is the bulk normal strain component and k is the bulk modulus. This material *first* starts to yield when $cs = f_y$, hence

$$\varepsilon_{11} = \frac{f_y}{c}.$$

To compute the stress $\sigma_{11} = \sigma_0$ at which this occurs, combine the last two equations to obtain

$$\sigma_0 = \frac{\pi\delta^4 f_y}{10}.$$

After repeating these steps for 2D and 1D and using (2.26), we arrive at

$$\sigma_0 = \begin{cases} 3\sigma_y/5 & \text{if } D = 3, \\ 3\sigma_y/4 & \text{if } D = 2, \\ \sigma_y & \text{if } D = 1. \end{cases}$$

Thus, the fact that bonds do not all yield at the same value of bulk strain effectively results in a strain hardening response in which the fully hardened stress exceeds the initial yield stress (Figure 2.5).

Microplastic materials models suffer from the same limitation as microelastic models: they imply severe restrictions on the material behavior that can be modeled, notably $\nu = 1/4$. Additionally, the plastic response in a microplastic model allows permanent change in volume (dilatation) and therefore is not well suited to modeling metals. However, microplastic models have been found to be effective in application to some geological materials in which irreversible volume change can occur.

2.3.2 LPS material with plasticity

The linear peridynamic solid (LPS) model is easily modified to include plasticity. To do this, simply modify (2.22) so that the deviatoric response, which is essentially bond-based, allows bonds to yield similarly to the microplastic material:

$$\underline{\mathbf{T}}\langle\xi\rangle = \left(\frac{-3k\vartheta\underline{\omega}\langle\xi\rangle|\xi|}{m} + \underline{t}^d\langle\xi\rangle \right) \mathbf{M}, \tag{2.27}$$

where

$$\underline{t}^d\langle\xi\rangle = \begin{cases} \alpha\underline{\omega}\langle\xi\rangle\underline{\dot{e}}^d\langle\xi\rangle, & \text{if } |\underline{t}^d\langle\xi\rangle| < t_y \text{ or } \underline{t}^d\langle\xi\rangle\underline{\dot{e}}^d\langle\xi\rangle < 0, \\ 0 & \text{otherwise} \end{cases}$$

where $\underline{e}^{\mathrm{d}}$ is the deviatoric extension state defined in (2.21) and t_y is a bond yield force density similar to f_y in the microplastic model. The calibrations in (2.26) carry over to the present material by setting $f_y = 2t_y$. (The factor of 2 appears because the LPS model is state-based; therefore, each bond carries half the force of its bond based analog. The resulting calibration of t_y in terms of the flow stress σ_y, after all the bonds have yielded, is as follows:

$$t_y = \begin{cases} 3\sigma_y/\pi\delta^4 & \text{if } D = 3, \\ 6\sigma_y/\pi h_2\delta^3 & \text{if } D = 2, \\ \sigma_y/h_1\delta^2 & \text{if } D = 1. \end{cases} \tag{2.28}$$

An additional enhancement of the LPS model allows nonlinear pressure-volume response:

$$\underline{\mathbf{T}}\langle\xi\rangle = \left(\frac{-p(\vartheta)\underline{\omega}\langle\xi\rangle|\xi|}{m} + \underline{t}^d\langle\xi\rangle \right) \mathbf{M}, \tag{2.29}$$

where m is given by (2.19) and $\underline{t}^d$ by (2.28).

Another approach to modeling plasticity in an ordinary state-based material model has been proposed by Mitchell [26]. In this approach, a yield surface is constructed in the space of deviatoric extension states, analogous to the standard concept of a yield surface in deviatoric strain space. In a yielded material, the gradient of this yield surface determines the flow "direction," that is, the relative rates of extension, among all the bonds in the family.

The correspondence materials reproduce any plastic response that is contained in the underlying classical material model $\sigma(\mathbf{F})$, which can include history-dependent variables. This is a key advantage of the correspondence models.

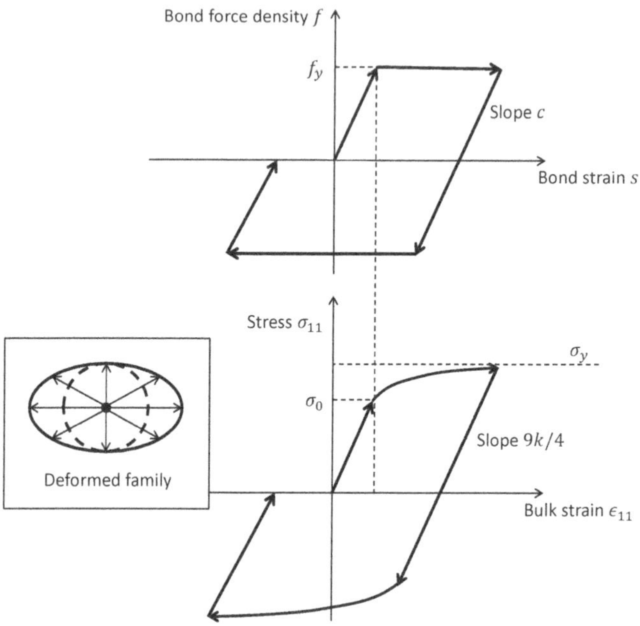

Figure 2.5: Microplastic material. Top: bond response. Bottom: bulk response under uniaxial strain. Inset: different bonds yield at different values of bulk strain, leading to an apparent strain hardening effect.

2.4 Damage and fracture

2.4.1 Damage in bond-based models

The simplest way to model damage in a peridynamic material model is through bond breakage. This means that after some bond damage criterion is met, a bond no longer can sustain nonzero force density. (A modification of this notion of bond breakage allows broken bonds to carry compressive but not tensile load.) For example, to apply bond breakage to the PM material (2.11), a history-dependent variable $\beta(t)$ is defined for each bond that changes suddenly and irreversibly from 1 to 0 when a bond breaks:

$$f = cs\beta, \qquad \beta(\xi,t) = \begin{cases} 1 & \text{if } s(t') < s_* \text{ for all } t' \leq t, \\ 0 & \text{otherwise} \end{cases} \tag{2.30}$$

where s_* is a constant called the *critical bond breakage strain*. The resulting material model is called the *prototype microelastic brittle* (PMB) material.

This material *nucleates* damage when the tensile strain in any bond exceeds s_*. At the nucleation site, the material effectively becomes weaker. The force density that the bond was carrying before it broke is transferred to its neighbors. This makes it more likely that these neighboring bonds will break, leading to a process of progressive damage. Broken bonds tend to occur along two-dimensional surfaces that we interpret as cracks. Neither the nucleation site nor crack trajectory and growth velocity are known in advance of modeling an application. These phenomena happen by themselves — in this sense, crack growth in peridynamics is *autonomous*.

It is useful to define the *net damage* field, which is a fraction of broken bonds connected to a point:

$$\varphi(\mathbf{x},t) = \frac{\int_{\mathcal{H}} (1-\beta(\xi,t))\, dV_\xi}{\int_{\mathcal{H}} dV_\xi}. \tag{2.31}$$

In the PMB model, the critical bond strain s_* can be calibrated from the measured critical energy release rate for a brittle material. To do this, observe that the bond micropotential at breakage is implied by (2.11) to be

$$w_* = \frac{f_* e_*}{2} = \frac{(cs_*)(\xi s_*)}{2} = \frac{c\xi s_*^2}{2}.$$

Identifying this as the work (per unit volume[2]) required to break a bond, it follows that the work per unit crack surface area G_* can be computed from (2.14) by setting $p = w_*$. In 3D, the result is

$$G_* = 2\pi \int_0^\delta \int_0^\xi \int_0^{\cos^{-1}(z/\xi)} w_* \xi^2 \sin\phi \, d\phi \, dz \, d\xi \, d\theta = \frac{\pi c s_*^2 \delta^5}{10}. \tag{2.32}$$

Carrying out similar computations for 2D and 1D and using (2.12) leads to

$$s_* = \begin{cases} \sqrt{5G_*/9k\delta} & \text{if } D = 3, \\ \sqrt{\pi G_*/3k'\delta} & \text{if } D = 2, \\ \sqrt{3G_*/E\delta} & \text{if } D = 1. \end{cases} \tag{2.33}$$

Since G_* is the energy consumed per unit area of crack growth, it can be identified with the critical energy release rate for a real material, G_{Ic}. In this way, the critical bond breakage strain s_* can be calibrated according to the energy release rate. The energy release rate for any mode of crack growth (I, II, or III) is the same.

When a bond breaks, what happens to the energy w_* that was stored in it? It dissappears. It is not converted to any other form of energy, such as kinetic or thermal energy. It represents energy that would be consumed by some crack tip process that is not explicitly modeled. However, in practice, there is a "snap-back" effect from the forces due to intact bonds that are no longer equilibrated by the bond that just broke. This imbalance creates accelerations that result in increased kinetic energy in the body, usually in the form of short-wavelength waves. These waves can adversely affect modeling of crack growth by causing fluctuations in the strain in bonds that are near breakage. It is therefore useful, in numerical modeling, to introduce damping forces that reduce these waves. One possible form of this damping in the PMB model takes the following form:

$$\mathbf{f}_{\text{damp}}(\mathbf{q},\mathbf{x},t) = \frac{\gamma_{\text{damp}} c \bar{\varphi} \dot{e}(\mathbf{q},\mathbf{x},t)\mathbf{M}}{v_0}, \qquad \bar{\varphi} = \frac{\varphi(\mathbf{x},t)+\varphi(\mathbf{q},t)}{2} \tag{2.34}$$

where γ_{damp} is a dimensionless constant, c is the spring constant in (2.12), v_0 is the dilatational wave speed for the undamaged material, $\dot{e}$ is the rate of bond extension, $\mathbf{M}$ is the deformed bond direction unit vector, and φ is the net damage defined in (2.31). This damping force density (per unit volume^2) is added to the bond forces from the constitutive model. For purposes of energy accounting, the damping force dissipates energy at a rate given by

$$\dot{E}_{\text{damp}} = \int_{\mathcal{B}} \dot{\mathbf{y}}(\mathbf{x},t) \cdot \left[\int_{\mathcal{H}_{\mathbf{x}}} \mathbf{f}_{\text{damp}}(\mathbf{q},\mathbf{x},t)\, dV_{\mathbf{q}} \right] dV_{\mathbf{x}}.$$

This energy should be interpreted as being stored in the body in inaccessible form, like thermal energy, which is not treated explicitly in this mechanical theory. This dissipated energy should not be considered as contributing to G_*.

Using the numerical method described later in this chapter, the PMB model predicts quasi-static tensile crack growth as illustrated in Figure 2.6. In this model, a preexisting crack is present in a brittle plate. The lower ends of the plate are pulled apart, causing the crack to grow. The figure shows the computed "lost energy" E_{lost} as a function of crack tip position, where

$$E_{\text{lost}} = \text{external work} - (\text{strain energy} + \text{kinetic energy} + \text{damping energy}).$$

Figure 2.7 illustrates the same problem with loading at a higher rate. The graph of lost energy versus crack tip position increases in slope when there are two branches simultaneously growing, because each consumes energy at the same rate. Ha and Bobaru [12, 13] discuss dynamic fracture modeling in peridynamics in more depth. Bobaru and Hu studied the effect of changing the horizon in a peridynamic model of dynamic fracture [2]. See also [4]. Chapters 9, 10 and 13 of this book contain more detailed discussions and results for dynamic fracture.

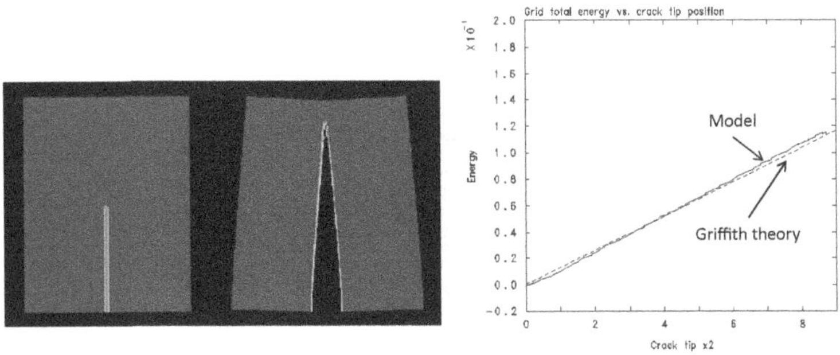

Figure 2.6: Quasi-static crack growth in a PMB plate. Left: initial damage. Center: final damage. Right: lost energy as a function of crack tip position compared with the Griffith theory (energy release rate times crack area).

Modeling crack propagation with the PMB model in practice can entail some difficulties. One such issue is the fact that the calculation of G_* above assumes that each crack is responsible for all the broken bonds that initially connected material particles that ended up on opposite sides of a crack. This assumption is not valid if there is already damage present, for example from an earlier crack. If such earlier damage is present, the new crack would dissipate less energy than the value of G_* computed above. A simple but approximate way to correct for this effect takes into account the *net damage* at any point $\mathbf{x}$ defined in (2.31). In the vicinity of a single crack, the net damage does not exceed 0.5, because a point on the crack surface still has unbroken bonds connected to other points on its own side of the crack. Therefore, if the damage is near or above 0.5, it makes sense to increase the critical bond breakage strain s_*. One possible form for this dependence is to assume, for the bond connecting a pair of points $\mathbf{x}$ and $\mathbf{q}$,

$$s_* = s_{*0}R(\bar{\varphi}), \qquad s_{*0} = \sqrt{5G_*/9k\delta},$$

$$R(\bar{\varphi}) = \begin{cases} 1 & \text{if}\, 0 \leq \bar{\varphi} \leq 0.5 \\ 0.5/(1-\min(0.95,\bar{\varphi})) & \text{if}\, 0.5 < \bar{\varphi} < 1, \end{cases}$$

where $\bar{\varphi}$ is the mean net damage in the bond defined in (2.34).

2.4.2 Damage in ordinary state-based material models

Most state-based material models use an influence (weighting) function $\underline{\omega}\langle\xi\rangle$ that controls how much each bond affects the force state, which represents the response of the material to the collective deformation of the family. An example is the LPF (2.19), (2.20), which computes a net volume change in the family according to a weighted average of the bond elongations. To implement bond breakage in such a model, the easiest approach is to set $\underline{\omega}\langle\xi\rangle = 0$ when the bond ξ breaks, thereby

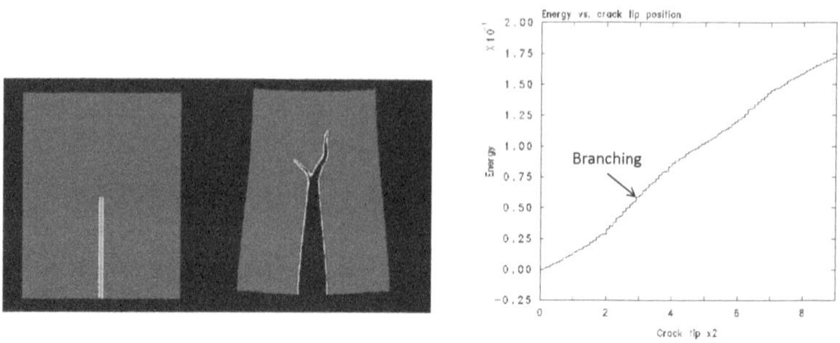

Figure 2.7: Dynamic crack growth in a PMB plate. Left: initial damage. Center: final damage. Right: lost energy as a function of crack tip position, showing increased energy dissipation due to branching.

effectively excluding the bond from affecting the response. This also has the effect of setting the bond force to zero in the broken bond. In modeling a fluid, and even some solids, it may be appropriate to allow broken bonds to "heal" if they go into compression after failure in tension.

Recall from the previous section that in the bond based PMB material, a critical bond breakage strain s_* can be derived from the measured critical energy release rate for a material. The crucial ingredient in this derivation is that a definite amount of work (per unit volume^2) w_* is supplied to a given bond when it attains this critical strain. In state-based materials, because each bond responds to the collective deformation of the entire family, there is no one-to-one relation between bond strain and work done on the bond. So, the concept of a critical bond breakage strain may not apply.

Instead, Foster [11] proposed that in a state-based material, damage can be characterized in terms of a critical bond work w_*. The relation (2.32) still applies. As the deformation evolves over time, the work done on each bond ξ is computed:

$$w(\xi,t) = \int_0^t \underline{\mathbf{T}}\langle\xi\rangle \cdot \dot{\underline{\mathbf{Y}}}\langle\xi\rangle \, dt$$

and bond breakage occurs when

$$w(\xi,t) = w_*/2$$

where w_* is determined from (2.32). The factor of 1/2 appears because in a state-based model, if the bond ξ originates at $\mathbf{x}$, there is another bond $-\xi$ connected to the endpoint $\mathbf{x}+\xi$ that has consumed the same amount of work by the time it breaks.

2.4.3 Damage in correspondence material models

In the bond-based theory, in which a bond is the basic unit of material modeling and each bond responds independently of all the others, the concept of bond breakage is completely compatible with the material model. In this sense, bond breakage is *native* to the bond-based theory. The same is true of the ordinary state-based models such as the LPS that are essentially bond-based but with some measure of the collective volume change included. In contrast, correspondence material models map a stress tensor, which is determined using an approximate deformation gradient tensor, onto bond forces. Each bond directly affects the forces in all the others in the family.

Keeping this in mind, the use of simple bond breakage based on a critical bond breakage strain creates a certain dissonance. One manifestation of this is the mechanical implication of breaking large numbers of bonds in evaluating $\bar{\mathbf{F}}$. Suppose damage is implemented by setting $\underline{\omega}\langle\xi\rangle = 0$ for each broken bond ξ. This has the effect of zeroing out the forces in the broken bonds, and it modifies the dependence of W on these bonds in an energetically consistent way. However, because $\mathbf{K}$ is being reduced by this bond breakage, $\mathbf{K}^{-1}$ increases in such a way that a uniform deformation of the family still results in the same value of W, even though some bonds are now broken. This actually *increases* the force density in the remaining intact bonds. This poses a conceptual as well as a practical problem: damage in some bonds increases the forces in the remaining non-broken bonds.

In (2.23), we typically start with a "healthy" looking, continuous, and spherically symmetric influence function $\underline{\omega}$. In modeling an application with a lot of damage, it often occurs that 95% or even 100% of the bonds in a family break. Suddenly our healthy looking influence function starts to look very unhealthy: the shape tensor $\mathbf{K}$ may become singular, preventing the evaluation of $\mathbf{K}^{-1}$ altogether.

An effective — but completely *ad hoc* — workaround for these difficulties is to reduce the stress tensor σ as the net damage increases so that it vanishes in any state with, say, 80% or more net damage. The case of singular, or nearly singular, $\mathbf{K}$ is another pathological situation that must be dealt with by setting $\sigma = \mathbf{0}$. The conceptual problem here is that unlike the bond-based model in which bond breakage is a self-contained approach to damage, we are now introducing an additional modification to the material model that is similar to a continuum damage model in the standard theory. A practical problem with this type of fix is that by modifying σ, we also modify the energy required to break bonds, altering the energy balance for crack growth.

From the above considerations, evidently simple bond breakage is not native to the correspondence approach to material modeling, as it is to bond-based materials models. So, is there a different approach that is native to correspondence materials? This is currently an open question. One avenue of research is to adapt a continuum type damage model from the standard theory similarly to the way the stress-strain model is adapted. In this approach, suppose there is a known failure model for a material that can be expressed in the form

$$\psi(\mathbf{F}) = \psi_0.$$

That is, a material point starts out with $\psi(\mathbf{1}) = 0$, and fails whenever the above condition is met. The function ψ represents a family of surfaces in the space of tensors. To adapt this to correspondence materials, let $\underline{\phi}\langle\xi\rangle$ be a *damage state* that, for each bond ξ, varies continuously from 0 to 1 as a bond undergoes damage. Then express an evolution law for this damage state with the general form

$$\dot{\underline{\phi}} = \begin{cases} 0 & \text{if } \psi(\bar{\mathbf{F}}) < \psi_0 \\ \psi_{\mathbf{F}}(\bar{\mathbf{F}})\bar{\mathbf{F}}_{\underline{\mathbf{Y}}} \bullet \dot{\underline{\mathbf{Y}}} & \text{otherwise,} \end{cases}$$

where $\psi_{\mathbf{F}}$ is the tensor gradient of ψ and $\bar{\mathbf{F}}_{\underline{\mathbf{Y}}}$ is the Fréchet derivative of $\bar{\mathbf{F}}$. The effect of this evolution law is that the bonds that contribute the most to ψ become damaged the fastest.

Another promising approach to modeling damage in non-ordinary state models has been proposed by Tupek, Rimoli, and Radovitzky [39]. This approach also uses a continuum damage model as a criterion for bond damage. In this case, bond damage occurs according to the status of the continuum damage at both endpoints of the bond.

2.4.4 Nucleation strain

In the bond damage model proposed in Section 2.4.1, bonds break when their strain s exceeds a critical value s_* that is calibrated according to the material energy release rate. This implies that s_* is the nucleation strain in the model: damage starts where the macroscopic strain (over a family) exceeds s_*. The problem here is that, according to (2.33), s_* depends strongly on δ and cannot be specified independently of G_*. It would be better to be able to reproduce both G_* and some macroscopic nucleation strain that will be denoted ε_*.

To do this in the PMB model, we can add a tail to the tensile response of a bond, as shown in Figure 2.8. The peak of the curve is set to the material nucleation strain ε_*. The tail of the curve extends out to a bond strain s_t chosen so that the energy release rate calculation (2.32) agrees with G_*. In this calculation, set

$$w_* = \frac{c\xi\varepsilon_* s_t}{2}.$$

This means that (2.32) and (2.33) still hold, provided we replace s_*^2 by $\varepsilon_* s_t$. The results are

$$G_* = \frac{\pi c\varepsilon_* s_t \delta^5}{10},$$

$$s_t = \begin{cases} 5G_*/9\varepsilon_* k\delta & \text{if } D=3, \\ \pi G_*/3\varepsilon_* k'\delta & \text{if } D=2, \\ 3G_*/\varepsilon_* E\delta & \text{if } D=1. \end{cases}$$

This assumes that $s_t \geq \varepsilon_*$, which implies that δ must be chosen sufficiently small to ensure this, according to the above equations. For steel, assuming a nucleation strain of $\varepsilon_* = 0.0025$, this works out to $\delta > 9$ cm. However, for weaker materials, much smaller values of δ are sometimes obtained. The significance of this is that it restricts the mesh spacing.

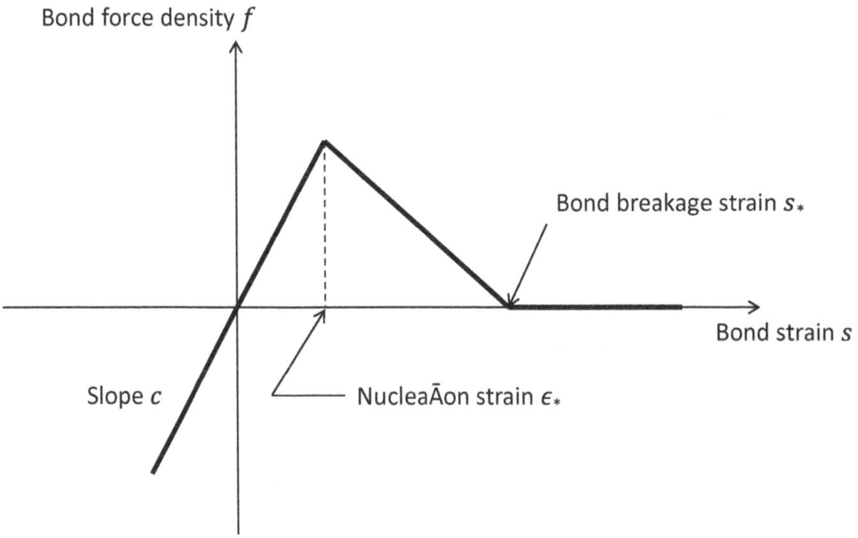

Figure 2.8: Modification of the PMB model to allow nucleation strain and critical energy release rate to be specified independently.

For a given elastic material model, a material stability condition can be derived that detects deformations in which a jump discontinuity can form spontaneously and grow. This type of instability can be interpreted as a crack nucleation condition [37]. Of course, not all material models admit such discontinuities. When a discontinuity nucleates and grows as a crack, everything happens autonomously and evolves according to the same set of equations. There is no supplementary relation that controls the nucleation and growth of cracks.

2.5 Treatment of boundaries and interfaces

2.5.1 Bond-based materials

For the PM and PMB materials, the derivation (2.12) of the constant c assumed that the point $\mathbf{x}$ is in the interior of a homogeneous body, sufficiently far from any boundaries that its family is completely contained within the body. If the family of $\mathbf{x}$ does contain a boundary, then the computation of W in (2.7) must be modified such that the volume of integration is restricted to $\mathcal{H}_{\mathbf{x}}$. Under isotropic expansion, the strain energy density found by summing the peridynamic bond energies is required to match the corresponding value in the standard theory, whether or not the family fully occupies the sphere of radius δ. This leads to the following correction for c in the PM material model:

$$c = \frac{18k\lambda_b}{\pi\delta^4}, \qquad \lambda_b = \frac{2V_0}{V(\mathbf{x}) + V(\mathbf{q})}, \qquad V_0 = \frac{4\pi\delta^3}{3},$$

where $V(\mathbf{x})$ and $V(\mathbf{q})$ are the volumes of $\mathcal{H}_\mathbf{x}$ and $\mathcal{H}_\mathbf{q}$, respectively (this only includes volume where there is material present, not empty space). This expression is for 3D, but similar corrections are easily derived for 2D and 1D. With this correction, bonds are stiffer near the boundaries. With microplastic materials, the bond force density at yield, f_y, would also be increased by the same correction factor λ_b. Other approaches to boundary correction factors include those of Oterkus [27] and Liu and Hong [21].

The properties of bonds connecting two different materials, denoted (1) and (2), must be assigned taking into account the properties of both materials. The micromodulus can be set using the idea of two springs connected end-to-end ("in series") representing the properties of each. Suppose the bond occupies length $\xi_{(1)}$ in material (1) and $\xi_{(2)}$ in material (2). Also let $c_{(1)}$ and $c_{(2)}$ denote the spring constant in the PM or PMB material, given by (2.12). The resulting value of c for the interface bond in the PM or PMB model is then

$$c = \frac{\xi_{(1)} + \xi_{(2)}}{\dfrac{\xi_{(1)}}{c_{(1)}} + \dfrac{\xi_{(2)}}{c_{(2)}}}.$$

If it is not practical to compute the partial bond lengths $\xi_{(1)}$ and $\xi_{(2)}$, the approximation

$$\xi_{(1)} = \xi_{(2)} = \frac{\xi}{2}, \qquad c = \bar{c} := \frac{2}{\dfrac{1}{c_{(1)}} + \dfrac{1}{c_{(2)}}} \tag{2.35}$$

usually gives satisfactory results, where ξ is the total length of the interface bond. The harmonic average is used here because the two partial bonds can be thought of as two springs "in series" rather than "in parallel." For microplastic materials, it makes sense to assume that whichever of these end-to-end bonds yields first determines the yield force density in the entire interface bond, therefore

$$f_y = \min\{f_{y(1)}, f_{y(2)}\}.$$

Under the assumption (2.35), interface fracture energy G_* may be computed using (2.32), provided $\bar{c}$ is used in place of c. This permits the critical bond strain for interface bonds to be computed directly in terms of the energy release rate for interface cracks, if this quantity is known. If it is not known, it is reasonable to set it to

$$G_* = \min\{G_{*(1)}, G_{*(2)}\}.$$

We emphasize that in the bond-based theory, the bond itself is the basic unit of material properties. A common software bug is to assign different properties to an interface bond depending on whether the acceleration at point $\mathbf{x}$ or $\mathbf{q}$ is being computed.

This situation would result in an imbalance of linear momentum, leading to bad results.

An example of interfacial crack growth is shown in Figure 2.9. In this calculation, an edge crack grows between two hard inclusions in a plate. The interface is weaker than either material. The crack attaches to the interface or detaches seamlessly according to the conditions that prevail near the crack tip. Agwai, Guven, and Madenci [1] present additional results and validation on interface crack modeling and dynamic fracture. Kilic and Madenci [17] discuss computation of crack path curvature and branching in glass plates with a temperature gradient.

Figure 2.9: Model of crack growth along the surfaces of hard inclusions in a brittle plate.

2.5.2 State-based materials

As discussed previously, the correspondence materials account automatically for boundaries, so no correction factor is necessary to account for partially filled families. For points near an interface, it is usually not satisfactory in practice to treat the material model as though the family contains only one material. For example, at an interface between air and steel, we anticipate much larger deformations on the air side. Therefore, in computing the force state for a point on the steel side, the model would produce unrealistically large stresses and force densities if it pretended that the bonds to air were bonds to steel.

One approach to treating interface bonds in a correspondence model is to alter the influence function $\underline{\omega}$ to reflect the relative stiffness of the materials. For example, using the end-to-end partial bond idea discussed in the previous section, the influence function for an interface bond would be reassigned according to

$$\underline{\bar{\omega}}\langle\xi\rangle = \underline{\omega}\langle\xi\rangle k_{\text{air}}/k_{\text{steel}}.$$

(It is not required that the influence function be radially symmetric or that its value be the same at each endpoint of a bond.)

A better approach, in practice, is to apply the correspondence model only to the subset of a family that contains the same material as **x**. That is, set $\omega\langle\xi\rangle = 0$ whenever ξ is an interface bond. The interface bonds are treated using the bond-based theory. The PM or PMB models are suitable for these bonds, with properties assigned as described in the previous section.

2.6 Emu numerical method

The earliest and simplest numerical method for approximate solution of the peridynamic equations was proposed in [33]. I will refer to this as "the Emu method," after the first peridynamic software that implemented it. However, it uses only well-known and standard techniques for evaluating the peridynamic integral, and several other researchers have developed similar approaches independently and in some cases improved upon it. A related method, including a quasi-static solver, is discussed by Kilic and Madenci [18].

The Emu method relies on midpoint quadrature in a Lagrangian spatial discretization. Each node i represents a fixed volume V_i in the reference configuration, which includes the dimension factor h_D. The peridynamic integral in (2.2) is replaced by a finite sum:

$$\rho_i \ddot{\mathbf{u}}_i^n = \sum_{j \in \mathcal{H}_i} \mathbf{f}(\mathbf{x}_j, \mathbf{x}_i, t^n) V_j + \mathbf{b}_i^n, \tag{2.36}$$

where $\mathcal{H}_i$ is the discretized family of node i, n is the time step number, and **u** is displacement. The acceleration is approximated by a finite difference:

$$\ddot{\mathbf{u}}_i^n = \frac{\mathbf{u}_i^{n+1} - 2\mathbf{u}_i^n + \mathbf{u}_i^{n-1}}{\Delta t^2},$$

where Δt is the time step size. The pairwise bond force $\mathbf{f}(\mathbf{x}_j, \mathbf{x}_i, t^n)$ is evaluated from the material model using the displacements at time step n. As discussed in [33], this method has a stability restriction on the time step size similar to the Courant–Friedrichs–Levy (CFL) condition in numerical solutions of the standard hyperbolic PDEs of continuum mechanics. However, in the peridynamic model, the horizon takes the place of the mesh spacing in the CFL stability limit.

One source of error in this discretized model is the treatment of nodes j whose volume V_j is partially within the horizon of node i and partially outside. This error can be reduced by replacing the volume V_j in (2.36) by a factor R_{ij} that accounts for the fraction that is outside the horizon. This factor would have different values for different i. In applying this type of correction, it is required that $R_{ji} = R_{ij}$, otherwise linear momentum might fail to be balanced. Yu, Xin, and Lease [42] and Seleson [28] have provided detailed analysis of the geometry of the overlap between discretized node volumes and spherical horizons, and related integration methods for the discretized equations. The use of a micromodulus function or influence function that smoothly approaches zero for bonds near the horizon tends to reduce the errors from this overlap effect.

Another problem that can occur with the Emu method is interpenetration. After a large number of bonds have broken, there is nothing to prevent nodes from moving in between each other. In most simulations, this is an undesirable effect. It can be avoided by permitting compressive forces in broken bonds, which tends to preserve the initial ordering of the nodes. It can also be controlled with contact forces, which create repulsive forces between nodes that get too close together. These contact forces are added to any bond forces that result from other sources, such as the material model and damping. An example of an expression for contact forces is given by

$$\mathbf{f}_{\text{contact}}(\mathbf{q},\mathbf{x},t) = \gamma_{\text{contact}} c \min\{0, s - s_{\text{contact}}\}\mathbf{M}$$

where γ_{contact} is a dimensionless constant of order 1 and s_{contact} is the compressive bond strain at which the contact force kicks in, typically 0.9. Tupek [38] addresses interpenetration as an aspect of more basic properties of peridynamic material models, rather than a side effect of discretization, and proposes an alternative strain measure as a way to avoid the problem. Note that in some applications, such as diffusion, interpenetration could be desirable.

Use of this simple numerical method has the advantage of allowing material parameters to be "calibrated in place." In this method, we disregard the continuum model calibration such as (2.12) entirely. Instead, each node has its parameters determined individually within the discretized material model. For example, for the linear peridynamic fluid (2.19), (2.20), we use consistent approximations for ϑ_i and m_i:

$$\vartheta_i = \frac{3}{m_i} \sum_{j \in \mathcal{H}_i} \omega_{ij} |\xi_{ij}| e_{ij} \Delta V_j, \qquad m_i = \sum_{j \in \mathcal{H}_i} \omega_{ij} |\xi_{ij}|^2 \Delta V_j.$$

It is easily confirmed that under a small isotropic expansion of the family of node i, in which $e_{ij} = \varepsilon |\xi_{ij}|$, the discretized model exactly reproduces the value of strain energy predicted by the local model:

$$W_i = \frac{k\vartheta_i^2}{2} = \frac{9k\varepsilon^2}{2}.$$

The point is that m_i is calibrated for each node "on the fly," rather than using the continuum expression in (2.19). This method automatically accounts for boundaries as well as discretization errors. It seems possible to re-calibrate the model in the same way to account for damage as it progresses, thus avoiding the softening effect of damage on the effective elastic properties at a node. However, I know of no one who has tried this in a code.

The calibration of material properties in place as described above ensures only that the strain energy density agrees with the value obtained from the classical theory in the special case of small isotropic deformation. This method is therefore not completely general, although it does greatly reduce boundary effects in many applications. A more sophisticated way of correcting for boundary effects is discussed in [24]. A method that, for linear elastic solids, gives the correct strain energy density for deformations that are not isotropic, as well as those that are, is proposed in [25]. A variety of other methods for correcting response at surfaces have been proposed. It is sometimes remarked that surface correction methods have an *ad hoc* character, but this objection is based implicitly on the unsupportable view that nonlocal material response should obey mathematical expressions that are independent of position.

The Emu numerical method can credibly make a surprising claim. The claim is that, in equilibrium, it is exact, provided the body is modeled as a collection of discrete points. The analyst is free to declare that such a set of points is the way he or she chooses to model some system, even if that system is really continuous. After making that decision, the numerical method described above provides an exact solution of the model. This is a claim that cannot be made by a finite element model of the standard PDEs, because the finite element equations provide only an approximate, rather than exact, solution of the PDEs. The observation here is mainly philosophical: in peridynamics, we can shift the nature of the approximation from the numerical solution of the model to the model itself as a representation of a real system. However, this point could have implications for the way software is verified, since verification traditionally involves confirming that a numerical solution is an accurate approximation of a mathematical model. If there is no approximation, perhaps verification is an easier task.

Since the introduction of the Emu method, a number of more sophisticated and accurate methods have been proposed. Emmrich and Weckner [7, 8] have investigated quadrature methods for the integral equation. One complication of application of higher order quadrature methods is that in many problems of interest, the fields being integrated are discontinuous, in some cases negating the benefits of higher order quadrature that are obtained for continuous functions. Chen and Gunzburger developed a discontinuous Galerkin method for peridynamics [6]. Wang and Tian [40] have investigated a novel, efficient solver for the dense matrices that arise in peridynamic models. Bobaru et al. [3] discuss numerical convergence and mesh adaptivity in the Emu method. Chapter 4 of this book contains further discussion of numerical methods for peridynamics, and Chapter 5 describes available peridynamic software.

2.7 Conclusions

In this chapter we have tried to convey the basic concepts of the mechanical theory of peridynamics, with an emphasis on material models that are used in practice. These and other models are discussed further in Chapter 6 of this book. Some handy formulas are given that are not available or hard to find in the literature. We have not touched on the thermodynamic theory or diffusive transport, which is discussed in Chapter 16. We have also given some specifics of how a simple numerical method based on midpoint quadrature can be applied to the peridynamic equations, along with some pitfalls of applying this method and techniques to avoid them.

References

[1] A. Agwai, I. Guven, and E. Madenci, Predicting crack propagation with peridynamics: A comparative study, *International Journal of Fracture*, **171**: 65–78, 2011.

[2] F. Bobaru and W. Hu, The meaning, selection, and use of the peridynamic horizon and its relation to crack branching in brittle materials, *International Journal of Fracture*, **176**: 215–222, 2012.

[3] F. Bobaru, M. Yang, L. Alves, S. A. Silling, E. Askari, and J. Xu, Convergence, adaptive refinement, and scaling in 1D peridynamics, *International Journal for Numerical Methods in Engineering*, **77**: 852–877, 2009.

[4] F. Bobaru and G. Zhang, Why do cracks branch? A peridynamic investigation of dynamic brittle fracture, *International Journal of Fracture*, **196**(1): 59–98, 2015.

[5] M. S. Breitenfeld, P. H. Geubelle, O. Weckner, and S. A. Silling, Non-ordinary state-based peridynamic analysis of stationary crack problems, submitted to *Computer Methods in Applied Mechanics and Engineering*, **272**: 233–250, 2014.

[6] X. Chen and M. Gunzburger, Continuous and discontinuous finite element methods for a peridynamics model of mechanics, *Computer Methods in Applied Mechanics and Engineering*, **200**: 1237–1250, 2011.

[7] E. Emmrich and O. Weckner, The peridynamic equation and its spatial discretization, *Mathematical Modelling and Analysis*, **12**: 17–27, 2007.

[8] E. Emmrich and O. Weckner, Analysis and numerical approximation of an integro-differential equation modeling non-local effects in linear elasticity, *Mathematics and Mechanics of Solids*, **12**: 363–384, 2007.

[9] E. Emmrich and O. Weckner, On the well-posedness of the linear peridynamic model and its convergence towards the Navier equation of linear elasticity, *Communications in Mathematical Sciences*, **5**: 851–864, 2007.

[10] G. I. Evangelatos and P. D. Spanos, A collocation approach for spatial discretization of stochastic peridynamic modeling of fracture, *Journal of Mechanics of Materials and Structures*, **6**: 1171–1195, 2011.

[11] J. Foster, S. A. Silling, and W. Chen, An energy based failure criterion for use with peridynamic states, *International Journal for Multiscale Computational Engineering*, **9**: 675–688, 2011.

[12] Y. D. Ha and F. Bobaru, Studies of dynamic crack propagation and crack branching with peridynamics, *International Journal of Fracture*, **162**: 229–244, 2010.

[13] Y. D. Ha and F. Bobaru, Characteristics of dynamic brittle fracture captured with peridynamics, *Engineering Fracture Mechanics*, **78**: 1156–1168, 2011.

[14] W. Hu, Y. D. Ha, and F. Bobaru, Peridynamic model for dynamic fracture in unidirectional fiber-reinforced composites, *Computer Methods in Applied Mechanics and Engineering*, **217–220**: 247–261, 2012.

[15] Y. Hu, Y. Yu, and H. Wang, Peridynamic analytical method for progressive damage in notched composite laminates, *Composite Structures*, **108**: 801–810, 2014.

[16] B. Kilic, A. Agwai, and E. Madenci, Peridynamic theory for progressive damage prediction in center-cracked composite laminates, *Composite Structures*, **90**: 141–151, 2009.

[17] B. Kilic and E. Madenci, Prediction of crack paths in a quenched glass plate by using peridynamic theory, *International Journal of Fracture*, **156**: 165–177, 2009.

[18] B. Kilic and E. Madenci, An adaptive dynamic relaxation method for quasi-static simulations using the peridynamic theory, *Theoretical and Applied Fracture Mechanics*, **53**: 194–204, 2010.

[19] I. A. Kunin, *Elastic Media with Microstructure. I-One-dimensional models*, New York: Springer, 1982.

[20] R. Lehoucq and S. A. Silling, Force flux and the peridynamic stress tensor, *Journal of the Mechanics and Physics of Solids*, **56**: 1566-1577, 2008.

[21] W. Liu and J.-W. Hong, Discretized peridynamics for linear elastic solids, *Computational Mechanics*, **50**: 579–590, 2012.

[22] D. J. Littlewood, K. Mish, and K. Pierson, Peridynamic simulation of damage evolution for structural health monitoring, ASME paper IMECE2012-86400, 2012.

[23] E. Madenci and E. Oterkus, *Peridynamic Theory and Its Applications,* New York: Springer, 2013.

[24] E. Madenci and E. Oterkus, *op. cit.,* pp. 104–114.

[25] J. M. Mitchell, S. A. Silling, and D. J. Littlewood, A position-aware linear solid constitutive model for peridynamics, *Journal of Mechanics of Materials and Structures*, **10**: 539-557, 2015.

[26] J. A. Mitchell, A nonlocal, ordinary, state-based plasticity model for peridynamics, Sandia National Laboratories report SAND2011-3166, 2011.

[27] E. Oterkus, Peridynamic theory for modeling three-dimensional damage growth in metallic and composite structures, Ph.D. thesis, University of Arizona, 2010.

[28] P. Seleson, Improved one-point quadrature algorithms for two-dimensional peridynamic models based on analytical calculations, *Computer Methods in Applied Mechanics and Engineering*, **282**: 184–217, 2014.

[29] P. Seleson, M. L. Parks, M. Gunzburger, and R. B. Lehoucq, Peridynamics as an upscaling of molecular dynamics, *Multiscale Modeling and Simulation*, **8**(1): 204–227, 2009.

[30] P. Seleson and M. L. Parks, On the role of the influence function in the peridynamic theory, *International Journal for Multiscale Computational Engineering*, **9**(6): 689–706, 2011.

[31] S. A. Silling, Reformulation of elasticity theory for discontinuities and long-range forces, *Journal of the Mechanics and Physics of Solids*, **48**(1): 175–209, 2000.

[32] S. A. Silling, Linearized theory of peridynamic states, *Journal of Elasticity*, **99**: 85–111, 2010.

[33] S. A. Silling and E. Askari, A meshfree method based on the peridynamic model of solid mechanics, *Computers and Structures*, **83**: 1526–1535, 2005.

[34] S. A. Silling, M. Epton, O. Weckner, J. Xu, and E. Askari, Peridynamic states and constitutive modeling, *Journal of Elasticity* **88**: 151–184, 2007.

[35] S. A. Silling and R. Lehoucq, Convergence of peridynamics to classical elasticity theory, *Journal of Elasticity*, **93**: 13–37, 2008.

[36] S. A. Silling and R. Lehoucq, Peridynamic theory of solid mechanics, *Advances in Applied Mechanics*, **44**: 74–167, 2010.

[37] S. A. Silling, O. Weckner, E. Askari, and F. Bobaru, Crack nucleation in a peridynamic solid, *International Journal of Fracture*, **162**: 219–227, 2010.

[38] M. R. Tupek, Extension of the peridynamic theory of solids for the simulation of materials under extreme loadings, Ph.D. thesis, Massachusetts Institute of Technology, 2013.

[39] M. R. Tupek, J. J. Rimoli, and R. Radovitzky, An approach for incorporating classical continuum damage models in state-based peridynamics, *Computer Methods in Applied Mechanics and Engineering*, **263**: 20–26, 2013.

[40] H. Wang and H. Tian, A fast Galerkin method with efficient matrix assembly and storage for a peridynamic model, *Journal of Computational Physics*, **231**: 7730-7738, (2012).

[41] J. Xu, A. Askari, O. Weckner, and S. Silling, Peridynamic analysis of impact damage in composite laminates, *Journal of Aerospace Engineering*, **21**: 187–194, 2008.

[42] K. Yu, X. J. Xin, and K. Lease, A new adaptive integration method for the peridynamic theory, *Modeling and Simulation in Materials Science and Engineering*, **19**: 045003, 2011.

Part II

Mathematics, Numerics, and Software Tools of Peridynamics

Chapter 3

Nonlocal Calculus of Variations and Well-Posedness of Peridynamics

Qiang Du

Columbia University

CONTENTS

3.1 Introduction 63
3.2 A brief review of well-posedness results 66
3.3 Nonlocal balance laws and nonlocal vector calculus 67
3.4 Nonlocal calculus of variations — an illustration 72
3.5 Nonlocal calculus of variations — further discussions 79
3.6 Summary 81
References 82

3.1 Introduction

Recent studies of peridynamic models in various applications, as documented in other parts of this handbook, have motivated the development of mathematical theory for nonlocal models [18]. Similar to the mathematical study of local PDE models, a systematic and rigorous mathematical treatment of nonlocal integral equations such as those represented by peridymamic models can offer theoretical insight in the modeling process. It can also provide guidance to numerical simulations and serve as a basis for verification and validation studies.

To gain better mathematical understanding of peridynamic models, there are a number of popular, and at the same time fundamental, mathematical questions to be answered first, such as

- What are suitable function spaces and topologies in which to define the solutions?

- How to incorporate suitable nonlocal boundary conditions or nonlocal constraints?
- Are the peridynamic models well-posed in the sense that solutions exist which are unique and are dependent continuously on data?
- What are the relations between nonlocal peridynamic models, differential equation models in classical continuum mechanics, dynamics of interacting particles and/or molecular dynamics?
- What are the characteristics of solutions to peridynamic models? What are their regularity properties? What singular behavior may solutions develop?

In the last few years, these questions have received much attention from the mathematics community. There has been considerable progress made towards answering these questions in some specialized cases. Moreover, the new development of mathematical theory for peridynamics has also helped to resolve various issues in other related subject areas such as nonlocal diffusion and stochastic jump processes. In this chapter, our main objective is to review some of the progress that has been made on establishing the well-posedness of nonlocal peridynamic models. We choose to emphasize on discussing results that have been established within the framework of nonlocal vector calculus and nonlocal calculus of variations. Materials are presented here with both general, less-technical descriptions and illustrative examples so as to offer readers a glimpse of the key ingredients of the nonlocal vector calculus framework and the theory of nonlocal calculus of variations, along with a demonstration of how they can be applied to formulate and analyze peridynamic models. Discussions of relations between nonlocal and local models are given in the next chapter, together with relations between their discretizations. Qualitative studies of how solutions to peridynamic models change with respect to materials parameters and external conditions are of great importance in practice. In particular, the characterization of their regularity properties is a much desirable work to be carried out. But other than the minimal regularity guaranteed by the well-posedness theory, no further discussions on the regularities are made here.

With a few exceptions, most of the materials presented in this chapter are associated with linear equations and variational problems. As pointed out by Silling in Section 2.2.8 of the handbook, under the assumption of small deformation, the peridynamic equation of motion may be approximated by a linear integro-differential equation given in (2.25), which is restated below as

$$\rho \partial_{tt} \mathbf{u}(\mathbf{x},t) = \int \mathrm{C}(\mathbf{x}',\mathbf{x})(\mathbf{u}(\mathbf{x}',t) - \mathbf{u}(\mathbf{x},t))d\mathbf{x}' + \mathbf{b}(\mathbf{x}) \tag{3.1}$$

where ρ represents a constant density and $\mathbf{b}(\mathbf{x})$ denotes the external force.

Factors that affect properties of solutions to the above linearized model include the tensor C, which defines the type of nonlocal interactions in the materials system, and appropriate nonlocal boundary conditions and initial conditions.

For a linear bond-based model, C is a rank-one tensor of the form,

$$\mathrm{C}(\mathbf{x}',\mathbf{x}) = \omega(|\mathbf{x}'-\mathbf{x}|)\left[\alpha(\mathbf{x}',\mathbf{x})\otimes\alpha(\mathbf{x}',\mathbf{x})\right] \tag{3.2}$$

where

$$\alpha(\mathbf{x}',\mathbf{x}) = \frac{\mathbf{x}'-\mathbf{x}}{|\mathbf{x}'-\mathbf{x}|^2}$$

denotes a scaled vector along the bond $\mathbf{x}$ to $\mathbf{x}'$.

For the state-based model [41], C is given by

$$\begin{aligned}
\mathrm{C}(\mathbf{x}',\mathbf{x}) &= \eta\,\omega(|\mathbf{x}'-\mathbf{x}|))\left[\alpha(\mathbf{x}',\mathbf{x})\otimes\alpha(\mathbf{x}',\mathbf{x})\right] \\
&+\sigma\int\Big(\underline{\omega}(|\mathbf{x}-\mathbf{z}|)\,\omega(|\mathbf{z}-\mathbf{x}'|)\,\alpha(\mathbf{x},\mathbf{z})\otimes\alpha(\mathbf{z},\mathbf{x}') \\
&-\underline{\omega}(|\mathbf{x}-\mathbf{x}'|)\,\omega(|\mathbf{x}'-\mathbf{z}|)\,\alpha(\mathbf{x},\mathbf{x}')\otimes\alpha(\mathbf{x}',\mathbf{z}) \\
&+\underline{\omega}(|\mathbf{x}-\mathbf{z}|)\,\omega(|\mathbf{x}-\mathbf{x}'|)\,\alpha(\mathbf{x},\mathbf{z})\otimes\alpha(\mathbf{x},\mathbf{x}')\Big)\,d\mathbf{z}.
\end{aligned} \tag{3.3}$$

The additional terms appearing in the tensor above but not in the tensor for the bond-based model signify the indirect interactions of materials points $\mathbf{x}'$ and $\mathbf{x}$ through an intermediate point $\mathbf{z}$ [41].

Assuming that the nonlocal interaction kernel $\omega(|\mathbf{x}'-\mathbf{x}|)$ is supported in a spherical neighborhood of radius δ (the so-called peridynamic horizon), then in the bond-based model, we see that the nonlocal interaction exists only when the distance between $\mathbf{x}$ and $\mathbf{x}'$ is smaller than δ. However, the range of interactions becomes broader in the state-based model due to the indirect interactions [41].

Let us recall some of the comments made on linearized peridynamic models in Section 2.2.8. First, it is pointed out that it is easier to study issues like the well-posedness and convergence of numerical approximations in the context of linearized equations. Moreover, Silling notes that linearized peridynamics also retains some of the complexities of the full nonlinear theory when one is concerned with practical numerical computations. Thus, mathematical analysis of linear equations of the form (3.1) and their corresponding steady-state equations serves as an interesting first step towards a better mathematical understanding of more general nonlinear peridynamic models. This is not to discount the greater practical significance of nonlinear theory. Indeed, extending the linear theory to full nonlinear models will surely be a major undertaking and is currently under active investigation.

The rest of the chapter is organized as follows: we first present in Section 3.2 a brief overview of recent mathematical studies on the well-posedness of peridynamic models. This is followed by an introduction of the notion of nonlocal flux, nonlocal balance laws, and the nonlocal vector calculus in Ssection 3.3. Then, in Section 3.4, we offer a step-by-step guided tour of the key elements of nonlocal calculus of variations, using a nonlocal constrained value problem associated with a linear bond-based model as the illustrative example. The mathematical framework established here can be used to analyze other nonlocal problems. It also lays the foundation for a rigorous analysis of numerical approximations. The mathematical development mirrors the standard calculus of variations theory for local PDEs, which helps to add some familiarity to the discussion. Yet, whenever possible, we point out the differences between the local version and the nonlocal counterpart. We focus on presenting the main ingredients without getting into greater technical details so as to make the exposition more accessible. Only basic variational principles and function spaces are required. Through simple but detailed illustrations, it is hoped that researchers outside the mathematics community can also benefit from a good understanding of the basic mathematical theory which may be important for verification and validation of the peridynamic models and simulations. We then continue with various extensions in section 3.5 and finally offer some remarks in section 3.6.

3.2 A brief review of well-posedness results

One of the first theoretical questions asked about a mathematical equation is often the existence and uniqueness of its solutions. We mention in this section a few selected works in the literature concerning the well-posedness of peridynamic models. This is not intended to be a comprehensive review since there are a large number of studies on related nonlocal models which are not discussed here. We instead focus on several earlier works in this direction that help set the stage for more detailed discussions given in the next section.

As with the development of peridynamic models, rigorous mathematical studies of the well-posedness of peridynamics have started with bond-based peridynamic models that are special cases of state-based peridynamic models. The existence and uniqueness of solutions to the initial value problem associated with a linear bond-based peridynamic equation of motion were first demonstrated by Emmrich and Weckner in [23]. They studied a one-dimensional infinite peridynamic bar with an integrable nonlocal interaction kernel. The resulting solution space is given by the standard L^p Lebesgue function space, a space of functions whose pth powers are integrable ($p \geq 1$). The well-posedness results were extended by Du and Zhou [20] to allow more general kernels that have finite second moments but may not be integrable. These more general kernels lead to a wider class of solution spaces.

For nonlinear initial value problems, well-posedness studies include the work of Erbay, Erkip and Muslu [24] in one space dimension, and the work of Du, Kamm, Lehoucq, and Parks [17] for one-dimensional scalar nonlocal conservation laws. More recently, Emmrich and Puhst analyzed nonlinear bond-based models [22] with globally Lipschitz continuous pairwise force functions. For nonlinear variational problems, a recent existence result was given in [5]. A study related to nonlinear variational problems associated with the more general state-based models was made in [36].

Zhou and Du [47] extended the analysis of initial value problems to study variational problems associated with nonlocal boundary conditions and considered two-dimensional bond-based peridynamic systems as special cases. Analysis of a scalar nonlocal variational problem in multiple dimensions has also been studied by Aksoylu and Parks in [2].

For further review of early literatures on the subject of well-posedness of peridynamic models, we refer to the paper by Silling and Lehoucq [39] and the paper by Lehoucq, Emmrich and Puhst [21].

A more systematic effort to develop a mathematical theory for nonlocal problems in parallel to that for local PDEs has been made in a series of works by Du, Gunzburger, Lehoucq and Zhou [13, 14, 15] resulting in the framework of nonlocal vector calculus developed in [15]. It is then applied to nonlocal and fractional or anomalous diffusion, nonlocal transport, and nonlocal mechanics problems such as peridynamics [13]. The paper [13] provides demonstrations of how nonlocal equations are well-posed on bounded domains with various nonlocal boundary conditions (constraints on domains of nonzero volume). It also shows a compelling analogy that can be drawn between nonlocal models and classical PDEs, including that of a conforming finite element discretization. In [14], a rigorous variational formulation of a linear nonlocal state-based PD system, which is called a peridynamic Navier equation, is presented using basic operators of the nonlocal vector calculus. The paper also demonstrates that, in free space, the local limit of the nonlocal peridynamic Navier operator recovers the conventional Navier operator of linear elasticity with the entire range of allowable elastic constants, e.g., the full range of Poisson ratio in the interval $(-1, 1/2)$. Moreover, the linear peridynamic Navier equation is shown to be well-posed over the space of square integrable functions for a locally integrable tensor kernel. A constraint on the displacement field over a volume of nonzero measure is applied to avoid using the trace operator that is not well defined on this space.

A series of follow-up works by Mengesha and Du further developed the framework of nonlocal calculus of variations that allows one to study the well-posedness of linear and nonlinear peridynamic models under more general assumptions. The papers [33, 34, 35, 36] generalized the analysis of linear nonlocal models to include a nonintegrable tensor kernel and to cases where the kernel may change signs. This latter generalization, suggested by Silling, means that both repulsive and attractive forces can be present which is necessary for dealing with material instability within the peridynamic mechanics theory [40, 44]. Nonlinear variational problems associated with a variety of volumetric constraints are also studied. In addition, [33, 34, 35, 36] documented basic ingredients for the functional analysis of nonlocal energy spaces and nonlocal variational problems with general kernels for both scalar equations and for systems.

3.3 Nonlocal balance laws and nonlocal vector calculus

In continuum mechanics, physical balance laws are often written mathematically as systems of partial differential equations. On the other hand, peridynamic models introduced by Silling in [40] can be viewed as nonlocal balance laws by defining a nonlocal flux in terms of interactions between regions having positive measures and possibly not sharing a common boundary.

Nonlocal flux For a typical peridynamic equation of motion, such as the equation (2.25) with a kernel given by (3.2), a *nonlocal flux* functional can be defined for any pair of domains Ω_1 and Ω_2 by

$$\mathcal{F}_\ell(\Omega_1,\Omega_2) = \int_{\Omega_1}\int_{\Omega_2} \mathrm{C}(\mathbf{x}',\mathbf{x})(\mathbf{u}(\mathbf{x}')-\mathbf{u}(\mathbf{x}))dV_{\mathbf{x}}' dV_{\mathbf{x}} \tag{3.4}$$

for a tensor $\mathrm{C}(\mathbf{x}',\mathbf{x})$ which is symmetric with respect to $\mathbf{x}'$ and $\mathbf{x}$.

For more general state-based peridynamic models [42] of the form:

$$\rho\partial_{tt}\mathbf{y}(\mathbf{x},t) = \int \{\underline{\mathrm{T}}[\mathbf{x}] < \mathbf{x}'-\mathbf{x} > -\underline{\mathrm{T}}[\mathbf{x}'] < \mathbf{x}-\mathbf{x}' >\} dV_{\mathbf{x}'} + \mathbf{b}(\mathbf{x}), \tag{3.5}$$

with $\underline{\mathrm{T}}[\mathbf{x}] < \mathbf{x}'-\mathbf{x} >$ and $\underline{\mathrm{T}}[\mathbf{x}'] < \mathbf{x}-\mathbf{x}' >$ being the peridynamic force states, a similar notion of nonlocal flux can be defined, as in [15], by

$$\mathcal{F}(\Omega_1,\Omega_2) = \int_{\Omega_1}\int_{\Omega_2} \{\underline{\mathrm{T}}[\mathbf{x}] < \mathbf{x}'-\mathbf{x} > -\underline{\mathrm{T}}[\mathbf{x}'] < \mathbf{x}-\mathbf{x}' >\} dV_{\mathbf{x}'} dV_{\mathbf{x}}. \tag{3.6}$$

A defining characteristic of the nonlocal flux is that the integrands in (3.4) and (3.6) are anti-symmetric with respect to an interchange of $\mathbf{x}$ and $\mathbf{x}'$. This is a necessary and sufficient condition for the resulting balance law to satisfy an action-reaction principle. The nonlocal flux is an additive form, in the sense that

$$\mathcal{F}(\Omega_1,\Omega_2) + \mathcal{F}(\Omega_1,\Omega_3) = \mathcal{F}(\Omega_1,\Omega_2\cup\Omega_3), \quad \forall \Omega_2\cap\Omega_3 = \emptyset,$$

and an alternating form in the sense that

$$\mathcal{F}(\Omega_1,\Omega_2) = -\mathcal{F}(\Omega_2,\Omega_1), \quad \forall \Omega_1,\Omega_2.$$

Moreover, nonlocal flux can be well-defined for two domains Ω_1 and Ω_2 without letting Ω_1 and Ω_2 be in direct contact. Thus, it broadens the notion of local fluxes. These discussions and axiomatic derivations of various properties of nonlocal flux functionals can be found in [15]. The study of nonlocal flux there also influenced how nonlocal operators are introduced in the nonlocal vector calculus framework.

Nonlocal balance laws By integrating the equation of motion (3.5) over a spatial domain $\Omega \subset \mathbb{R}^n$, the nonlocal balance law corresponding to the state-based peridynamic model (3.5) takes on the form:

$$\frac{d}{dt}\int_{\Omega} \rho\dot{\mathbf{y}}\, dV_{\mathbf{x}} = \mathcal{F}(\Omega,\mathbb{R}^n\setminus\Omega) + \int_{\Omega} \mathbf{b}\, dV_{\mathbf{x}}.$$

In other words, peridynamic equations may be seen as continuum models that represent nonlocal balance laws involving nonlocal fluxes and nonlocal operators; they avoid classical notions like the local flux and deformation gradient, thus offering the opportunity to work with more singular solutions to better characterize materials defects.

The nonlocal vector calculus developed in [15] introduces some basic nonlocal operators to systematically represent nonlocal flux and nonlocal balance laws. The theory also helps to elucidate the often puzzling issue of boundary conditions, or, more properly speaking, nonlocal solution constraints for nonlocal equations.

Nonlocal operators Various simple nonlocal operators have been defined as the basic building blocks of nonlocal vector calculus. For example, for any two points $\mathbf{x}'$ and $\mathbf{x}$ in $\mathbb{R}^n$, we pick $\alpha = \alpha(\mathbf{x}',\mathbf{x})$ as a given two-point vector such as the unit vector $\alpha(\mathbf{x}',\mathbf{x}) = (\mathbf{x}'-\mathbf{x})/|\mathbf{x}'-\mathbf{x}|$ or, to be used here, the rescaled unit vector $\alpha(\mathbf{x}',\mathbf{x}) = (\mathbf{x}'-\mathbf{x})/|\mathbf{x}'-\mathbf{x}|^2$. With respect to α, we have the nonlocal divergence operator $\mathscr{D}$ defined by

$$(\mathscr{D}\psi)(\mathbf{x}) = \int_{\mathbb{R}^n} \big(\psi(\mathbf{x},\mathbf{x}') + \psi(\mathbf{x}',\mathbf{x})\big)\cdot \alpha(\mathbf{x},\mathbf{x}')\,d\mathbf{x}'$$

for any 2-point tensor-valued function $\psi\colon \mathbb{R}^n \times \mathbb{R}^n \to \mathbb{R}^{n\times m}$. Its adjoint operator $\mathscr{D}^*$ is defined by

$$(\mathscr{D}^*\mathbf{v})(\mathbf{x},\mathbf{x}') = \big(\mathbf{v}(\mathbf{x}') - \mathbf{v}(\mathbf{x})\big)\otimes \alpha(\mathbf{x}',\mathbf{x})$$

for any vector-valued function $\mathbf{v}\colon \mathbb{R}^n \to \mathbb{R}^m$. We note the matrix and vector product identities that

$$(\beta\otimes\alpha)\cdot\alpha = \beta\alpha^{\mathrm{T}}\alpha = |\alpha|^2\beta, \quad (\alpha\otimes\alpha)\cdot\beta = \alpha^{\mathrm{T}}\beta\alpha, \quad \forall\beta\in\mathbb{R}^m. \tag{3.7}$$

Unlike classical divergence operators that map between vector and scalar fields of the same variable, operators $\mathscr{D}$ and $\mathscr{D}^*$ are maps between vector and tensor fields with different numbers of variables. This can be better understood through the exterior calculus interpretation of differential operators. Viewing vector and tensor fields with different numbers of variables as nonlocal forms of different orders, the nonlocal operators mimic exterior differentiations of local differential forms having different orders.

$\mathscr{D}$ and $\mathscr{D}^*$ are adjoint operators to each other in the sense that

$$\int_{\mathbb{R}^n} \mathbf{v}(\mathbf{x})\cdot(\mathscr{D}\psi)(\mathbf{x})d\mathbf{x} = \int_{\mathbb{R}^n}\int_{\mathbb{R}^n} (\mathscr{D}^*\mathbf{v})(\mathbf{x},\mathbf{x}')\cdot\psi(\mathbf{x},\mathbf{x}')d\mathbf{x}'d\mathbf{x}.$$

This duality property may also be seen as an instance of the Fubini theorem if the integrands satisfy suitable regularity assumptions. Symbolically speaking, and to draw analogy to classical vector calculus, it can also be interpreted as a nonlocal integration by parts formula.

For notational convenience, we introduce as in [15] that

$$(\mathfrak{D}\varphi)(\mathbf{x}) = \int_{\mathbb{R}^n} \big(\varphi(\mathbf{x},\mathbf{x}') + \varphi(\mathbf{x}',\mathbf{x})\big)\cdot\alpha(\mathbf{x},\mathbf{x}')\,d\mathbf{x}', \quad \forall\varphi\colon \mathbb{R}^n\times\mathbb{R}^n\to\mathbb{R}^n.$$

Its adjoint is then given by

$$(\mathfrak{D}^*v)(\mathbf{x},\mathbf{x}') = \big(v(\mathbf{x}') - v(\mathbf{x})\big)\alpha(\mathbf{x}',\mathbf{x}), \quad \forall v\colon \mathbb{R}^n\to\mathbb{R}.$$

Note that $\mathscr{D}^*\mathbf{v}$ is a symmetric 2-point tensor-valued function, that is $(\mathscr{D}^*\mathbf{v})(\mathbf{x},\mathbf{x}') = (\mathscr{D}^*\mathbf{v})(\mathbf{x}',\mathbf{x})$. Meanwhile, for any 2-point tensor function $\psi = \psi(\mathbf{x},\mathbf{y})$, $\mathscr{D}\psi$ depends only on the symmetric part of ψ with respect to its two variables $\mathbf{x}$ and $\mathbf{y}$. In addition, we see that the integrand in the definition of $\mathscr{D}\psi$ is an anti-symmetric 2-point function so that the identity

$$\int_{\Omega} (\mathscr{D}\psi)(\mathbf{x})d\mathbf{x} = -\int_{\Omega^c} (\mathscr{D}\psi)(\mathbf{x})d\mathbf{x}$$

always holds. This can be interpreted as a nonlocal version of the classical divergence theorem [15], since the term on the right-hand side can be precisely characterized as a nonlocal flux

$$-\mathcal{F}(\Omega^c,\Omega) = -\mathcal{F}(\Omega,\Omega^c)$$

that replaces the classical local flux defined on the boundary of Ω.

Nonlocal extensions to Stokes theorem and other integral identities can also be established with the definition of nonlocal curl operators.

We note that in the definitions of nonlocal operators, while the space dimensions m and n can be different in general, the cases most relevant to peridynamics are $m = 1$ and $m = n$.

Formulating nonlocal models via nonlocal operator The nonlocal vector calculus allows us to reformulate linear peridynamic models in ways similar to local linear PDE models expressed in terms of basic divergence, gradient, and curl operators.

For example, with a given kernel $\omega_\delta = \omega_\delta(|\mathbf{x}' - \mathbf{x}|)$, for any function $\mathbf{u} = \mathbf{u}(\mathbf{x})$ defined in $\mathbb{R}^n$, using (3.7) and the notation $\underline{\omega}_\delta(|\mathbf{y}-\mathbf{x}|) = \omega_\delta(|\mathbf{y}-\mathbf{x}|)|\alpha(\mathbf{x},\mathbf{y})|^2$, a composition of above defined operators $\mathscr{D}$ and $\mathscr{D}^*$ gives

$$\mathcal{L}_\delta(\mathbf{u})(\mathbf{x}) = -\mathscr{D}(\omega_\delta\, \mathscr{D}^*\mathbf{u})(\mathbf{x}) = \int \underline{\omega}_\delta(|\mathbf{y}-\mathbf{x}|)(\mathbf{u}(\mathbf{y}) - \mathbf{u}(\mathbf{x}))\,d\mathbf{y} \tag{3.8}$$

which is often called a *nonlocal Laplacian* or a nonlocal diffusion operator.

In the context of peridynamic models, the kernel function ω_δ has a compact support over $|\mathbf{y}-\mathbf{x}| \le \delta$ with $\delta > 0$ being referred to as the horizon parameter. The kernel is suitably scaled so that

$$-\mathscr{D}(\omega_\delta\, \mathscr{D}^*) \to -\nabla\cdot(K\nabla), \quad \text{as} \quad \delta \to 0$$

for a coefficient matrix denoted by the tensor K. More detailed discussions are given later; an intuitive example of such a limit is when $\underline{\omega}_\delta(|\mathbf{z}|)$ approaches a constant multiple of the Dirac delta function at $\mathbf{x} = \mathbf{0}$. This is why $\mathcal{L}_\delta$ may be viewed as an nonlocal analog of classical diffusion (Laplacian) operator.

Peridynamic Navier operator More closely related to peridynamic models is the so-called peridynamic Navier operator

$$\mathcal{L}_s = -\eta\mathfrak{D}(\omega_\delta\,\mathfrak{D}^*) - \sigma\mathfrak{D}_{\omega_\delta}(\mathfrak{D}^*_{\omega_\delta}) \tag{3.9}$$

where $\mathbf{I}$ is the identity tensor. Moreover, given

$$\tilde{\omega}_\delta(\mathbf{x},\mathbf{y}) = \frac{n}{m(\mathbf{x})}\omega_\delta(|\mathbf{x}-\mathbf{y}|) \quad \text{with} \quad m(\mathbf{x}) = \int_\Omega \omega_\delta(|\mathbf{x}-\mathbf{y}|)d\mathbf{y},$$

the operators $\mathfrak{D}_{\omega_\delta}$ and $\mathfrak{D}^*_{\omega_\delta}$ that map (1-point) functions to themselves are defined by

$$\mathfrak{D}_{\omega_\delta}(v)(\mathbf{x}) = \mathfrak{D}\big(\tilde{\omega}_\delta(\mathbf{x},\mathbf{y})v(\mathbf{x})\big)(\mathbf{x}), \quad \forall v\colon \mathbb{R}^n \to \mathbb{R}, \tag{3.10}$$

$$\mathfrak{D}^*_{\omega_\delta}(\mathbf{u})(\mathbf{x}) = \int_{\mathbb{R}^n} \mathfrak{D}^*(\mathbf{u})(\mathbf{x},\mathbf{y})\,\tilde{\omega}_\delta(|\mathbf{x}-\mathbf{y}|)\,d\mathbf{y}, \quad \forall \mathbf{u}\colon \mathbb{R}^n \to \mathbb{R}^n. \tag{3.11}$$

The expression given in (3.9) offers a concise formulation of the integral operator associated with the tensor kernel given in (3.3). Moreover, since the nonlocal operators $\mathfrak{D}_{\omega_\delta}$ and $\mathfrak{D}^*_{\omega_\delta}$ have close correspondence to the classical divergence and gradient operator in the local limit [15, 37, 36], we have as $\delta \to 0$,

$$-\eta\mathfrak{D}(\omega_\delta\mathfrak{D}^*) - \sigma\mathfrak{D}_{\omega_\delta}(\mathfrak{D}^*_{\omega_\delta}) \to -\mu\nabla\cdot\nabla - \mu\nabla\nabla\cdot - \lambda\nabla\nabla\cdot,$$

where the limit is the classical, local Navier operator of linear elasticity [10].

For the special case of $\sigma = 0$, we get the nonlocal vector calculus formulation of the bond-based linear peridynamic operator

$$\mathcal{L}_b = -\mathfrak{D}(\omega_\delta\mathfrak{D}^*)$$

whose local limit is given by

$$\mu\nabla\cdot\nabla + \mu\nabla\nabla\cdot,$$

a special form of the Navier operator for homogeneous materials with a Poisson ratio $1/4$. More discussions will be provided in Chapter 4.

The weighted nonlocal operators The weighted nonlocal operators $\mathfrak{D}_{\omega_\delta}$ and $\mathfrak{D}^*_{\omega_\delta}$ defined in (3.10) and (3.11) have been introduced in [15], as they are more closely aligned with conventional differential operators defined on vector fields than operators $\mathscr{D}$, $\mathscr{D}^*$, $\mathfrak{D}$, and $\mathfrak{D}^*$.

Indeed, it has been shown via Fourier analysis that, in the whole space, the local limits as $\delta \to 0$ of $\mathfrak{D}_{\omega_\delta}$ and $\mathfrak{D}^*_{\omega_\delta}$ correspond to classical gradient and divergence operators [15], subject to suitable choices of the kernels ω_δ. Recently, the analysis has been further extended in [36, 37] to offer more detailed characterizations of the related function spaces and topologies on bounded domains.

Weighted nonlocal operators also have natural physical interpretations. In fact, for a displacement field $\mathbf{u} = \mathbf{u}(\mathbf{x})$, the quantity $\mathfrak{D}^*_{\omega_\delta}(\mathbf{u})(\mathbf{x})$ gives the linearized nonlocal dilatation and

$$\mathfrak{D}^*(\mathbf{u})(\mathbf{x},\mathbf{x}') - \frac{1}{n}\mathfrak{D}^*_{\omega_\delta}(\mathbf{u})(\mathbf{x})$$

gives the deviatoric part of the nonlocal elastic strain.

Symbolically, one may recover $\mathscr{D}^*$ and $\mathfrak{D}^*$ from $\mathscr{D}^*_{\omega_\delta}$ by picking a special weight ω_δ in the form of a Dirac delta measure.

Nonlocal vector calculus and Green's identities In [15], more extensive discussions have been given on $\mathscr{D}$, $\mathscr{D}^*$, $\mathscr{D}_\omega$, $\mathscr{D}^*_{\omega_\delta}$, along with nonlocal Green's identities such as

$$\int_{\mathbb{R}^n} \mathbf{u}\left(\mathscr{D}(\omega(\mathscr{D}^*\mathbf{v}))\right)d\mathbf{x} - \int_{\mathbb{R}^n} \left(\mathscr{D}(\omega(\mathscr{D}^*\mathbf{u}))\right)\mathbf{v}d\mathbf{x} = 0,$$

$$\int_{\mathbb{R}^n} \mathbf{u}\,\mathscr{D}_{\omega_\delta}(\mathscr{D}^*_{\omega_\delta}\mathbf{v})dx = -\int_{\mathbb{R}^n} (\mathscr{D}^*_{\omega_\delta}\mathbf{u}\mathscr{D}^*_{\omega_\delta}\mathbf{v})d\mathbf{x} = \int_{\mathbb{R}^n} \mathscr{D}_{\omega_\delta}(\mathscr{D}^*_{\omega_\delta}\mathbf{u})\mathbf{v}d\mathbf{x},$$

and

$$\int_{\mathbb{R}^n} \mathbf{u}\,\mathfrak{D}_{\omega_\delta}(\mathfrak{D}^*_{\omega_\delta}\mathbf{v})dx = -\int_{\mathbb{R}^n} (\mathfrak{D}^*_{\omega_\delta}\mathbf{u}\mathfrak{D}^*_{\omega_\delta}\mathbf{v})d\mathbf{x} = \int_{\mathbb{R}^n} \mathfrak{D}_{\omega_\delta}(\mathfrak{D}^*_{\omega_\delta}\mathbf{u})\mathbf{v}d\mathbf{x}.$$

These identities are nonlocal analogs of the classical counterparts that are stated formally in [15] and they can be rigorously justified in suitably defined function spaces. The latter are discussed in the next section.

In [15], other nonlocal operators such as nonlocal curls have also been introduced, In the context of peridynamics, nonlocal curls have not been used in the analysis or formulation but it is conceivable that there is some potential value when electromagnetic interactions are coupled to mechanical and thermodynamic models.

3.4 Nonlocal calculus of variations — an illustration

Building upon the foundation of the nonlocal vector calculus, we may further develop the theory of nonlocal calculus of variations that can be used to analyze nonlocal steady-state variational problems and time-dependent problems associated with peridynamic equations. To make the presentation accessible to a broader audience, some of the key ingredients are presented here through an illustrative example, namely a variational problem associated with a linear bond-based peridynamic model with a homogeneous condition imposed on the displacement field on part of the material domain. We assume only minimal knowledge of the variational principles for PDEs and function spaces for most of the discussion.

A steady-state problem We consider a linear steady-state problem related to the nonlocal bond-based peridynamic model

$$-\mathcal{L}_\delta\mathbf{u}(\mathbf{x}) = \mathbf{b}(\mathbf{x}), \quad \forall \mathbf{x} \in \Omega \tag{3.12}$$

where Ω is a bounded domain in $\mathbb{R}^n$ and the nonlocal bond-based peridynamic operator $\mathcal{L}_\delta$ is defined by

$$\begin{aligned}\mathcal{L}_\delta \mathbf{u}(\mathbf{x}) &= -\mathfrak{D}(\omega_\delta \mathfrak{D}^*(\mathbf{u}))(\mathbf{x}) \\ &= \int_{\Omega\cup\Omega_I} \omega_\delta(|\mathbf{y}-\mathbf{x}|)\alpha(\mathbf{y},\mathbf{x})\otimes\alpha(\mathbf{y},\mathbf{x})\cdot\big(\mathbf{u}(\mathbf{y})-\mathbf{u}(\mathbf{x})\big)\,d\mathbf{y} \\ &= \int_{\Omega\cup\Omega_I} \omega_\delta(|\mathbf{y}-\mathbf{x}|)\frac{\mathbf{y}-\mathbf{x}}{|\mathbf{y}-\mathbf{x}|^2}\otimes\frac{\mathbf{y}-\mathbf{x}}{|\mathbf{y}-\mathbf{x}|^2}\cdot\big(\mathbf{u}(\mathbf{y})-\mathbf{u}(\mathbf{x})\big)\,d\mathbf{y} \quad (3.13)\end{aligned}$$

where Ω_I denotes the interaction domain on which the solution may be constrained. Without loss of generality, Ω and Ω_I are assumed to have no interior intersection, and we use $\tilde{\Omega} = \Omega\cup\Omega_I$ to denote the whole domain. The above bond-based model corresponds to the steady-state equation of (3.1) with the tensor C being a scalar multiple of the outer product of the unit vector along the bond direction, so that the nonlocal force is a linear Hookean spring force aligned with the bond direction.

Based on the definitions of the nonlocal operator $\mathcal{L}_\delta$, we see that the definition of $\mathcal{L}_\delta\mathbf{u}(\mathbf{x})$ in Ω is related to how $\mathbf{u}$ is defined on Ω_I. This leads to the issue of nonlocal constraints.

Nonlocal boundary conditions and nonlocal constraints For nonlocal models such as peridynamics, one important issue is the proper understanding of the nonlocal analog of boundary conditions as those specified for PDEs. In [13], the notion of volume constrained problems is discussed. Some remarks along this direction are given here.

First of all, one may define a notion of essential constraints and natural or variational conditions for nonlocal problems by drawing analogy to the classical variational problems for PDEs that are defined with respect to suitable boundary conditions. At the same time, for nonlocal models there is a conceptual change to the notion of boundary. Indeed, the underlying problem is dictated by the form and the domain of nonlocal interactions so that when a materials point $\mathbf{x}$ is located close to a material boundary, it is perceivable that a new physical principle must be adopted to account for the presence of the boundary. Thus, either additional constraints are imposed or changes to laws of nonlocal interactions that offer similar effects to that of the local boundary conditions on PDEs should be applied. It is possible to distinguish these different types of constraints or changes through the framework of nonlocal calculus of variations.

For example, if Ω_I is an empty set, then $\mathcal{L}_\delta$ can be well defined all over $\tilde{\Omega} = \Omega$ without the need to impose additional conditions on $\mathbf{u}$. In this case, the term b may represent either soft or hard loading with the distinction only showing up through the dependence on the horizon parameter. This case may be viewed as a nonlocal analog of natural boundary conditions for PDEs. In our earlier works, we also have tried to give a more symbolically similar formulation by taking a non-empty Ω_I in the definitions of the nonlocal operators as that given by (3.13), and imposing the Equation (3.12) on both Ω that may represent the nonlocal equation, and Ω_I that may represent the domain of nonlocal constraints or nonlocal boundary conditions.

Again, viewed in terms of $\tilde{\Omega}$, such a separation of the equation domain and the constraint domain becomes superfluous mathematically, though they can be subject to different physical interpretations.

On the other hand, Ω_I can be a constraint set of nonzero measure (i.e., it is not a lower-dimensional boundary) on which essential, rather than variational or natural, conditions on the solutions are imposed. For example, we may impose the condition $\mathbf{u} = 0$ on Ω_I. This is an essential condition that we must build explicitly into the variational principle. A typical example is

$$\Omega_I = \Omega_\delta = \{\mathbf{x} \in \Omega^c : \mathrm{dist}(\mathbf{x}, \partial\Omega) < \delta\},$$

which represents a δ-layer surrounding Ω with $\partial\Omega$ being its inner boundary.

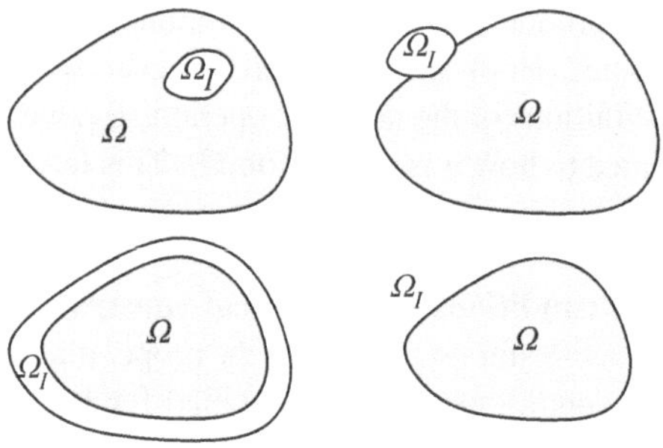

Figure 3.1: Four of the possible configurations for the domain Ω and interaction set Ω_I.

As seen in later discussions, for interaction kernels that may have stronger singularities than those with infinite first-order moments, the solutions of nonlocal problems may enjoy better spatial regularities so that constraints on a lower-dimensional boundary could be well defined. For such cases, essential conditions on the boundary can also be imposed, though we do not intend to pursue further discussions here.

Nonlocal constrained value problem Related to Equation (3.12), we consider the following nonlocal constrained value problem.

$$\begin{cases} -\mathcal{L}_\delta \mathbf{u}(\mathbf{x}) = \mathbf{b}(\mathbf{x}), & \forall \mathbf{x} \in \Omega, \\ \mathbf{u}(\mathbf{x}) = 0, & \forall \mathbf{x} \in \Omega_\delta. \end{cases} \tag{3.14}$$

The rest of the discussion provides a systematic study of the well-posedness of the above problem and it largely follows from the materials presented in [34].

Energy functional A step towards a variational formulation of the nonlocal problem is to establish the corresponding energy functional corresponding to (3.12) that can be defined as

$$E(u) = \int_{\tilde{\Omega}}\int_{\tilde{\Omega}} \frac{1}{2}\omega_\delta(|\mathbf{y}-\mathbf{x}|)\left(\mathfrak{D}^*\mathbf{u}(\mathbf{x},\mathbf{y})\right)^2 d\mathbf{y}d\mathbf{x} - \int_{\tilde{\Omega}} \mathbf{b}(\mathbf{x})\mathbf{u}(\mathbf{x})d\mathbf{x}. \tag{3.15}$$

The energy includes the nonlocal elasticity energy and the contribution from the external force $\mathbf{b} = \mathbf{b}(\mathbf{x})$.

Another step following the definition of the energy functional is the computation of the energy variation

$$\begin{aligned}\frac{d}{dt}E(\mathbf{u}+t\mathbf{v})\,|_{t=0} &= \int_{\tilde{\Omega}}\int_{\tilde{\Omega}} \omega_\delta(\mathfrak{D}^*\mathbf{u})(\mathfrak{D}^*\mathbf{v})\,d\mathbf{y}d\mathbf{x} - \int_{\tilde{\Omega}} \mathbf{b}(\mathbf{x})\mathbf{v}(\mathbf{x})d\mathbf{x} \\ &= \int_{\tilde{\Omega}} \mathfrak{D}\left(\omega_\delta(\mathfrak{D}^*\mathbf{u})\right)\mathbf{v}(\mathbf{x})d\mathbf{x} - \int_{\tilde{\Omega}} \mathbf{b}(\mathbf{x})\mathbf{v}(\mathbf{x})d\mathbf{x} \\ &= \int_{\tilde{\Omega}} \left(-\mathcal{L}_\delta\mathbf{u}(\mathbf{x}) - \mathbf{b}(\mathbf{x})\right)\mathbf{v}(\mathbf{x})d\mathbf{x},\end{aligned}$$

where the second equation is obtained formally through the duality or the definition of the nonlocal operators for functions that make the operations well defined. This calculation is reminiscent of that in the principle of virtual works of classical mechanics and it can be used to establish the relation between the strong form and the weak form of the variational problem when the functions are chosen from suitable spaces.

Assumptions on the nonlocal interaction kernel The properties of the linear nonlocal model are very much dependent on the nonlocal interaction kernel $\underline{\omega}_\delta$ used to define the nonlocal operator. It is therefore helpful to state clearly the type of kernels that are amendable to the nonlocal calculus of variations frameworks. We note that ω_δ is often assumed to satisfy

1. $$\omega_\delta(r) \geq 0, \forall r \in (0,\delta), \tag{3.16}$$
2. $$\text{supp}(\omega_\delta) \subset B_\delta(0), \quad \text{i.e.,} \quad \underline{\omega}_\delta(r) = 0, \forall r \geq \delta, \tag{3.17}$$
3. $$\int_{\mathbb{R}^n} \omega_\delta(|\mathbf{x}|)d\mathbf{x} = m < \infty \tag{3.18}$$

where the last normalization condition on ω_δ implies that $\underline{\omega}_\delta$ has finite second-order moments (which is necessary for well-defined elastic moduli [40]). The latter is also equivalent to the requirement that the energy is finite for a linear displacement field. A consequence is that for any square integrable vector-valued function that has square integrable first-order partial derivatives, the energy remains finite.

We note that the non-negativity assumption $\omega_\delta(r) \geq 0$ can be relaxed [33]. It is also possible to consider interaction kernels that are not radial nor translational invariant. More discussions are given in the next section.

Solution space The canonical function space in which to look for a solution of the variational problem is one that includes functions that have well-defined energy and satisfy specified constraints.

For the problem under consideration, we first define the so-called energy space which contains square integrable functions (i.e., functions in $L^2(\tilde{\Omega})$) with a finite energy

$$V = \{\mathbf{u} \in L^2(\tilde{\Omega}) \mid \int_{\tilde{\Omega}} \int_{\tilde{\Omega}} \omega_\delta(|\mathbf{x}-\mathbf{y}|)\,(\mathfrak{D}^*\mathbf{u})^2 d\mathbf{y}d\mathbf{x} < \infty\}.$$

The space V is an inner product space corresponding to the norm

$$\|\mathbf{u}\|_v = (|\mathbf{u}|_v^2 + \|\mathbf{u}\|^2_{L^2(\tilde{\Omega})})^{1/2}$$

where the semi-norm $|\mathbf{u}|_v$ on V is defined as

$$|\mathbf{u}|_v^2 = \int_{\tilde{\Omega}} \int_{\tilde{\Omega}} \omega_\delta(|\mathbf{x}-\mathbf{y}|)\,(\mathfrak{D}^*\mathbf{u})^2 d\mathbf{y}d\mathbf{x}.$$

For the nonlocal interaction kernels satisfying the assumptions made earlier, we note that in general V contains the space $H^1(\tilde{\Omega})$, which is the subspace of $L^2(\tilde{\Omega})$ with functions having all of their first-order partial derivatives also in $L^2(\tilde{\Omega})$. Meanwhile, for a kernel $\underline{\omega}_\delta$ that is integrable, it has been shown that $V = L^2(\tilde{\Omega})$.

The solution space of the variational problem is simply a subspace of V with functions satisfying the imposed constraints. That is, for the problem under consideration, we have

$$V_{c,\delta} = \{\mathbf{u} \in V \mid \mathbf{u} = 0 \text{ in } \Omega_I\} \subset V. \tag{3.19}$$

Various properties of the function spaces V and $V_{c,\delta}$ can be studied in both specific spaces identified with either L^2 spaces or fractional Sobolev spaces for specific kernels [15] and general abstract solution spaces [34]. These studies lead to results like

- V, equipped with the norm $\{\|\mathbf{u}\|^2_{L^2} + |\mathbf{u}|_v^2\}^{1/2}$, is a Hilbert space.
- The embeddings $H^1 \hookrightarrow V \hookrightarrow L^2$ are continuous.
- The solution space $V_{c,\delta}$ is a subspace of V that is closed and dense in L^2 (thus separable), containing a dense subspace that is made of all functions in $H_0^1(\Omega)$ with zero extensions on Ω_I.

Additional properties of V and $V_{c,\delta}$ can be established, such as compact embedding properties and extension theorems [34], which require more technical discussions and are beyond the scope of our presentation here.

We note that the discussion of the function spaces associated with the nonlocal models is not only of interest for mathematical analysis and well-posedness studies but also important for designing numerical approximation schemes, in particular, conforming finite element approximations. For example, when the kernel becomes integrable, then $V = L^2$ so that discontinuous finite elements are in fact conforming approximations. This is in sharp contrast to the conforming approximation of conventional second-order elliptic equations that only allow continuous finite element spaces.

Variational principle We can now consider the constrained minimization problem

$$\min_{\mathbf{u}\in V_{c,\delta}} E(\mathbf{u}) .$$

Based on the calculation of the variations of the energy functional given earlier, we see that a necessary and sufficient condition for the solution $\mathbf{u} \in V_{c,\delta}$ of the variational problem, if it exists, is that

$$\frac{d}{dt}E(\mathbf{u}+t\mathbf{v}) \mid_{t=0}= 0 \tag{3.20}$$

for any $\mathbf{v}$ also in $V_{c,\delta}$. The condition (3.20) is commonly referred to as the Euler–Lagrange equation for the minimization problem.

Bilinear form and weak form Let us first define

$$B_\delta(\mathbf{u},\mathbf{v}) = \int_{\tilde{\Omega}}\int_{\tilde{\Omega}} \omega_\delta(|\mathbf{x}-\mathbf{y}|)(\mathfrak{D}^*\mathbf{u})(\mathfrak{D}^*\mathbf{v})\,dyd\mathbf{x}, \tag{3.21}$$

which is a symmetric continuous bilinear form defined in $V \times V$. The symmetry of $B_\delta(\cdot,\cdot)$ means simply that

$$B_\delta(\mathbf{u},\mathbf{v}) = B_\delta(\mathbf{v},\mathbf{u})$$

for any $\mathbf{u},\mathbf{v} \in V$.

Let $(\cdot,\cdot)$ denote the standard L^2 inner product on $\tilde{\Omega}$. The weak form of the variational problem is then given by: for $\mathbf{b} \in V^*_{c,\delta}$ (the dual space of $V_{c,\delta}$, which is the space of all bounded linear functionals on $V_{c,\delta}$), find $\mathbf{u} \in V_{c,\delta}$ such that

$$B_\delta(\mathbf{u},\mathbf{v}) = (\mathbf{b},\mathbf{v}), \quad \forall\, \mathbf{v} \in V_{c,\delta} \tag{3.22}$$

which is equivalent to (3.20) or rather $\mathbf{u}$ is the minimizer of the energy functional in the solution space $V_{c,\delta}$. This can also be viewed as a consequence of the nonlocal version of the principle of virtual work.

Let Π denote the set of infinitesimally rigid displacements given by

$$\Pi = \{\mathbf{v} : \mathbf{v}(\mathbf{x}) = \mathbb{Q}\mathbf{x}+\mathbf{p}, \mathbb{Q} \in \mathbb{R}^{d\times d}, \mathbb{Q}^T = -\mathbb{Q}, \mathbf{p} \in \mathbb{R}^d\}.$$

An important property that ensures the unique solvability of the variational problem is that

$$Z = \{\mathbf{v} \in V \mid |\mathbf{u}|_v^2 = B_\delta(\mathbf{v},\mathbf{v}) = 0\} = \Pi ,$$

that is, the set Z contains only rigid body displacement. This particular result has been verified in [14] for integrable kernels and in more general solution spaces for more general kernels [34].

Consequently, we have

$$V_{c,\delta} \cap Z = V_{c,\delta} \cap \Pi = \{0\} . \tag{3.23}$$

This equation (3.23) is important not only mathematically but also physically as it refers to the fact that there is no nontrivial zero energy mode associated with the energy functional defined by (3.15) in the absence of the external force, that is, if $\mathbf{b}(\mathbf{x}) = 0$ in Ω.

Nonlocal Poincare's inequality For the well-posedness of the variational problem, we need to show that the bilinear form is coercive in the solution space $V_{c,\delta}$.

Similar to the local counterpart for PDEs, the coercivity of the bilinear form is a consequence of a nonlocal Poincare's inequality. That is, there is a constant $c > 0$, such that for any $\mathbf{v} \in V_{c,\delta}$,

$$|\mathbf{v}|_v \geq c\|\mathbf{v}\|_{L^2}. \tag{3.24}$$

Inequalities like (3.24) have been established for both scalar models and systems with various kernels in, for example, [20, 13, 15, 33]. In general, one can first establish the result for kernels that are integrable, that is, kernels like $\underline{\omega}_\delta(|\mathbf{x}'-\mathbf{x}|) = |\mathbf{x}'-\mathbf{x}|^{-2}\omega_\delta(|\mathbf{x}'-\mathbf{x}|)$, and then get the desired inequality for ω_δ itself by noticing that

$$4\,\mathrm{diam}(\Omega)^2\underline{\omega}_\delta(|\mathbf{x}' \quad \mathbf{x}|) \geq \omega_\delta(|\mathbf{x}'-\mathbf{x}|)$$

where $\mathrm{diam}(\Omega)$ denotes the diameter of the domain Ω.

For an integrable kernel $\underline{\omega}_\delta(|\mathbf{x}'-\mathbf{x}|)$, the nonlocal Poincare's inequality can be established by utilizing certain compactness of sequence of functions in L^2 with unit L^2 norms but vanishing $|\cdot|_v$ semi-norms. Such a sequence must have a weakly convergent subsequence in L^2, with the weak limit in the kernel space Z. This in turn implies that the weak limit has to be 0 by (3.23).

Given that $\underline{\omega}_\delta$ is integrable and that convolution operators associated with such kernels are compact, the vanishing seminorm in $|\cdot|_v$ then implies the convergence in L^2 to 0 with the help of the fact

$$\int_{\tilde{\Omega}}\int_{\tilde{\Omega}} \underline{\omega}_\delta(|\mathbf{x}'-\mathbf{x}|)d\mathbf{x}'d\mathbf{x} \geq \beta > 0$$

for some positive constant $\beta > 0$. The convergence of a subsequence in L^2 norm to 0 contradicts with the fact that all the functions in the sequence have unit L^2 norms. This contradiction shows the validity of the nonlocal Poincare's inequality. We refer to [32, 34] for more detailed discussions.

The nonlocal Poincare's inequality implies that the semi-norm $|\cdot|_V$ defines a norm on the solution space $V_{c,\delta}$ and the bilinear form $B_\delta(\cdot,\cdot)$ is coercive, that is

$$B_\delta(\mathbf{v},\mathbf{v}) \geq \frac{c^2}{1+c^2}(|\mathbf{v}|_v^2 + \|\mathbf{v}\|_{L^2}^2) \quad \forall \mathbf{v} \in V_{c,\delta}. \tag{3.25}$$

In addition, we have that the symmetric bilinear form $B_\delta(\mathbf{u},\mathbf{v})$ defines an equivalent inner product on the solution space.

Well-posedness of the variational problem By the properties established for the function spaces and the bilinear form, we are led to the well-posedness of the variational problem or rather the existence, uniqueness, and stability of the weak solution in the sense that for any given $\mathbf{b} \in V_{c,\delta}^*$, there exists a unique solution $\mathbf{u} \in V_{c,\delta}$, such that

$$B_\delta(\mathbf{u},\mathbf{v}) = (\mathbf{b},\mathbf{v}) \quad \forall \mathbf{v} \in V_{c,\delta},$$

and

$$|\mathbf{u}|_v \leq \|\mathbf{b}\|_* .$$

Consequently, by the nonlocal Poincare's inequality (3.24), we have

$$\|\mathbf{u}\|_v = (|\mathbf{u}|_v^2 + \|\mathbf{u}\|_{L^2}^2)^{1/2} \leq \frac{\sqrt{1+c^2}}{c} \|\mathbf{b}\|_*$$

with $\|\cdot\|_*$ being the norm in the dual space $V^*_{c,\delta}$.

The above well-posedness result also implies that the nonlocal operator $\mathcal{L}_\delta$ is a continuous and self-adjoint operator from $V_{c,\delta}$ to its dual space with a bounded inverse.

The well-posedness also provides support to the formal derivation of the nonlocal Green identity and the calculation of the variations of the energy that can in fact be rigorous for functions in the desired function spaces. We refer to [33, 35] for more details on suitable interpretations of the operators and identities especially for cases invoking non-integrable interaction kernels.

Note that in the special case that we have an integrable nonlocal interaction kernel, the space V and its dual V^* are both equivalent to the conventional Hilbert space $L^2(\tilde{\Omega})$. In this case, the weak solutions are also functions in the same space so that the nonlocal equation holds also in L^2. For nonlocal interaction kernels that are not integrable, the fact that $\mathbf{b} \in V^*_{c,\delta}$ can belong to a weaker space than L^2 implies that it is possible to have the weak solution of the nonlocal model even for external forces that are not integrable functions. Moreover, the solutions are expected to have better regularities than the data in such a context, a property that is common for standard elliptic PDEs, but not shared by the case associated with an integrable kernel [20, 13].

3.5 Nonlocal calculus of variations — further discussions

The framework outlined in the above section is presented for a linear steady-state bond-based peridynamic system. Yet, as demonstrated in [33, 34, 35, 36], various extensions can be made.

For example, well-posedess has been shown for linear scalar equations with kernels with changing signs [33]. In [34], the theory has been applied to perturbations of the linear bond-based models that allow both repulsive and attractive nonlocal interactions. Results have also been presented on the well-posedness of variational problems involving nonlinear perturbations, that is, the energy functional can take on the form

$$E(u) = \int_{\tilde{\Omega}} \int_{\tilde{\Omega}} \frac{1}{2} \omega_\delta(|\mathbf{y}-\mathbf{x}|) (\mathfrak{D}^*\mathbf{u}(\mathbf{x},\mathbf{y}))^2 d\mathbf{y} d\mathbf{x} + \int_{\tilde{\Omega}} \Psi(\mathbf{u})(\mathbf{x}) d\mathbf{x} - \int_{\tilde{\Omega}} \mathbf{b}(\mathbf{x})\mathbf{u}(\mathbf{x}) d\mathbf{x},$$

which contains an additional term in comparison to (3.15). In [34], the cases where

$\Psi = \Psi(\mathbf{u})$ is either a convex map or a more general continuous map are considered. For the latter case, extra conditions on the ω_δ are imposed.

The theory or nonlocal calculus of variations is also applicable to the more general state-based systems. A convenient reformulation of the nonlocal energy in this case is given by [14, 35]

$$\begin{aligned} E(u) &= \int_{\tilde{\Omega}}\int_{\tilde{\Omega}} \frac{\eta}{2}\omega_\delta(|\mathbf{y}-\mathbf{x}|)\left(\mathfrak{D}^*\mathbf{u}(\mathbf{x},\mathbf{y}) - \frac{1}{n}\mathfrak{D}^*_{\omega_\delta}\mathbf{u}(\mathbf{x})\right)^2 d\mathbf{y}d\mathbf{x} \\ &+ \int_{\tilde{\Omega}} \frac{\tau}{2}\left(\mathfrak{D}^*_{\omega_\delta}\mathbf{u}(\mathbf{x})\right)^2 d\mathbf{x} - \int_{\tilde{\Omega}} \mathbf{b}(\mathbf{x})\mathbf{u}(\mathbf{x})d\mathbf{x}. \end{aligned}$$

By computing the variations of the above energy, we can get the linear peridynamic Navier operator given in (3.9), which also matches with the nonlocal operator in (3.1) with a tensor C given in (3.3).

The study in [35] has shown, in particular, that the same energy space can be used for both bond-based and state-based systems associated with the same nonlocal interaction kernel. In other words, the spaces V and $V_{c,\delta}$ defined in the previous section can be used to define energy and solution spaces for the linear state-based system as well.

In a recent study [36], we have also established some basic theory for nonlinear variational problems associated with the following more general energy functional:

$$\begin{aligned} E(u) &= \int_{\tilde{\Omega}}\int_{\tilde{\Omega}} \omega_\delta(|\mathbf{y}-\mathbf{x}|)\,\zeta\left(\mathfrak{D}^*\mathbf{u}(\mathbf{x},\mathbf{y}) - \frac{1}{n}\mathfrak{D}^*_{\omega_\delta}\mathbf{u}(\mathbf{x})\right) d\mathbf{y}d\mathbf{x} \\ &+ \int_{\tilde{\Omega}} \phi\left(\mathfrak{D}^*_{\omega_\delta}\mathbf{u}(\mathbf{x})\right) d\mathbf{x} - \int_{\tilde{\Omega}} \mathbf{b}(\mathbf{x})\mathbf{u}(\mathbf{x})d\mathbf{x}, \end{aligned}$$

where ζ and ϕ are functions that are assumed to have suitable growth and convexity properties. The results cover variational problems associated with a variety of volumetric conditions.

Time-dependent peridynamic models are of greater interests in practice. Naturally, studies of steady-state models and nonlocal operators lead also to results on the well-posedness of time-dependent peridynamic equation of motion [33] such as the one given by

$$\rho u_{tt}(\mathbf{x},t) + B(t)u_t(\mathbf{x},t) + \mathcal{L}_\delta u(\mathbf{x},t) = b(\mathbf{x},t), \quad t \in (0,T), \mathbf{x} \in \Omega \tag{3.26}$$

with initial value $u(\cdot,0) = u_0$ and $u_t(\cdot,0) = v_0$ and nonlocal constraint $u(\cdot,t) = 0$ on Ω_I. Here, ρ is mass density and for each $t \in [0,T]$ and $B(t)$ is a bounded linear operator on $L^2(\Omega)$ that represents the damping effect. An example of such an operator used is

$$B(t)w(\mathbf{x}) = \int_\Omega \frac{h(\mathbf{x}'-\mathbf{x})}{|\mathbf{x}'-\mathbf{x}|}(w(\mathbf{x}') - w(\mathbf{x}))d\mathbf{x}'$$

which appears in [11]. Another form of $B(t)$ would be to take the local operator

$$B(t)w(x) = \beta(x,t)w(x)$$

for a smooth and nonnegative function β. The existence, uniqueness, and continuous dependence of the time-dependent solution to (3.26) are established for initial data u_0 in the constrained solution space similar to $V_{c,\delta}$ and $v_0 \in L^2$.

Similar analysis can also be applied to recently studied models of nonlocal diffusion and convection presented in [16] where the relation between time-dependent nonlocal models and stochastic jump processes has also been explored.

3.6 Summary

With increasing attention being given to peridynamics in materials modeling and simulations, the development of rigorous mathematical theory becomes more and more important. Peridynamics is proposed as an alternative to classical PDE-based continuum models [40]. By avoiding the explicit use of spatial derivatives, it adopts an integral formulation to incorporate nonlocal interactions and to allow singular solutions that potentially can describe the morphology of cracks and fractures more effectively. These advantages in modeling come with some added cost: nonlocal interactions are more difficult to efficiently evaluate and singular solutions are harder to resolve numerically. Given such complexities, it is essential to offer sound verification and validation of modeling and simulations based on peridynamics. Rigorous mathematical analysis, in combination with physical experiments, can play a central role in verification and validation processes.

In recent years, new mathematical results about peridynamics have started to appear [18]. A systematic and rigorous mathematical framework for nonlocal models is also emerging. This chapter presented an overview of the well-posedness study of peridynamic equations, mostly through the framework of nonlocal calculus of variations developed for nonlocal models [13, 14, 33, 34, 35, 36]. Such a framework was built upon the basic elements of the nonlocal vector calculus [13, 15]. A step-by-step illustration was provided here for a variational problem associated with a linear bond-based peridynamic model with minimal background in advanced mathematics.

The development of mathematical theory for peridynamics is related to many studies of other nonlocal models. One may consult, for example, additional references on nonlocal diffusion [3], papers on nonlocal operators in image analysis [7, 25, 28], fractional PDEs [9, 12, 31, 29, 30], nonlocal dispersal [27], and stochastic processes [4, 8]. The basic tool sets developed for the mathematical analysis of peridynamics will likely find broader applications.

There are obviously many more mathematical questions than what have been discussed here. Going back to the questions raised in the introduction section, we see that our focus has been on addressing the first three questions. Nonlocal calculus of variations provides us a rigorous basis with which to study the mathematical properties of simplified linearized models and more general nonlinear peridynamic models. Besides the application to well-posedness theory, it can also be applied to study many other interesting mathematical questions. For example, the framework can help us explore the consistency between the nonlocal models and their local limits when the latter are valid. This addresses the fourth question raised in the introduction, which we will revisit in Chapter 4. Another important application is to analyze numerical approximations of peridynamic models such as the Asymptotically Compatible schemes developed in [45] and [46]. Detailed discussions are also given in Chapter 4. Concerning the last question posed in the introduction, we note that characterizing the regularity properties of solutions to peridynamic models is a very important and challenging task. Other than the minimal regularity guaranteed by the well-posedness theory, we do not address the subject here. More future research efforts in this direction are naturally called for. Coupling the analysis closely with constitutive modeling involving nonlinear interactions should also be given high priority.

In summary, deeper mathematical understanding of peridynamic models in their full generality and in broad applications is very much in need. It is important to have modelers, simulation code developers, mathematicians, and experimentalists working together to formulate the important mathematical questions and find their effective solutions.

Acknowledgments Much of the material presented here is either taken from or inspired by joint work with Max Gunzburger, Rich Lehoucq, Kun Zhou, Tadele Mengesha, Zhan Huang, and Xiaochuan Tian. The author's research is supported in part by the US NSF grant DMS-1318586, US DOE grant DE-SC0005346, and AFOSR MURI center for material failure prediction through peridynamics.

References

[1] B. Aksoylu and T. Mengesha. Results on nonlocal boundary value problems. *Numerical Functional Analysis and Optimization*, 31(12): 1301–1317, 2010.

[2] B. Aksoylu and M.L. Parks. Variational theory and domain decomposition for nonlocal problems. *Applied Mathematics and Computation*, 217(14): 6498–6515, 2011.

[3] F. Andreu, J. M. Mazón, J. D. Rossi, and J. Toledo. *Nonlocal Diffusion Problems*, volume 165 of *Mathematical Surveys and Monographs*. American Mathematical Society, 2010.

[4] R. Bass, M. Kassmann, and T. Kumagai. Symmetric jump processes: Localization, heat kernels and convergence. *Annales de l'Institut Henri Poincaré - Probabilités et Statistiques*, 46: 59–71, 2010.

[5] J. Bellido and C. Mora-Corral. Existence for nonlocal variational problems in peridynamics. *SIAM Journal on Mathematical Analysis*, 46(1): 890–916, 2014.

[6] J. Bourgain, H. Brezis, and P. Mironescu. *Another look at Sobolev spaces*, pages 439–455. IOS Press, Amsterdam, 2001.

[7] A. Buades, B. Coll, and J. M. Morel. Image denoising methods. A new nonlocal principle. *SIAM Review*, 52(1): 113–147, 2010.

[8] N. Burch and R.B. Lehoucq. Classical, nonlocal, and fractional diffusion equations on bounded domains. *International Journal for Multiscale Computational Engineering*, 9(6): 661–674, 2011.

[9] L. Caffarelli and L. Silvestre. An extension problem related to the fractional Laplacian. *Communications in Partial Differential Equations*, 32: 1245–1260, 2007.

[10] P. Ciarlet. *Three-dimensional elasticity*, volume 1. Elsevier, 1988.

[11] K. Dayal and K. Bhattacharya. Kinetics of phase transformations in the peridynamic formulation of continuum mechanics. *Journal of the Mechanics and Physics of Solids*, 54:1811–1842, 2006.

[12] O. Defterli, M. D'Elia, Q. Du, M. Gunzburger, R. Lehoucq, and M. Meerschaert. Fractional diffusion in bounded domains. *Fractional Calculus and Applied Analysis*, 18: 342–360, 2015..

[13] Q. Du, M. Gunzburger, R.B. Lehoucq, and K. Zhou. Analysis and approximation of nonlocal diffusion problems with volume constraints. *SIAM Review*, 56: 676–696, 2012.

[14] Q. Du, M. Gunzburger, R.B. Lehoucq, and K. Zhou. Analysis of the volume-constrained peridynamic navier equation of linear elasticity. *Journal of Elasticity*, 113: 193–217, 2013.

[15] Q. Du, M. Gunzburger, R.B. Lehoucq, and K. Zhou. A nonlocal vector calculus, nonlocal volume-constrained problems, and nonlocal balance laws. *Mathematical Models and Methods in Applied Sciences*, 23: 493–540, 2013.

[16] Q. Du, Z. Huang, and R. Lehoucq. Nonlocal convection-diffusion volume-constrained problems and jump processes. *Discrete and Continuous Dynamical Systems – Series B*, 19: 373–389, 2014.

[17] Q. Du, J.R. Kamm, R. Lehoucq, and M.L. Parks. A new approach to nonlocal nonlinear conservation laws. *SIAM Journal on Applied Mathematics*, 72: 464–487, 2012.

[18] Q. Du and R. Lipton. Peridynamics, fracture, and nonlocal continuum models. *SIAM News*, 47(3), 2014.

[19] Q. Du and X. Tian. Robust discretization of nonlocal models related to peridynamics. *Lecture Notes in Computational Science and Engineering*, 2014.

[20] Q. Du and K. Zhou. Mathematical analysis for the peridynamic nonlocal continuum theory. *Mathematical Modelling and Numerical Analysis*, 45: 217–234, 2011.

[21] E. Emmrich, R.B. Lehoucq, and D. Puhst. Peridynamics: A nonlocal continuum theory. In Michael Griebel and Marc Alexander Schweitzer, editors, *Meshfree Methods for Partial Differential Equations VI*, volume 89 of *Lecture Notes in Computational Science and Engineering*, pages 45–65. Springer Berlin Heidelberg, 2013.

[22] E. Emmrich and D. Puhst. Well-posedness of the peridynamic model with Lipschitz continuous pairwise force function. *Communications in Mathematical Sciences*, 11: 1039–1049, 2013.

[23] E. Emmrich and O. Weckner. On the well-posedness of the linear peridynamic model and its convergence towards the Navier equation of linear elasticity. *Communications in Mathematical Sciences*, 5: 851–864, 2007.

[24] H.A. Erbay, A. Erkip, and G.M. Muslu. The Cauchy problem for a one-dimensional nonlinear elastic peridynamic model. *Journal of Differential Equations*, 252:4392–4409, 2012.

[25] G. Gilboa and S. Osher. Nonlocal operators with applications to image processing. *Multiscale Modeling Simulation*, 7: 1005–1028, 2008.

[26] M. Gunzburger and R.B. Lehoucq. A nonlocal vector calculus with application to nonlocal boundary value problems. *Multiscale Modeling & Simulation*, 8(5): 1581–1598, 2010.

[27] C.Y. Kao, Y. Lou, and W. Shen. Random dispersal vs. nonlocal dispersal. *Discrete and Continuous Dynamical Systems – Series A* , 26: 551–596, 2010.

[28] Y. Lou, X. Zhang, S. Osher, and A. Bertozzi. Image recovery via nonlocal operators. *Journal of Scientific Computing*, 42: 185–197, 2010.

[29] M. Meerschaert, D. Benson, and B. Bäumer. Multidimensional advection and fractional dispersion. *Physical Review E*, 59(5): 5026, 1999.

[30] M. Meerschaert, J. Mortensen, and S. Wheatcraft. Fractional vector calculus for fractional advection-dispersion. *Physica A: Statistical Mechanics and its Applications*, 367: 181–190, 2006.

[31] M. Meerschaert and A Sikorskii. *Stochastic models for fractional calculus*, volume 43. Walter de Gruyter, 2011.

[32] T. Mengesha. Nonlocal korn-type characterization of sobolev vector fields. *Communications in Contemporary Mathematics*, 14: 1250028, 2012.

[33] T. Mengesha and Q. Du. Analysis of a scalar nonlocal peridynamic model with a sign changing kernel. *Discrete and Continuous Dynamical Systems – Series B*, 18: 1415–1437, 2013.

[34] T. Mengesha and Q. Du. The bond-based peridynamic system with Dirichlet type volume constraint. *Proceedings of the Royal Society of Edinburgh, A*, 144: 161–186, 2014.

[35] T. Mengesha and Q. Du. Nonlocal constrained value problems for a linear peridynamic Navier equation. *Journal of Elasticity*, 116: 27–51, 2014.

[36] T. Mengesha and Q. Du. On the variational limit of a class of nonlocal functionals related to peridynamics. *Nonlinearity*, 28: 3999–4035, 2015.

[37] T. Mengesha and D. Spector. Localization of nonlocal gradients in various topologies. *Calculus of Variations and PDEs*, 52: 253–279, 2015.

[38] A. Ponce. An estimate in the spirit of Poincaré's inequality. *Journal of the European Mathematical Society*, 6(1): 1–15, 2004.

[39] S. Silling and R.B. Lehoucq. Peridynamic theory of solid mechanics. *Advances in Applied Mechanics*, 44: 73–168, 2010.

[40] S.A. Silling. Reformulation of elasticity theory for discontinuities and long-range forces. *Journal of the Mechanics and Physics of Solids*, 48(1):175–209, 2000.

[41] S.A. Silling. Linearized theory of peridynamic states. *Journal of Elasticity*, 99: 85–111, 2010.

[42] S.A. Silling, M. Epton, O. Weckner, J. Xu, and E. Askari. Peridynamic states and constitutive modeling. *Journal of Elasticity*, 88(2): 151–184, 2007.

[43] S.A. Silling and R.B. Lehoucq. Convergence of peridynamics to classical elasticity theory. *Journal of Elasticity*, 93(1): 13–37, 2008.

[44] S.A. Silling, O. Weckner, E. Askari, and F. Bobaru. Crack nucleation in a peridynamic solid. *International Journal of Fracture*, 162(1): 219–227, 2010.

[45] X. Tian and Q. Du. Analysis and comparison of different approximations to nonlocal diffusion and linear peridynamic equations. *SIAM Journal of Numerical Analysis*, 51: 3458–3482, 2013.

[46] X. Tian and Q. Du. Asymptotically compatible schemes and applications to robust discretization of nonlocal models. *SIAM Journal of Numerical Analysis*, 52: 1641–1665, 2014.

[47] K. Zhou and Q. Du. Mathematical and numerical analysis of linear peridynamic models with nonlocal boundary conditions. *SIAM Journal of Numerical Analysis*, 48(5): 1759–1780, 2010.

Chapter 4

Local Limits and Asymptotically Compatible Discretizations

Qiang Du

Columbia University

CONTENTS

4.1	Introduction	87
4.2	Local PDE limits of linear peridynamic models	90
4.3	Discretization schemes and discrete local limits	97
4.4	Asymptotically compatible schemes for peridynamics	100
4.5	Summary	103
	References	104

4.1 Introduction

In this chapter, we continue the mathematical analysis of peridynamic models based on the nonlocal calculus of variations framework presented in Chapter 3.

In practical applications, verification and validation are important aspects of peridynamic modeling and simulations. It is clear that one objective of peridynamic modeling and simulation is to provide an alternative to conventional continuum mechanical models represented by partial differential equations when the latter can neither capture the underlying physics nor have meaningful mathematical solutions due to the presence of the singularities. Yet, as a popular benchmark test, it has been a common practice to reproduce solutions of peridynamic models that are consistent with classical continuum models for the same given data when the latter are physically valid and mathematically well defined.

Addressing such a consistency issue for peridynamics has played a prominent role in both modeling and code development efforts [4, 5, 6, 9, 20, 23, 24, 27, 34, 36, 35, 38, 40, 45, 44, 46, 48, 49, 50, 51, 53].

This chapter is concerned with, in the context of linear peridynamic models and in particular those corresponding to steady-state variational problems, the convergence of nonlocal models to their local limits as well as the behavior of the numerical solutions to nonlocal models in the local limits. Once again, as the rationale to choose linear models for illustrations, we quote the statement from the earlier introduction to peridynamics in this handbook by Silling that *linearized peridynamics is not as much of a simplification as one might expect over the full nonlinear theory* in practical numerical computations. The comment is particularly fitting to the current discussion, and we will see that the study of linearized equations provides a clear picture of the various convergence properties and limiting behaviors of interest. The picture is largely put together based on the theory of nonlocal calculus of variations outlined in the previous chapter.

More specifically, we let δ denote the horizon parameter used in peridynamics and u_δ be the solution of the corresponding nonlocal equation. Assuming that the peridynamic model is solved numerically, with h being a discretization parameter that either measures the grid size or particle spacing used in the numerical schemes, we use u_δ^h to denote the numerical solution. The discussions here are centered around answering the following questions:

- What are the relations between the peridynamic model with solution u_δ and the limiting classical PDE model with solution u_0 as the horizon parameter δ goes to zero?

- What are the behaviors of the numerical scheme of the peridynamic model and its solution u_δ^h when δ goes to zero? Do we end up with a numerical scheme of the limiting local model? Does u_δ^h converge to u_0^h that is the numerical approximation of u_0?

- What are the behaviors of numerical approximations u_δ^h of the peridynamic model when both δ and h go to zero? Does u_δ^h converge to u_0?

The above questions are illustrated in the diagram given in Figure 4.1 that highlights the different paths connecting the different solutions and limits.

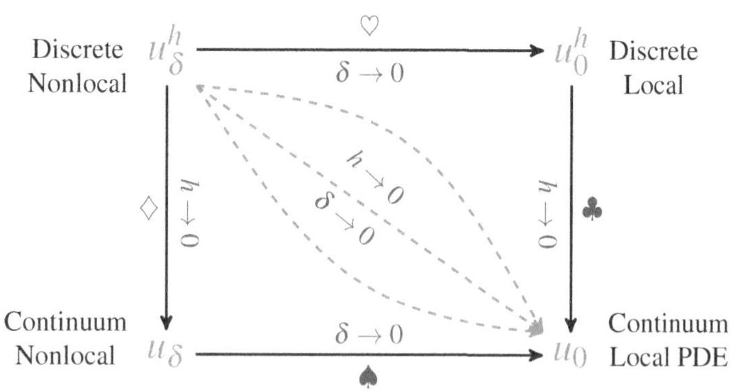

Figure 4.1: A diagram of possible paths between u_δ, u_δ^h, u_0^h, and u_0 via various limits.

The paths along the boundary of the diagram (marked with symbols $\diamondsuit$, $\heartsuit$, $\clubsuit$, $\spadesuit$) are for limits of one parameter with the other being fixed: the $\spadesuit$ path connects solutions of local and nonlocal models, which is desirable from a modeling perspective; the $\clubsuit$ path is very much a well-understood subject of numerical PDEs and we do not expect much complication there; the $\diamondsuit$ path assures the convergence of the discretization to the nonlocal problem that should be achievable for schemes that are designed for such a purpose. The $\heartsuit$ path is the most intriguing one among the four, because it is not clear whether the local limit of the numerical schemes for nonlocal problems would remain to be an effective scheme for local limit problems. If the answer is affirmative, then we get a nice *commutative diagram*, as it is commonly called in mathematics. It means that we can take either the path through those marked with $\diamondsuit$ and $\spadesuit$ or one marked with $\heartsuit$ and $\clubsuit$ to get the convergence of u_δ^h to u_0.

Naturally, there are important issues for practical numerical simulations concerning other possible paths. For example, the linear diagonal path signifies the situation that h and δ go to zero proportionally. While the curved path above the diagonal means $\delta = o(h)$, the path below the diagonal can represent the case that $h = o(\delta)$. The former case tends to give more sparsity in discrete nonlocal interactions as the horizon shrinks, while the latter corresponds to an increased level of mesh refinement or particle density as the horizon gets reduced so that a loss of sparsity becomes more evident. Our main objective in this chapter is to introduce mathematical approaches that help determine if and in what sense the various limits in the above diagram hold. In particular, we advocate the so-called asymptotically compatible schemes that can get the convergence of u_δ^h to u_0 no matter which path one takes from the diagram. Such schemes are thus most robust for problems involving different scales and requiring varying level of numerical resolutions. Linearized model problems are used as illustrative examples.

The rest of the chapter is organized as follows. In section 4.2, we discuss the local limits of peridynamic models using a very simple one-dimensional linear variational problem as an illustration. We present different techniques to relate the nonlocal integral operators and local partial differential operators and we also show the convergence of solutions to the nonlocal models to the solutions of the local PDEs for the same given set of data and domain. In section 4.3, we discuss numerical discretizations of nonlocal models. We first demonstrate that it is easy to construct numerical schemes that are convergent when solving the nonlocal models; however, in the local limit, depending on the relations between the horizon δ and the mesh size h, these schemes may lead to numerical solutions that are consistent with a wrong local limit. Such observations further motivate us to present in section 4.4 the notion of asymptotically compatible schemes that are more robust for multiscale simulations in the sense that they are less sensitive with respect to various couplings of the horizon and mesh spacing. Finally, a summary of this chapter together with brief discussions of other numerical analysis issues are given in section 4.5.

4.2 Local PDE limits of linear peridynamic models

We begin with a special case of the variational problem (3.14) where the linear bond-based peridynamic operator is defined in (3.13).

One-dimensional model For illustration, we use a constant interaction kernel, a choice often used in many existing studies [40], and consider the problem defined for a finite one dimensional elastic bar. The model problem defined on $\mathbb{R}$ takes on the following form

$$\begin{cases} -\mathcal{L}_\delta u_\delta(x) = -2\displaystyle\int_{-\delta}^{\delta} (u_\delta(x+s) - u_\delta(x))c_\delta ds = b(x), \ \forall x \in (0,1), \\ u_\delta(x) = 0, \quad \forall x \in (-\delta, 0) \cup (1, 1+\delta), \end{cases} \tag{4.1}$$

where c_δ is a constant to be specified. For convenience, we let $\Omega = (0,1)$ be the spatial domain where the nonlocal equation is imposed, $\Omega_\delta = (-\delta,0) \cup (1,1+\delta)$ be the domain where the solution is constrained to be zero, and $\tilde{\Omega} = (-\delta, 1+\delta)$ be the whole spatial domain.

The nonlocal operator $\mathcal{L}_\delta$ can be viewed as a special case of the bond-based operator given in (3.13) or the one-dimensional version of the nonlocal diffusion operator defined in (3.8) with the kernel $\underline{\omega}_\delta(r)$ being a constant $2c_\delta$ for $|r| \leq \delta$ and 0 otherwise.

The corresponding elastic energy takes on the form

$$E_1(v) = \int_\Omega e_1(v)(x)dx = \int_\Omega \int_{-\delta}^{\delta} \frac{1}{2} c_\delta \big(v(x+s) - v(x)\big)^2 dsdx \tag{4.2}$$

where $e_1(v)(x)$ denotes the nonlocal energy density at x associated with the displacement field $v = v(x)$.

Following the discussions in section 3.4, the energy space in this case is in fact $V = L^2$ and the solution space is given by L_0^2, which is a subspace of L^2 containing functions that are square integrable on $(0,1)$ and vanish outside the unit interval.

Note that the natural space for the external force $b = b(x)$, as demonstrated in the well-posedness study, is also L^2, that is, we assume that b is a square integrable function in $(0,1)$.

We apply the standard patch test to compute the nonlocal energy density for a linear displacement field $v_l(x) = x$. That is, we substitute v_l in (4.2) to get that for $x \in (0,1)$,

$$e_1(v_l)(x) = \frac{1}{3} c_\delta \delta^3 .$$

To match with the elastic energy density of the same displacement field of a unit linear elastic bar with a unit elastic constant, we thus set

$$c_\delta = \frac{3}{2\delta^3} . \tag{4.3}$$

Local limit We now examine the local limit of (4.1). Based on the patch test, we expect that the local limiting model is given by the two point boundary value problem of a differential equation defined in $(0,1)$ given by

$$\begin{cases} -\mathcal{L}_0 u_0(x) = -\dfrac{d^2}{dx^2} u_0(x) = b(x), \quad \forall x \in (0,1), \\ u_0(0) = u_0(1) = 0. \end{cases} \tag{4.4}$$

Showing that (4.1) is consistent to (4.4) as $\delta \to 0$ is an almost trivial investigation for this particular example. However, we present the details of this investigation using several different techniques that have been used to study the local limit of much more general nonlocal peridynamic models. The objective is to elucidate what are the assumptions needed and conclusions that different techniques may lead to.

Convergence to local limit The techniques used to investigate the local limit of nonlocal peridynamic models include Taylor expansions like those used in [20, 46], Fourier transform like [19, 12], consistency of weak forms like [13], and nonlocal calculus of variations like [28, 29, 30].

For easy reference, we formulate the weak form of the nonlocal and local problems, respectively.

Define the bilinear form for the nonlocal problem as in (3.21):

$$B_\delta(u,v) = \int_\Omega \int_{-\delta}^{\delta} c_\delta \big(u(x+s) - u(x)\big)\big(v(x+s) - v(x)\big) ds dx$$

for any $u, v \in L_0^2$.

We then have the weak form for (4.1): find $u \in L_0^2$ such that

$$B_\delta(u_\delta, v) = (b, v), \; \forall\, v \in L_0^2, \tag{4.5}$$

where $(\cdot,\cdot)$ denotes the standard inner product on $(0,1)$.

For (4.4), we have: find $u \in H_0^1$ such that

$$B_0(u_0, v) = \int_0^1 \frac{du_\delta}{dx}(x) \frac{dv}{dx}(x) dx = (b, v), \; \forall\, v \in H_0^1. \tag{4.6}$$

We note that weak solutions of the local limit are in H_0^1, that contains square integrable functions with the first-order derivative also square integrable and function values vanishing at $x = 0$ and $x = 1$. In one dimension, weak solutions in H_0^1 are necessarily continuous in $(0,1)$. In comparison, weak solutions of nonlocal models for a given δ are in L_0^2 only; thus, they are allowed to have discontinuities.

Let us make a remark here. The comparison of the solution spaces shows that for the same external force that is square integrable, the solution of a nonlocal model may exhibit more singular behavior (like discontinuity) than the solution of the local limit. More singular solutions generally demand more computational effort, a possible price to pay with the nonlocal models, but a worthy one since one might be trying to capture such singular behavior to characterize the physical reality, which is not feasible via the local PDE models. At the same time, for problems where there are no expected singularities and local PDEs have solutions that provide good or adequate descriptions of nature, one would like to see solutions of nonlocal models be consistent with solutions of limiting PDE models.

Consistency via Taylor expansion We now consider the technique based on Taylor expansions that has been among the most popular approaches to exploring the connection between nonlocal models and local limits [20, 46]. Largely, the goal is to show that for a smoothly defined function $u = u(x)$, or say $u \in C_c^\infty$, the space of functions that have continuous derivatives of all orders with a compact support, we have for any x inside its support,

$$\mathcal{L}_\delta u(x) \to \mathcal{L}_0 u(x) \quad \text{as } \delta \to 0.$$

We apply the Taylor expansion

$$u(x \pm s) = u(x) \pm \frac{du}{dx}(x)s + \frac{1}{2}\frac{d^2u}{dx^2}(x)s^2 \pm \ldots$$

in $\mathcal{L}_\delta u$ to get as $\delta \to 0$ that

$$\begin{aligned}
\mathcal{L}_\delta u(x) &= c_\delta \int_{-\delta}^{\delta} \big(u(x+s) - 2u(x) + u(x-s)\big)ds \\
&= c_\delta \int_0^\delta s^2 ds \frac{d^2u}{dx^2}(x) + 2c_\delta \sum_{k=2}^{\infty} \frac{1}{(2k)!} \int_0^\delta s^{2k} ds \frac{d^{2k}u}{dx^{2k}}(x) \\
&= \frac{d^2u}{dx^2}(x) + \sum_{k=2}^{\infty} \frac{3}{(2k)!} \delta^{2(k-1)} \frac{d^{2k}u}{dx^{2k}}(x) \\
&= \mathcal{L}_0 u(x) + O(\delta^2).
\end{aligned}$$

Hence

$$\mathcal{L}_\delta u(x) - \mathcal{L}_0 u(x) \to 0, \quad \forall x \in \Omega = (0,1). \tag{4.7}$$

This establishes formally the point-wise consistency of the nonlocal and local operators and also the second order of convergence, in δ, for smooth functions. Moreover, if one uses a mean-value theorem, then we may also get

$$\mathcal{L}_\delta u(x) = \mathcal{L}_0 u(\theta)$$

for some θ with $|\theta - x| \le \delta$, which implies that for a given $u \in C_c^\infty$,

$$\sup_{x\in(0,1)} |\mathcal{L}_\delta u(x)| \text{ is also uniformly bounded with respect to } \delta. \tag{4.8}$$

One may further work with the weak form, as in [13], to show that for both u and v in H_0^1 with zero extension outside the unit interval,

$$B_\delta(u,v) \to B_0(u,v), \quad \text{as } \delta \to 0,$$

which gives the consistency of the weak forms. Formally, for smooth functions u and v, we can see this from the following calculation,

$$
\begin{aligned}
B_\delta(u,v) &= \int_\Omega \int_{-\delta}^{\delta} c_\delta \big(u(x+s)-u(x)\big)\big(v(x+s)-v(x)\big)dsdx \\
&= \frac{3}{2}\int_\Omega \int_{-1}^{1} \Big(\frac{u(x+\delta z)-u(x)}{\delta}\Big)\Big(\frac{v(x+\delta z)-v(x)}{\delta}\Big)dzdx \\
&\to \frac{3}{2}\int_\Omega \int_{-1}^{1} \frac{du}{dx}(x)\frac{dv}{dx}(x)z^2 dzdx \\
&= B_0(u,v)\,.
\end{aligned}
$$

Since it is of particular interest to study discontinuous solutions in peridynamic models in practice, techniques based on the Taylor expansion alone do not provide a complete characterization of the local limit of possibly non-smooth u_δ.

Fourier analysis and dispersion relation In the whole space (or for problems with periodic solutions), Fourier analysis can be an effective tool to analyze linear nonlocal models and to explore local limits [19, 12]. This type of analysis has also been widely used to study the dispersion relations of the nonlocal models [19, 49].

Assuming that $\mathcal{L}_\delta v(x)$ in (4.1) is defined for any $x \in \mathbb{R}$, then by taking the Fourier transform, we get

$$
\begin{aligned}
\widehat{-\mathcal{L}_\delta v}(\xi) &= c_\delta \int_{\mathbb{R}} \int_{-\delta}^{\delta} \big(2v_\delta(x)e^{ix\xi} - v_\delta(x+s)e^{ix\xi} - v_\delta(x-s)e^{ix\xi}\big)dsd\xi \\
&= c_\delta \left(\int_{-\delta}^{\delta} (2-2\cos(s\xi))ds\right)\hat{v}_\delta(\xi) \\
&= \frac{6}{\delta^2}\left(1-\frac{1}{\delta\xi}\sin(\delta\xi)\right)\hat{v}_\delta(\xi) = \lambda_\delta(\xi)\hat{v}_\delta(\xi),
\end{aligned}
$$

which gives a dispersion relation for the nonlocal model with $\delta > 0$, and $\{\lambda_\delta(\xi), \xi \in \mathbb{R}\}$ gives the point spectrum (eigenvalues) of the nonlocal operator $-\mathcal{L}_\delta$.

The plot of $\lambda_\delta(\xi)$ is given in Figure 4.2. A few observations are in order. First of all, for a given $\delta > 0$, since $|\sin(y)| \le |y|$ for any $y \in \mathbb{R}$, we see that

$$
|\lambda_\delta(\xi)| \le \frac{12}{\delta^2}, \quad \forall \xi \in \mathbb{R}.
$$

On the other hand,

$$
\xi^2 - \lambda_\delta(\xi) = 2\sin(\frac{\theta}{2})^2\xi^2
$$

for any $\xi \in \mathbb{R}$ and some $|\theta| \le \delta|\xi|$. Thus, for any given $\xi \in \mathbb{R}$, it is easy to see that

$$
\lambda_\delta(\xi) = \xi^2 + O(\delta^2), \quad \text{as } \delta \to 0.
$$

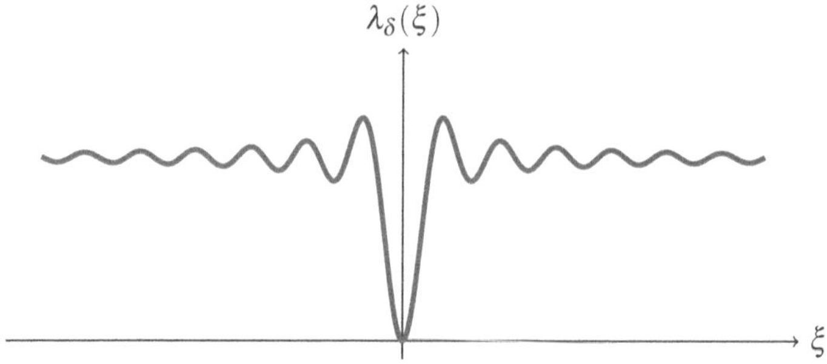

Figure 4.2: Dispersion relation for nonlocal model with a constant interaction kernel.

Furthermore, direct calculation based on the Fourier transform shows, for any function $v \in H^{s+2}(\mathbb{R})$ (with $s \geq -1$), that

$$\|\mathcal{L}_0 v - \mathcal{L}_\delta v\|_{H^s} \leq C\delta^2 \|v\|_{H^{s+2}} \tag{4.9}$$

where the H^s norms are either the standard Sobolev space norms or the dual norms. We refer to [19, 12] for more details of such estimates.

Fourier analysis is effective wherever it is applicable but it is generally not well-suited to deal with variational problems on bounded domains with general nonlocal constraints, so its range of application is limited.

Fourier analysis, point spectrum, stiffness matrix conditioning Fourier analysis of a linear nonlocal operator is not only useful to the study of its local limit but also allows us to study additional properties of the operator spectrum.

The uniform bound on the point spectrum $\lambda_\delta(\xi)$, with respect to ξ, is an interesting feature of linear peridynamic models with integrable kernels, in contrast to that for the local limit [19, 49]. Note that the uniform upper bound is dependent on δ and grows like $1/\delta^2$. Thus the bound becomes infinity as $\delta \to 0$, which is as expected given the convergence of nonlocal operators to the local differential operators having unbounded spectra as $\delta \to 0$.

In general, for non-integrable kernels that satisfy the assumptions given in Section 3.4, it is possible to see the growth of $|\lambda_\delta(\xi)|$ with respect to $|\xi|$, but the finite second-order moment assumption implies that the growth is always below the quadratic order (corresponding to the local limit). In fact, for such more general kernels, an upper bound is given by a constant multiple of

$$\max\left\{\frac{1}{\delta^2}, |\xi|^2\right\},$$

for any $\delta > 0$ and $\xi \in \mathbb{R}$. Sharper bounds can be derived [19, 54] for more specialized kernels.

The above analysis on eigenvalues $\lambda_\delta(\xi)$ in essence also determines the spectra of stiffness matrices generated from the discretization of nonlocal operators. Typically, with a finite dimensional approximation using a parameter h as the mesh or particle spacing, we only need to consider those wave numbers that are no bigger than order $O(h)$ in magnitude. Thus, adopting similar analysis to the discrete case [54], we are led to an upper bound on the condition number of discrete nonlocal stiffness matrices given by

$$\min\left\{O(\frac{1}{\delta^2}), O(\frac{1}{h^2})\right\}$$

for models with an integrable kernel, and bounds like

$$\min\left\{O(\frac{1}{\delta^{2-\beta}h^{\beta}}), O(\frac{1}{h^2})\right\}$$

for nonlocal interaction kernels typically behaving like $1/|\mathbf{x}-\mathbf{x}'|^{n+\beta}$ for materials points $\mathbf{x}$ and $\mathbf{x}'$ close to each other in $\mathbb{R}^n$ with an exponent $\beta \in (0,2)$. We refer to [54, 10, 3] for more discussions.

Convergence of weak solutions For both the Taylor expansion technique and the Fourier analysis technique, the relations we have looked at are mostly between nonlocal and local operators, rather than between solutions of the corresponding nonlocal and local variational problems.

Using the nonlocal calculation of variations, we can investigate the convergence of the nonlocal solutions u_δ to the local limit u_0 more directly. One of the important ingredients for such an investigation is to have a more precise (sharp) bound on the (constant of the) nonlocal Poincare's inequality as stated in (3.24). That is, there is a constant $M > 0$, independent of δ, such that

$$|v|_v^2 = B_\delta(v,v) \ge M\|v\|_{L^2}^2, \quad \forall v \in L_0^2.$$

We wish to establish that the constant M is independent of δ. The argument is similar in spirit to the one used to establish the standard nonlocal Poincare's inequality (3.24) given in section 3.4 but uses additional techniques given in [7, 29, 37]. Indeed, if we cannot find such a constant, it implies that we can find a sequence of $\delta_k \to 0$ as $k \to \infty$ and $v_k \in L_0^2$ such that

$$\|v_k\|_{L^2} = 1, \quad B_{\delta_k}(v_k,v_k) \to 0, \quad \text{as } k \to \infty.$$

This further gives a convergent subsequence in L^2 with a limit point v_* in H_0^1 and with unit L^2 norm. Moreover, one can argue that $B_0(v_*,v_*)$ is bounded by any small fixed positive constant [7, 29, 37]. Note that for v_* in H_0^1, we have from the standard Poincare's inequality on the unit interval that

$$B_0(v_*,v_*) \ge \pi^2\|v_*\|_{L^2}^2 = 1.$$

This thus leads to a contradiction. Hence, we can have the desirable independence of M on δ.

Now, with the sharp characterization of the constant M, for a given $b = b(x) \in L^2$, we get the uniform bound of the solutions u_δ with respect to δ. In other words, there is a constant $D > 0$, independent of δ, such that for

$$\int_\Omega \left\{ u_\delta^2(x) + \int_{-\delta}^{\delta} c_\delta \big(u_\delta(x+s) - u_\delta(x)\big)^2 ds \right\} dx \le D.$$

Next, the uniform bound on the sequence $\{u_\delta\}$ implies its relative compactness in L_0^2 with its limiting points in H_0^1, a technical result that follows from [7, 37] as quoted before. See also [28, 51] for more discussions.

Let u_* denote a limiting point; then for any $v \in C_c^\infty$, using the nonlocal Green's identity and the classical Green's identity (integration by parts), we have

$$(b + \mathcal{L}_0 u_*, v) = (-\mathcal{L}_\delta u_\delta + \mathcal{L}_0 u_*, v) = -(u_\delta, \mathcal{L}_\delta v - \mathcal{L}_0 v) + (u_* - u_\delta, \mathcal{L}_0 v).$$

The calculations done earlier based on Taylor expansions now become handy to easily deduce that $\mathcal{L}_\delta v - \mathcal{L}_0 v \to 0$ in L^2. Combined with the uniform bound of $\mathcal{L}_0 v$ and $\|u_\delta\|_{L^2}$ with respect to δ and the convergence of the subsequence of u_δ to u_*, we find that

$$(b + \mathcal{L}_0 u_*, v) = 0,$$

which implies that $u_* = u_0$. Consequently, we can deduce that $u_\delta \to u_0$ in L^2 based on the uniqueness of the local limiting solution u_0.

It is important to note that the above conclusions make no additional assumptions on the properties of solutions of nonlocal models and local models than those implied from the basic well-posedness theory. The also point out an interesting fact about the local limit of linear nonlocal models: for solutions of the nonlocal models that are not necessarily smooth or perhaps are even discontinuous, as $\delta \to 0$, they will approach solutions of the local PDE model that enjoys better regularity. In the one-dimensional example, this means that the possible discontinuities in the nonlocal solutions will disappear in the local limit.

More general systems and nonlocal constraints Discussions in this section are focused on a simple one-dimensional model to present the techniques and conclusions without heavy technical jargon. They illustrate the validity of the local PDE limits of the nonlocal peridynamic models, at least in the linear and variational settings. Similar studies can be made for more general systems, like the nonlocal peridynamic Navier equation corresponding to the linear state-based peridynamic models. As alluded to in Chapter 3, there is a natural connection to the classical Navier equation, that is,

$$\mathcal{L}_s = -\eta \mathfrak{D}(\omega_\delta \mathfrak{D}^*) - \sigma \mathfrak{D}_{\omega_\delta}(\mathfrak{D}^*_{\omega_\delta}) \to -\mu \nabla \cdot \nabla - (\mu + \lambda) \nabla \nabla \cdot \,,$$

as $\delta \to 0$. This has been pointed out based on Taylor expansions as in [46], Fourier analysis as in [11], and compactness via nonlocal calculus of variations as in [30]. In [31], γ-convergence of the nonlocal variational problems to the local limit has also been established We note that for the compactness argument to work, additional assumptions about nonlocal interaction kernels have to be imposed, such as the non-increasing property of the interaction as the distance increases, a physically reasonable condition. We present some of the more technical discussions in the next section as part of the framework for the analysis of discrete schemes and their local limits.

4.3 Discretization schemes and discrete local limits

There have been a lot of efforts towards developing effective numerical methods for peridynamic models. One of the key questions to be addressed is whether these methods are effective for treating the nonlocal interactions while providing consistent results with classical continuum mechanical models given by PDEs when the latter are mathematically well defined and physically valid. Issues of consistency and limiting behavior are highlighted in the diagram in Figure 4.1.

We present the study again using the one-dimensional linear problem (4.1) as an illustration. The results given in the previous section provide the link between the solutions of the nonlocal and local models at the bottom of the diagram. Questions about limits and paths outlined in other parts of the diagram remain to be answered.

To begin we note first that nonlocal operators can often be viewed as discrete operators in continuum forms. Taking for example the operator $\mathcal{L}_\delta$ given in (4.1), we have

$$\mathcal{L}_\delta u(x) = \frac{3}{\delta^3}\int_0^\delta \big(u(x+s) - 2u(x) + u(x-s)\big)ds = \frac{3}{\delta^3}\int_0^\delta D_s^2 u(x)ds \qquad (4.10)$$

where $D_s^2 u(x) = u(x+s) - 2u(x) + u(x-s)$ is a standard second-order finite difference. As stated in [18], the operator $\mathcal{L}_\delta$ can be viewed as an averaged (continuum) difference operator. The fact that the average is taken over the range of $s \in (0,\delta)$ tells us that rather than a finite difference, we in fact have a *continuum of differences* supported over the range $(0,\delta)$. This means that the linear nonlocal model is a continuum combination of discrete schemes.

With (4.10), it is intuitively clear that we may obtain a fully-discrete finite dimensional algebraic system by approximating the *continuum difference* $\mathcal{L}_\delta$ by discrete finite differences through suitable quadrature approximations (thus leading to the *quadrature based finite difference* discretization as named in [50]).

For example, following the discussions given in [50, 18], we take a mesh spacing parameter h for discrete quadrature points and assume that $h = 1/N = \delta/r$ for some positive integers r and N. Let us then consider a scheme based on the discrete operator $\mathcal{L}_\delta^h$, as defined in [50], given by: for $i = 1,\dots N$,

$$-\mathcal{L}_\delta^h u_i = -\frac{3h}{\delta^3}\sum_{m=1}^{r}(u_{i-m} - 2u_i + u_{i+m}) = b(x_i), \qquad (4.11)$$

where $\{u_i\}$ are approximations of $\{u(x_i)\}$ at nodal points $\{x_i = ih\}_{i=-r}^{N+r}$.

As illustrated in [50], the discrete operator $\mathcal{L}_\delta^h$ in (4.11) is obtained from a simple Riemann sum quadrature of the integral in (4.10). Thus, for any given $\delta > 0$, we easily see the consistency of the discretization. In fact, since the Riemann sum quadrature is of first-order accuracy, one can check that for the solution of (4.1), we have

$$\mathcal{L}_\delta^h u_\delta(x_i) + b(x_i) = \frac{3h}{\delta^3} \sum_{m=1}^{r} (u_\delta(x_{i-m}) - 2u_\delta(x_i) + u_\delta(x_{i+m})) + O(h).$$

Moreover, the numerical discretization (4.11) also leads to a symmetric M-matrix with a bandwidth $2r+1$ as its stiffness matrix. This can further lead to the stability of the approximation for any given δ as $h \to 0$. Combining stability with consistency, we see that (4.11) gives a convergent first-order discretization of (4.1).

A further observation can be made, following discussions presented in [50], by considering a special case with $r = 1$ in (4.11). This gives a center difference scheme

$$-\frac{3}{h^2}(u_{i-1} - 2u_i + u_{i+1}) = b_i,$$

which is a convergent discretization of the differential equation

$$-3\frac{d^2u}{dx^2}(x) = b(x)$$

as $h = \delta \to 0$, instead of (4.4).

In other words, if we set h and δ to zero proportionally, the numerical solution of the discrete scheme for the nonlocal problem yields a convergent approximation to a local limit that, unfortunately, is associated with a consistently over-estimated elastic constant! The over-estimated constant is off the correct limit by a factor of 3. One can do similar calculations for other choices of r to get the elastic constant coming from the discretization to be equal to

$$\frac{3}{r^3}\sum_{m=1}^{r} m^2 = 1 + \frac{3}{2r} + \frac{1}{2r^2}. \tag{4.12}$$

Obviously, we see that the value is getting closer to be unity as $r \to \infty$, but for any finite r, it is off the mark by a constant factor. The factor is still more than 150%, even for $r = 3$ that has been a popular choice.

While the convergence of numerical solutions is often a desirable feature, the possibility of convergence to a wrong limit is alarming. Without prior knowledge, such convergence might be mistakenly used to verify or disapprove numerical simulation results, and we see the risks involved due to the wrong local limits produced by some discrete solutions to nonlocal models.

To recap, it is known in practice that to obtain consistency to the desirable local PDE models, the mesh or quadrature points spacing may have to be reduced at a faster pace than the reduction of the horizon parameter [6, 5]. Otherwise, there could be potential complications when the horizon parameter is coupled proportionally with the discretization parameter. The latter scenario has been encountered in existing studies when modeling and simulating complex multiscale problems and thus deserves much careful study. Although this discrepancy is illustrated only via a simple example here, it has been shown to be a generic feature of the type of discretizations represented by (4.11) and many other schemes such as the piecewise constant Galerkin finite element approximations, for scalar nonlocal models and general state-based peridynamic systems [50, 51].

Similar phenomena have been observed by other research groups and sometimes they are viewed as issues mostly attributed to quadrature approximations. A remedy suggested by some is to normalize the discrete scheme by a constant factor as a way to numerically calibrate the discrete stiffness matrix depending on the relation between h and δ, which is ad hoc at best and certainly neither robust nor effective in practice when more complex peridynamic models are in consideration.

On a more positive note, the complications due to the use of discrete schemes like (4.11) can be resolved through other means. For example, it is proposed in [50] that an alternative formulation similar to the one below works much more robustly by suitably adjusting the weights for the second-order difference: for $i = 1, \ldots N$,

$$-\mathcal{L}^h_{\delta,1} u_i = -\frac{1}{r\delta^2} \sum_{m=1}^{r} \frac{m^3 - (m-1)^3}{m^2} (u_{i-m} - 2u_i + u_{i+m}) = b(x_i) . \tag{4.13}$$

The elastic constant corresponding to (4.13) is given by

$$\frac{1}{r\delta^2} \sum_{m=1}^{r} \frac{m^3 - (m-1)^3}{m^2} m^2 h^2 = \frac{r^3}{r^3} = 1 , \tag{4.14}$$

independently of r! Hence, as shown in [50], (4.13) gives a scheme that is convergent to nonlocal models for any fixed δ as $h \to 0$ and it is convergent to the correct local limit whenever $\delta \to 0$ and $h \to 0$, regardless how the two parameters are coupled. Moreover, by taking into consideration that δ may not need to be an integer multiple of h, a generalized form of (4.13) in fact gives a convergent scheme for nonlocal problems that approaches to a convergent scheme for the correct local limit models as $\delta \to 0$. We note that the modified choice of weights of second-order differences over the simple arithmetic average is not made at random nor by a posteriori calibration. Naturally, it is interesting to characterize how such schemes can be constructed in general.

A straightforward interpretation of schemes like (4.13) for one-dimensional scalar equations was given in [50]. The discussion was further extended in [51] to much more general settings. In particular, the concept and theory of asymptotically compatible schemes are introduced for general parametrized problems as a way to provide robust approximations to nonlocal peridynamics that are guaranteed to be consistent in the local limits.

4.4 Asymptotically compatible schemes for peridynamics

In this section, we review the theory of asymptotically compatible schemes for peridynamic models. The discussions are considerably more technical than those given in the earlier sections. To avoid unnecessary complications, we present a definition for some special nonlocal peridynamic models and their local limits. More precise and general definitions can be found in [51], but the exposition here is written with less technical jargon.

The analysis presented here may be accessible to those who had some basic exposure to functional analysis and Sobolev spaces that have been used to analyze PDEs. We will make an attempt in the end of the section to explain the main findings in simpler terms.

Definition of asymptotically compatible schemes We adopt notations previously introduced in the Chapter 3 and also in earlier sections of this chapter.

A family of approximations $\{u_{\delta,h}\}$ for nonlocal problems parametrized by the horizon δ and mesh size h is said to be asymptotically compatible (AC) to the solution u_0 of the local limit if, for any sequence $\delta_n \to 0$ and $h_n \to 0$, we have $\|u_{\delta_n,h_n} - u_0\|_{L^2} \to 0$.

With both δ and h as parameters, there are various limiting processes as outlined in Figure 4.1. Among these processes of interest, the limit of discrete approximations to local PDE problems, as mesh is refined ($h \to 0$) and marked by the symbol ♣, has been most widely studied as part of numerical PDEs. The limit as $\delta \to 0$ of the nonlocal model to the local limit, marked by the symbol ♠, is described in the previous section. The other limiting cases all involve discretizations of nonlocal models.

Based on earlier examples, we see that some natural and popular choices for discretizing nonlocal models may not be asymptotically compatible.

Key properties and assumptions Analytically, the keys to establishing AC schemes for linear nonlocal variational problems are the uniform embeddings and the asymptotically compact embedding properties associated with the solution spaces.

The framework of AC schemes can be developed for problems exemplified by (4.1) as well as those associated with more general nonlocal operators like those in (3.13). For the latter cases, given a finite $\delta_0 > 0$ (say $\delta_0 = 1/4$ for example), an important assumption about the nonlocal interaction kernel ω_δ, in addition to those specified in (3.16)-(3.18), is needed [50, 51]:

$$\omega_\delta(r) = \frac{1}{\delta^n}\omega_1(\frac{r}{\delta}), \quad \frac{d}{dr}(\frac{1}{r^2}\omega_\delta(r)) \le 0, \quad \int \omega_1(|\mathbf{x}|)d\mathbf{x} = 1. \tag{4.15}$$

For the piecewise constant kernel used in (4.1), we have $\omega_1(r) = 3r^2$ for $|r| \le 1$ and 0 otherwise so as to get that $\underline{\omega}_\delta(r) = 2c_\delta$ for $|r| \le \delta$ and 0 otherwise.

The condition (4.15) assures that the continuous embedding constant $H^1 \hookrightarrow V \hookrightarrow$

L^2 as well as the constant in the nonlocal Poincare's inequality (3.24) are all independent of δ for $0 < \delta \leq \delta_0$.

Another useful property is the *asymptotically compact embedding* property: given the above ω_δ, any function sequence $\{u_\delta\}$ uniformly bounded in $V_{c,\delta}$ with respect to $\delta \to 0$ is also relatively compact in L^2 with limit point in H_0^1.

One can obtain the convergence of u_δ to u_0 in L^2 under the above assumptions as demonstrated for a special kernel in the previous section.

Conforming Galerkin discretization We now turn our attention to the discretization of nonlocal problems. The following Galerkin approximation associated with the bilinear form $B_\delta(\cdot,\cdot)$ is considered: find the discrete solution $u_\delta^h \in W_{\delta,h} \subset V_{c,\delta}$ such that

$$B_\delta(u_\delta^h, v^h) = (b, v^h), \; \forall\, v^h \in W_{\delta,h}. \tag{4.16}$$

We note that in [50], the equivalence and similarities between Galerkin approximations of the weak form of a nonlocal problem and standard difference or collocation/quadrature approximations of the strong form have been investigated.

We need to specify some assumptions about $\{W_{\delta,h}\}$ that include, first of all, conditions ensuring the convergence of approximations to elements of the solution space $V_{c,\delta}$ as $h \to 0$ for each fixed $\delta \in (0, \delta_0]$, and the second part is concerned with the limiting behavior as both $h \to 0$ and $\delta \to 0$ at the same time.

i) For each $\delta \in (0, \delta_0]$, the family $\{W_{\delta,h}, h \in (0, h_0]\}$ is dense in the nonlocal solution space in the sense that, $\forall v \in V_{c,\delta}$, there exists a sequence $\{v_n \in W_{\delta,h_n}\}$ with $h_n \to 0$ as $n \to \infty$, such that

$$\|v - v_n\|_v \to 0 \quad \text{as} \quad n \to \infty. \tag{4.17}$$

ii) $\{W_{\delta,h}, \delta \in (0, \delta_0], h \in (0, h_0]\}$ is *asymptotically dense* in H_0^1 (the local solution space), i.e., $\forall v \in H_0^1$, there exists a sequence $\{v_n \in W_{\delta_n,h_n}\}_{h_n \to 0,\, \delta_n \to 0}$ as $n \to \infty$, such that

$$\|v - v_n\|_{H^1} \to 0 \quad \text{as} \quad n \to \infty. \tag{4.18}$$

Note that assumption i) is a standard one to assure a good approximation for the nonlocal problem with a fixed $\delta > 0$, which is often satisfied by most discrete spaces. On the other hand, assumption ii) imposes additional restrictions that, for instance, rule out the choice of discontinuous piecewise constant finite element space.

Best approximation and asymptotic compatibility Under the assumptions specified, some typical approximation and convergence results can be established. They include the best approximation property

$$|u_\delta - u_\delta^h|_v \leq \inf_{v^h \in W_{\delta,h}} |u_\delta - v^h|_v, \qquad \text{as } h \to 0$$

for nonlocal problems, which shows the consistency of the path marked with $\diamondsuit$ in the diagram in Figure 4.1. More careful a priori error estimates given in [54] for some special nonlocal variational problems have shown the optimal error orders with respect to the mesh size as $h \to 0$ for a fixed $\delta > 0$; see also related discussions in [10].

Meanwhile, discretizations satisfying the assumptions stated here are also automatically AC schemes so that the rest of the links in Figure 4.1 are also valid mathematically. In particular, the local limit of the AC schemes becomes a discretization of the local PDE limit as indicated by the path marked with $\heartsuit$ in the diagram [51].

A recap of AC schemes via examples As seen from the previous discussion and more detailed studies given in [50], there are popular discretization schemes of nonlocal peridynamics that are not AC. In particular, if δ is taken to be proportional to h, then as $h \to 0$, piecewise constant conforming finite element solutions actually converge to the incorrect limit, similarly to those based on simple Riemann sum quadrature approximations to nonlocal operators. Given such unfortunate reality, it is thus of practical interest to provide some constructive remedies to avoid the undesirable effects.

Indeed, in [51], it has been shown that as long as the condition $h = o(\delta)$ is met as $\delta \to 0$, then we are able to obtain the correct local limit even for discontinuous piecewise constant finite element approximations when they are of the conforming type. This result is consistent to the simple calculation done in (4.12); as long as $\delta/h = r \to \infty$ together with h or δ going to zero, we expect to have the correct local limit. Practically speaking, this implies that a mild growth of bandwidth is needed as the mesh is refined to recover the correct local limit for Riemann sum-like quadratures or piecewise constant finite element schemes.

Meanwhile, it has been shown in [51] that all conforming Galerkin approximations of the nonlocal models containing continuous piecewise linear functions are automatically AC. This means that they can recover the correct local limit as long as both δ and h are diminishing, even if nonlocality measure δ is reduced at a much faster pace than the mesh spacing h. While the analysis is highly technical, an intuitive explanation is that even with h larger than δ, the nonlocal features are still encoded in the stiffness matrices to ensure the correct local limit due to the incorporation of higher order (than constant) basis functions. A more mathematical interpretation is that continuous piecewise linear functions provide a space large enough to give good approximations of the local limit in the local solution space (that is in the space of H^1), whereas discontinuous piecewise constant functions can not.

Naturally, schemes using higher order basis functions tend to provide higher order accuracy to the nonlocal problems as well. At the moment, our theory on AC schemes does not offer any estimate of the order of convergence with respect to the different coupling of h and δ. Preliminary numerical experiments in [51] offer some insight to the balance of modeling and discretization errors, but much more theoretical analysis remains to be carried out.

Yet, it is important for us to emphasize that the advantages of having higher order schemes are not limited to the gain in accuracy (that is generally applicable only with smooth functions). It is more significant that they offer the robustness to get consistent solutions with the desirable physical models regardless of how the modeling and discretization parameters are coupled. The latter is one of the main findings of our recent studies on the discretization of peridynamic models.

4.5 Summary

In this chapter, our discussion touches upon two important aspects of peridynamic modeling and simulations: the consistency with the local PDE limit and the convergence of numerical approximation in both the nonlocal setting and in the local limit. While the illustration is done for simple linear variational problems, the analysis can also be generalized to time-dependent problems as done in [19]. Other nonlocal constrained value problems can also be considered, and extensions to nonlinear variations have also started to emerge in recent works based on the theory of Gamma convergence (see [24] for example). The details are beyond the scope of the current discussion and will be provided in future publications.

Concerning the numerical schemes for peridynamics, a robust method has the best chance of providing effective approximations in a multiscale setting where the mesh sizes and modeling parameters may have to be chosen in more flexible manners. While it is important to debate which particular model (continuum or discrete) provides best descriptions of the physical reality, it is also critical to understand what mathematical models the computational codes are truly simulating. Asymptotically compatible schemes, being either conforming Galerkin type approximations of weak forms [51], nonconforming DG approximations [52] or collocation and quadrature based approximations of strong forms [17], offer the potential to solve the desirable models with different choices of parameter to gain efficiency and to avoid the pitfall of reaching inconsistent limits. There is also the flexibility to adapt the theory for other forms of approximations such as those based on meshfree methods and reproducing kernel spaces [8, 25, 26], and Smoothed Particle Hydrodynamics (SPH) [32, 33].

There are certainly many other challenging computational issues to be explored, such as those related to numerical quadratures and scalable solvers. Conditioning of stiffness matrices for nonlocal problems has been a concern for which we have reached a good theoretical understanding, as alluded in the Fourier analysis of the point spectrum for the nonlocal operators given here; see [54] and [1, 2, 3, 10, 41] for more discussions. Adaptivity is another interesting direction to pursue. The paper [14] contains an a posteriori error analysis of conforming finite element methods for solving linear nonlocal diffusion and peridynamic mechanics models. A general abstract framework is developed for a posteriori error analysis of nonlocal volume-constrained problems for scalar equations. The reliability and efficiency of the estimators are proved, and relations between nonlocal and classical local a posteriori error estimations are also studied. In [16], a convergent adaptive finite element algorithm for the numerical solution of scalar nonlocal models has been developed. For problems involving certain non-integrable kernel functions, the convergence of the adaptive algorithm is rigorously derived with the help of several basic ingredients, such as an upper bound of the estimator, the estimator reduction, and the orthogonality property. This represents the first time in the literature that the convergence of adaptive finite element methods for nonlocal problems is rigorously proved. How these estimators and methods work in the local limit and for general time-dependent and nonlinear peridynamic models is under current investigation.

Acknowledgments Much of the material presented here is either taken from or inspired by joint work with Max Gunzburger, Rich Lehoucq, Kun Zhou, Tadele Mengesha, Zhan Huang, and Xiaochuan Tian. The author also would like to thank Michael Parks, Stewart Silling, Florin Bobaru, John Foster, Erdogan Madenci, and John Mitchell for their discussions on the subject. The author's research was supported in part by the US NSF grant DMS-1318586, US DOE grant DE-SC0005346, and AFOSR MURI center for material failure prediction through peridynamics.

References

[1] B. Aksoylu and T. Mengesha. Results on nonlocal boundary value problems. *Numerical Functional Analysis and Optimization*, 31(12): 1301–1317, 2010.

[2] B. Aksoylu and M.L. Parks. Variational theory and domain decomposition for nonlocal problems. *Applied Mathematics and Computation*, 217(14): 6498–6515, 2011.

[3] B. Aksoylu and Z. Unlu. Conditioning analysis of nonlocal integral operators in fractional Sobolev spaces. *SIAM Journal on Numerical Analysis*, 52(2): 653–677, 2015.

[4] E. Askari, F. Bobaru, R.B. Lehoucq, M.L. Parks, S.A. Silling, and O. Weckner. Peridynamics for multiscale materials modeling. *Journal of Physics: Conference Series*, 125: 012078, 2008.

[5] F. Bobaru and M. Duangpanya. The peridynamic formulation for transient heat conduction. *International Journal of Heat and Mass Transfer*, 53: 4047–4059, 2010.

[6] F. Bobaru, M. Yang, L.F. Alves, S.A. Silling, E. Askari, and J. Xu. Convergence, adaptive refinement, and scaling in 1d peridynamics. *International Journal for Numerical Methods in Engineering*, 77(6): 852–877, 2009.

[7] J. Bourgain, H. Brezis, and P. Mironescu. *Another look at Sobolev spaces*, pages 439–455. IOS Press, Amsterdam, 2001.

[8] J.S. Chen, C.T. Wu, and T. Belytschko. Regularization of material instabilities by meshfree approximations with intrinsic length scales. *International Journal for Numerical Methods in Engineering*, 47(7): 1303–1322, 2000.

[9] X. Chen and M. Gunzburger. Continuous and discontinuous finite element methods for a peridynamics model of mechanics. *Computer Methods in Applied Mechanics and Engineering*, 200: 1237–1250, 2011.

[10] Q. Du, M. Gunzburger, R. B. Lehoucq, and K. Zhou. Analysis and approximation of nonlocal diffusion problems with volume constraints. *SIAM Review*, 56: 676–696, 2012.

[11] Q. Du, M. Gunzburger, R. B. Lehoucq, and K. Zhou. Analysis of the volume-constrained peridynamic Navier equation of linear elasticity. *Journal of Elasticity*, 113: 193–217, 2013.

[12] Q. Du, M. Gunzburger, R. B. Lehoucq, and K. Zhou. A nonlocal vector calculus, nonlocal volume-constrained problems, and nonlocal balance laws. *Mathematical Models and Methods in Applied Sciences*, 23: 493–540, 2013.

[13] Q. Du, Z. Huang, and R. B. Lehoucq. Nonlocal convection-diffusion volume-constrained problems and jump processes. *Discrete and Continuous Dynamics Systems – Series B*, 19: 373–389, 2014.

[14] Q. Du, L. Ju, L. Tian, and K. Zhou. A posteriori error analysis of finite element method for linear nonlocal diffusion and peridynamic models. *Mathematics of Computation*, 82: 1889–1922, 2013.

[15] Q. Du and R. Lipton. Peridynamics, fracture, and nonlocal continuum models. *SIAM News*, 47(3), 2014.

[16] Q. Du, L. Tian, and X. Zhao. A convergent adaptive finite element algorithm for nonlocal diffusion and peridynamic models. *SIAM Journal on Numerical Analysis*, 51: 1211–1234, 2013.

[17] Q. Du and X. Tian. Asymptotically compatible schemes for peridynamics based on numerical quadratures. *Proceedings of the ASME 2014 International Mechanical Engineering Congress and Exposition*, pages IMECE2014–39620, 2014.

[18] Q. Du and X. Tian. Robust discretization of nonlocal models related to peridynamics. In *Meshfree Methods for Partial Differential Equations VII*, edited by Michael Griebel and Marc Alexander Schweitzer, Lecture Notes in Computational Science and Engineering, 2014.

[19] Q. Du and K. Zhou. Mathematical analysis for the peridynamic nonlocal continuum theory. *Mathematical Modelling and Numerical Analysis*, 45: 217–234, 2011.

[20] E. Emmrich and O. Weckner. On the well-posedness of the linear peridynamic model and its convergence towards the Navier equation of linear elasticity. *Communications in Mathematical Sciences*, 5(4): 851–864, 2007.

[21] L. González F. Maciá, M. Antuono, and A. Colagrossi. Theoretical analysis of the no-slip boundary condition enforcement in sph methods. *Progress of Theoretical Physics*, 125: 1091–1121, 2011.

[22] M. Gunzburger and RB Lehoucq. A nonlocal vector calculus with application to nonlocal boundary value problems. *Multiscale Modeling & Simulation*, 8(5): 1581–1598, 2010.

[23] B. Kilic and E. Madenci. Coupling of peridynamic theory and the finite element method. *Journal of Mechanics of Materials and Structures*, 5(5): 707–733, 2010.

[24] R. Lipton. Dynamic brittle fracture as a small horizon limit of peridynamics. *Journal of Elasticity*, 114: 1–30, 2014.

[25] W.K, Liu, Y. Chen, S. Jun, J.S. Chen, T. Belytschko, C Pan, R.A. Uras, and C.T. Chang. Overview and applications of the reproducing kernel particle methods. *Archives of Computational Methods in Engineering*, 3(1): 3–80, 1996.

[26] W.K. Liu, S. Li, and T. Belytschko. Moving least-square reproducing kernel methods (i) methodology and convergence. *Computer Methods in Applied Mechanics and Engineering*, 143(1): 113–154, 1997.

[27] R. Macek and S. Silling. Peridynamics via finite element analysis. *Finite Elements in Analysis and Design*, 43(15): 1169–1178, 2007.

[28] T. Mengesha and Q. Du. Analysis of a scalar nonlocal peridynamic model with a sign changing kernel. *Discrete and Continuous Dynamics Systems – Series B*, 18: 1415–1437, 2013.

[29] T. Mengesha and Q. Du. The bond-based peridynamic system with Dirichlet type volume constraint. *Proceedings of the Royal Society of Edinburgh, A*, 144: 161–186, 2014.

[30] T. Mengesha and Q. Du. Nonlocal constrained value problems for a linear peridynamic Navier equation. *Journal of Elasticity*, 116: 27–51, 2014.

[31] T. Mengesha and Q. Du. On the variational limit of a class of nonlocal functionals related to peridynamics. *Nonlinearity*, 28: 3999–4035, 2015.

[32] J. Monaghan. Smoothed particle hydrodynamics. *Annual Review of Astronomy and Astrophysics*, 30: 543–574, 1992.

[33] J. Monaghan. Smoothed particle hydrodynamics. *Reports on Progress in Physics*, 68(8): 1703–1759, 2005.

[34] E. Oterkus and E. Madenci. Peridynamic analysis of fiber-reinforced composite materials. *Journal of Mechanics of Materials and Structures*, 7(1): 45–84, 2012.

[35] M. Parks, D. Littlewood, J. Mitchell, and S. A. Silling. Peridigm users' guide v1.0.0,. *Sandia Rep. 2012-7800, Sandia National Laboratories, Albuquerque, New Mexico*, 2012.

[36] M. Parks, S. Plimpton, R. B. Lehoucq, and S. A. Silling. Peridynamics with LAMMPS: A user guide. *Sandia National Laboratory Report, SAND2008-0135, Albuquerque, New Mexico*, 2008.

[37] A. C. Ponce. An estimate in the spirit of Poincaré's inequality. *Journal of the European Mathematical Society*, 6(1): 1–15, 2004.

[38] R. Rahman and J. T. Foster. Implementation of elastic-plastic model in PDLAMMPS. *preprint*, `http://lammps.sandia.gov/doc/PDF/PDLammps_EPS.pdf`, 2014.

[39] E. M. Ryan, A. M. Tartakovsky, and C. Amon. A novel method for modeling Neumann and Robin boundary conditions in smoothed particle hydrodynamics. *Computer Physics Communications*, 181(12): 2008–2023, 2010.

[40] P. Seleson and M. Parks. On the role of the influence function in the peridynamic theory. *International Journal for Multiscale Computational Engineering*, 9(6): 689–706, 2011.

[41] P. Seleson, M. Parks, M. Gunzburger, and R. B. Lehoucq. Peridynamics as an upscaling of molecular dynamics. *Multiscale Modeling & Simulation*, 8: 204–227, 2009.

[42] S. A. Silling. Reformulation of elasticity theory for discontinuities and long-range forces. *Journal of the Mechanics and Physics of Solids*, 48(1): 175–209, 2000.

[43] S.A. Silling. Linearized theory of peridynamic states. *Journal of Elasticity*, 99: 85–111, 2010.

[44] S. A. Silling and E. Askari. A meshfree method based on the peridynamic model of solid mechanics. *Computers and Structures*, 83(17): 1526–1535, 2005.

[45] S. A. Silling, M. Epton, O. Weckner, J. Xu, and E. Askari. Peridynamic states and constitutive modeling. *Journal of Elasticity*, 88(2): 151–184, 2007.

[46] S. A. Silling and R. B. Lehoucq. Convergence of peridynamics to classical elasticity theory. *Journal of Elasticity*, 93(1): 13–37, 2008.

[47] S. A. Silling and R. B. Lehoucq. Peridynamic theory of solid mechanics. *Advances in Applied Mechanics*, 44: 73–168, 2010.

[48] S. A. Silling, O. Weckner, E. Askari, and F. Bobaru. Crack nucleation in a peridynamic solid. *International Journal of Fracture*, 162(1): 219–227, 2010.

[49] S. A. Silling, M. Zimmermann, and R. Abeyaratne. Deformation of a peridynamic bar. *Journal of Elasticity*, 73: 173–190, 2003.

[50] X. Tian and Q. Du. Analysis and comparison of different approximations to nonlocal diffusion and linear peridynamic equations. *SIAM Jounral on Numerical Analysis*, 51: 3458–3482, 2013.

[51] X. Tian and Q. Du. Asymptotically compatible schemes and applications to robust discretization of nonlocal models. *SIAM Journal on Numerical Analysis*, 52: 1641–1665, 2014.

[52] X. Tian and Q. Du. Nonconforming discontinuous Galerkin methods for nonlocal variational problems. *SIAM Journal on Numerical Analysis*, 53: 762–781, 2015.

[53] H. Wang and H. Tian. A fast Galerkin method with efficient matrix assembly and storage for a peridynamic model. *Journal of Computational Physics*, 240: 49–57, 2012.

[54] K. Zhou and Q. Du. Mathematical and numerical analysis of linear peridynamic models with nonlocal boundary conditions. *SIAM Journal on Numerical Analysis*, 48(5): 1759–1780, 2010.

Chapter 5

Roadmap for Software Implementation

David Littlewood

Sandia National Laboratories

CONTENTS

5.1	Introduction	109
5.2	Evaluating the internal force density	112
5.3	Bond damage and failure	115
5.4	The tangent stiffness matrix	115
5.5	Modeling contact	119
5.6	Meshfree discretizations for peridynamics	122
5.7	Proximity search for identification of pairwise interactions	124
5.8	Time integration	125
	5.8.1 Explicit time integration for transient dynamics	126
	5.8.2 Estimating the maximum stable time step	128
	5.8.3 Implicit time integration for quasi-statics	129
5.9	Example simulations	131
	5.9.1 Fragmentation of a brittle disk resulting from impact	131
	5.9.2 Quasi-static simulation of a tensile test	132
5.10	Summary	134
	Acknowledgements	136
	References	136

5.1 Introduction

Peridynamics has advanced the state of the art for modeling material failure and fragmentation within a computational simulation code. To be effective in modeling real-world systems, the underlying theory must be accompanied by an effective software implementation. The goal of this chapter is to provide a roadmap for implementing peridynamics within an analysis code. This is motivated in part by the gaps that exist between much of the available literature on peridynamics, which is often focused on theoretical considerations, and implementation challenges that can arise in practice.

This chapter focuses exclusively on the meshfree approach of Silling and Askari [33]. To date, the vast majority of peridynamic simulations have utilized this approach. The decision to focus on this particular computational strategy is not meant to imply that it is superior to alternative approaches. Applying the finite element technique to peridynamics, for example, may provide superior results in some cases. Nonetheless, the meshfree approach of Silling and Askari has proven to be a reliable and efficient strategy for addressing problems in solid mechanics with pervasive material failure and fracture. Advantages include ease of implementation, computational efficiency, and the presence of a natural mechanism for material separation. The method of Silling and Askari is a direct discretization of the strong form of the peridynamic balance of linear momentum [34],

$$\rho(\mathbf{x})\,\ddot{\mathbf{u}}(\mathbf{x},t) = \int_{\mathcal{H}_{\mathbf{x}}} \left(\underline{\mathbf{T}}[\mathbf{x},t]\langle \mathbf{q}-\mathbf{x}\rangle - \underline{\mathbf{T}}[\mathbf{q},t]\langle \mathbf{x}-\mathbf{q}\rangle \right) dV_{\mathbf{q}} + \mathbf{b}(\mathbf{x},t), \tag{5.1}$$

where ρ is the material density, $\ddot{\mathbf{u}}$ is acceleration, t is time, $\mathbf{x}$ and $\mathbf{q}$ are material points, $\mathcal{H}_{\mathbf{x}}$ is a spherical neighborhood of radius δ centered at $\mathbf{x}$, $\underline{\mathbf{T}}$ denotes a peridynamic force state, $dV_{\mathbf{q}}$ denotes the infinitesimal volume associated with $\mathbf{q}$, and $\mathbf{b}$ is a body force density. For problems in three-dimensional space, the approach of Silling and Askari requires evaluation of a three-dimensional integral. In contrast, solution of the weak form of the peridynamic balance of linear momentum, for example using a finite element discretization, requires evaluation of a six-dimensional integral. For this reason, solution strategies for peridynamic models based on the weak form entail significant geometric complexity and computational expense relative to strong-form strategies.

The goal of a peridynamic simulation is to model material response within a body or a set of bodies under prescribed conditions over a specified period of time. This is accomplished numerically by discretizing the governing equations in both space and time. Following Silling and Askari [33], the physical system is discretized into a finite number of nodes, each having a (scalar) volume associated with it. A co-locational approach is employed, meaning that the nodes are used for tracking field variables such as displacement and velocity, and also as material points for constitutive model evaluations. This strategy is comparable to a one-point Gauss quadrature scheme in which all fields are assumed constant over the domain of integration and the Gauss point is located at the centroid of the domain (akin to midpoint integration).

The response of a peridynamic system as a whole is determined by solving the following semidiscrete equation of motion over time for each material point in the discretization,

$$\rho(\mathbf{x})\,\ddot{\mathbf{u}}(\mathbf{x},t) = \sum_{\mathcal{H}_{\mathbf{x}}} \left(\underline{\mathbf{T}}[\mathbf{x},t]\langle \mathbf{q}-\mathbf{x}\rangle - \underline{\mathbf{T}}[\mathbf{q},t]\langle \mathbf{x}-\mathbf{q}\rangle \right) \Delta V_{\mathbf{q}} + \mathbf{b}(\mathbf{x},t), \tag{5.2}$$

where the integral in Equation (5.1) has been replaced by a summation over the set of neighbors for material point $\mathbf{x}$. The force states $\underline{\mathbf{T}}[\mathbf{x},t]$ and $\underline{\mathbf{T}}[\mathbf{q},t]$ are determined through evaluation of the constitutive models at points $\mathbf{x}$ and $\mathbf{q}$, respectively. Contact interactions, if any, provide additional nodal force densities that must be included in Equation (5.2). In general, simulations may involve multiple bodies, multiple materials, and any number of specified initial and boundary conditions. Equation (5.2) holds for both bond-based and state-based constitutive models, as bond-based models can be considered a subset of the more general family of state-based models.

The progression of a simulation is driven by a time integration routine. The duration of the simulation is divided into a finite number of time steps or load steps at which Equation (5.2) is evaluated. In the case of the explicit time integration scheme described in Section 5.8.1, Equation (5.2) is applied to solve directly for nodal accelerations. The accelerations are then applied at each node to advance the simulation from one time step to the next. In contrast, the implicit time integration scheme for quasi-static simulations outlined in Section 5.8.3 treats the nodal displacements as the unknowns. Equation (5.2) is applied to solve for the displacement values that yield zero acceleration for those nodal degrees of freedom not subjected to kinematic boundary conditions. Regardless of the time integration strategy, simulations proceed by prescribing a set of known values at a given time step, e.g., positions or velocities over a subset of the domain, followed by determination of the remaining field and material state data through application of the governing peridynamic equations. The net result is a complete description of kinematic and material state data for all material points (nodes) in the discretization for every time step in the simulation, from which any number of secondary quantities of interest may be determined.

At a high level, a computational peridynamics code contains several key components: a time integrator, a method for evaluating the internal force, routines for evaluating bond damage laws, routines for applying initial and boundary conditions, a means to evaluate and apply contact forces, and input/output routines for storing and retrieving data. For transient dynamic simulations, the critical time step must be estimated. Simulations employing implicit time integration generally require the implementation of a nonlinear solver. One standard approach for solving a nonlinear system of equations is Newton's method, which requires the construction of a tangent stiffness matrix and a method for solving a linear system of equations.

Peridynamics has been implemented in stand-alone software, and also as a component of larger, more comprehensive simulation codes. *EMU*, written by Stewart Silling, was the first computational peridynamics code [33]. Examples of preexisting software packages in which peridynamics has been implemented include *LAMMPS* and *Sierra/SolidMechanics* [25, 30]. It has also been demonstrated that peridynamics can be implemented in the commercial finite element code *Abaqus* [22]. A more recent implementation of peridynamics is the open-source code *Peridigm* [26, 2].

There is potential for confusion with regard to the nomenclature applied to peridynamic discretizations. In this chapter, the terms *node*, *nodal volume*, and *material point* are used to describe a discretized peridynamic body. The term *nodal volume* refers directly to the pointwise position and volume data utilized in the discretization method of Silling and Askari. Nodes are used both as a means to track model geometry and as the locations for material model evaluations. The terms *node* and *material point* are used in reference to these distinct roles and provide alignment with standard computational mechanics terminology. The approach of Silling and Askari is typically referred to as *meshless* or *meshfree*, owning to the fact that nodes and nodal connectivities defining physical volumetric elements (as in standard finite element approaches) are not considered. A primary source of confusion with regard to this terminology stems from the fact that, in practice, geometric data defining volumetric elements are often utilized in the discretization process as a means to compute nodal volumes (cf. Henke and Shanbhag [15]). Further, the geometries of the regions associated with each node have been utilized to improve the accuracy of peridynamic quadrature by taking into account the partial intersections of these entities with the spheres that define peridynamic neighborhoods [7, 28, 29].

The following sections address the primary components of a computational peridynamics code. Where it is illustrative to do so, the *Peridigm* code is used as an example. The discussion assumes that simulations are three-dimensional, although concepts are generally applicable to one- and two-dimensional simulations as well. The majority of the content is focused on serial code. While implementation of a parallel code adds complexity, there are no fundamental changes required to the core peridynamic algorithms. Software implementation of a peridynamic constitutive model is presented first, followed by sections covering bond failure, the tangent stiffness matrix, contact modeling, meshfree discretizations, the proximity search for identifying peridynamic bonds, and time integration. Finally, example simulations demonstrating explicit transient dynamics and quasi-statics are presented.

5.2 Evaluating the internal force density

The core kernels of a peridynamic code are the routines for computing the internal force density. The internal force density is determined via constitutive laws that compute pairwise force densities between bonded material points based on kinematic data and, optionally, material state variables. Software routines associated with a constitutive model include initialization, calculation of the internal force density, and, optionally, calculation of the tangent stiffness matrix. Each of these routines has access to initial positions, nodal volumes, neighbor lists, values for the current time and the time step size, and basic kinematic data that are updated throughout the simulation by the time integration routine (*e.g.*, velocities and current positions). In addition, constitutive models may define material-specific data fields that are tracked and updated over the course of a simulation.

Constitutive model routines are intimately tied to the time integrator. The goal of a time integrator is to advance a simulation from a given time step or load step, n, to the next step, $n+1$. For this reason, values corresponding to both step n and step $n+1$ are tracked for many fields. Tracking nodal positions at steps n and $n+1$, for example, enables a constitutive model to compute rates of deformation. The storing of solutions at both step n and $n+1$ also facilitates the evaluation of trial configurations for implicit time integration. In this case, the solution at step n must be available for reuse in the evaluation of multiple trial configurations.

The initialization and internal force density routines for the *linear peridynamic solid* constitutive model [34] with a Gaussian influence function are presented in Algorithms 1 and 2, respectively. Here, $\mathbf{x}$ denotes nodal coordinates in the undeformed configuration, $\boldsymbol{\xi}$ is a bond vector in the undeformed configuration, $\mathbf{u}$ denotes displacement, η denotes relative displacement of bonded material points, δ is the peridynamic horizon, $\underline{\omega}$ is an influence function value, $\underline{e}$ denotes extension state, $\underline{e}^d$ denotes deviatoric extension state, m denotes weighted volume, ΔV is a nodal volume, θ denotes dilatation, $\underline{t}$ is a scalar force state, $\underline{\mathbf{M}}$ is the unit vector aligned with a bond in the deformed configuration, and $\mathbf{f}$ denotes internal force density. The material parameters μ and k are the shear modulus and bulk modulus, respectively. Note that in the internal force density routine, the force state corresponding to each material point is computed only once. After the force state for a given point, $\mathbf{x}$, is computed, it is immediately applied to the bonds $\langle \mathbf{q}_i - \mathbf{x} \rangle$ and $\langle \mathbf{x} - \mathbf{q}_i \rangle$ that correspond to each neighbor, $\mathbf{q}_i$, of $\mathbf{x}$. The resulting pairwise force density vectors are summed directly into the global force density vector for both $\mathbf{x}$ and $\mathbf{q}$ in accordance with Equation (5.2). For simplicity, Algorithm 2 is presented with the assumption that the nodal volumes, ΔV_i, are the full nodal volumes associated with node i. As discussed in Section 5.7, the physical volumetric region associated with a node may lie partially inside and partially outside a given peridynamic neighborhood. The accuracy of Algorithm 2 may be improved by modifying the nodal volumes to account for partial intersections (cf. Bobaru and Ha [7], Seleson [28], and Seleson and Littlewood [29]).

Algorithm 1 The initialization routine for a *linear peridynamic solid* material with a Gaussian influence function.

procedure LINEAR PERIDYNAMIC SOLID INITIALIZATION
    ▷ Compute the weighted volume for each node.
    for each node i **do**
        $m_i \leftarrow 0$
        for each node j in neighbor list for node i **do**
            $\boldsymbol{\xi} \leftarrow \mathbf{x}_j - \mathbf{x}_i$
            $\underline{\omega} \leftarrow \exp\left(-\frac{|\boldsymbol{\xi}|^2}{\delta^2}\right)$
            $m_i \leftarrow m_i + \underline{\omega}\,|\boldsymbol{\xi}|^2\,\Delta V_j$
        end for
    end for
end procedure

Algorithm 2 Routine for calculation of the internal force density for a *linear peridynamic solid* material with a Gaussian influence function.

procedure LINEAR PERIDYNAMIC SOLID INTERNAL FORCE
    ▷ Initialize the global force density vector to zero.
    for each node i **do**
        $\mathbf{f}_i \leftarrow 0$
    end for
    ▷ Compute the dilatation for each node.
    for each node i **do**
        $\theta_i \leftarrow 0$
        for each node j in neighbor list for node i **do**
            $\boldsymbol{\xi} \leftarrow \mathbf{x}_j - \mathbf{x}_i$
            $\boldsymbol{\eta} \leftarrow \mathbf{u}_j - \mathbf{u}_i$
            $\underline{\omega} \leftarrow \exp\left(-\frac{|\boldsymbol{\xi}|^2}{\delta^2}\right)$
            $\underline{e} \leftarrow |\boldsymbol{\xi}+\boldsymbol{\eta}| - |\boldsymbol{\xi}|$
            $\theta_i \leftarrow \theta_i + \frac{3}{m_i}\,\underline{\omega}\,|\boldsymbol{\xi}|\,\underline{e}\,\Delta V_j$
        end for
    end for
    ▷ Compute the pairwise contributions to the global force density vector.
    for each node i **do**
        for each node j in neighbor list for node i **do**
            $\boldsymbol{\xi} \leftarrow \mathbf{x}_j - \mathbf{x}_i$
            $\boldsymbol{\eta} \leftarrow \mathbf{u}_j - \mathbf{u}_i$
            $\underline{\omega} \leftarrow \exp\left(-\frac{|\boldsymbol{\xi}|^2}{\delta^2}\right)$
            $\underline{e} \leftarrow |\boldsymbol{\xi}+\boldsymbol{\eta}| - |\boldsymbol{\xi}|$
            $\underline{e}^d \leftarrow \underline{e} - \frac{\theta_i\,|\boldsymbol{\xi}|}{3}$
            $\underline{t} \leftarrow \frac{3}{m_i}\,k\,\theta_i\,\underline{\omega}\,|\boldsymbol{\xi}| + \frac{15\mu}{m_i}\,\underline{\omega}\,\underline{e}^d$
            $\underline{\mathbf{M}} \leftarrow \frac{\boldsymbol{\xi}+\boldsymbol{\eta}}{|\boldsymbol{\xi}+\boldsymbol{\eta}|}$
            $\mathbf{f}_i \leftarrow \mathbf{f}_i + \underline{t}\,\underline{\mathbf{M}}\,\Delta V_j$
            $\mathbf{f}_j \leftarrow \mathbf{f}_j - \underline{t}\,\underline{\mathbf{M}}\,\Delta V_i$
        end for
    end for
end procedure

5.3 Bond damage and failure

Peridynamic bonds provide a natural framework for modeling material damage and failure. Following Silling and Askari [33], a scalar damage value is assigned to each bond in a peridynamic body. One approach is to define a damage variable, d_{ij}, that tracks the damage between the bonded material points $\mathbf{x}_i$ and $\mathbf{x}_j$. The damage variable is restricted to values between 0.0 and 1.0, where $\mathrm{d}_{ij} = 0.0$ denotes an undamaged bond and $\mathrm{d}_{ij} = 1.0$ denotes complete bond failure. Bond damage values are updated throughout the course of a simulation and incorporated directly into the calculation of internal force density. For example, the influence function that defines a weighting between the bonded material points $\mathbf{x}_i$ and $\mathbf{x}_j$ in a state-based constitutive law may be modified by the factor $(1.0 - \mathrm{d}_{ij})$. In this way, the influence of damaged bonds is reduced, and bonds that have failed completely are effectively excluded from the simulation. The accumulation of broken bonds leads to material separation and the formation of cracks.

Pseudocode for the *critical stretch* bond failure law [33] is presented in Algorithm 3. The bond damage values, d_{ij}, are initialized to 0.0 at the onset of the simulation. At each time step or load step, the bond stretch, s, defined as the change in bond length divided by the undeformed bond length, is computed for each pair of bonded material points in the body. The stretch values are then checked against a specified critical value, s_o. If the stretch exceeds the critical value, the bond is irreversibly broken (d_{ij} is set to 1.0).

As shown by Silling and Askari [33], the critical stretch value is directly related to the fracture energy associated with the creation of free surfaces,

$$s_o = \sqrt{\frac{5G_o}{9k\delta}}, \tag{5.3}$$

where G_o is the fracture energy per unit area, k is the bulk modulus, and δ is the peridynamic horizon. Additional connections between peridynamic bond failure laws and standard concepts in classical mechanics, including the formulation of a work-based bond failure criterion, are given by Foster, et al. [12].

5.4 The tangent stiffness matrix

The tangent stiffness matrix is required when applying many nonlinear solver strategies for implicit time integration, such as Newton's method. It is comprised of the derivatives of nodal forces with respect to nodal displacements. Alternative formulations are also possible, for example one in which the nodal velocities are treated as the cardinal unknowns. A common approach for construction of the tangent stiffness matrix is to first consider the internal force (i.e., the constitutive model response),

Algorithm 3 Routine for evaluation of the *critical stretch* bond failure law. Bond damage values, d_{ij}, are initialized to zero at the beginning of the simulation and set to a value of one if the bond stretch exceeds the specified critical value.

procedure CRITICAL STRETCH BOND FAILURE
  for each node i **do**
    ▷ Evaluate the stretch of each bond.
    for each node j in neighbor list for node i **do**
      $\boldsymbol{\xi} \leftarrow \mathbf{x}_j - \mathbf{x}_i$
      $\boldsymbol{\eta} \leftarrow \mathbf{u}_j - \mathbf{u}_i$
      $s = \frac{|\boldsymbol{\xi}+\boldsymbol{\eta}| - |\boldsymbol{\xi}|}{|\boldsymbol{\xi}|}$
      ▷ Check the bond stretch against the critical value.
      if $s \geq s_o$ **then**
        $\mathrm{d}_{ij} = 1.0$
      end if
    end for
  end for
end procedure

after which the matrix is modified to reflect boundary conditions. For general nonlinear problems, the tangent stiffness matrix is a function of the chosen linearization point and must be re-evaluated whenever the nodal displacements are updated. It is important to note that several so-called linear peridynamic constitutive models impose a nonlinear relationship between force and displacement, and therefore require the application of a nonlinear solver. An example is the *linear peridynamic solid*, in which pairwise forces are linearly related to bond stretch but nonlinearly related to nodal displacements.

The tangent stiffness matrix is defined as

$$K_{ij} = \frac{\partial f_i^{int}(\mathbf{u})}{\partial u_j}, \tag{5.4}$$

where f_i^{int} is a component of the internal force vector, $\mathbf{u}$ is the displacement vector, and u_j is a component of $\mathbf{u}$. The tangent stiffness matrix relates an infinitesimal perturbation of a displacement degree of freedom to the resulting change in the global internal force vector. The tangent stiffness matrix for a nonlocal model contains nonzero entries for each pair of nodes that interact directly, and is inherently more dense than that of a local model. For a general state-based peridynamic model, direct interactions between nodes can take two forms: interactions between nodes that are bonded to each other, and interactions between nodes that share a common neighbor.

Methods for construction of the tangent stiffness matrix include both analytical and computational approaches. Analytical approaches may be developed using the concept of a peridynamic *modulus state*, as presented by Silling [32]. Examples of peridynamic constitutive models for which a modulus state has been derived include the *linear peridynamic solid* [34, 23], and the ordinary state-based viscoelasticity and plasticity models developed by Mitchell [23, 24]. When available, the modulus state provides a direct path to construction of the tangent stiffness matrix and is generally preferred over other approaches for reasons of accuracy and efficiency. Computational approaches for construction of the tangent stiffness matrix include finite difference methods, described below, as well as methods that utilize software engineering techniques such as automatic differentiation [14] and the complex-step approach [9]. The *Peridigm* code includes a finite difference scheme that is applicable to the full set of available constitutive models, as well as several material-model-specific routines that utilize the *Sacado* [3, 27] software package for automatic differentiation.

A finite difference scheme for construction of the tangent stiffness matrix is presented in Algorithm 4. Finite difference strategies are generic in that specifics of the constitutive model formulation are not considered; all that is required is application of the constitutive model for evaluation of the internal force. Under the approach outlined in Algorithm 4, the partial derivatives in Equation (5.4) are approximated using a central difference scheme,

$$K_{ij} \approx \frac{f_i^{int}\left(\mathbf{u}+\epsilon^j\right) - f_i^{int}\left(\mathbf{u}-\epsilon^j\right)}{2\varepsilon}. \tag{5.5}$$

Here, ϵ^j is a vector containing a single nonzero entry, ε, at the position corresponding to the jth degree of freedom in the discretization. The notation used in Algorithm 4 to denote perturbed quantities is defined as follows: $\underline{\mathbf{T}}^{\varepsilon+}$ is a force state evaluated under a positive perturbation of a displacement degree of freedom, $\underline{\mathbf{T}}^{\varepsilon-}$ is the force state evaluated under the analogous negative perturbation, $\underline{\mathbf{T}}^{\varepsilon+}\langle \mathbf{x}_k - \mathbf{x}_i \rangle$ and $\underline{\mathbf{T}}^{\varepsilon-}\langle \mathbf{x}_k - \mathbf{x}_i \rangle$ are the force densities per unit volume resulting from the application of the perturbed force states to the bond $\langle \mathbf{x}_k - \mathbf{x}_i \rangle$, $\mathbf{f}^{\varepsilon+}$ and $\mathbf{f}^{\varepsilon-}$ are the corresponding nodal forces, and $\mathbf{f}^{\text{diff}}$ is the difference between $\mathbf{f}^{\varepsilon+}$ and $\mathbf{f}^{\varepsilon-}$. For an accurate approximation, the magnitude of ε should be chosen to be small relative to the discretization, but not so small that the limits of machine precision become a significant factor. The default perturbation size in *Peridigm* is $1.0e^{-6}$ times the nodal spacing.

Finite difference schemes for constructing the tangent stiffness matrix involve perturbing individual displacement degrees of freedom and examining the resulting change in each component of the global force vector. To minimize computational expense, the evaluation of the internal force under a perturbation of a single displacement degree of freedom should be restricted to include only those components of the internal force vector that are directly affected. As described above, this subset of the internal force vector contains, at most, the union of the neighbor list of the perturbed node and the neighbor lists of each neighbor of the perturbed node.

The *Peridigm* algorithm for evaluating the tangent stiffness matrix by finite difference operates by traversing each node in the discretization and evaluating the force state at that node under a perturbation of a displacement degree of freedom. Perturbations must be considered for the node itself, as well as for each node in the node's neighbor list. Each time the force state is evaluated, it is applied to each bond in the neighbor list and multiplied by the volumes connected by that bond. The resulting force values are then divided by the perturbation length and assembled into the proper locations in the tangent stiffness matrix. An assumption implicit to this approach is that the peridynamic force state at a given material point can be determined based solely on the constitutive model data (and history) at that point, plus basic kinematic data (*e.g.*, position, velocity) at each neighbor. The approach utilized in *Peridigm* minimizes the number of internal force calculations required for construction of the tangent stiffness matrix by central finite difference. Alternative approaches, such as those based on graph coloring, provide additional strategies for evaluating Equation (5.5). It should be noted that the volumes ΔV_i and ΔV_j in Algorithm 4 must be treated in a manner consistent with evaluation of the constitutive model; if a correction for partial neighbor intersections is used in the constitutive model, it should be applied for construction of the tangent stiffness matrix as well.

Algorithm 4 Construction of the tangent stiffness matrix by central finite difference.

---

procedure TANGENT STIFFNESS MATRIX
    ▷ Initialize the tangent stiffness matrix to zero.
    $\mathbf{K} \leftarrow \mathbf{0}$
    ▷ Traverse each node in the discretization.
    for each node i **do**
        $\{traversal\ list\} \leftarrow$ node i and all neighbors of node i
        for each node j in $\{traversal\ list\}$ **do**
            ▷ Evaluate the force state at $\mathbf{x}_i$ under perturbations of displacement.
            for each displacement degree of freedom r at node j **do**
                $\underline{\mathbf{T}}^{\varepsilon+} \leftarrow \underline{\mathbf{T}}[\mathbf{x}_i](\mathbf{u} + \epsilon^r)$
                $\underline{\mathbf{T}}^{\varepsilon-} \leftarrow \underline{\mathbf{T}}[\mathbf{x}_i](\mathbf{u} - \epsilon^r)$
                ▷ Evaluate pairwise forces under perturbations of displacement.
                for each node k in neighbor list of node i **do**
                    $\mathbf{f}^{\varepsilon+} \leftarrow \underline{\mathbf{T}}^{\varepsilon+}\langle \mathbf{x}_k - \mathbf{x}_i \rangle \, \Delta V_i \, \Delta V_k$
                    $\mathbf{f}^{\varepsilon-} \leftarrow \underline{\mathbf{T}}^{\varepsilon-}\langle \mathbf{x}_k - \mathbf{x}_i \rangle \, \Delta V_i \, \Delta V_k$
                    $\mathbf{f}^{\text{diff}} \leftarrow \mathbf{f}^{\varepsilon+} - \mathbf{f}^{\varepsilon-}$
                    for each degree of freedom s at node k **do**
                        $\mathrm{K}_{sr} \leftarrow \mathrm{K}_{sr} + \frac{f_s^{\text{diff}}}{2\varepsilon}$
                    end for
                end for
            end for
        end for
    end for
end procedure

5.5 Modeling contact

Modeling contact is important for many applications of peridynamics. Impact, penetration, and fragmentation simulations, for example, require the ability to capture multi body contact interactions. This includes bodies that are disconnected at the onset of a simulation, and also those that become disconnected as a result of material failure. Contact modeling in peridynamics has received little attention relative to contact modeling for classical finite element analysis, which has been addressed extensively in the literature (see, for example, Laursen [18]). To date, the most common approach to modeling contact in peridynamics is the *short-range force* approach of Silling and Askari [33]. The *short-range force* approach models contact interactions between bodies using a method similar to techniques employed in molecular dynamics. At each step in the simulation, spring-like repulsive forces are applied between nodes that are in close proximity to one another. This approach produces acceptable results in many cases. Alternatively, it has been demonstrated that a conventional (local) contact algorithm may be applied within a peridynamic simulation [19].

The basic components of a contact model are the detection algorithm and the enforcement algorithm. The detection algorithm performs a proximity search in the current (deformed) configuration. The proximity search required for contact is not different in principle from the proximity search required for the creation of neighbor lists. The *Peridigm* code executes the proximity search for contact in three phases: geometry-based parallel decomposition, identification and communication of off-processor data, and execution of an on-processor proximity search. The first two steps are required only for parallel simulations. Their goal is to reduce the parallel search to a set of serial searches. As described in Section 5.6, this is achieved by creating geometry-based parallel partitions and expanding the partitions in each direction by the contact search distance, thus collecting on each partition the necessary data to identify contact interactions for the on-processor nodes. A serial proximity search may then be carried out using a number of approaches, such as those based on quadtree or *k*-d tree data structures. Because nodes travel a finite distance over a time step, it is occasionally necessary to consider node paths as opposed to simply examining the node positions at the end of the step (*e.g.*, in hypervelocity simulations).

Short-range force algorithms consist of pairwise repulsive forces that increase in magnitude as the distance between nodes decreases. An extremely straightforward version of a *short-range force* calculation is illustrated in Algorithm 5. Here, $\mathbf{y}$ denotes nodal coordinates in the deformed configuration, ℓ is the distance between two nodes in the deformed configuration, ℓ_o is a threshold distance (*e.g.*, kick-in distance for short-range forces), C is a user-defined constant for controlling the magnitude of the short-range forces, f_c is a contact force magnitude per unit volume squared, $\underline{\mathbf{M}}$ is the vector between two nodes in the deformed configuration, and $\mathbf{f}^{\text{contact}}$ denotes contact force density. Strengths of the *short-range force* approach include ease of implementation and relatively low computational cost. If desired, friction effects may be included through extensions to the *short-range force* model. The kinematic

Algorithm 5 A *short-range force* contact algorithm.

procedure SHORT-RANGE FORCE CONTACT ALGORITHM
▷ Initialize the contact force densities to zero.
for each node i **do**
$\mathbf{f}_i^{\text{contact}} \leftarrow \mathbf{0}$
end for
▷ Examine the results of the global proximity search for each node.
for each node i **do**
$\mathbf{y}_i \leftarrow \mathbf{x}_i + \mathbf{u}_i$
▷ Compute pairwise contact force densities.
for each node j in proximity of node i **do**
$\mathbf{y}_j \leftarrow \mathbf{x}_j + \mathbf{u}_j$
$\ell \leftarrow |\mathbf{y}_j - \mathbf{y}_i|$
if $\ell < \ell_o$ **then**
$\mathrm{f}_c \leftarrow \left(\frac{9C}{\pi\,\delta^4}\right)\left(\frac{\ell_o - \ell}{\delta}\right)$
$\underline{\mathbf{M}} \leftarrow \frac{\mathbf{y}_j - \mathbf{y}_i}{\ell}$
$\mathbf{f}_i^{\text{contact}} \leftarrow \mathbf{f}_i^{\text{contact}} - \mathrm{f}_c\,\Delta V_j\,\underline{\mathbf{M}}$
$\mathbf{f}_j^{\text{contact}} \leftarrow \mathbf{f}_j^{\text{contact}} + \mathrm{f}_c\,\Delta V_i\,\underline{\mathbf{M}}$
end if
end for
end for
end procedure

information required for a Coulomb friction model, for example, may be determined by examining the relative motion of any two nodes interacting via contact.

An additional extension of the *short-range force* model concerns peridynamic bonds. In its most simple form, the *short-range force* model does not take peridynamic bonds into account. This results in the possibility that material points may interact through the material model and the contact model simultaneously, for instance when a body is subjected to large hydrostatic compression. This effectively results in rapid material stiffening, often well beyond the stiffness prescribed by the constitutive model. This scenario can be avoided by considering neighbor list information in the contact model and disallowing contact forces between bonded nodes. For further discussion specific to correspondence constitutive models, in particular the possibility of material interpenetration, see Tupek and Radovitzky [36].

The computational expense of detecting and enforcing contact interactions often comprises a significant portion of the overall simulation cost. One strategy of reducing this expense is to limit the range of possible contact interactions to include only specific subregions or node sets. Further, so-called self-contact may be disabled by designating that contact interactions may not occur between any two nodes that are within the same subregion.

A final consideration in contact modeling is the contact model's effect on the maximum stable time step for explicit time integration. The short-range force model, for example, typically results in a reduction of the critical time step due to what is effectively an increase in material stiffness in regions where contact is occurring. This can be problematic if the critical time step is sharply reduced when advancing the simulation from time t^n to t^{n+1} (see Section 5.8.1). Care must be taken to ensure that the time step applied at t^n is sufficiently small to guarantee stability at t^{n+1}.

The *short-range force* algorithm does not perform well under all conditions. One source of difficulty is the tracking of surfaces. The *short-range force* model effectively represents surfaces as a collection of nodes, a strategy that breaks down as the distance between adjacent nodes grows large under finite deformation. This scenario is exacerbated in cases where the peridynamic discretization is derived from a nonuniform hexahedral or tetrahedral mesh (see Section 5.6). In this case, slender elements are replaced by nodes that yield poor representations of continuous surfaces. Poor surface approximations may lead to unphysical interpenetration and subsequent numerical difficulties. A possible solution is the construction of explicit surface representations, for example using a level-set approach (cf. Chi et al. [10]).

Contact modeling in peridynamic simulations remains an open area of research. To date, contact has been applied primarily within explicit transient dynamic peridynamic simulations. A penalty- or constraint-based approach is more likely to be successful when implicit time integration is utilized. Furthermore, general issues related to nonlocality in contact have not been explored. For example, the interaction distance for the *short-range force* contact model (ℓ_o in Algorithm 5) is typically assigned a value approximately equal to the node spacing in the undeformed discretization, whereas the interaction distance for the constitutive model is equal to the horizon. In this scenario, the contact model is effectively a local model, in contrast to the constitutive law, which is nonlocal. The application of a local contact model within a peridynamic simulation is tantamount to applying local boundary conditions to a nonlocal model. A nonlocal contact model that produces force densities acting over volumetric regions may be more appropriate for peridynamic simulations.

5.6 Meshfree discretizations for peridynamics

The meshfree approach of Silling and Askari requires that the domain of interest be discretized into a set of nodal volumes, each of which is defined by its initial coordinates, (x, y, z), and volume, ΔV. Nodal volumes may be grouped into blocks, each tied to a specific material model and set of constitutive parameters. Standard *Peridigm* simulations operate on a discretization defined by tuples of the form $(x, y, z, \Delta V, block\_id)$. Discretizations of this type may be created internally within the code, read from a file, or created by processing a hexahedral or tetrahedral mesh (*e.g.*, a mesh created for use in a classical finite element code). Several options for generating peridynamic discretizations are discussed in the recent work by Henke and Shanbhag [15], including the use of a Cartesian grid and a more sophisticated approach based on centroidal Voronoi tessellation. Each node in a discretization may be assigned a unique global ID. In the case of a parallel code, local (on-processor) IDs may also be assigned, and a mapping maintained between the local and global IDs. Unique global IDs facilitate the definition of node sets, to which initial and boundary conditions are applied. The use of node sets to facilitate the specification of initial and boundary conditions is commonplace in classical finite element codes. A caveat regarding peridynamic models is that constraints are generally specified over a volumetric region, as opposed to a two-dimensional surface, which may complicate the specification of node sets in practice.

The use of blocks to define sets of nodal volumes facilitates simulations involving multiple materials and enables the application of specialized constitutive laws to bonds that cross material boundaries. Direct application of Equation (5.2) results in an averaging of constitutive responses when computing net pairwise interactions across material boundaries. Alternatively, a constitutive law that is specific to a material interface may be applied. Material damage along grain boundaries in a polycrystalline material, for instance, may be captured with a specialized bond failure law. In related work by Katiyar et al. [17], the harmonic mean was shown to be an appropriate interface condition for modeling nonlocal diffusion.

Mesh generation tools developed for finite element analyses can be applied to create discretizations for peridynamic simulations. Mesh generation software typically operates on a geometric description of the domain and produces a mesh of solid hexahedral or tetrahedral elements. Converting a mesh of this type into a meshfree discretization for a peridynamic simulation is straightforward: solid elements are converted to a $(x, y, z, \Delta V, block\_id)$ tuple, where (x, y, z) is the centroid of the original element, ΔV is the volume of the original element, and the $block\_id$ is transferred directly from the original element to the peridynamic nodal volume. Calculation of the

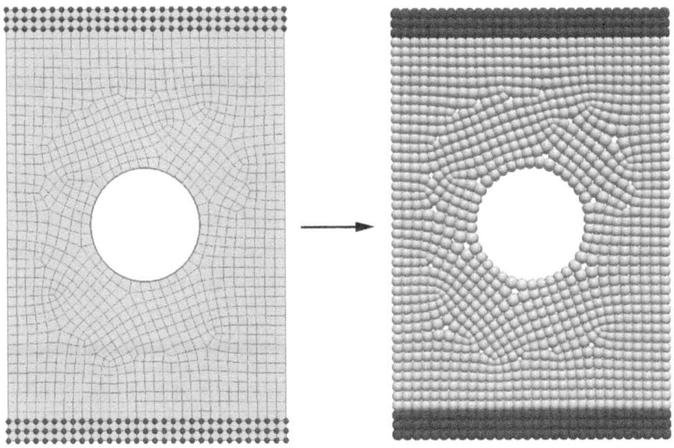

Figure 5.1: Conversion of a standard finite element mesh to a meshfree discretization. Nodes belonging to a node set are highlighted near the top and bottom of the domain shown. Nodal volumes in the meshfree discretization are visualized as spheres having volumes equal to the (scalar) nodal values.

volume and centroid of a tetrahedron is trivial; for details regarding geometric calculations on hexahedral elements, see Grandy [13]. Conversion of a finite element mesh in this fashion yields a one-to-one relationship between the elements in the original finite element mesh and the nodal volumes in the meshfree peridynamic discretization. The nodes in the original mesh are not preserved, however, which complicates the transfer of node sets defined on the original mesh to nodes in the peridynamic discretization.

It is important to recognize several differences between the requirements for a viable peridynamic discretization and those for a classical finite element mesh. Boundary conditions in the classical local theory are applied over a two-dimensional surface, whereas nonlocal simulations require that constraints be applied over a three-dimensional volume. This complicates the definition of node sets for peridynamic simulations due to the fact that mesh-generation tools do not typically include functionality for defining three-dimensional layers along domain boundaries. The use of highly graded nodal spacing, which is common in classical finite element analyses, is an additional source of difficulty. The use of a highly graded discretization in peridynamic simulations is problematic because the ratio of the nodal spacing to the horizon varies wildly over the domain. Computational expense can increase dramatically if the node spacing is reduced relative to the horizon size, which occurs in the highly refined regions. A final point on the requirements for a viable peridynamic discretization concerns the chosen length scale. In many peridynamic constitutive models, the value of the horizon is raised to the fourth power, for example in the *prototype microelastic brittle* material [33]. For this reason, the unit of length should be chosen such that the value of the horizon is of order one to avoid potential numerical difficulties.

5.7 Proximity search for identification of pairwise interactions

A proximity search is required to identify the neighbors of the material points in a peridynamic body. A so-called neighbor list is a record of these points, for example a list of node IDs for all the nodes that comprise a given neighborhood. As a consequence of nonlocality, neighbor lists extend beyond the set of nearest neighbors to include all material points lying within a sphere centered at a given material point and having a radius equal to the peridynamic horizon. Note that, because the meshfree approach of Silling and Askari is a Lagrangian approach, neighbor lists do not change as a result of deformation and may be created and stored at the onset of a simulation [33].

The creation of neighbor lists, while straightforward in theory, presents several challenges in practice. For a general unstructured discretization, a global proximity search must be executed to find all material points within the sphere that defines the neighborhood of each material point in the model. For simplicity, the discussion below is restricted to the case in which a single horizon value is assigned to the entire computational domain. For a treatment of a variable horizon, see Bobaru and Ha [7] and Silling et. al [35]. The *Peridigm* code utilizes the *Zoltan* software package for the creation of neighbor lists [8]. The *Zoltan* package contains routines with a variety of applications to parallel partitioning, load balancing, and data management, a number of which facilitate efficient proximity searches. Search algorithms based on quadtree or k-d tree data structures provide additional alternatives. In the case of a parallel code, proximity searches require an initial global decomposition phase. One strategy for this initial step is to create a geometrically based parallel decomposition, for example using the Recursive Coordinate Bisection (RCB) algorithm [6]. When constructing neighbor lists by examining the distance between material points, expanding the bounding box for a given partition by a length equal to the horizon in each dimension allows for the identification of all potential off-processor neighbors to the material points in that partition.

Examining the distance between material points is not sufficient for quadrature schemes in which partial neighbor intersections are considered [7, 28, 29]. Following the approach of Seleson [28] and Seleson and Littlewood [29], for example, the domain is discretized into a set of subvolumes, each of which has a single peridynamic material point located at its centroid. The division of the domain into a set of subvolumes is analogous to the meshing procedure for classical finite element analysis. In the case of finite element analysis, the subvolumes are used directly as elements in the numerical simulation. In the case of a peridynamic simulation, the subvolumes are used only during pre-processing as a means to determine accurately the extent to which any pair of material points interacts. The pivotal case is the one in which the subvolume for a material point lies only partially within the sphere that defines the neighborhood for another material point. In this case, the intersection of the subvolume and the sphere may be computed and used in the calculation of pairwise internal forces (*e.g.*, through modification to ΔV_i and ΔV_j in Algorithm 2). Consideration of partial intersections adds significant complexity to the creation of neighbor lists. For general unstructured grids, the proximity search must take subvolume geometry into account. This complicates both the serial proximity search and the geometric decomposition required for parallel codes. The serial proximity search must be capable of identifying and computing the intersections of subvolumes and spheres. The geometric decomposition required for parallel codes must employ an expanded bounding box that allows for identification of all potential off-processor partial intersections. Further, the results of the partial neighbor intersection calculations must be stored for use throughout the simulation, in addition to the standard $(x, y, z, \Delta V, block\_id)$ data.

Care must be taken when constructing neighbor lists in the vicinity of small geometric features. The danger is that bonds may pass across features that are small relative to the horizon size, for example small holes or notches, in a way that is unphysical. One possible solution is to include in the proximity search process a set of geometric entities through which bonds may not pass. For example, a finite plane may be used to define a pre-crack. Any bond that would pass through the pre-crack plane is excluded from the model. This approach can be made quite general in the case of a discretization created from a standard finite element mesh. Shell elements, for instance, provide a means to define planes that can be used in the proximity search process [30]. An alternative, simpler approach to controlling the construction of neighbor lists is to explicitly allow or disallow bond creation between specific material subregions.

5.8 Time integration

Numerical time integration plays a central role in engineering analysis codes. Conceptually, it is the time integrator that drives the simulation and orchestrates the passing of information to and from various computational kernels. Time integration schemes for peridynamic models do not differ significantly from those of classical, local models, for which a vast literature exists. Standard approaches for engineering codes can be found in the excellent books by Hughes [16] and by Belytschko et al. [5]. The time integration strategies for peridynamic models presented below were adopted directly from these texts.

Time integration schemes fall into two broad categories: explicit and implicit. Explicit time integration is typically utilized for transient simulations in which inertial effects play an important role. It is extremely resilient, and as such is well suited for peridynamic simulations involving large deformations and pervasive material failure. The great drawback of explicit time integration is that it is only conditionally stable, which limits the maximum allowable size of the time step. Implicit methods constitute a second class of time integration schemes. They are unconditionally stable, allowing for much larger time steps than those permitted by explicit approaches. Implicit time integration, however, requires the solution of a system of equations involving both the current state and a future state of the system. This adds considerable complexity and computational expense.

A discussion of explicit time integration for transient dynamics and implicit time integration for nonlinear quasi-static analysis is presented below. Far from a comprehensive treatment of numerical time integration, the intent is to highlight the most salient aspects of explicit and implicit methods for peridynamics. For details regarding other common time integration strategies, such as Newmark–Beta schemes for implicit dynamics, see Belytschko et al. [5].

Note that, in some cases, it may be advantageous to apply multiple time integrators, in sequence, within a single simulation. An example is mechanical or thermal preloading, which may be simulated as a quasi-static process, followed by the simulation of a dynamic process such as impact.

5.8.1 Explicit time integration for transient dynamics

Generally, explicit time integration schemes determine system response over the course of a simulation by advancing through a large number of small time steps. At each step, known quantities in the current state of the system are used to determine the future state of the system. An explicit time integration scheme well suited for application to peridynamics is outlined below. It is a central difference scheme in which velocities are defined at the midpoints of the time intervals [5].

1. Initialize $n=0, t=0, \mathbf{u}=\mathbf{0}$, and $\mathbf{a}=\mathbf{0}$. Set initial conditions for $\mathbf{v}$ and initialize material state variables (if any).

2. Estimate the maximum stable time step, Δt_{crit}, and determine the simulation time step, Δt.

3. Update the times t^{n+1} and $t^{n+\frac{1}{2}}$: $\quad t^{n+1}=t^n+\Delta t, \quad t^{n+\frac{1}{2}}=\frac{1}{2}\left(t^n+t^{n+1}\right)$.

4. First partial velocity update, $\mathbf{v}^{n+\frac{1}{2}}=\mathbf{v}^n+\left(t^{n+\frac{1}{2}}-t^n\right)\mathbf{a}^n$.

5. Enforce velocity boundary conditions by setting $\mathbf{v}^{n+\frac{1}{2}}$ for nodes with prescribed velocities. Displacement boundary conditions can be converted to velocities and applied here, or they can be applied directly after Step 7.

6. Update nodal displacements, $\mathbf{u}^{n+1}=\mathbf{u}^n+\mathbf{v}^{n+\frac{1}{2}}\Delta t$.

7. Evaluate the internal force by means of the constitutive model and damage model for $\mathbf{u}^{n+1}$ and $\mathbf{v}^{n+\frac{1}{2}}$. Update material state variables accordingly. Optionally, recompute the estimate for the critical time step, Δt_{crit}, and update the simulation time step, Δt.

8. Evaluate contact forces for $\mathbf{u}^{n+1}$ and $\mathbf{v}^{n+\frac{1}{2}}$. Update the critical time step, Δt_{crit}, if required by the contact algorithm.

9. Compute acceleration $\mathbf{a}^{n+1} = \mathbf{M}^{-1}\mathbf{f}^{n+1}$, where $\mathbf{M}$ is the mass matrix and $\mathbf{f}^{n+1} = \mathbf{f}^{n+1}_{int} + \mathbf{f}^{n+1}_{ext} + \mathbf{f}^{n+1}_{contact}$.
10. Second partial velocity update, $\mathbf{v}^{n+1} = \mathbf{v}^{n+\frac{1}{2}} + \left(t^{n+1} - t^{n+\frac{1}{2}}\right)\mathbf{a}^{n+1}$.
11. Evaluate energy balance as a check on stability.
12. If the simulation is not complete, update $n \leftarrow n+1$, and go to Step 3.

The first step in the time integration scheme is initialization. The time, t, and time step, n, as well as nodal values for displacement, $\mathbf{u}$, and acceleration, $\mathbf{a}$, are set to zero. The nodal velocities, $\mathbf{v}$, are set to reflect initial conditions on the velocity field. Variables for representing material states within a constitutive model must also be initialized in Step 1. The maximum stable time step, also referred to as the critical time step, Δt_{crit}, is determined in Step 2, as described below. The time step, Δt, is typically determined by multiplying the critical time step by a safety factor. To guard against numerical instability, the safety factor must be less than 1.0; typical values fall in the range of 0.70 to 0.95. A constant value of Δt may be applied throughout the course of the simulation, or, alternatively, the time step may be recomputed as part of Steps 7 or 8 to reflect evolving model geometry, material state, and contact interactions.

The time-stepping process begins with Step 3, in which both the time at end of the step, t^{n+1}, and the mid-step time, $t^{n+\frac{1}{2}}$, are computed. The midstep velocity is then determined in Step 4 using the current values for velocity and acceleration. Step 5 concerns the application of prescribed displacements and velocities, which are typically assigned to node sets that define volumetric subregions within the computational domain. Prescribed velocities may be applied directly. Prescribed displacements may be converted to velocities through the relation

$$\mathbf{v}^{n+\frac{1}{2}} = \frac{1}{\Delta t}\left(\mathbf{u}^{n+1} - \mathbf{u}^{n}\right). \tag{5.6}$$

The midstep velocities, $\mathbf{v}^{n+\frac{1}{2}}$, are then applied to determine the displacements at the end of the time step, $\mathbf{u}^{n+1}$, in Step 6.

The constitutive model, damage model, and contact model are invoked in Steps 7 and 8 for evaluation of the nodal forces, $\mathbf{f}^{n+1}$. The inputs to these models are the displacements at the end of the time step, $\mathbf{u}^{n+1}$, and the mid-step velocities, $\mathbf{v}^{n+\frac{1}{2}}$, as well as the material state variables stored at step n. The damage law may be applied first, to determine the current state of damage for each bond, followed by evaluation of the constitutive model. Contact algorithms, if any, are generally not tied directly to the constitutive model and may be evaluated independently. Contact forces, constitutive model forces, and prescribed body forces are summed together to determine the net force on each node.

Following determination of nodal forces, the nodal accelerations, $\mathbf{a}^{n+1}$, are computed by dividing by the mass associated with each node. The resulting acceleration values are then applied to determine the velocities at the end of the time step, $\mathbf{v}^{n+1}$, in Step 10. The time stepping process then returns to Step 3 and repeats until the simulation is complete.

Prior to incrementing to the next time step, it may be advantageous to carry out a check on global energies to ensure that the simulation remains numerically stable (Step 12). A check on the global energy balance typically involves integration of the kinetic energy, external work, and stored elastic energy over the computation domain (cf. Equation (62) in Silling et al. [34]). An unphysical increase in the total energy of the system generally indicates an unstable simulation, for example resulting from an inaccurate estimate of the critical time step. Note that simulations involving crack formation necessarily absorb energy through the creation of new surfaces. Care must be taken in this case because numerical instability may manifest as the creation of additional (unphysical) cracks, as opposed to an increase in kinetic energy or stored elastic energy.

5.8.2 Estimating the maximum stable time step

A means for estimating the critical time step for the *prototype microelastic brittle* model was published by Silling and Askari [33]. At each material point in the discretization, a value for the maximum stable time step is determined as follows,

$$\Delta t_{\text{crit}} = \sqrt{\frac{2\rho}{\sum_p \Delta V_p C_p}}, \tag{5.7}$$

where ρ is the density, p iterates over all the neighbors of the given material point, ΔV_p is the volume associated with neighbor p, and C_p is the micromodulus between the given material point and neighbor p. The minimum value over all material points in the discretization is taken as the critical time step for the simulation. Equation (5.7) was derived for the one-dimensional case. Its application in two- and three-dimensional simulations results in a conservative estimate of the time step. (Effectively, it assumes that all bonds for a given material point are collinear.)

It has been demonstrated that Equation (5.7) may provide a reasonable estimate of the maximum stable time step for other constitutive models, for example the state-based *linear peridynamic solid* [21]. In this case, the micromodulus, C_p, appearing in the denominator of Equation (5.7) is replaced by an effective micromodulus,

$$C_p^{\text{eff}} = \frac{1}{|\boldsymbol{\xi}|} \frac{18k}{\pi\delta^4}, \tag{5.8}$$

where $|\boldsymbol{\xi}|$ is the distance between the given material point and neighbor p in the reference configuration, k is the bulk modulus, and δ is the horizon. The second term in Equation (5.8) is taken from Silling and Askari [33]; it is the spring constant for a *prototype microelastic brittle* material having bulk modulus k and horizon δ.

Another option for estimating the maximum critical time step is the well-known Courant–Friedrichs–Lewy (CFL) approach [11],

$$\Delta t_{\text{crit}} = \frac{h}{c}, \tag{5.9}$$

where h is a characteristic length associated with the discretization and c is the wave speed,

$$c = \sqrt{\frac{k}{\rho}}. \tag{5.10}$$

Equation (5.9) generally yields a very conservative estimate of the critical time step for peridynamic models when h is set equal to the node spacing. This is because wave propagation in peridynamic models is largely dictated by the size of the horizon. The CFL limit is commonly applied in classical finite element analyses, where nodes interact directly only when connected by a common element. In contrast, nodes in a peridynamic simulation may interact directly when separated by as much as twice the horizon. A natural question regarding Equation (5.9) is, can the horizon be used as the value for h? In preliminary numerical experiments, this approach yielded a nonconservative critical time step [21].

5.8.3 Implicit time integration for quasi-statics

Quasi-static analysis is appropriate for simulations in which relatively large load steps are desirable and in which dynamic effects are negligible. The governing equation for quasi-static simulations is obtained by setting the acceleration term to zero in Equation (5.2),

$$\sum_{\mathcal{H}_{\mathbf{x}}} \left(\underline{\mathbf{T}}[\mathbf{x},t] \langle \mathbf{q}-\mathbf{x} \rangle - \underline{\mathbf{T}}[\mathbf{q},t] \langle \mathbf{x}-\mathbf{q} \rangle \right) \Delta V_{\mathbf{q}} + \mathbf{b}(\mathbf{x},t) = \mathbf{0}. \tag{5.11}$$

A quasi-static simulation consists of a series of load steps at which static equilibrium is enforced. Dynamic (inertial) phenomena, such as wave propagation, are not considered. At the onset of each load step, the boundary conditions are updated. In contrast to explicit transient dynamics, this update constitutes an increase in the applied load, as opposed to an increment of a physically meaningful time parameter. Updating the boundary conditions generally results in an unequilibrated system configuration. The challenge is then to determine the configuration of the degrees of freedom not subjected to kinematic boundary conditions such that equilibrium is re-established.

The equilibrium configuration for a given load step is determined by minimizing the nodal forces. This process focuses on the residual vector, $\mathbf{r}$, which is defined as the nodal force vector with the degrees of freedom subject to kinematic boundary conditions removed or set to zero. In a perfectly equilibrated configuration, the residual vector is equal to $\mathbf{0}$. A scalar-valued residual, r, is defined as a means to track the incremental progress of the nonlinear solver as $\mathbf{r}$ approaches $\mathbf{0}$. The scalar-valued residual may be defined, for example, as the ℓ_2-norm or ∞-norm of $\mathbf{r}$. A load step is complete when r has dropped below a specified threshold value. The chosen form of r and the specified threshold value define the convergence criterion for the nonlinear solver. For additional discussion of convergence criteria for the nonlinear solution process, see Belytschko et al. [5].

The solution procedure outlined below for nonlinear quasi-static simulations is valid for both linear and nonlinear problems. For linear problems, however, the iterative procedure for reduction of the residual is not required and may be replaced with a single solve of the global linear system, given in Step 5b, below. In the linear case, a single Newton step, $\Delta\mathbf{u}$, is guaranteed to re-establish static equilibrium.

The solution process for general, nonlinear quasi-static peridynamic simulations using Newton's method as the nonlinear solver may be written as follows:

1. Initialize $n = 0$, $\mathbf{u} = \mathbf{0}$, and $\mathbf{v} = \mathbf{0}$. Initialize material state variables.
2. Update the load step $n \leftarrow n + 1$ and pseudo time t. Update the boundary conditions.
3. Evaluate the residual vector, $\mathbf{r}$, and residual r. Determine the convergence criterion for the load step.
4. Assign an initial guess to the trial displacement $\mathbf{u}_{\text{trial}}$ (for example, $\mathbf{u}_{\text{trial}} = \mathbf{u}^n$).
5. Apply Newton's method to minimize the residual.
   (a) Construct the tangent stiffness matrix, $\mathbf{K}$, for the configuration $\mathbf{u}_{\text{trial}}$ (recompute, reuse, or modify).
   (b) Solve the linear system $\mathbf{K}\Delta\mathbf{u} = -\mathbf{r}$ for the Newton step, $\Delta\mathbf{u}$.
   (c) Set $\mathbf{u}_{\text{trial}} = \mathbf{u}_{\text{trial}} + \Delta\mathbf{u}$.
   (d) Evaluate the residual vector, $\mathbf{r}$, and the residual, r, for the updated configuration $\mathbf{u}_{\text{trial}}$.
   (e) If the convergence criterion is not met, go to Step 5a.
6. Set $\mathbf{u}^{n+1} = \mathbf{u}_{\text{trial}}$. Set material state data to their trial values.
7. If the simulation is not complete, go to Step 2.

The simulation begins with field initialization, as well as the initialization of material state variables associated with the constitutive models. It is assumed that the system is in equilibrium at the onset of the simulation. Load stepping begins with Step 2, in which an increment of the boundary conditions is applied. This may take the form of updated displacement values in the case of kinematic boundary conditions, or updated external force values in the case of prescribed body forces. The extent to which the system is no longer in equilibrium is determined through the residual evaluation in Step 3. As discussed above, the residual is a function of the forces corresponding to degrees of freedom not subjected to kinematic boundary conditions. (The forces corresponding to degrees of freedom at which kinematic boundary conditions are applied are generally nonzero and are referred to as reaction forces.) The convergence criterion for a given load step is typically computed relative to the initial residual value, although an absolute criterion may also be used.

Solution of the nonlinear problem begins with Step 4. The goal is to determine a displacement vector increment, $\Delta\mathbf{u}$, that will reduce the residual such that the convergence criterion is met. While many techniques exist for solving nonlinear problems,

Newton's method is perhaps the most straightforward. Here, the residual is minimized by solving a series of linear problems. The linear problems are defined as

$$\mathbf{K}\Delta\mathbf{u} = -\mathbf{r}, \tag{5.12}$$

where $\mathbf{K}$ is the tangent stiffness matrix, $\mathbf{r}$ is the residual vector, and $\Delta\mathbf{u}$ is an (unknown) increment in the nodal displacements. The tangent matrix is a linearization of the system about the current trial displacement, $\mathbf{u}_{\text{trial}}$, as described in Section 5.4. Prior to solving the linear system in Equation (5.12), the tangent stiffness matrix and right-hand side vector are typically modified to reflect kinematic boundary conditions. One strategy is to set the rows and columns of the stiffness matrix corresponding to kinematic boundary conditions to zero, with a scalar on the diagonal (*e.g.*, the ℓ_2-norm of the unmodified diagonal). The entries in the residual vector corresponding to kinematic boundary conditions are also set to zero. The result of these modifications is that the solution of Equation (5.12) yields a zero increment in displacement for degrees of freedom corresponding to kinematic boundary conditions.

A quasi-static load step is deemed to be converged when the residual drops below a specified threshold value. The solution procedure then advances to Step 6, in which both the field variables and the material state variables are set equal to their trial values. The load stepping process is then repeated until the simulation is complete.

5.9 Example simulations

Results of an impact and fragmentation simulation and the results of a tensile bar simulation are presented in the following sections. Taken together, they are realizations of the numerical and computational strategies discussed above for evaluation of peridynamic constitutive models, calculation of the tangent stiffness matrix, contact modeling, and explicit and implicit time integration. The first is an explicit transient dynamics simulation of fracture that results from a projectile impacting a brittle disk. It is an initial value problem in which an initial velocity is assigned to the projectile. The second is a quasi-static simulation of a bar loaded in tension. In this case, the simulation is driven by prescribed displacement boundary conditions applied to the end portions of the tensile specimen. Both simulations are fully three-dimensional and were carried out using *Peridigm* [26, 2].

5.9.1 Fragmentation of a brittle disk resulting from impact

A peridynamic simulation of a fragmenting brittle disk is illustrated in Figure 5.2. It is a modified version of the brittle fragmentation simulation presented by Silling and Askari [33], variations of which were later carried out by Parks et al. [25], and Henke and Shanbhag [15]. The simulation is presented for the purpose of demonstrating explicit time integration for transient dynamics (Section 5.8.1), contact (Section 5.5), and crack formation via bond failure (Section 5.3). For a more complete treatment of

Table 5.1: Constitutive parameters for the disk, modeled using the *linear peridynamic solid* constitutive model and the *critical stretch* bond failure law.

Parameter	Value
Horizon	1.17 mm
Density	2.20 g/cm^3
Bulk Modulus	14.90 GPa
Shear Modulus	8.94 GPa
Critical Stretch	0.0005

Table 5.2: Constitutive parameters for the projectile, modeled using the *linear peridynamic solid* constitutive model. No bond failure law was assigned to the projectile.

Parameter	Value
Horizon	1.17 mm
Density	7.70 g/cm^3
Bulk Modulus	160.00 GPa
Shear Modulus	78.30 GPa

modeling brittle fragmentation with peridynamics, including an investigation of convergence with respect to nodal spacing, horizon size, and time step size, see Henke and Shanbhag [15].

The disk and projectile were modeled using the *linear peridynamic solid* constitutive model [34], and material failure in the disk was modeled with the *critical stretch* bond failure law [33]. Constitutive model parameters are given in Tables 5.1 and 5.2. The geometry of the disk is a cylinder of radius 37.0 mm and height 2.5 mm. The discretization for the disk was created by converting an unstructured hexahedral mesh, created with the *Cubit* mesh generation code [1], into a meshfree discretization containing approximately 180.000 nodes. The geometry of the projectile is a sphere with radius 5.0 mm. The discretization for the projectile was created by converting a tetrahedral mesh, created with *Cubit*, into a meshfree representation containing approximately 18.000 nodes. The node spacing for both the disk and the projectile was set to roughly one third of the peridynamic horizon. Contact was modeled with a *short-range force* contact model, as described in Section 5.5. The contact radius was set to 0.775 mm and the spring constant to 1000.0 GPa. The disk was initially at rest, and the projectile was assigned an initial velocity of 100.0 m/s. The simulation was run to a final simulation time of 400.0 µs. A constant time step of 0.0416 µs (0.7 times the estimated critical time step) was used throughout the simulation. The fragmenting disk simulation was run in parallel using ten 3.0 GHz Intel Xeon processors, requiring a total run time of approximately 2.8 hours.

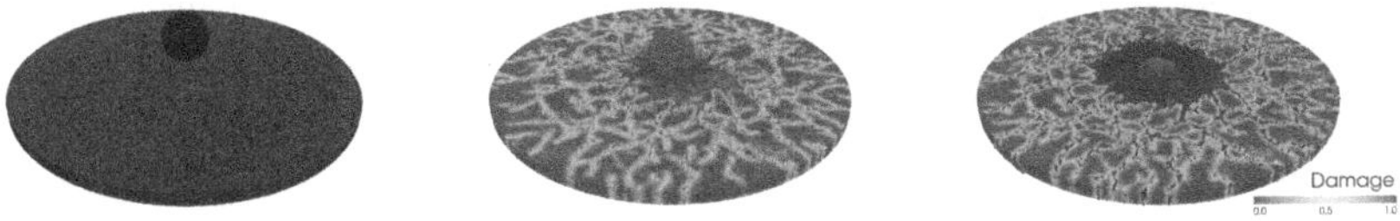

(a) 0 μs. (b) 44 μs. (c) 131 μs.

Figure 5.2: Explicit dynamic simulation of brittle fracture resulting from impact. The different shades (colors for the electronic version) indicate percentage of broken bonds, or damage, at a particular material point.

Table 5.3: Constitutive parameters for the tensile bar. The projectile was modeled using a linear elastic correspondence constitutive law.

Parameter	Value
Horizon	0.095 25 cm
Density	8.00 g/cm^3
Young's Modulus	180.00 GPa
Poisson's Ratio	0.30
Stabilization Coefficient	0.02

5.9.2 Quasi-static simulation of a tensile test

A peridynamic simulation of a tensile test experiment is presented in Figure 5.3. The simulation replicates a standard laboratory test for material characterization [4]. The simulation was carried out using implicit quasi-static time integration (Section 5.8.3). The bar was modeled using an elastic correspondence constitutive model [34]. As described by Littlewood [20], and Tupek and Radovitzky [36], the formulation for correspondence constitutive models given by Silling et al. [34] may exhibit unphysical low-energy modes of deformation. An additional stabilization term was applied in the tensile test simulation to mitigate the impact of these low-energy modes on the solution [20]. Constitutive model parameters for the bar are given in Table 5.3. The length of the tensile bar is 10.16 cm, with a maximum width of 1.27 cm, a minimum width of 0.635 cm, and a thickness of 0.315 cm. The discretization was created by converting an unstructured hexahedral mesh, created using *Cubit*, to a meshfree representation containing approximately 99.000 nodes. The simulation was divided into four load steps. At each load step, the nonlinear solver was applied to reduce the residual by approximately nine orders of magnitude. The tensile bar simulation was run in parallel using ten 3.0 GHz Intel Xeon processors, requiring a total run time of approximately 1.3 hours.

Boundary conditions were applied in the form of prescribed displacements over volumes at the ends of the bar. The volumes over which prescribed displacements were applied were defined as spanning from each end of the bar to a distance equal to two times the horizon inward from the end of the bar. Nodes within these volumes were assigned an initial displacement of zero and a final displacement equal to $0.005x$, where x is the distance from the center of the bar (the origin). The result is a prescribed linear displacement field corresponding to an engineering strain of 0.5% applied to the end portions of the bar. This displacement field is an approximation (educated guess that disregards the variable cross section of the bar) of the true displacement field that would occur in a physical tensile bar experiment. It is expected that, due to the approximate nature of the boundary conditions, unphysical artifacts will be present in and around the end segments of the bar. These artifacts are not expected to significantly pollute the solution in the narrow section of the bar.

The tensile bar simulation was designed to mimic a standard engineering tensile test experiment for material characterization. Young's modulus and Poisson's ratio may be determined experimentally by monitoring the elongation of small segments of the bar using strain gauges and tracking the total force applied to the bar using a load cell. These data can then be used to compute the engineering strain (change in length of the strain gauge divided by its initial length) in both the lateral and longitudinal directions as well as the engineering stress (total force divided by the initial cross-sectional area of the narrow section of the bar). The Young's modulus is computed as the ratio of the engineering stress to the engineering strain in the longitudinal direction. Poisson's ratio is computed as the ratio of the lateral strain to the longitudinal strain. This procedure was modeled within the peridynamic simulation by tracking the displacements of pairs of nodes as well as the net reaction forces in the end segments of the bar. One pair of nodes was chosen to approximate the positions at which the ends of a 2.54 cm strain gauge would be located, assuming the strain gauge were aligned in the longitudinal direction and centered along the length of the bar. A second pair of nodes was chosen to approximate a lateral strain gauge spanning a narrow 0.54 cm section of the bar centered in the longitudinal direction. The resulting engineering stress-strain curve is presented in Figure 5.3(b). The value of Young's modulus calculated in this manner was 181.5 GPa, which differs from the prescribed value of 180.0 GPa by 0.83%. The Poisson's ratio was calculated to be 0.309, which differs from the prescribed value of 0.30 by 3.0%.

5.10 Summary

The application of peridynamics for computational simulation depends critically on an efficient and robust software implementation. In the preceding sections, a roadmap was presented for implementation of the meshfree approach of Silling and Askari [33], which is the numerical technique used for the vast majority of engineering peridynamic simulations to date. Codes developed at Sandia National Laboratories, including *EMU*, *Peridigm*, *LAMMPS*, and *Sierra/SolidMechanics*, contain peridynamics functionality based on this approach.

The state of computational peridynamics codes reflects the level of maturity of peridynamics as a whole. While great strides have been made since the introduc-

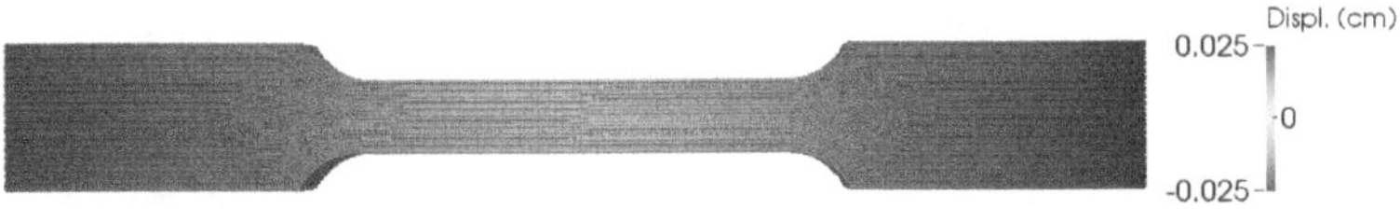

(a) Displacement in the loading direction.

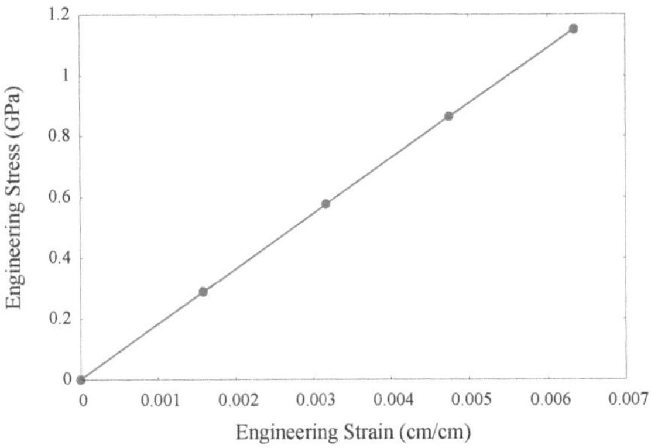

(b) Computed engineering stress-strain curve. The recovered elastic modulus is 181.5 GPa.

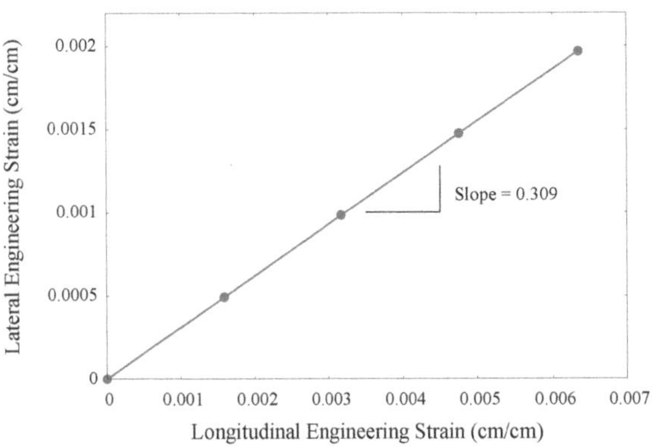

(c) Computed lateral and longitudinal strains. The recovered Poisson's ratio is 0.309.

Figure 5.3: Quasi-static simulation of a tensile test.

tion of peridynamics by Silling in 2000 [31], numerous open research areas remain. With respect to computational simulation, alternatives to the meshfree approach of Silling and Askari should be explored, including alternative discretization techniques, methods for higher-order quadrature, and Eulerian or Arbitrary Lagrangian–Eulerian (ALE) formulations. Constitutive models, contact models, and bond-failure models remain significantly limited, particularly in the context of implicit time integration. In addition to theoretical and numerical considerations, widespread adoption of peridynamics as a tool for engineering analysis requires improvements in the performance and usability of computational peridynamics codes. Practitioners accustomed to mainstream finite element analysis tools would benefit from pre- and postprocessing tools designed specifically for peridynamic models, for example for the specification of volumetric regions for the application of boundary conditions.

Nonlocality pervades many aspects of a peridynamic simulation code. This is apparent within the internal force evaluation, where constitutive model routines must traverse a neighbor list data structure, and in the tangent stiffness matrix, which is significantly more dense than its counterpart in the local theory. The material in the preceding sections draws from experience implementing both stand-alone peridynamics codes and peridynamics functionality within larger software packages. It is hoped that the material presented will provide clarity on the implementation of peridynamic theory, as presented in the literature to date, in a computational simulation code.

Acknowledgements

Sandia National Laboratories is a multi-program laboratory managed and operated by Sandia Corporation, a wholly owned subsidiary of Lockheed Martin Corporation, for the U.S. Department of Energy's National Nuclear Security Administration under contract DE-AC04-94AL85000. SAND 2014-1530P.

References

[1] *Cubit* mesh generation code. http://cubit.sandia.gov.

[2] *Peridigm* peridynamics code. http://peridigm.sandia.gov.

[3] *Sacado* software package. http://trilinos.org/packages/sacado/.

[4] ASTM Standard E8 / E8M-13a. Standard test methods for tension testing of metallic materials. ASTM International, West Conshohocken, PA, 2013.

[5] T. Belytschko, W. K. Liu, and B. Moran. *Nonlinear finite elements for continua and structures*. John Wiley & Sons, Ltd., Chichester, England, 2000.

[6] M. Berger and S. Bokhari. A partitioning strategy for nonuniform problems on multiprocessors. *IEEE Transactions on Computers*, C-36(5): 570–580, 1987.

[7] F. Bobaru and D. Y. Ha. Adaptive refinement and multiscale modeling in 2D peridynamics. *International Journal for Multiscale Computational Engineering*, 9(6): 635–660, 2011.

[8] E. G. Boman, U. V. Catalyurek, C. Chevalier, and K. D. Devine. The Zoltan and Isorropia parallel toolkits for combinatorial scientific computing: partitioning, ordering, and coloring. *Scientific Programming*, 20(2): 129–150, 2012.

[9] M. D. Brothers, J. T. Foster, and H. R. Millwater. A comparison of different methods for calculating tangent-stiffness matrices in a massively parallel computational peridynamics code. *Computer Methods in Applied Mechanics and Engineering*, 279:247–267, 2014.

[10] S.-W. Chi, C.-H. Lee, J.-S. Chen, and P.-C. Guan. A level set enhanced natural kernel contact algorithm for impact and penetration modeling. *International Journal for Numerical Methods in Engineering*, 102: 839–866, 2015.

[11] R. Courant, K. O. Friedrichs, and H. Lewy. Über die partiellen differenzensleichungen der mathematischen physik. *Mathematische Annalen*, 100(1):32–74, 1928.

[12] J. T. Foster, S. A. Silling, and W. Chen. An energy based failure criterion for use with peridynamic states. *Journal for Multiscale Computational Engineering*, 9: 675–687, 2011.

[13] J. Grandy. Efficient computation of volume of hexahedral cells. Technical Report UCRL-ID-128886, Lawrence Livermore National Laboratory, Livermore, CA, 1997.

[14] A. Griewank and A. Walther. *Evaluating derivatives: Principles and techniques of algorithmic differentiation*, 2nd ed. Frontiers in Applied Mathematics. SIAM, Philadelphia, PA, 2008.

[15] S. F. Henke and S. Shanbhag. Mesh sensitivity in peridynamic simulations. *Computer Physics Communications*, 185: 181–193, 2014.

[16] T. J. R. Hughes. *The finite element method: Linear static and dynamic finite element analysis*. Englewood Cliffs, NJ: Prentice-Hall, 1987.

[17] A. Katiyar, J. T. Foster, H. Ouchi, and M. M. Sharma. A peridynamic formulation of pressure driven convective fluid transport in porous media. *Journal of Computational Physics*, 261: 209–229, 2014.

[18] T. A. Laursen. *Computational contact and impact mechanics: Fundamentals of modeling interfacial phenomena in nonlinear finite element analysis*. Springer-Verlag, Heidelberg, Germany, 2002.

[19] D. J. Littlewood. Simulation of dynamic fracture using peridynamics, finite element modeling, and contact. In *Proceedings of the ASME 2010 International Mechanical Engineering Congress and Exposition (IMECE)*, Vancouver, British Columbia, Canada, 2010.

[20] D. J. Littlewood. A nonlocal approach to modeling crack nucleation in AA 7075-T651. In *Proceedings of the ASME 2011 International Mechanical Engineering Congress and Exposition (IMECE)*, Denver, Colorado, 2011.

[21] D. J. Littlewood, J. D. Thomas, and T. R. Shelton. Estimation of the critical time step for peridynamic models. Presented at the SIAM Conference on Mathematical Aspects of Materials Science, Philadelphia, Pennsylvania, 2013.

[22] R. W. Macek and S. A. Silling. Peridynamics via finite element analysis. *Finite Elements in Analysis and Design*, 43: 1169–1178, 2007.

[23] J. A. Mitchell. A nonlocal, ordinary, state-based plasticity model for peridynamics. Report SAND2011-3166, Sandia National Laboratories, Albuquerque, NM and Livermore, CA, 2011.

[24] J. A. Mitchell. A nonlocal, ordinary-state-based viscoelasticity model for peridynamics. Report SAND2011-8064, Sandia National Laboratories, Albuquerque, NM and Livermore, CA, 2011.

[25] M. L. Parks, R. B. Lehoucq, S. J. Plimpton, and S. A. Silling. Implementing peridynamics within a molecular dynamics code. *Computer Physics Communications*, 179(11): 777–783, 2008.

[26] M. L. Parks, D. J. Littlewood, J. A. Mitchell, and S. A. Silling. Peridigm Users' Guide v1.0.0. Report SAND2012-7800, Sandia National Laboratories, Albuquerque, NM and Livermore, CA, 2012.

[27] E. Phipps and R. Pawlowski. Efficient expression templates for operator overloading-based automatic differentiation. In S. Forth, P. Hovland, E. Phipps, J. Uke, and A. Walther, editors, *Recent Advances in Algorithmic Differentiation*, volume 87 of *Lecture Notes in Computational Science and Engineering*. Springer, 2012.

[28] P. D. Seleson. Improved one-point quadrature algorithms for two-dimensional peridynamic models based on analytical calculations. *Computer Methods in Applied Mechanics and Engineering*, 282: 184–217, 2014.

[29] P. D. Seleson and D. J. Littlewood. Convergence studies in meshfree peridynamic simulations. *Computers and Mathematics with Applications*, 77(11): 2432–2448, 2016.

[30] SIERRA Solid Mechanics Team. Sierra/SolidMechanics 4.40 User's Guide. Report SAND2016-2707, Sandia National Laboratories, Albuquerque, NM and Livermore, CA, 2016.

[31] S. A. Silling. Reformulation of elasticity theory for discontinuities and long-range forces. *Journal of the Mechanics and Physics of Solids*, 48: 175–209, 2000.

[32] S. A. Silling. Linearized theory of peridynamic states. *Journal of Elasticity*, 99: 85–111, 2010.

[33] S. A. Silling and E. Askari. A meshfree method based on the peridynamic model of solid mechanics. *Computers and Structures*, 83: 1526–1535, 2005.

[34] S. A. Silling, M. Epton, O. Weckner, J. Xu, and E. Askari. Peridynamic states and constitutive modeling. *Journal of Elasticity*, 88: 151–184, 2007.

[35] S. A. Silling, D. J. Littlewood, and P. Seleson. Variable horizon in a peridynamic medium. *Journal of Mechanics of Materials and Structures*, 10(5): 591–612, 2015.

[36] M. R. Tupek and R. Radovitzky. An extended constitutive correspondence formulation of peridynamics based on nonlinear bond-strain measures. *Journal of the Mechanics and Physics of Solids*, 65: 82–92, 2014.

Part III

Material Models and Links to Atomistic Models

Chapter 6

Constitutive Modeling in Peridynamics

John T. Foster

The University of Texas at Austin

CONTENTS

6.1	Introduction	144
6.2	Kinematics, momentum conservation, and terminology	146
6.3	Linear peridynamic isotropic solid	150
6.3.1	Plane elasticity	152
6.3.1.1	Plane stress	153
6.3.1.2	Plane strain	156
6.3.2	"Bond-based" theories as a special case	156
6.3.3	On the role of the influence function	158
6.3.4	Other elasticity theories	159
6.4	Finite Deformations	159
6.4.1	Invariants of peridynamic scalar-states	162
6.5	Correspondence models	163
6.5.1	Non-ordinary correspondence models for solid mechanics	164
6.5.2	Ordinary correspondence models for solid mechanics	165
6.6	Plasticity	166
6.6.1	Yield surface and flow rule	167
6.6.2	Loading/unloading and consistency	167
6.6.3	Discussion	168
6.7	Non-ordinary models	169
6.7.1	A non-ordinary beam model	169
6.7.2	A non-ordinary plate/shell model	171
6.7.3	Other non-ordinary models	171
6.8	Final Comments	172
	References	173

6.1 Introduction

The peridynamic theory of solid mechanics was originated by Silling [53, 60] as a reformulation of classical elasticity * primarily motivated by the desire to replace the spatial derivatives in the governing equations with an integral functional that provides mathematical consistency with the nature of displacement discontinuities (e.g., cracks) in an otherwise continuously deforming body. Numerical simulations conducted with discrete peridynamic equations have demonstrated a unique capability to easily capture many interesting and challenging problems related to crack propagation, e.g., crack branching [19, 8], oscillatory crack path instabilities in membranes [57], pervasive failure of brittle materials [55], and complex fracture patterns in composite laminates [25].

The early research in peridynamics focused on the theory's attractiveness for modeling fracture, while utilizing simple constitutive models. It is a common misconception that the peridynamic theory as introduced in [53] was presented solely as a Navier one-parameter isotropic elasticity model [34]. A careful rereading of Section 15 of the original paper shows an outline of a generalization that was further developed and presented in full detail in [58]. This chapter presented models for peridynamic fluids and isotropic elastic solids that recover their classical analog as limiting cases. It also introduced some new terminology and notation for handling the mathematics of *peridynamic states*, which are essentially generalizations of tensors. This terminology has led many researchers to use the adjectives *bond-based* and *state-based* when referring to peridynamics, where "bond-based" refers to elasticity models of the one-parameter type, as they derive from a central force-potential between material points similar to a simple molecular *bond*, and "state-based" refers to the generalized theory †. While detailing the peridynamic theory in [58] the authors introduced additional terminology related to constitutive modeling such as *ordinary* and *non-ordinary* materials.

Non-ordinary materials are interesting because they allow for the development of material constitutive response models that have no analog in the classical theory of elasticity, e.g. materials that inherently resist rotation (including discontinuous rotations) between pairs of material points. These types of models have been used to develop peridynamic models for plates and shells [40, 41]. Non-ordinary models have found the most widespread use as a way to incorporate any classical stress tensor constitutive model into a computational code that solves the discrete peridynamic equations [64, 14, 9]; however, this can be accomplished in a way that results in an ordinary model as well [63]. These types of implementations are commonly called *correspondence models* in peridynamics because they derive from assuming a correspondence of strain-energy density functionality between the classical and

* "Classical elasticity" refers to the momentum balance law developed by Cauchy [34].

† As the former is simply a special case of the latter, I will dispense with these unnecessary adjectives as it can lead to confusion and the perception that there are two theories of peridynamics that are somehow disjoint.

peridynamic models. Additionally, for a particular choice of meshfree discretization when combined with a non-ordinary correspondence model, direct connections can be made to classical meshfree methods [5].

Ordinary material models include the isotropic elastic model presented in [58] and have been extended to include plasticity and viscoelastic behavior [36, 37]. They have been proposed as an upscaling of molecular dynamics both for central potential models [50] and more complex models such as the embedded atom model [51]. It's important to note that these are upscaled models that retain essential features of the molecular models, e.g., wave dispersion, while allowing for continuous discretizations such as the Galerkin finite element approximation and can benefit from a massive reduction in degrees of freedom over the full-atom counterparts. The mathematical form of peridynamics allows for it to recover the equations of molecular dynamics as a special case and has therefore been proposed as a multiscale modeling method for atomistic-to-continuum coupling [48, 45] and as a bridging technique between nonlocal and local continuum elasticity [47, 4].

Given the integral nature of the peridynamic conservation of linear momentum equation, an inherent length scale is present in the model. Influence from different physics present in a constitutive material model can be localized via kernel (i.e. attenuation) functions to recover different characteristics of material and geometric length scales [65, 54] present in real materials. Also, owing to the nonlocal formulation, peridynamics can recover virtually all of the weakly and strongly nonlocal models presented in literature including Kröner [26], Eringen [13], and Kunin [27] as special cases. These models are typically recovered by first *restricting the kinematics to smooth fields* such that spatial derivatives can be evaluated or through linearizations of the geometrically nonlinear finite deformation peridynamic theory.

Finally, it's worth mentioning that, given the etymology of the word *, peridynamics strictly refers to a theory of mechanical deformation; however, the usefulness of the mathematics and notation have been extended to nonlocal diffusion processes to develop theories of heat transfer [6] and flow in a porous media [24]. Both recover the classical theories as a special case, but are additionally capable of modeling anomalous diffusion processes such as Lévy motions and recover the fractional-Laplacian models as special cases.

The primary purpose of this chapter is to present a state-of-the-art review of constitutive modeling in peridynamics while presenting the case for the theory as a unification of other generalized theories of elasticity. Secondary, it is a goal that this chapter serve as a reference and standardization of the new and vast terminology that has arisen in the literature devoted to peridynamics. The rest of this chapter is organized as follows. Section 6.2 will derive the peridynamic momentum equation from Hamilton's principle and discuss the generalized kinematic measures in the theory. Section 6.3 will review the isotropic elasticity model presented in [58] and show how the simple constitutive model of [53] arises as a special case. Section 6.4 will present

*The word "peridynamic" is an adjective with etymology as the contraction of the Greek word "peri" meaning near or surrounding and the French word "dyne" which derived from the Ancient Greek "dunamis" meaning force. "Peridynamics" is a noun referring to the peridynamic theory of solid mechanics.

a general elasticity model valid for finite deformations. Section 6.5 will show examples of constitutive correspondence models of both ordinary and non-ordinary type. Section 6.6 will demonstrate extensions of the elastic models to plasticity. Section 6.7 will show further examples of non-ordinary materials. Finally, Section 6.8 will provide conclusions and a discussion of challenges and opportunities.

6.2 Kinematics, momentum conservation, and terminology

Hamilton's principle states that among admissible motions, the actual motion of a body is such that

$$\int_{t_1}^{t_2} (\delta(T - V) + \delta W)\,\mathrm{d}t = 0, \tag{6.1}$$

where δ is the variation operator from the calculus-of-variations, T and V are the kinetic and potential energies, respectively, and δW is the virtual work. t_1 and t_2 are fixed times in which the configuration of the body is assumed to be prescribed. For a bounded region $\mathcal{B}$ defined in the reference configuration at t_1, with a mass density ρ, a peridynamic body has kinetic energy

$$T = \int_{\mathcal{B}} \frac{1}{2}\rho\dot{\mathbf{u}}\cdot\dot{\mathbf{u}}\,\mathrm{d}V,$$

where $\dot{\mathbf{u}}$ is the time-derivative of displacement $\mathbf{u} = \mathbf{x}(\mathbf{X},t) - \mathbf{X}$, i.e., the velocity of a point $\mathbf{x}$ at time t. A standard convention from continuum mechanics literature, that lower case boldface $\mathbf{x}$ indicates a position vector of a material point in the current (i.e. deformed) configuration and an uppercase boldface $\mathbf{X}$ indicates a position vector in the reference configuration, is used. The first integral in (6.1) then evaluates to

$$\begin{aligned}
\int_{t_1}^{t_2} \delta T\,\mathrm{d}t &= \int_{t_1}^{t_2} \frac{\partial T}{\partial \dot{\mathbf{u}}}\cdot\delta\dot{\mathbf{u}}\,\mathrm{d}t, \\
&= \left[\frac{\partial T}{\partial \dot{\mathbf{u}}}\cdot\delta\mathbf{u}\right]_{t_1}^{t_2} - \int_{t_1}^{t_2} \frac{\mathrm{d}}{\mathrm{d}t}\left(\frac{\partial T}{\partial \dot{\mathbf{u}}}\right)\cdot\delta\mathbf{u}\,\mathrm{d}t, \\
&= -\int_{t_1}^{t_2}\int_{\mathcal{B}} \rho\ddot{\mathbf{u}}\cdot\delta\mathbf{u}\,\mathrm{d}V\,\mathrm{d}t,
\end{aligned} \tag{6.2}$$

where integration-by-parts is used between the first and second steps and $\delta\mathbf{u}(t_1) = \delta\mathbf{u}(t_2) = 0$ is used to eliminate the bracketed term between the second and third steps due to the nature of the variations being fixed at $[t_1, t_2]$. For the potential energy, I will assume a stored energy functional of the form

$$\psi = \psi(\underline{\mathbf{Y}}),$$

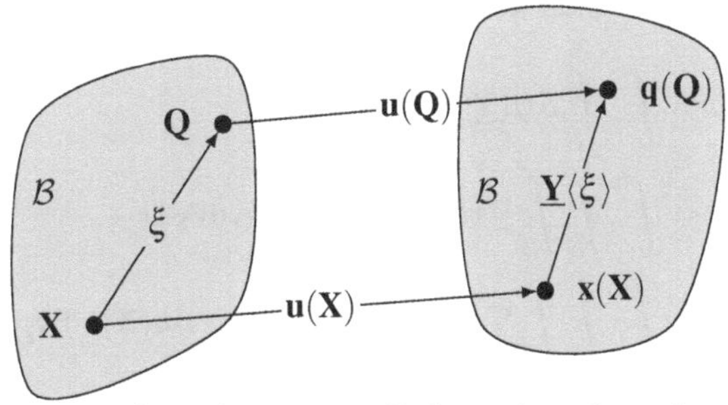

Figure 6.1: Schematic description of peridynamic kinematics.

where $\underline{\mathbf{Y}}$ is the *deformation vector-state*, i.e.

$$\underline{\mathbf{Y}} = \underline{\mathbf{Y}}(\mathbf{X})\langle\xi\rangle = \mathbf{u}(\mathbf{Q}) - \mathbf{u}(\mathbf{X}) + \xi \tag{6.3}$$

which is a generalized tensor-like operator that maps all vectors ξ emanating from $\mathbf{X}$ into their deformed images in the current configuration. With

$$\xi = \mathbf{Q} - \mathbf{X}.$$

$\mathbf{Q}$ is a material point separated from $\mathbf{X}$ by a finite distance in the reference configuration. The $\langle\cdot\rangle$ brackets are used when needed for clarity to indicate which vector the functional operates on. As the displacements $\mathbf{u}(\mathbf{X})$ and $\mathbf{u}(\mathbf{Q})$ can, in general, be nonlinear and discontinuous, the deformation state represents a relaxation in kinematics compared with the classical theory. A schematic of the kinematics described above is shown in Figure 6.1.

The potential energy term in (6.1) can then be evaluated as

$$\begin{aligned}
\int_{t_1}^{t_2} \delta V \,\mathrm{d}t &= \int_{t_1}^{t_2} \int_{\mathcal{B}} \delta \psi(\underline{\mathbf{Y}}(\mathbf{X})\langle \xi \rangle) \mathrm{d}V_{\mathbf{X}} \,\mathrm{d}t \\
&= \int_{t_1}^{t_2} \int_{\mathcal{B}} \int_{\mathcal{B}} \underline{\nabla} \psi(\mathbf{X})\langle \xi \rangle \cdot \delta \underline{\mathbf{Y}} \,\mathrm{d}V_{\mathbf{Q}} \,\mathrm{d}V_{\mathbf{X}} \,\mathrm{d}t \\
&= \int_{t_1}^{t_2} \int_{\mathcal{B}} \int_{\mathcal{B}} \underline{\nabla} \psi(\mathbf{X})\langle \xi \rangle \cdot (\delta \mathbf{u}(\mathbf{Q}) - \delta \mathbf{u}(\mathbf{X})) \,\mathrm{d}V_{\mathbf{Q}} \,\mathrm{d}V_{\mathbf{X}} \,\mathrm{d}t \\
&= \int_{t_1}^{t_2} \left(\int_{\mathcal{B}} \int_{\mathcal{B}} \underline{\nabla} \psi(\mathbf{X})\langle \xi \rangle \cdot \delta \mathbf{u}(\mathbf{Q}) \,\mathrm{d}V_{\mathbf{Q}} \,\mathrm{d}V_{\mathbf{X}} \right. \\
&\quad \left. - \int_{\mathcal{B}} \int_{\mathcal{B}} \underline{\nabla} \psi(\mathbf{X})\langle \xi \rangle \cdot \delta \mathbf{u}(\mathbf{X}) \,\mathrm{d}V_{\mathbf{Q}} \,\mathrm{d}V_{\mathbf{X}} \right) \mathrm{d}t \\
&= \int_{t_1}^{t_2} \left(\int_{\mathcal{B}} \int_{\mathcal{B}} \underline{\nabla} \psi(\mathbf{Q})\langle -\xi \rangle \cdot \delta \mathbf{u}(\mathbf{X}) \,\mathrm{d}V_{\mathbf{Q}} \,\mathrm{d}V_{\mathbf{X}} \right. \qquad (6.4) \\
&\quad \left. - \int_{\mathcal{B}} \int_{\mathcal{B}} \underline{\nabla} \psi(\mathbf{X})\langle \xi \rangle \cdot \delta \mathbf{u}(\mathbf{X}) \,\mathrm{d}V_{\mathbf{Q}} \,\mathrm{d}V_{\mathbf{X}} \right) \mathrm{d}t \\
&= \int_{t_1}^{t_2} \int_{\mathcal{B}} \left(\int_{\mathcal{B}} \underline{\nabla} \psi(\mathbf{Q})\langle -\xi \rangle - \underline{\nabla} \psi(\mathbf{X})\langle \xi \rangle \,\mathrm{d}V_{\mathbf{Q}} \right) \cdot \delta \mathbf{u}(\mathbf{X}) \,\mathrm{d}V_{\mathbf{X}} \,\mathrm{d}t, \qquad (6.5)
\end{aligned}$$

where the dummy variables-of-integration have been added as subscripts to $\mathrm{d}V$ for clarity. The integral over $\mathrm{d}V_{\mathbf{Q}}$ in the second line of (6.5) arises due to properties of the Fréchet derivative of functions of peridynamic states [58] as an inner product on the function spaces of $\underline{\nabla}\psi$ and $\delta\underline{\mathbf{Y}}$. $\underline{\nabla}$ is the Fréchet derivative operator. It is defined for functions of states, if it exists, as the function $\underline{\nabla}\psi(\underline{\mathbf{Y}})$ that satisfies

$$\psi(\underline{\mathbf{Y}}\langle \xi \rangle + \Delta \underline{\mathbf{Y}}\langle \xi \rangle) = \psi(\underline{\mathbf{Y}}\langle \xi \rangle) + \int \underline{\nabla} \psi(\underline{\mathbf{Y}}\langle \xi \rangle) \cdot \Delta \underline{\mathbf{Y}}\langle \xi \rangle \,\mathrm{d}V_{\xi} + \mathcal{O}(|\Delta \underline{\mathbf{Y}}\langle \xi \rangle|^2).$$

Sequential change-of-variable substitutions $\mathbf{X} \leftrightarrow \mathbf{Q}$ followed by a change in order-of-integration was used in the manipulations on the first term between the fourth and fifth of (6.5). Finally, the virtual work term in (6.1) is simply

$$\int_{t_1}^{t_2} \delta W \,\mathrm{d}t = \int_{t_1}^{t_2} \int_{\mathcal{B}} \rho \mathbf{b} \cdot \delta \mathbf{u} \,\mathrm{d}V \,\mathrm{d}t, \qquad (6.6)$$

where $\mathbf{b}$ is an external body force.

Substituting equations (6.2), (6.5), and (6.6) into (6.1) and utilizing the fundamental lemma of the calculus of variations, the final form of the peridynamic momentum equation is

$$\rho \ddot{\mathbf{u}}(\mathbf{X}) = \int_{\mathcal{B}} (\underline{\mathbf{T}}(\mathbf{X})\langle \xi \rangle - \underline{\mathbf{T}}(\mathbf{Q})\langle -\xi \rangle) \,\mathrm{d}V_{\mathbf{Q}} + \rho \mathbf{b}(\mathbf{X}), \qquad (6.7)$$

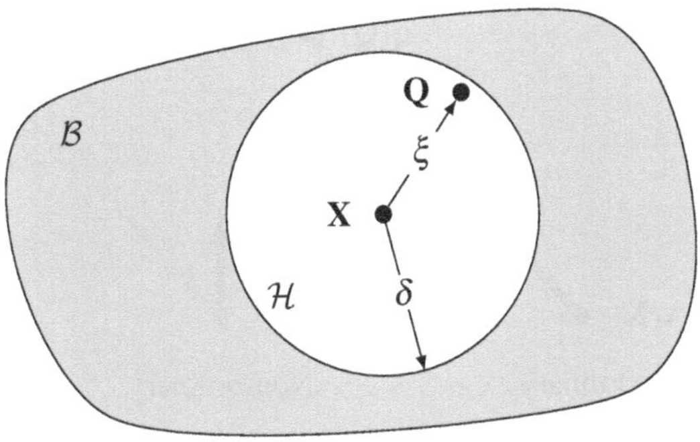

Figure 6.2: Schematic of peridynamic body with a horizon.

where the definition of a *force vector-state* is introduced as

$$\underline{\mathbf{T}}\langle\xi\rangle = \underline{\nabla}\psi(\underline{\mathbf{Y}}\langle\xi\rangle).$$

The particular choice of the strain-energy functional in a hyperelastic setting or otherwise force vector-state defines the constitutive response of the material. It is often physically justified due to the attenuation of nonlocal interactions at a distance to specify a characteristic length scale in a peridynamic model δ such that

$$\underline{\mathbf{T}}\langle\xi\rangle = 0 \quad \forall \quad |\xi| < \delta.$$

In the literature on peridynamics, δ is often called the material's *horizon* and $\mathcal{H} = \{\mathbf{Q}\,|\,\mathbf{X}\in\mathcal{B}, \mathbf{Q}\in\mathcal{B}, |\xi| < \delta\}$ is identified as the *family of* **X**. An example of a peridynamic body with a horizon is shown in Figure 6.2.

There are two broad classifications of constitutive models in peridynamics, *ordinary* materials and *non-ordinary* materials. They are defined based on the directional relationship of the force vector-state to the deformation vector-state. Ordinary materials have force vector-states that are co-linear with the deformation vector-state, whereas non-ordinary materials have force vector-states that are in general not co-linear with the deformation vector-state. This is most easily understood via an illustration as shown in Figure 6.3. Ordinary materials possess the property that they automatically conserve angular momentum [58] without additional restrictions. Non-ordinary materials do not inherently possess this property, but they can be shown to conserve angular momentum with additional considerations. An examination and classification of a variety of both ordinary and non-ordinary material constitutive models will be the primary concern of the next few sections.

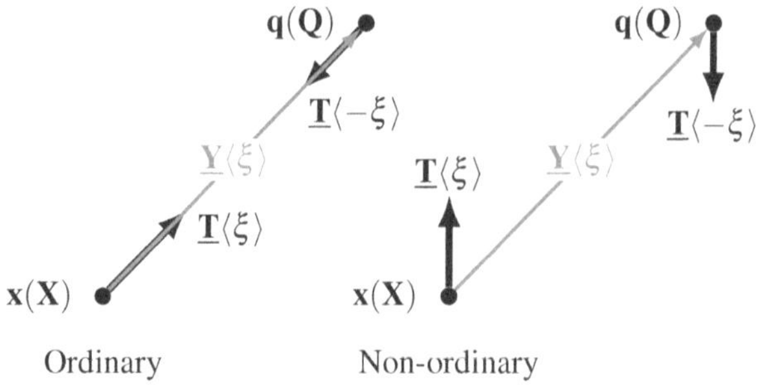

Figure 6.3: Ordinary vs. non-ordinary materials.

6.3 Linear peridynamic isotropic solid

In [58], Silling et al. proposed an ordinary peridynamic material model for linear elastic isotropic solids. Given that ordinary materials have force vector-states that act in the direction of the deformation vector-state (as illustrated in Figure 6.3), they can be expressed as

$$\underline{\nabla}\psi = \underline{\mathbf{T}} = \underline{t}\,\frac{\underline{\mathbf{Y}}}{\sqrt{\underline{\mathbf{Y}}\cdot\underline{\mathbf{Y}}}} = \underline{t}\,\underline{\mathbf{M}}$$

where it should be clear that $\underline{\mathbf{M}}$ is a *unit-state* in the direction of $\underline{\mathbf{Y}}$. $\underline{t}$ is then a peridynamic *force scalar-state* which arises from a change in a multiplicative factor of $\underline{\mathbf{Y}} = \underline{y}\underline{\mathbf{M}}$, i.e. $\underline{y}$ termed the *deformation scalar-state*. If $\underline{\mathbf{Y}}$ is used as the fundamental measure of deformation in composing a strain-energy density functional for an ordinary peridynamic material and it is assumed that $\underline{y}$ and $\underline{\mathbf{M}}$ are independent arguments of a strain-energy density functional ψ, it can be verified [58] that

$$\underline{\nabla}_{\underline{\mathbf{M}}}\psi = \underline{\mathbf{0}} \qquad \text{and} \qquad \underline{\nabla}_{\underline{y}}\psi = \underline{t}$$

where the subscripts on the $\underline{\nabla}$s represent Fréchet differentiation with respect to the subscripted states. Now, using a decomposition of the deformation scalar-state into components involving the net volume change θ and the remainder after separating out the volume change $\underline{e}^d$, analogous to the *spherical* and *deviatoric* components of strain in the classical theory, a strain-energy density functional for an isotropic elastic solid parameterized by horizon δ centered at $\mathbf{X}$ with family $\mathcal{H}$ is

$$\psi(\theta, \underline{e}^d) = \frac{K\theta^2}{2} + \frac{\alpha}{2}\int_{\mathcal{H}} (\underline{\omega}\,\underline{e}^d)\cdot \underline{e}^d \,\mathrm{d}V_\xi, \tag{6.8}$$

where K is the material's bulk modulus, and α is a material property representing the materials resistance to shearing. θ is a measure of dilatation and is defined mathematically as

$$\theta = \frac{3}{m}\int_{\mathcal{H}} (\underline{\omega}\,\underline{x})\cdot \underline{e}\,\mathrm{d}V_\xi,$$

with

$$\underline{x} = \sqrt{\xi \cdot \xi}, \quad \underline{y} = \sqrt{\underline{\mathbf{Y}} \cdot \underline{\mathbf{Y}}}, \quad \text{and} \quad \underline{e} = \underline{y} - \underline{x}.$$

$\underline{\omega}$ is an influence function * whose role will be discussed in the sequel. $\underline{e}$ is a measure of the extension of the vector defined by ξ in the reference configuration. It is defined as the *extension scalar-state*. m is a weighted volume defined by

$$m = \int_{\mathcal{H}} (\underline{\omega}\underline{x}) \cdot \underline{x} \mathrm{d}V_{\xi}.$$

Returning to (6.8) it's now possible to define the remainder of the extension after subtracting away the dilatation component

$$\underline{e}^d = \underline{e} - \frac{\theta \underline{x}}{3}.$$

$\underline{e}^d$ is called the *deviatoric extension scalar-state.*

Assuming that $\underline{\omega}$ is spherical †, then it is possible to develop an analytic expression for α in (6.8) such that the change in elastic strain-energy under a deformation that preserves the net volume of $\mathcal{H}$ is equivalent to the change in elastic strain-energy undergoing an infinitesimal isochoric deformation in the classical theory. First, assume a continuously differentiable displacement field and perform a Taylor expansion of (6.3) about $\mathbf{Q} = \mathbf{X}$

$$\begin{aligned} \underline{\mathbf{Y}}(\mathbf{X})\langle\xi\rangle &\approx (\nabla\mathbf{u} + \mathbf{I})\,\xi + \frac{1}{2}\nabla^2\mathbf{u}\,\xi^2 + \ldots, \\ &\approx \mathbf{F}\,\xi + \frac{1}{2}\nabla^2\mathbf{u}\,\xi^2 + \ldots, \end{aligned} \tag{6.9}$$

where ∇ is the gradient operator taken with respect to $\mathbf{X}$. $\mathbf{F}$ is the deformation gradient, which is the fundamental deformation measure in the classical theory ‡. Under these assumptions, for small deformations, i.e. $|\nabla\mathbf{u}| << 1$, the deformation vector-state is well-approximated by the first term in (6.9), using this to evaluate the extension scalar-state

$$\underline{e} = \sqrt{(\mathbf{F}\,\xi) \cdot (\mathbf{F}\,\xi)} - \sqrt{\xi \cdot \xi} = \frac{\xi \cdot \varepsilon \cdot \xi}{|\xi|},$$

where the expression on the right comes from the classical kinematics for a small deformation of a vector ξ with the small strain tensor $\varepsilon_{ij} = (u_{i,j} + u_{j,i})/2$. Using this expression to evaluate (6.8) for a deformation with no net volume change in $\mathcal{H}$, i.e.

* Also called a attenuation function, kernel function, weight function, etc., in similar contexts in literature.

† This implies that $\mathbf{X}$ is greater than δ from any boundary of the material.

‡ Notice the deformation gradient is a local, linear approximation to $\underline{\mathbf{Y}}$.

$$
\begin{aligned}
&\underline{e} = \underline{e}^d \\
&\psi = \frac{\alpha}{2}\int_{\mathcal{H}} \underline{\omega}\langle\xi\rangle(\underline{e}^d\langle\xi\rangle)^2\,\mathrm{d}V_\xi \\
&\quad = \frac{\alpha}{2}\varepsilon_{ij}^d\varepsilon_{kl}^d\int_{\mathcal{H}} \underline{\omega}\langle\xi\rangle\frac{\xi_i\xi_j\xi_k\xi_l}{|\xi|^2}\,\mathrm{d}V_\xi \\
&\quad = \frac{\alpha}{2}\varepsilon_{ij}^d\varepsilon_{kl}^d\Bigg(\big(\delta_{ij}\delta_{lm}(1-\delta_{jl})+\delta_{ik}\delta_{jl}(1-\delta_{kj})+\delta_{il}\delta_{jk}(1-\delta_{ij})\big)\int_{\mathcal{H}} \underline{\omega}\langle\xi\rangle\frac{\xi_1^2\xi_2^2}{|\xi|^2}\,\mathrm{d}V_\xi \\
&\quad +\delta_{ij}\delta_{jl}\delta_{kj}\int_{\mathcal{H}} \underline{\omega}\langle\xi\rangle\frac{\xi_1^4}{|\xi|^2}\,\mathrm{d}V_\xi\Bigg) \\
&\quad = \frac{\alpha m}{30}\varepsilon_{ij}^d\varepsilon_{kl}^d\left(\delta_{ij}\delta_{kl}(1\quad\delta_{jl})+\delta_{ik}\delta_{jl}(1-\delta_{kj})+\delta_{il}\delta_{jk}(1-\delta_{ij})+3\delta_{ij}\delta_{jl}\delta_{kj}\right) \\
&\quad = \frac{\alpha m}{30}\varepsilon_{ij}^d\varepsilon_{kl}^d\left(\delta_{ij}\delta_{kl}+\delta_{ik}\delta_{jl}+\delta_{il}\delta_{jk}\right) \\
&\quad = \frac{\alpha m}{15}\varepsilon_{ij}^d\varepsilon_{ij}^d,
\end{aligned}
\tag{6.10}
$$

where the symmetry of the integrand has been exploited to evaluate only the nonzero combinations of indices in the integrals as shown in the third line of (6.10). The integration is then carried out over a volume of a sphere with radius * δ. Comparing the final expression in (6.10) with the strain-energy associated with an isochoric deformation from the classical theory, i.e. $\Omega = \mu\,\varepsilon_{ij}^d\varepsilon_{ij}^d$ where μ is the shear modulus, then apparently

$$
\alpha = \frac{15\mu}{m},
\tag{6.11}
$$

under these assumptions.

Evaluating the Fréchet derivative of (6.8) with respect to the arguments gives the force scalar-state for a isotropic peridynamic solid after substitution of (6.11)

$$
\underline{t} = \frac{3K\theta}{m}\underline{\omega}\,\underline{x} + \frac{15\mu}{m}\underline{\omega}\,\underline{e}^d.
\tag{6.12}
$$

This model will produce a change in strain-energy density, where the reference volume is the volume of $\mathcal{H}$, that is equivalent to a change in strain-energy density from the classical theory at a material point undergoing small deformations.

6.3.1 Plane elasticity

Following similar arguments as those above to match changes in strain energies, Le et al. [29] developed models for both plane stress and plane strain two-dimensional elasticity problems based on the linear peridynamic solid model.

*I have overloaded the symbol δ here indicating both a Kronecker delta function and the radius of integration or *peridynamic horizon*. However, the intended use should be clear as the Kronecker delta will always have indicial subscripts.

6.3.1.1 Plane stress

Reconsidering (6.8)

$$\psi(\theta,\underline{e}^d) = \frac{\beta(\gamma\theta)^2}{2} + \frac{\alpha}{2}\int_{\mathcal{H}} (\underline{\omega}\,\underline{e}^d)\cdot\underline{e}^d\,\mathrm{d}V_\xi, \tag{6.13}$$

where α, β and γ are parameters to be determined. Here the product $(\gamma\theta)$ represents the net volume change so that

$$\underline{e}^d = \underline{e} - \frac{(\gamma\theta)\underline{x}}{3}.$$

Evaluating the integral in the second term of (6.13) this time in cylindrical coordinates with $\mathrm{d}V_\xi = tr\mathrm{d}r\mathrm{d}\phi$ where t is a unit thickness, but is included explicitly for unit consistency, with $0 < r \leq \delta$ and $0 < \phi \leq 2\pi$, then following similar arguments from the previous three-dimensional case yields

$$\psi(\theta,\underline{e}^d) = \frac{\beta(\gamma\theta)^2}{2} + \frac{\alpha m}{16}\left(\left(\varepsilon^d_{kk}\right)^2 + 2\varepsilon^d_{ij}\varepsilon^d_{ij},\right); \tag{6.14}$$

however, in the three-dimensional case, the identity $\varepsilon^d_{kk} = 0$ was used as $k = \{1,2,3\}$ for further simplification. In two dimensions $i = j = k = \{1,2\}$ and

$$\varepsilon^d_{11} + \varepsilon^d_{22} + \varepsilon^d_{33} = 0,$$

or

$$\begin{aligned} \varepsilon^d_{11} + \varepsilon^d_{22} &= -\varepsilon^d_{33}, \\ \varepsilon^d_{kk} &= -\varepsilon^d_{33}, \end{aligned} \tag{6.15}$$

for $k = \{1,2\}$. Substituting (6.15) into (6.14) gives

$$\begin{aligned} \psi(\theta,\underline{e}^d) &= \frac{\beta(\gamma\theta)^2}{2} + \frac{\alpha m}{16}\left(\varepsilon^d_{33}\right)^2 + \frac{\alpha m}{8}\varepsilon^d_{ij}\varepsilon^d_{ij}, \\ &= \frac{\beta(\gamma\theta)^2}{2} + \frac{\alpha m}{16}\left(\varepsilon_{33} - \frac{1}{3}\varepsilon_{vol}\right)^2 + \frac{\alpha m}{8}\varepsilon^d_{ij}\varepsilon^d_{ij}, \end{aligned} \tag{6.16}$$

using the standard definition of deviatoric strain with ε_{vol} as the volumetric strain, i.e.

$$\varepsilon_{vol} = \varepsilon_{11} + \varepsilon_{22} + \varepsilon_{33}. \tag{6.17}$$

Using Generalized Hooke's law

$$\varepsilon_{ij} = \frac{1}{2\mu}\sigma_{ij} + \left(\frac{1}{9K} - \frac{1}{6\mu}\right)\sigma_{kk}\delta_{ij},$$

under the assumptions of plane stress $\sigma_{33} = \sigma_{13} = \sigma_{31} = \sigma_{23} = \sigma_{32} = 0$ along with (6.17) to solve for ε_{33} in terms of ε_{vol} gives

$$\varepsilon_{33} = \frac{2\mu - 3K}{6\mu}\varepsilon_{vol}. \tag{6.18}$$

Substituting (6.18) into (6.16) and recognizing for sufficiently small deformations $(\gamma\theta) \approx \varepsilon_{vol}$ then

$$\psi = \left(\frac{\alpha K^2 m}{64\mu^2} + \frac{\beta}{2}\right)\varepsilon_{vol}^2 + \frac{\alpha m}{8}\varepsilon_{ij}^d \varepsilon_{ij}^d. \tag{6.19}$$

It is easily verified that the classical strain-energy density in plane stress [29] is

$$W = \left(\frac{K(2\mu + K)}{4\mu}\right)\varepsilon_{vol}^2 + \mu\,\varepsilon_{ij}^d \varepsilon_{ij}^d. \tag{6.20}$$

Equating coefficients of the strain terms between (6.19) and (6.20) yields two equations that can be solved for the peridynamic parameters in plane stress

$$\alpha = \frac{8\mu}{m}, \qquad \beta = K + \frac{K^2}{4\mu}.$$

Thus far, I have not defined $(\gamma\theta)$ except to indicate that $(\gamma\theta) \to \varepsilon_{vol}$ in the limit of small deformations. Considering the case of a transversely isotropic deformation, where $\varepsilon_{11} = \varepsilon_{22} = \varepsilon_0$ in plane stress, substituting this and (6.18) into (6.17) and solving,

$$\varepsilon_{vol} = \frac{12\mu}{3K + 4\mu}\varepsilon_0.$$

Likewise, the peridynamic deformation vector-state for this deformation is $\underline{\mathbf{Y}} = (1 + \varepsilon_0)\underline{\mathbf{X}}$. Evaluating θ and m as defined in Section 6.3 in cylindrical coordinates results in $\theta = 3\varepsilon_0$, therefore

$$\gamma = \frac{4\mu}{3K + 4\mu}.$$

Following [58] and evaluating a change in ψ to determine the force vector-state (I follow a different approach than [29], leading to an equivalent but slightly simplified result, therefore the details are presented *),

$$
\begin{aligned}
\Delta \psi &= \beta\gamma\theta\underline{\nabla}_{\underline{e}}(\gamma\theta)\bullet\Delta\underline{e}+\alpha\left(\underline{\omega}\,\underline{e}^{d}\right)\bullet\Delta\underline{e}^{d}\\
&= \beta\gamma\theta\underline{\nabla}_{\underline{e}}(\gamma\theta)\bullet\Delta\underline{e}+\alpha\left(\underline{\omega}\,\underline{e}^{d}\right)\bullet\left(\Delta\underline{e}-\frac{\underline{x}}{3}\underline{\nabla}_{\underline{e}}(\gamma\theta)\bullet\Delta\underline{e}\right)\\
&= \left(\left(\beta\gamma\theta-\frac{\alpha}{3}\left(\underline{\omega}\,\underline{e}^{d}\right)\bullet\underline{x}\right)\underline{\nabla}_{\underline{e}}(\gamma\theta)+\alpha\underline{\omega}\,\underline{e}^{d}\right)\bullet\Delta\underline{e}\\
&= \left(\left(\beta\gamma\theta-\frac{\alpha}{3}\left(\underline{\omega}\,\underline{e}-\underline{\omega}\frac{\gamma\theta\underline{x}}{3}\right)\bullet\underline{x}\right)\underline{\nabla}_{\underline{e}}(\gamma\theta)+\alpha\underline{\omega}\,\underline{e}^{d}\right)\bullet\Delta\underline{e}\\
&= \left(\left(\beta\gamma\theta-\frac{\alpha}{3}\left(\frac{m\theta}{3}-\frac{\gamma m\theta}{3}\right)\right)\underline{\nabla}_{\underline{e}}(\gamma\theta)+\alpha\underline{\omega}\,\underline{e}^{d}\right)\bullet\Delta\underline{e}\\
&= \left(\left(\beta\gamma\theta-\frac{(1-\gamma)\alpha m\theta}{9}\right)\underline{\nabla}_{\underline{e}}(\gamma\theta)+\alpha\underline{\omega}\,\underline{e}^{d}\right)\bullet\Delta\underline{e}\\
&= \left(3\left(\beta\gamma-\frac{8\mu(1-\gamma)}{9}\right)\frac{\gamma\theta}{m}\underline{\omega}\,\underline{x}+\alpha\underline{\omega}\,\underline{e}^{d}\right)\bullet\Delta\underline{e}\\
&= \left(\frac{4K\mu}{3K+4\mu}\frac{\theta}{m}\underline{\omega}\,\underline{x}+\frac{8\mu}{m}\underline{\omega}\,\underline{e}^{d}\right)\bullet\Delta\underline{e}\\
&= \underline{t}\bullet\Delta\underline{e},
\end{aligned}
\tag{6.21}
$$

where the notation has been introduced such that given two states $\underline{a}$ and $\underline{b}$

$$
\underline{a}\langle\xi\rangle\bullet\underline{b}\langle\xi\rangle=\int_{\mathcal{H}}\underline{a}\langle\xi\rangle\cdot\underline{b}\langle\xi\rangle\,\mathrm{d}V_{\xi}.
$$

The force scalar-state for a plane stress linear isotropic solid is then

$$
\underline{t}=\frac{4K\mu}{3K+4\mu}\frac{\theta}{m}\underline{\omega}\,\underline{x}+\frac{8\mu}{m}\underline{\omega}\,\underline{e}^{d}.
$$

While following essentially the same derivation procedure, the results presented here are different from [29], owing to the desire to use consistent definitions of θ and m between this plane strain model and the full three-dimensional model presented in Section 6.3 (only the bounds of integration are different as one is two-dimensional and one is three-dimensional). The authors in [29] leave the resulting force scalar-state in a form that is not identically decomposed into co-isotropic and co-deviatoric scalar-states, i.e. $\underline{t}^{i}$ and $\underline{t}^{d}$, respectively. Additionally, the authors use three elastic parameters (only two are independent) in their final result which looks peculiar.

*I would like to acknowledge helpful discussions with Drs. Quang Le and Florin Bobaru regarding this model.

6.3.1.2 Plane strain

For plane strain, we'll return to (6.13) to determine α, β and γ under the assumptions of plane strain. Namely, $\varepsilon_{33} = 0$, so (6.13) becomes

$$\begin{aligned} \psi(\theta, \underline{e}^d) &= \frac{\beta(\gamma\theta)^2}{2} + \frac{\alpha m}{16}\left(-\frac{1}{3}\varepsilon_{vol}\right)^2 + \frac{\alpha m}{8}\varepsilon_{ij}^d\varepsilon_{ij}^d \\ &= \left(\frac{\beta}{2} + \frac{\alpha m}{144}\right)\varepsilon_{vol}^2 + \frac{\alpha m}{8}\varepsilon_{ij}^d\varepsilon_{ij}^d, \end{aligned} \tag{6.22}$$

again, with $(\gamma\theta) \approx \varepsilon_{vol}$ for sufficiently small deformations. The classical strain-energy density for plane strain assumptions [29] is

$$W = \left(\frac{K}{2} + \frac{\mu}{9}\right)\varepsilon_{vol}^2 + \mu\,\varepsilon_{ij}^d\varepsilon_{ij}^d. \tag{6.23}$$

Equating coefficients between (6.22) and (6.23) to solve for α and β gives

$$\alpha = \frac{8\mu}{m}, \qquad \beta = K + \frac{\mu}{9},$$

and consideration of a small transversely isotropic deformation under plane strain assumptions gives

$$\gamma = \frac{2}{3}.$$

Substituting these parameters into the second-to-last step in (6.21) the peridynamic scalar force-state for an isotropic solid under plane strain assumptions becomes

$$\underline{t} = \frac{4(3K - \mu)\,\theta}{9}\frac{}{m}\underline{\omega}\,\underline{x} + \frac{8\mu}{m}\underline{\omega}\,\underline{e}_d.$$

6.3.2 "Bond-based" theories as a special case

Using the relationship of shear modulus to Poisson's ratio and bulk modulus, evaluated at a Poisson's ratio of $\nu = 1/4$

$$\mu = \left[\frac{3K(1-2\nu)}{2(1+\nu)}\right]_{\nu=1/4} = \frac{3K}{5}.$$

Substituting this into (6.12) and simplifying

$$\begin{aligned} \underline{t} &= \frac{3K\theta}{m}\underline{\omega}\,\underline{x} + \frac{9K}{m}\underline{\omega}\left(\underline{e} - \frac{\theta\underline{x}}{3}\right), \\ &= \frac{9K}{m}\underline{\omega}\,\underline{e}. \end{aligned} \tag{6.24}$$

(6.24) is a so-called "bond-based" model because the force vector-state acting on each ξ (recall $\underline{\mathbf{T}}\langle\xi\rangle = \underline{t}\underline{\mathbf{M}}\langle\xi\rangle$) is dependent only on the extension of each independent vector identified by ξ in the reference position. The vector ξ is commonly referred to as a *bond* in the literature on peridynamics. Furthermore, if $\underline{\omega} = 1/|\xi|$ is used and I evaluate m in spherical coordinates with radius δ to get $m = \pi\delta^4$, and substitute these expressions into (6.24) and then into the peridynamic equation of motion, the integrand of (6.7) becomes

$$\begin{aligned}\mathbf{f}(\mathbf{X},\mathbf{Q}) &= \underline{\mathbf{T}}(\mathbf{X})\langle\xi\rangle - \underline{\mathbf{T}}(\mathbf{Q})\langle-\xi\rangle,\\ &= \frac{18K}{\pi\delta^4}\frac{\underline{e}}{|\xi|},\\ &= cs,\end{aligned}$$

with

$$c = \frac{18K}{\pi\delta^4}, \quad \text{and} \quad s = \frac{\underline{e}}{|\xi|},$$

which is identical to the *prototype microelastic brittle* (PMB) material model of [56]. c is known as the *micropotential* and s is the *bond-strain*.

Ganzenmüller et al. [16] recognized that the PMB model is only accurate for peridynamic materials in the bulk of a continuum body (away from the surfaces) because the integrals involved are evaluated in closed-form in polar coordinates. In a real body of interest, especially near surfaces, the values of the micropotential will be inaccurate with respect to classical strain-energies where $\mathcal{H}$ is not spherical. They proposed an *improved prototype microelastic brittle* model which has normalization terms that are computed with numerical quadrature. I will show here that their model is a special case of the LPS model with $\mu = 1/4$, $\underline{\omega} = 1/|\xi|^2$ leaving m to be evaluated by quadrature. Substituting this choice of $\underline{\omega}$ into (6.24) and assuming that m is evaluated with a single-point collocation quadrature rule, the integrand of (8.2) becomes

$$\begin{aligned}\mathbf{f}(\mathbf{X},\mathbf{Q}) &= \underline{\mathbf{T}}(\mathbf{X})\langle\xi\rangle - \underline{\mathbf{T}}(\mathbf{Q})\langle-\xi\rangle,\\ &= \left(\frac{9K}{\sum_{j\in\mathcal{H}_i} V_j} + \frac{9K}{\sum_{k\in\mathcal{H}_j} V_k}\right)\frac{\underline{e}}{|\xi|^2},\\ &= \frac{1}{2}\left(\frac{18K}{\sum_{j\in\mathcal{H}_i} V_j} + \frac{18K}{\sum_{k\in\mathcal{H}_j} V_k}\right)\frac{s}{|\xi|},\\ &= \frac{(c_i + c_j)}{2}\frac{s}{|\xi|},\end{aligned}$$

which is identical to the improved PMB model of [16]. It should be noted that in most computational codes, it's commonplace to evaluate m numerically. In the open source peridynamics code *Peridigm* [44], for example, the *improved PMB* model has been implicitly available since its initial release by setting the Poisson's ratio to $1/4$ and setting $\underline{\omega} = 1/|\xi|^2$ in the linear peridynamic solid material model.

6.3.3 On the role of the influence function

It should be clear from the previous examples that the influence function $\underline{\omega}$ is *part of the constitutive response*, i.e. a change to the influence function is a change to the material model. It's most prevalent use in peridynamic literature includes an *indicator function* $\underline{\omega}\langle\xi\rangle = \underline{\mu}\langle\xi\rangle\underline{\hat{\omega}}\langle\xi\rangle$ where $\underline{\mu}\langle\xi\rangle$ takes on the value of either 0 or 1 depending on some criteria, and $\underline{\hat{\omega}}\langle\xi\rangle$ describes the strength of the force-density as a function of ξ, e.g. brittle fracture can be described with

$$\underline{\omega}\langle\xi\rangle = \frac{\underline{\mu}(s)\langle\xi\rangle}{|\xi|} = \begin{cases} 0 & s \geq \sqrt{\frac{5G_c}{9K\delta}}, \\ 1/|\xi| & \text{otherwise,} \end{cases}$$

for the PMB model described in Section 6.3.2 where G_c is a critical strain-energy release-rate. This is often referred to as a *bond-breakage criterion*. This expression can be derived by considering the total energy required to "break all bonds" that cross a fracture surface of unit area [56]. Similar criteria based on energy considerations have been proposed for more complex constitutive responses [15].

A question related to peridynamics that is often encountered is, "How do you choose the characteristic length-scale, δ?" The question is so common that Bobaru and Hu [7] wrote an entire article attempting to provide an answer, at least in the context of brittle fracture. Using the specific example of crack branching in Homalite-100 they demonstrate that under rapid loading conditions along a crack face [46], neither the onset of crack branching nor the speed of propagation were dependent on the size of δ; however, in loading conditions where stress waves reflecting from sample boundaries interacted with a propagating crack before it branched, then the branching phenomenon was dependent on the size of the δ. They argued that wave dispersion was the cause of this and made the case that the "peridynamic horizon can be understood as an *effective nonlocal interaction* that captures an effective length-scale or nonlocal behavior induced by the material microstructure and/or the type of dynamic loading." See also Chapter 10 in this handbook. One could additionally argue that sample geometry would add additional length-scale considerations to the problem, since it was actually wave *reflections* interacting with the crack tip that caused crack branching behavior to differ.

In the work of Bobaru and Hu, only the characteristic interaction length δ was considered; however, Seleson [49] investigated the role of the *strength of locality* of the influence function and its effects on wave dispersion and dynamic crack propagation while holding the characteristic length δ fixed. They did this by defining a family of influence functions of the form

$$\underline{\omega} = \underline{\mu}(s)\langle\xi\rangle\left(\frac{1}{|\xi| + \delta}\right)^p$$

whereby increasing p had the effect of localizing the influence function. They analytically show that with $p = 20$ (for $K = 1$, $\rho = 1$, and $\delta = 5$) the dispersion relations are very near the linear dispersion relation for the classical (local) wave equation. Interestingly, they show qualitative comparisons for complex fracture patterns that manifest in high-resolution three-dimensional simulations of a hard spherical projectile impacting a brittle disk modeled as a PMB material with the p-dependent influence functions shown above. Vastly different fracture patterns are observed for different values of p. Their results seem to elucidate a better question regarding peridynamic material models, i.e. "How do you determine the influence function in a peridynamic material?" As influence functions could be proposed to localize at any characteristic length-scale δ, the former question is implicitly contained in the later. The answer to this question is an area of research in peridynamic material modeling, but a generic answer would be that the influence function should be determined from a rigorous homogenization of any microstructure and/or microarchitecture that is being excluded at the resolution of the simulation. Weckner and Silling [65] offer a numerical approach to this type of homogenization by matching phonon dispersion curves from *ab initio* lattice-dynamics simulations for single crystal silicon. An approach that starts from classical micro-mechanical arguments and exposes the influence function as a sort of Green's function that arises from a proper homogenization of a composite laminate is presented in [54].

6.3.4 Other elasticity theories

Other elasticity theories have been proposed within the framework of ordinary peridynamic solids. References [3, 1, 2] propose an isotropic linear elastic solid that extends from the linearization of a generic hyperelastic peridynamic constitutive model (discussed in the sequel). Their model curiously has four material constants, three independent and one arbitrary. It is not clear what advantages this model has over the linear peridynamic solid model presented in Section 6.3 or its direct linearization. The works in [23, 18, 42] present models for transversely isotropic material behavior and failure that are useful in the modeling of certain composite laminates.

6.4 Finite Deformations

Examining the special case of a large volumetric compression, or $\underline{\mathbf{Y}} \to \underline{\mathbf{0}}$, where $\underline{\mathbf{0}}$ is the zero vector-state, yields the following limiting cases

$$\begin{aligned} \underline{e} &\to -\underline{x}, \\ \theta &\to -3, \\ \underline{e}^d &\to 0. \end{aligned}$$

Substituting these limiting values into (6.12) gives

$$\underline{t} \to -\frac{9K}{m}\underline{x},$$

which is a constant. For a continuously increasing compression, a monotonically increasing resistance, i.e. force scalar-state, would be expected if material inversion or matter interpenetration is disallowed. The kinematic measure $\underline{\mathbf{Y}}$ is apparently not restrictive enough for the physically observed deformations of continuum solids. * To prevent matter interpenetration, there must be an additional kinematic constraint $\underline{\mathbf{Y}} \neq \underline{\mathbf{0}}$.

To enforce this constraint and deal with the possibility of matter interpenetration for large compressive deformations, Tupek [62] introduced a nonlinear bond-strain measure inspired by the Seth–Hill [52, 21] family of strain tensors. The nonlinear bond-strain measure has the form

$$\underline{\varepsilon}_{(m)}\langle\xi\rangle = \frac{1}{2m}\left(\underline{c}\langle\xi\rangle^m - 1\right),$$

where

$$\underline{c}\langle\xi\rangle = \frac{\underline{\mathbf{Y}}\langle\xi\rangle \cdot \underline{\mathbf{Y}}\langle\xi\rangle}{\xi \cdot \xi},$$

with $m \in \mathbb{R}$. This nonlinear bond-strain measure can then be used to propose an elastic strain-energy density function, e.g.

$$\psi = \frac{\bar{\lambda}}{2}\theta_{(m)}^2 + \bar{\mu}\,\mathrm{tr}_{\underline{\omega}}\left(\underline{\varepsilon}_{(m)}^2\right) \tag{6.25}$$

where $\theta_{(m)} = \mathrm{tr}_{\underline{\omega}}\left(\underline{\varepsilon}_{(m)}\right)$ is a nonlinear measure of the volumetric dilatation, and $\bar{\lambda}$ and $\bar{\mu}$ are the peridynamic Lamé constants. The $\underline{\omega}$-trace operator is defined for a scalar-state as

$$\mathrm{tr}_{\underline{\omega}}\left(\underline{a}\right) = \int_{\mathcal{H}} \underline{\omega}\langle\xi\rangle\,\underline{a}\langle\xi\rangle\,\mathrm{d}V_{\xi}.$$

For consistency with the trace of the identity matrix in three dimensions, the influence function is constrained to satisfy

$$\mathrm{tr}_{\underline{\omega}}\left(\underline{1}\right) = \int_{\mathcal{H}} \underline{\omega}\langle\xi\rangle\,\underline{1}\langle\xi\rangle\,\mathrm{d}V_{\xi} = 3$$

where $\underline{1}\langle\xi\rangle = 1$ is the *identity scalar-state*. Computing the Fréchet derivative of (6.25) gives a force vector-state for the nonlinear elastic solid as

$$\underline{\mathbf{T}}_{(m)}\langle\xi\rangle = \left(\bar{\lambda}\theta_{(m)} + 2\mu\underline{\varepsilon}_{(m)}\langle\xi\rangle\right)\underline{\omega}\langle\xi\rangle\frac{\underline{c}\langle\xi\rangle^{m-1}}{|\xi|^2}\underline{\mathbf{Y}}\langle\xi\rangle.$$

By inspection, any $m \leq 0$ will lead to $\underline{\varepsilon}_{(m)}\langle\xi\rangle \to -\infty$, $\theta_{(m)} \to -\infty$, $\underline{\mathbf{T}}\langle\xi\rangle \to -\infty$, and $\psi \to \infty$ for finite compressions. Or in other words, it will require an infinite amount of strain energy to collapse the material onto itself, and therefore this material model for $m \leq 0$ has the desired kinematic behavior.

*While not restrictive enough in this case, the generalized kinematics contained in $\underline{\mathbf{Y}}$ could prove useful in other settings, e.g. quantum teleportation.

It is also possible to write the strain-energy density in terms of the sum of volumetric and deviatoric components as

$$\psi = \frac{\bar{K}}{2}\theta_{(m)}^2 + \bar{\mu}\,\mathrm{tr}_{\underline{\omega}}\left(\underline{\varepsilon}_{(m)}^d \underline{\varepsilon}_{(m)}^d\right), \tag{6.26}$$

where the $\underline{\omega}$-traceless deviatoric scalar-state $\underline{\varepsilon}_{(m)}^d$ is defined as

$$\underline{\varepsilon}_{(m)}^d = \underline{\varepsilon}_{(m)} - \frac{1}{3}\mathrm{tr}_{\underline{\omega}}\left(\underline{\varepsilon}_{(m)}\right)\underline{1}$$

where $\bar{K} = \bar{\lambda} + \frac{2}{3}\bar{\mu}$. Notice that for the special case of $m = 1/2$, the nonlinear bond-strain is $\underline{\varepsilon}_{(1/2)} = \underline{e}\underline{x}$, where $\underline{e}$ is the extension scalar-state defined in Section 6.3. Using this, and a unnormalized influence function of $\underline{\omega} = \frac{1}{3}\gamma\underline{\hat{\omega}}|\xi|^{-2}$, $\bar{K} = K$, and $\alpha = 5\bar{\mu}/\gamma$ and substituting into (6.26)

$$\begin{aligned}
\psi &= \frac{\bar{K}}{2}\theta_{(1/2)}^2 + \bar{\mu}\,\mathrm{tr}_{\underline{\omega}}\left(\underline{\varepsilon}_{(1/2)}^d \cdot \underline{\varepsilon}_{(1/2)}^d\right), \\
&= \frac{K}{2}\theta_{(1/2)}^2 + \gamma\frac{\alpha}{6}\int_{\mathcal{H}}\frac{3}{\gamma}\underline{\hat{\omega}}|\xi|^2\left(\underline{\varepsilon}_{(1/2)}\langle\xi\rangle - \frac{1}{3}\theta_{(1/2)}\right)^2 \mathrm{d}V_{\xi}, \\
&= \frac{K}{2}\theta^2 + \frac{\alpha}{2}\int_{\mathcal{H}}\underline{\hat{\omega}}\left(\underline{e}\langle\xi\rangle - \frac{\theta\underline{x}}{3}\right)^2 \mathrm{d}V_{\xi}, \\
&= \frac{K}{2}\theta^2 + \frac{\alpha}{2}\int_{\mathcal{H}}\left(\underline{\hat{\omega}}\underline{e}^d\langle\xi\rangle\right)\cdot\underline{e}^d\langle\xi\rangle\,\mathrm{d}V_{\xi}.
\end{aligned} \tag{6.27}$$

Comparing (6.27) to (6.8) shows they are identical, which demonstrates that the nonlinear bond-strain theory is a generalization of the model presented in §6.3. Using arguments similar to those shown in Section 6.3 the following relationships can be developed for the peridynamic constants and their classical analogs

$$\begin{aligned}
\bar{K} &= \bar{\lambda} + \frac{2}{3}\bar{\mu} = K, \\
\bar{\mu} &= \frac{5}{2}\mu, \\
\bar{\lambda} &= \lambda - \mu.
\end{aligned}$$

Briefly, it's easy to show that $\bar{\lambda} = 0$ reduces (6.25) to a single term that is only a function of individual nonlinear bond-strains and implies that $\lambda = \mu$ or that the Poisson's ratio $\nu = 1/4$, as expected for this type of material model.

6.4.1 Invariants of peridynamic scalar-states

Some interesting observations can be made using the formalism of the $\underline{\omega}$-trace operators introduced in the previous section. Rearranging (6.25) slightly,

$$\begin{aligned}\psi &= \frac{\bar{\lambda}}{2}\operatorname{tr}_{\underline{\omega}}\left(\underline{\varepsilon}_{(m)}\right)^2 + \bar{\mu}\operatorname{tr}_{\underline{\omega}}\left(\underline{\varepsilon}^2_{(m)}\right)\\ &= \frac{\bar{\lambda}}{2}\operatorname{tr}_{\underline{\omega}}\left(\underline{\varepsilon}_{(m)}\right)^2 + \bar{\mu}\operatorname{tr}_{\underline{\omega}}\left(\underline{\varepsilon}^2_{(m)}\right) + \bar{\mu}\operatorname{tr}_{\underline{\omega}}\left(\underline{\varepsilon}_{(m)}\right)^2 - \bar{\mu}\operatorname{tr}_{\underline{\omega}}\left(\underline{\varepsilon}_{(m)}\right)^2\\ &= \left(\frac{\bar{\lambda}}{2}+\bar{\mu}\right)\operatorname{tr}_{\underline{\omega}}\left(\underline{\varepsilon}_{(m)}\right)^2 - \bar{\mu}\left(\operatorname{tr}_{\underline{\omega}}\left(\underline{\varepsilon}_{(m)}\right)^2 - \operatorname{tr}_{\underline{\omega}}\left(\underline{\varepsilon}^2_{(m)}\right)\right). \end{aligned} \tag{6.28}$$

In this form, (6.28) is remarkably similar to a hyperelastic strain-energy density functional from classical theory written in terms of a Seth–Hill strain tensor $\mathbf{E}_{(m)}$, i.e.

$$\begin{aligned} W &= \left(\frac{\lambda}{2}+\mu\right)\operatorname{tr}\left(\mathbf{E}_{(m)}\right)^2 - \mu\left(\operatorname{tr}\left(\mathbf{E}_{(m)}\right)^2 - \operatorname{tr}\left(\mathbf{E}^2_{(m)}\right)\right),\\ &= \left(\frac{\lambda}{2}+\mu\right)I_1^2 - 2\mu I_2 \end{aligned}$$

where I_1 and I_2 are the traditional first and second invariants of the second-order strain tensor, $\mathbf{E}_{(m)}$. With this similarity in mind, and postulating the following peridynamic strain-energy density functional

$$\psi = \left(\frac{\bar{\lambda}}{2}+\bar{\mu}\right)\bar{I}_1^2 - 2\bar{\mu}\,\bar{I}_2,$$

with

$$\bar{I}_1 = \operatorname{tr}_{\underline{\omega}}\left(\underline{\varepsilon}_{(m)}\right), \quad \text{and} \quad \bar{I}_2 = \frac{1}{2}\left(\operatorname{tr}_{\underline{\omega}}\left(\underline{\varepsilon}_{(m)}\right)^2 - \operatorname{tr}_{\underline{\omega}}\left(\underline{\varepsilon}^2_{(m)}\right)\right),$$

as first and second invariants (in an infinite dimensional set) of peridynamic scalar-states. It's verified that for a given choice of m, the peridynamic invariants are in-fact the same as the first and second invariants of the Seth–Hill strain, e.g. for $m = 0$ and a uniform finite volume expansion which has the deformation gradient,

$$\mathbf{F} = J^{1/3}\mathbf{R}$$

where $\mathbf{R}$ is a unitary rotation tensor and J is the volume change between the reference and deformed configurations. The first invariant I_1 of $\mathbf{E}_{(0)}$ is then

$$\begin{aligned} I_1 &= \operatorname{tr}\left(\mathbf{E}_{(0)}\right)\\ &= \operatorname{tr}\left(\frac{1}{2}\log\left(\mathbf{F}^{\mathsf{T}}\mathbf{F}\right)\right)\\ &= \operatorname{tr}\left(\frac{1}{2}\log\left(J^{2/3}\mathbf{R}^{\mathsf{T}}\mathbf{R}\right)\right)\\ &= \operatorname{tr}\left(\frac{1}{3}\log(J)\mathbf{I}\right)\\ &= \log(J). \end{aligned} \tag{6.29}$$

For the same deformation, the peridynamic deformation vector-state is $\underline{\mathbf{Y}}\langle\xi\rangle = \mathbf{F}\xi = J^{1/3}\mathbf{R}\xi$, so the nonlinear bond-strain for $m = 0$ becomes

$$\underline{\varepsilon}_{(0)} = \log\left(J^{1/3}\right) = \frac{1}{3}\log(J).$$

Keeping in mind the normalized influence function

$$\bar{I}_1 = \mathrm{tr}_{\underline{\omega}}\left(\underline{\varepsilon}_{(0)}\langle\xi\rangle\right) = \frac{1}{3}\log(J)\int_{\mathcal{H}} \underline{\omega}\langle\xi\rangle\underline{1}\,\mathrm{d}V_{\xi} = \log(J), \tag{6.30}$$

comparing (6.29) to (6.30) completes the verification. In an effort to discover other possible invariants of peridynamic states, let us speculate that the third invariant may have the form

$$\bar{I}_3 = \frac{1}{6}\left(\mathrm{tr}_{\underline{\omega}}(\underline{\varepsilon})^3 - 3\mathrm{tr}_{\underline{\omega}}(\underline{\varepsilon})\,\mathrm{tr}_{\underline{\omega}}\left(\underline{\varepsilon}^2_{(m)}\right) + 2\mathrm{tr}_{\underline{\omega}}\left(\underline{\varepsilon}^3_{(m)}\right)\right), \tag{6.31}$$

based on the traditional third invariant $I_3 = \det(\mathbf{E})$ and the identity

$$\det(\mathbf{E}) = \frac{1}{6}\left(\mathrm{tr}(\mathbf{E})^3 - 3\mathrm{tr}(\mathbf{E})\mathrm{tr}(\mathbf{E}^2) + 2\mathrm{tr}(\mathbf{E}^3)\right). \tag{6.32}$$

It's verified that (6.31) and (6.32) are equivalent under finite volume expansions for a fixed value of m. Given that there are conceptually an infinite set of invariants to peridynamic states, then propose strain-energy functionality of the form

$$\psi = \psi(\bar{I}_1, \bar{I}_2, \bar{I}_3, \ldots),$$

which would have corresponding force vector states

$$\underline{\mathbf{T}}\langle\xi\rangle = \frac{\partial\psi}{\partial\bar{I}_1}\nabla I_1 + \frac{\partial\psi}{\partial\bar{I}_2}\nabla I_2 + \frac{\partial\psi}{\partial\bar{I}_3}\nabla I_3 + \ldots$$

providing immense generality in constitutive modeling within this framework.

6.5 Correspondence models

In the previously discussed models, peridynamic strain energy functionalities were proposed using arbitrary material constants multiplying individual bond-strain measures that are then related to their classical counterparts in the limit of vanishing nonlocality. It is also possible using a *correspondence of arbitrary strain-energy functionality* to develop peridynamic constitutive models that can utilize an arbitrary stress tensor from the classical theory.

There seems to be some inference in the literature that these types of models are always of the *non-ordinary* type of peridynamic material model [9, 64], and often what I will label simply *correspondence models* are truthfully but restrictively called *non-ordinary* material models. I will show in this section that correspondence models can be developed that are either ordinary or non-ordinary, and later in Section 6.7 I will discuss non-ordinary models that do not utilize a classical stress tensor for their constitutive response.

6.5.1 Non-ordinary correspondence models for solid mechanics

Return to the Taylor expansion in (6.9), keeping only the first-order term in ξ and labeling $\bar{\mathbf{F}} \approx \mathbf{F}$ as the approximate deformation gradient that minimizes the error of the Taylor-expansion in a least-squared sense over $\mathcal{H}$, i.e.

$$\begin{aligned} 0 &= \frac{\partial}{\partial \bar{F}_{kl}} \int_{\mathcal{H}} \underline{\omega} \left(\underline{Y}_i - \bar{F}_{ij}\xi_j\right)^2 \mathrm{d}V_\xi \\ &= \int_{\mathcal{H}} \underline{\omega} \left(-\underline{Y}_k \xi_l + \bar{F}_{kj}\xi_j\xi_l\right) \mathrm{d}V_\xi . \end{aligned}$$

Then solving for $\bar{\mathbf{F}}$ gives

$$\bar{\mathbf{F}} = \int_{\mathcal{H}} \underline{\omega} \underline{\mathbf{Y}}\langle\xi\rangle \otimes \xi \mathrm{d}V_\xi \left[\int_{\mathcal{H}} \underline{\omega} \xi \otimes \xi \mathrm{d}V_\xi\right]^{-1} . \tag{6.33}$$

Now assuming a correspondence of strain energy density from the classical theory defined as a function of $\bar{\mathbf{F}}$, $W = W(\bar{\mathbf{F}})$ and the peridynamic strain energy density ψ and investigating a change in strain energy over an increment of the deformation state $\Delta\underline{\mathbf{Y}}$

$$\begin{aligned} \Delta\psi &= \Delta W, \\ \Delta\psi &= \frac{\partial W(\bar{\mathbf{F}})}{\partial \bar{\mathbf{F}}} \Delta\bar{\mathbf{F}}, \\ \int_{\mathcal{B}} \underline{\nabla}_{\mathbf{Y}} \psi \cdot \Delta\underline{\mathbf{Y}} \mathrm{d}V_\xi &= \frac{\partial W(\bar{\mathbf{F}})}{\partial \bar{\mathbf{F}}} \int_{\mathcal{B}} \underline{\nabla}_{\mathbf{Y}} \bar{\mathbf{F}} \cdot \Delta\underline{\mathbf{Y}} \mathrm{d}V_\xi, \\ \int_{\mathcal{B}} \underline{\mathbf{T}} \cdot \Delta\underline{\mathbf{Y}} \mathrm{d}V_\xi &= \mathbf{S}(\bar{\mathbf{F}}) \int_{\mathcal{B}} \underline{\nabla}_{\mathbf{Y}} \bar{\mathbf{F}} \cdot \Delta\underline{\mathbf{Y}} \mathrm{d}V_\xi, \end{aligned}$$

where the definitions of the Fréchet derivative, the force vector-state, and the first Piola-Kirchhoff stress $\mathbf{S}$ have been utilized. Apparently,

$$\begin{aligned} \underline{\mathbf{T}}\langle\xi\rangle &= \mathbf{S}(\bar{\mathbf{F}}) \underline{\nabla}_{\mathbf{Y}} \bar{\mathbf{F}}, \\ &= \underline{\omega} \mathbf{S}(\bar{\mathbf{F}}) \mathbf{K}^{-1} \xi, \end{aligned}$$

with

$$\mathbf{K} = \int_{\mathcal{B}} \xi \otimes \xi \, \mathrm{d}V_\xi .$$

This type of correspondence model was first proposed in [58] and used in applications of plasticity in [64, 14]. It was quickly discovered that the model has several flaws that can manifest in computational implementations. Referring to Figure 6.1 and expanding the deformation state in (6.33)

$$\begin{aligned}\bar{\mathbf{F}} &= \int_{\mathcal{H}} (\underline{\omega}(\mathbf{q}(\mathbf{X}+\xi)-\mathbf{x}(\mathbf{X}))\otimes\xi)\,\mathrm{d}V_\xi \mathbf{K}^{-1},\\ &= \left(\int_{\mathcal{H}} \underline{\omega} q_i(\mathbf{X}+\xi)\xi_k \mathrm{d}V_\xi - x_i(\mathbf{X})\int_{\mathcal{H}} \underline{\omega}\xi_k \mathrm{d}V_\xi\right) K_{kj}^{-1}, \end{aligned} \tag{6.34}$$

assuming a spherical domain of integration $\mathcal{H}$ and influence function $\underline{\omega}$. Then the second term of (6.34) vanishes due to the anti-symmetry of the integrand and

$$\begin{aligned}\bar{\mathbf{F}} &= \int_{\mathcal{H}} \underline{\omega} q_i(\mathbf{X}+\xi)\xi_k \mathrm{d}V_\xi K_{kj}^{-1},\\ &= \int_{\mathcal{H}} \underline{\omega}\mathbf{q}(\mathbf{X}+\xi)\otimes\xi\,\mathrm{d}V_\xi \mathbf{K}^{-1}. \end{aligned} \tag{6.35}$$

(6.35) reveals that the approximate deformation gradient is independent of the deformed position of $\mathbf{X}$. Or in other words, there are an infinite number of positions of $\mathbf{x}$ that result in the same approximate deformation gradient. The problems associated with this shortcoming manifest as zero-energy deformation modes in computational implementations of peridynamics for which correction methods have been devised [32, 9]. A more complete analysis of other deficiencies related to kinematics of this constitutive formulation is presented in [62].

Despite the defects of this constitutive correspondence model, its use allows for the straightforward application of any classical constitutive model into peridynamics without modification. Likewise, the kinematics of large rotation formulations and related issues with objective stress-rates can be handled identically to any classical formulation. It is possible this constitutive correspondence formulation could be improved by using direct higher-order approximations for the deformation gradient such as those proposed in [10] or by direct differentiation of a higher-order displacement field approximation as discussed in [5].

6.5.2 Ordinary correspondence models for solid mechanics

Based on his ideas for nonlinear bond strain measures (presented in Section 6.4) Tupek introduced a tensorial measure of nonlocal strain [63, 62]

$$\bar{\mathbf{E}}_{(m)} = \int_{\mathcal{H}} \underline{\omega}\,\underline{\varepsilon}_{(m)}\langle\xi\rangle \underline{\bar{\mathbf{H}}}\langle\xi\rangle\,\mathrm{d}V_\xi, \tag{6.36}$$

where $\underline{\bar{\mathbf{H}}}$ is the *shape tensor-state* defined as

$$\underline{\bar{\mathbf{H}}}\langle\xi\rangle = \int_{\mathcal{H}} \frac{5}{2}\frac{\xi\otimes\xi}{\xi\cdot\xi} - \frac{1}{2}\underline{1}\,\mathrm{d}V_\xi,$$

and again the influence function is normalized such that $\mathrm{tr}_{\underline{\omega}}(\underline{1}) = 3$. This strain measure (6.36) is equivalent to the classical local infinitesimal strain $\varepsilon_{ij} = (u_{i,j}+u_{j,i})/2$

for any value of m under the standard assumption of small strain, i.e. $||u_{i,j}|| << 1$ [62]. Additionally, I will show that the strain measure exactly recovers the logarithmic strain, i.e. $m = 0$ under pure dilatation. Returning to second-to-last step of (6.29) to read off the evaluation of $\mathbf{E}_{(0)} = 1/3 \log(J)\mathbf{I}$, then verifying the nonlocal strain

$$\begin{aligned}\bar{\mathbf{E}}_{(0)} &= \frac{1}{3}\log(J)\int_{\mathcal{H}} \frac{5}{2}\frac{\xi\otimes\xi}{\xi\cdot\xi} - \frac{1}{2}\underline{1}\,\mathrm{d}V_{\xi},\\ \bar{\mathbf{E}}_{(0)} &= \frac{1}{3}\log(J)\mathbf{I},\\ \bar{\mathbf{E}}_{(0)} &= \mathbf{E}_{(0)},\end{aligned}$$

where the integral has been evaluated in polar coordinates and symmetry arguments exploited similar to (6.10) in the simplification. Tupek goes on to demonstrate other favorable properties of the nonlocal strain tensor regarding correct kinematic behavior in limiting conditions of finite compressions [62].

The force-vector state for this type of constitutive correspondence model can be evaluated by taking the Fréchet derivate of a classical strain energy functional proposed as a function of $\bar{\mathbf{E}}_{(m)}$, i.e.

$$\begin{aligned}\underline{\mathbf{T}}_{(m)}\langle\xi\rangle &= \underline{\nabla}_{\underline{\mathbf{Y}}} W(\bar{\mathbf{E}}_{(m)}),\\ &= \frac{\partial W}{\partial \bar{\mathbf{E}}_{(m)}} \underline{\nabla}_{\underline{\mathbf{Y}}} \bar{\mathbf{E}}_{(m)},\\ &= \left(\mathbf{S}_{(m)} : \underline{\bar{\mathbf{H}}}\langle\xi\rangle\right) \underline{\omega}\,\underline{c}\langle\xi\rangle^{m-1}|\xi|^{-2}\underline{\mathbf{Y}}\langle\xi\rangle\end{aligned}$$

where $\mathbf{S}_{(m)}$ is an appropriate work-conjugate stress measure to $\bar{\mathbf{E}}_{(m)}$. Notice that the force vector-state is aligned with $\underline{\mathbf{Y}}$ making it an *ordinary* material model.

6.6 Plasticity

A simple plasticity model that was proposed in [58] derives from an extension of the strain-energy density functional given in (6.8)

$$\psi(\theta, \underline{e}^d) = \frac{K\theta^2}{2} + \frac{\alpha}{2}\int_{\mathcal{H}} \left(\underline{\omega}\,\underline{e}^d - \underline{\omega}\,\underline{e}^{dp}\right)\cdot\left(\underline{e}^d - \underline{e}^{dp}\right), \tag{6.37}$$

where $\underline{e}^{dp}$ is an internal state variable called the *plastic deviatoric extension scalar-state* that will be defined via an associated flow rule in the sequel. The force vector-state associated with this constitutive model can be found by taking the Fréchet derivative of (6.37) with respect to the deformation-state. The result is an ordinary material model which, when evaluated, gives the force scalar-state, $\underline{t}$, as

$$\begin{aligned}\underline{t} &= \frac{3K\theta}{m}\underline{\omega}\,\underline{x} + \alpha\underline{\omega}(\underline{e}^d - \underline{e}^{dp}),\\ &= \underline{t}^i + \underline{t}^d.\end{aligned} \tag{6.38}$$

In (6.38) the two resulting terms were defined as the *isotropic force scalar-state* and the *deviatoric force scalar-state*, $\underline{t}^i$ and $\underline{t}^d$, respectively.

6.6.1 Yield surface and flow rule

In order to utilize the constitutive relation (6.38), a scalar-valued function, f, called the yield function is used to define the set of allowable deviatoric force scalar-states $E_{\underline{t}^d}$ as

$$E_{\underline{t}^d} = \{\underline{t}^d \in S^d \,|\, f(\underline{t}^d) \leq 0\},$$

where S^d is the set defined by

$$S^d = \{\underline{t}^d \in S \,|\, \int_{\mathcal{H}} \underline{t}^d \cdot \underline{x} \mathrm{d}V_\xi = 0\},$$

with S given as the space of scalar-states. The associated plastic flow rule is defined as

$$\dot{\underline{e}}^{dp} = \dot{\lambda} \underline{\nabla}^d f, \tag{6.39}$$

where $\underline{\nabla}^d f$ is the constrained Fréchet derivative of $f \in S$ on S^d and $\dot{\lambda}$ is the *consistency parameter* or plastic multiplier. A formal definition for the constrained Fréchet derivative is given in [58]. $\underline{\nabla}^d f$ produces functions that have no dilatation.

6.6.2 Loading/unloading and consistency

Following in analogy with classical plasticity theory, the evolution of peridynamic plastic deformations is governed by the consistency parameter $\dot{\lambda}$ and the yield function f.

When deviatoric force scalar-states $\underline{t}^d$ are such that $f(\underline{t}^d) < 0$, then it is assumed that there is no change in the plastic deformation and $\dot{\underline{e}}^{dp} = 0$, i.e. the material is elastic. The condition $f(\underline{t}^d) > 0$ is not allowed and the force scalar-state must be returned to the yield surface via the consistency condition such that $f(\underline{t}^d) = 0$.

The consistency parameter $\dot{\lambda}$ is a non-negative scalar that correlates with the rate of plastic deformation. When deformations are elastic, $\dot{\lambda} = 0$; otherwise, under plastic deformations $\dot{\lambda} > 0$. When taken together, the consistency parameter and the yield function are assumed to satisfy the Kuhn–Tucker complementary conditions

$$\dot{\lambda} \geq 0, \quad f(\underline{t}^d) \leq 0, \quad \dot{\lambda} f(\underline{t}^d) = 0.$$

Combining these conditions gives the consistency condition

$$\dot{\lambda} \dot{f}(t^d) = 0.$$

Using the consistency condition, a value for $\dot{\lambda}$ can be computed as follows

$$\begin{aligned} 0 =& \dot{f}(t^d) \\ =& \int_{\mathcal{H}} \underline{\nabla}^d f \cdot \underline{\dot{t}}^d \, \mathrm{d}V_\xi \\ =& \int_{\mathcal{H}} \underline{\nabla}^d f \cdot \alpha \underline{\omega} (\underline{\dot{e}}^d - \underline{\dot{e}}^{dp}) \, \mathrm{d}V_\xi \\ =& \int_{\mathcal{H}} \underline{\nabla}^d f \cdot \alpha \underline{\omega} (\underline{\dot{e}}^d - \dot{\lambda} \underline{\nabla}^d f) \, \mathrm{d}V_\xi . \end{aligned} \tag{6.40}$$

Equation (6.40) can be solved for $\dot{\lambda}$, which is given as

$$\dot{\lambda} = \frac{\int_{\mathcal{H}} \underline{\nabla}^d f \cdot \underline{\omega} \, \underline{\dot{e}}^d \, \mathrm{d}V_\xi}{\int_{\mathcal{H}} \underline{\nabla}^d f \cdot \underline{\omega} \, \underline{\nabla}^d f \, \mathrm{d}V_\xi} . \tag{6.41}$$

With $\dot{\lambda}$ in hand, the deviatoric plastic extension state can be evaluated via (6.39) and the force scalar-state updated with (6.38). Mitchell [36] developed an efficient backward-Euler implicit return algorithm (i.e. radial return algorithm) for the numerical integration of (6.41) and demonstrated the effectiveness of the model for a perfectly plastic constitutive response. He also developed a method for relating the *peridynamic yield value* to the classical yield stress. Lammi and Vogler extended the work of Mitchell to account for pressure dependence in a model that is analogous to a classical Drucker–Prager plasticity model and also introduced a nonassociative flow rule [28]. Recently Madenci and Oterkus extended peridynamic plasticity theory to include the effects of isotropic work hardening [35].

6.6.3 Discussion

While much of the work to date dealing with plasticity in peridynamics has been performed with the objective of reproducing the types of responses common in classical plasticity modeling, there are tremendous opportunities to develop extremely advanced constitutive responses within this framework. The reason for this is that the set S^d, that defines the peridynamic yield "surface", is of potentially infinite dimension, contrasted with the finite dimensional set of stress components where the classical yield-surface exists. The history-dependent internal state variable $\underline{e}^{dp}$ is evolved on each individual force vector-state associated with every ξ as opposed to each material point. This allows for the evolution of plasticity and material degradation with mind-blowing complexity. Reference [15, c.f. Section 3 and Fig. 4] gives a simple example of how a peridynamic failure criterion can be modified according to the accumulation of plastic work. If the plastic evolution occurred in an anisotropic or even discontinuous fashion as allowed for within this theory, then likewise extremely complex material failure could be predicted.

6.7 Non-ordinary models

It was shown in Section 6.5 that certain constitutive correspondence models result in material models that are of the *non-ordinary* type. However, this non-ordinary nature was a consequence of the formulation and not necessarily by design. Here, non-ordinary material models that intentionally exploit the property that the force vector-states do not align with the deformed position of ξ are explored.

6.7.1 A non-ordinary beam model

Consider the model illustrated in Figure 6.4 in which every vector ξ originating from a point is connected by a rotational spring to its opposite companion $-\xi$ originating from that same point. Defining the deformed angle between these bonds as ϕ, and

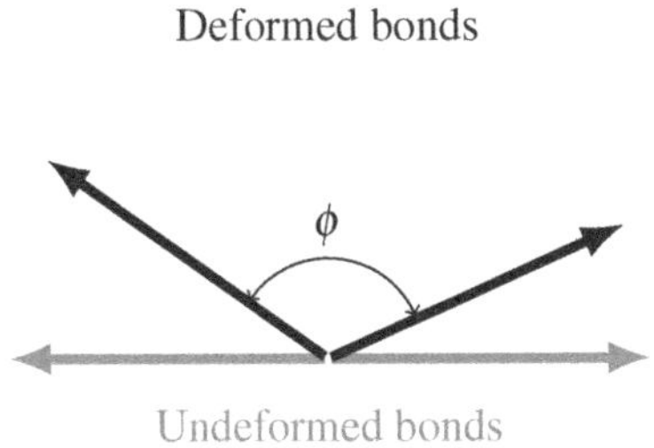

Figure 6.4: A bond-pair model.

choosing the potential energy of that spring to be $w(\xi) = \underline{\omega}\,\alpha(1+\cos(\phi))$ for the pair $\langle \xi, -\xi \rangle$, which is defined as a *bond-pair*, the non-ordinary force vector-state proposed by Silling et al. [58] is recovered by taking the Fréchet derivative.

$$
\begin{aligned}
\underline{\mathbf{T}}\langle \xi \rangle &= \underline{\nabla} w\left(\underline{\mathbf{Y}}\langle \xi \rangle\right) \\
&= \omega \frac{-\alpha}{|\underline{\mathbf{Y}}\langle \xi \rangle|} \frac{\underline{\mathbf{Y}}\langle \xi \rangle}{|\underline{\mathbf{Y}}\langle \xi \rangle|} \times \left(\frac{\underline{\mathbf{Y}}\langle \xi \rangle}{|\underline{\mathbf{Y}}\langle \xi \rangle|} \times \frac{\underline{\mathbf{Y}}\langle -\xi \rangle}{|\underline{\mathbf{Y}}\langle -\xi \rangle|} \right).
\end{aligned}
\tag{6.42}
$$

Though it looks complex, (6.42) indicates a bond force perpendicular to the deformed bond and in the plane containing both the deformed bond and its companion as illustrated in Figure 6.5. The force magnitude is proportional to the sine of the angle between the bonds divided by the length of the deformed bond. This response is consistent with the idea of a rotational spring between bonds as long as the change in angle is small. Because the potential energy and force states are functions of *pairs* of peridynamic bonds, Silling et al. [58] calls this type of material a *bond-pair model*. Other choices for the bond-pair potential function, such as $w = (\pi - \theta)^2$, are also possible, but result in more mathematically complex analysis.

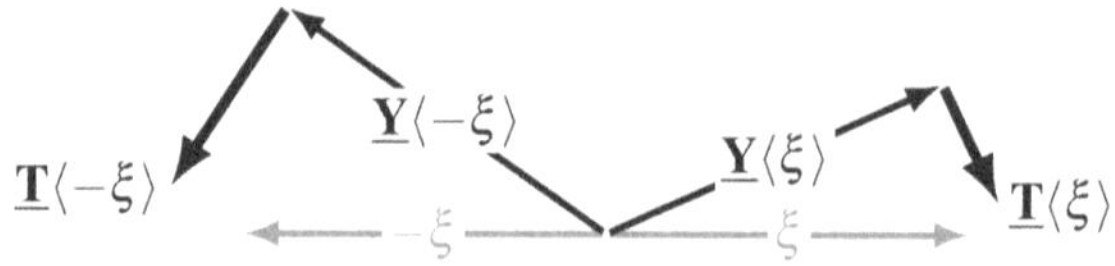

Figure 6.5: Deformation and force vector-states.

To determine an appropriate choice of α, the peridynamic model is assumed to have an equivalent strain energy density to a classical Euler–Bernoulli beam in the *local limit*, i.e. when the nonlocal length scale vanishes. Beginning with the assumptions from Euler beam theory, the length of the beam is much greater than thickness, vertical displacements are small, and rotations are small. For small vertical displacements (i.e. $\sin\phi \approx \phi$)

$$\phi(\underline{\mathbf{Y}}\langle\xi\rangle, \underline{\mathbf{Y}}\langle -\xi\rangle) \approx \pi - \frac{v(x+\xi) - 2v(x) + v(x-\xi)}{\xi}, \tag{6.43}$$

where v is the vertical displacement of material point. Momentarily assuming that v is continuous and using a Taylor series to expand the right-hand side of (6.43)

$$\begin{aligned} \phi(\underline{\mathbf{Y}}\langle\xi\rangle \underline{\mathbf{Y}}\langle -\xi\rangle) &\approx \pi - \xi\frac{\partial^2 v}{\partial x^2} + \mathcal{O}(\xi^3) \\ &\approx \pi - \xi\kappa + \mathcal{O}(\xi^3), \end{aligned} \tag{6.44}$$

with

$$\kappa = \frac{\partial^2 v}{\partial x^2}.$$

Substituting (6.44) into the equation for the strain-energy density of a single bond-pair,

$$\begin{aligned} w(\xi) &= \underline{\omega}\alpha\left(1 + \cos(\phi(\underline{\mathbf{Y}}\langle\xi\rangle, \underline{\mathbf{Y}}\langle -\xi\rangle))\right) \\ &\approx \underline{\omega}\alpha\frac{\xi^2}{2}(\kappa)^2 + \mathcal{O}(\xi^4). \end{aligned}$$

Using a influence function $\underline{\omega} = \underline{\omega}(|\xi|)$ and assuming that the $\underline{\omega}$ plays the role of a localization kernel, i.e. $\underline{\omega} = 0 \;\forall\; \xi > \delta$, the resulting strain-energy density, ψ, for any material point in the peridynamic beam is

$$\psi = \frac{\alpha}{2}\kappa^2 \int_{-\delta}^{\delta} \underline{\omega}\xi^2 \mathrm{d}\xi + \mathcal{O}(\delta^5).$$

Equating ψ with the classical Euler–Bernoulli beam strain-energy density, W, and taking the limit as $\delta \to 0$, one can solve for α

$$\begin{aligned} \lim_{\delta\to 0} \psi &= W, \\ \frac{\alpha}{2} m\kappa^2 &= \frac{EI}{2}\kappa^2, \\ \alpha &= \frac{EI}{m}, \end{aligned}$$

with

$$m = \int_{-\delta}^{\delta} \underline{\omega} \xi^2 \mathrm{d}\xi.$$

While this demonstrates the model's equivalence to a linearly elastic Euler beam, keeping an additional term from the Taylor series approximation of (6.43), a slightly more complex expression for change in angle that is an Euler beam governed by Eringen's model of nonlocal elasticity [13] is recovered. O'Grady and Foster [40] extend this model to show how the influence function can be used to represent plasticity, damage, and beams with different cross sections. They also present a numerical implementation that demonstrates how progressive material failure can be predicted with a model of this type.

6.7.2 A non-ordinary plate/shell model

Extending the beam model of the previous section into planar bond-pairs, O'Grady and Foster [41] present a non-ordinary model for plates that reproduces the behavior of Kirchhoff–Love plate theory in the local limit. A shortcoming of this approach is that there is a limitation to material responses with Poisson ratios of $\nu = 1/3$. To represent more general material responses, the strain energy is decomposed into isotropic and deviatoric components similar to the way the solid ordinary models of Section 6.3 were presented, these models are termed *bond-multiple models*. With the addition of in-plane forces of the PMB type (c.f. Section 6.3.2), a shell model is recovered that is demonstrated to have the correct in-plane tensile response to bending [41]. Crack propagation due to bending is also presented in [41].

6.7.3 Other non-ordinary models

Gerstle et al. [17] presented a peridynamic material model for concrete materials that is essentially non-ordinary in structure. His peridynamic model adds a separate conservation of angular momentum equation that includes the integration of pairwise moment functions. He calls the resulting formulation a *micropolar peridynamic material*. This work was performed before the formalism of peridynamic states existed. Within the formalism of vector-states, the two equations can be accounted for together within the material's constitutive response.

There remains a great deal of research into non-ordinary constitutive modeling to be performed as these models have no analog in classical theory. Perhaps these models can be used as a way to accurately represent the response of meta-materials and other homogenized structures that have unique material responses not accurately represented with the partial differential equations of the classical theory.

6.8 Final Comments

The focus of this chapter was to present the common constitutive models used in peridynamic modeling and provide a few extensions and generalizations. There were several constitutive models that were not discussed in the interest of brevity and because they did not fit broadly into the section topics covered here. For example, Moyer and Miraglia developed a Timeshenko beam theory for peridynamcs [38], and Taylor and Steigmann developed a theory of peridynamic thin-shells. Both of these models are ordinary models in contrast to the beam and shell models that were presented in Sections 6.7.1 and 6.7.2 primarily as motivation for the discussion of non-ordinary material models. We have not discussed in this chapter any of the numerical discretization techniques in peridynamics, which are covered thoroughly elsewhere in the literature [56, 11, 33, 44] and in Chapter 5 in this handbook. Also, no discussion was given on applications, for which there are many references available (e.g., [22, 25, 20, 43]), and which, quite frankly, are the highlight of peridynamics research and usefulness.

Finally, I would like to conclude this chapter with some open topics for future research within the peridynamic constitutive modeling field:

1. Analytic and/or computational homogenization procedures to determine the correct influence function for a given material's composition and microstructure along with the geometry and boundary values of the problem.

2. Finite deformation hyperelastic formulations in analogy with neo-Hookean, Ogden, etc. [39], models from the classical theory.

3. Eulerian force-vector states (i.e. described in the current configuration) and hypoelastic formulations.

4. Relating the regularization that occurs due to the nonlocal interactions in peridynamics plasticity models to strain-gradient and other nonlocal plasticity theories.

5. A multiplicative treatment of the elastic and plastic parts of the deformation vector-state for a more accurate theory of plasticity under finite deformations.

6. Exploiting the infinite dimensional space of functions that the peridynamic yield surface resides in to develop complex plasticity theories that can lead to accurate prediction of ductile fracture.

7. Rigorous constitutive treatment of material interfaces and delamination.

References

[1] A. R. Aguiar. On the determination of a peridynamic constant in a linear constitutive model. *Journal of Elasticity*, 122:27–39, 2016.

[2] A. R. Aguiar. Erratum to: On the Determination of a Peridynamic Constant in a Linear Constitutive Model. *Journal of Elasticity*, 122:41–42, 2016.

[3] A. R. Aguiar, and R. Fosdick. A constitutive model for a linearly elastic peridynamic body. *Mathematics and Mechanics of Solids*, 19(5): 502–523, 2014.

[4] Y. Azdoud, F. Han, and G. Lubineau. The morphing method as a flexible tool for adaptive local/non-local simulation of static fracture. *Computational Mechanics*, 54(3):711–722, 2014.

[5] M. Bessa, J. Foster, T. Belytschko, and W. K. Liu. A meshfree unification: Reproducing kernel peridynamics. *Computational Mechanics*, 53(6):1251–1264, 2014.

[6] F. Bobaru, and M. Duangpanya. The peridynamic formulation for transient heat conduction. *International Journal of Heat and Mass Transfer*, 53(19): 4047–4059, 2010.

[7] F. Bobaru, and W. Hu. The meaning, selection, and use of the peridynamic horizon and its relation to crack branching in brittle materials. *International Journal of Fracture*, 176(2): 215–222, 2012.

[8] F. Bobaru, and G. Zhang. Why do cracks branch? a peridynamic investigation of dynamic brittle fracture. *International Journal of Fracture*, 196(1): 59–98, 2015.

[9] M. Breitenfeld, P. Geubelle, O. Weckner, and S. A. Silling. Non-ordinary state-based peridynamic analysis of stationary crack problems. *Computer Methods in Applied Mechanics and Engineering*, 272: 233–250, 2014.

[10] J.-S. Chen, X. Zhang, and T. Belytschko. An implicit gradient model by a reproducing kernel strain regularization in strain localization problems. *Computer Methods in Applied Mechanics and Engineering*, 193(27): 2827–2844, 2004.

[11] X. Chen, and M. Gunzburger. Continuous and discontinuous finite element methods for a peridynamics model of mechanics. *Computer Methods in Applied Mechanics and Engineering*, 200(9): 1237–1250, 2011.

[12] Q. Du, M. Gunzburger, R. B. Lehoucq, and K. Zhou. Analysis of the volume-constrained peridynamic Navier equation of linear elasticity. *Journal of Elasticity*, 113(2): 193–217, 2013.

[13] A. C. Eringen and D. Edelen. On nonlocal elasticity. *International Journal of Engineering Science*, 10(3): 233–248, 1972.

[14] J. T. Foster, S. A. Silling, and W. Chen. Viscoplasticity using peridynamics. *International Journal for Numerical Methods in Engineering*, 81(10): 1242–1258, 2010.

[15] J. T. Foster, S. A. Silling, and W. Chen. An energy based failure criterion for use with peridynamic states. *International Journal of Multiscale Computational Engineering*, 9(6): 675–688, 2011.

[16] G. C. Ganzenmüller, S. Hiermaier, and M. May. Improvements to the prototype micro-brittle linear elasticity model of peridynamics. In *Meshfree Methods for Partial Differential Equations VIII*, M. Griebel and M. A. Schweitzer (eds), Lecture Notes in Computational Engineering 100, 163–184, 2015.

[17] W. Gerstle, N. Sau, and S. A. Silling. Peridynamic modeling of concrete structures. *Nuclear Engineering and Design*, 237(12-13): 1250–1258, 2007.

[18] M. Ghajari, L. Iannucci, and P. Curtis. A peridynamic material model for the analysis of dynamic crack propagation in orthotropic media. *Computer Methods in Applied Mechanics and Engineering*, 276: 431–452, 2014.

[19] Y. -D. Ha and F. Bobaru. Studies of dynamic crack propagation and crack branching with peridynamics. *International Journal of Fracture*, 162(1-2): 229–224, 2010.

[20] F. Han, Y. Azdoud, and G. Lubineau. Computational modeling of elastic properties of carbon nanotube/polymer composites with interphase regions. part ii: Mechanical modeling. *Computational Materials Science*, 81: 652–661, 2014.

[21] R. Hill. On constitutive inequalities for simple materials. *Journal of the Mechanics and Physics of Solids*, 16(4): 229–242, 1968.

[22] W. Hu, Y. D. Ha, and F. Bobaru. Modeling dynamic fracture and damage in fiber-reinforced composites with peridynamics. *International Journal for Multiscale Computational Engineering*, 9(6): 707–726, 2011.

[23] W. Hu, Y. D. Ha, and F. Bobaru. Peridynamic model for dynamic fracture in unidirectional fiber-reinforced composites. *Computer Methods in Applied Mechanics and Engineering*, 217: 247–261, 2012.

[24] A. Katiyar, J. T. Foster, H. Ouchi, and M. Sharma. A peridynamic formulation of pressure driven convective fluid transport in porous media. *Journal of Computational Physics*, 261: 209–229, 2014.

[25] B. Kilic, A. Agwai, and E. Madenci. Peridynamic theory for progressive damage prediction in center-cracked composite laminates. *Composite Structures*, 90(2): 141–151, 2009.

[26] E. Kröner. Mechanics of Generalized Continua. Springer-Verlag, 1968.

[27] I. A. Kunin. Elastic media with microstructure. I - One-dimensional models. *Springer Series on Solid State Sciences*, 26, Springer–Verlag, 1982.

[28] C. J. Lammi and T. J. Vogler. A nonlocal peridynamic plasticity model for the dynamic flow and fracture of concrete. Report SAND2014-18257, Sandia National Laboratories. 2014

[29] Q. Le, W. Chan, and J. Schwartz. A two-dimensional ordinary, state-based peridynamic model for linearly elastic solids. *International Journal for Numerical Methods in Engineering*, 98(8): 547–561, 2014.

[30] R. B. Lehoucq and S. A. Silling. Force flux and the peridynamic stress tensor. *Journal of the Mechanics and Physics of Solids*, 56(4): 1566–1577, 2008.

[31] R. Lipton. Cohesive dynamics and brittle fracture. *Journal of Elasticity*, 124(2): 143–191, 2016.

[32] D. J. Littlewood. Simulation of dynamic fracture using peridynamics, finite element modeling, and contact. *IMECE Conference Proceedings*, (44465): 209–217, 2010.

[33] D. J. Littlewood. Roadmap for peridynamic software implementation. Report SAND2015-9013, Sandia National Laboratories, 2015.

[34] A. E. H. Love. *A treatise on the mathematical theory of elasticity*. Cambridge University Press, 2013.

[35] E. Madenci and S. Oterkus. Ordinary state-based peridynamics for plastic deformation according to von Mises yield criteria with isotropic hardening. *Journal of the Mechanics and Physics of Solids*, 86: 192–219, 2016.

[36] J. A. Mitchell. A nonlocal, ordinary, state-based plasticity model for peridynamics. Sandia Report, SAND2011-3166, Sandia National Laboratories, 2011.

[37] J. A. Mitchell. A non-local, ordinary-state-based viscoelasticity model for peridynamics. Sandia Report, SAND2011-8064, Sandia National Laboratories, 2011.

[38] E. T. Moyer and M. J. Miraglia. Peridynamic solutions for Timoshenko beams. *Engineering*, 6(06): 304–317, 2014.

[39] R. W. Ogden. *Non-linear elastic deformations*. Courier Corporation, 1997.

[40] J. O'Grady and J. T. Foster. Peridynamic beams: A non-ordinary state-based model. *International Journal of Solids and Structures*, 51(18): 3177–3183, 2014.

[41] J. O'Grady. J. T. Foster. Peridynamic plates and flat shells: A non-ordinary state-based model. *International Journal of Solids and Structures*, 51(25–26): 4572–4579, 2014.

[42] E. Oterkus and E. Madenci. Peridynamic analysis of fiber-reinforced composite materials. *Journal of Mechanics of Materials and Structures*, 7(1): 45–84, 2012.

[43] H. Ouchi, J. York, J., A. Katiyar, J. T. Foster, and M. Sharma. A fully coupled porous flow and geomechanics model for fluid driven cracks: A peridynamics approach. *Computational Mechanics*, 55(3): 561–576, 2015.

[44] M. Parks, D. J. Littlewood, J. A. Mitchell, and S. A. Silling. Peridigm users' guide. Sandia Report, SAND2012-7800, Sandia National Laboratories, 2012.

[45] R. Rahman, J. T. Foster, and A. Haque. A multiscale modeling scheme based on peridynamic theory. *International Journal of Multiscale Computational Engineering*, 12(3): 223–248, 2014.

[46] K. Ravi-Chandar and W. Knauss. An experimental investigation into dynamic fracture: III. On steady-state crack propagation and crack branching. *International Journal of Fracture*, 26(2): 141–154, 1984.

[47] P. Seleson, S. Beneddine, and S. Prudhomme. A force-based coupling scheme for peridynamics and classical elasticity. *Computational Materials Science*, 66: 34–49, 2013.

[48] P. Seleson and M. Gunzburger. Bridging methods for atomistic-to-continuum coupling and their implementation. *Communications in Computational Physics*, 7(4): 831–876, 2010.

[49] P. Seleson and M. L. Parks. On the role of the influence function in the peridynamic theory. *International Journal of Multiscale Computational Engineering*, 9(6): 689–706, 2011.

[50] P. Seleson, M. L. Parks, M. Gunzburger, and R. B. Lehoucq. Peridynamics as an upscaling of molecular dynamics. *Multiscale Modeling & Simulation*, 8(1): 204–227, 2009.

[51] P. Seleson, M.L. Parks, and M. Gunzburger. Peridynamic state-based models and the embedded-atom model. *Communications in Computational Physics*, 15(1): 179–205, 2014.

[52] B. Seth. Generalized strain measure with applications to physical problems. Tech. rep., DTIC Document, 1961.

[53] S. A. Silling. Reformulation of elasticity theory for discontinuities and long-range forces. *Journal of the Mechanics and Physics of Solids*, 48(1): 175–209, 2000.

[54] S. A. Silling. Origin and effect of nonlocality in a composite. *Journal of Mechanics of Materials and Structures*, 9(2): 245–258, 2014.

[55] S. A. Silling and E. Askari. Peridynamic modeling of impact damage. In ASME/JSME 2004 Pressure Vessels and Piping Conference, American Society of Mechanical Engineers: 197–205, 2004.

[56] S. A. Silling and E. Askari. A meshfree method based on the peridynamic model of solid mechanics. *Computers and Structures*, 83(17-18): 1526–1535, 2005.

[57] S. A. Silling and F. Bobaru. Peridynamic modeling of membranes and fibers. *International Journal of Non-Linear Mechanics*, 40(2): 395–409, 2005.

[58] S. A. Silling, M. Epton, O. Weckner, J. Xu, and E. Askari. Peridynamic states and constitutive modeling. *Journal of Elasticity*, 88: 151–184, 2007.

[59] S. A. Silling and R. B. Lehoucq. Convergence of peridynamics to classical elasticity theory. *Journal of Elasticity*, 93(1): 13–37, 2008.

[60] S. A. Silling and R. B. Lehoucq. Peridynamic theory of solid mechanics. *Advances in Applied Mechanics*, 44: 73–168, 2010.

[61] S. A. Silling, D. J. Littlewood, and P. Seleson. Variable horizon in a peridynamic medium. *Journal of Mechanics of Materials and Structures*, 10(5): 591–612, 2015.

[62] M. R. Tupek. Extension of the peridynamic theory of solids for the simulation of materials under extreme loadings. PhD thesis, Massachusetts Institute of Technology, 2014.

[63] M. R. Tupek and R. Radovitzky. An extended constitutive correspondence formulation of peridynamics based on nonlinear bond-strain measures. *Journal of the Mechanics and Physics of Solids*, 65: 82–92, 2014.

[64] T. Warren, S. A. Silling, E. Askari, O. Weckner, M. Epton, and J. Xu. A non-ordinary state-based peridynamic method to model solid material deformation and fracture. *International Journal of Solids and Structures*, 46(5): 1186–1195, 2009.

[65] O. Weckner and S. A. Silling. Determination of nonlocal constitutive equations from phonon dispersion relations. *International Journal for Multiscale Computational Engineering*, 9(6): 623–634, 2011.

Chapter 7

Links between Peridynamic and Atomistic Models

Pablo Seleson

Oak Ridge National Laboratory

Michael L. Parks

Sandia National Laboratories

CONTENTS

7.1 Introduction 179
7.2 Molecular dynamics 181
7.3 A meshfree discretization of peridynamic models 182
7.4 Upscaling molecular dynamics to peridynamics 184
7.4.1 A one-dimensional nonlocal linear springs model * 184
7.4.2 A three-dimensional embedded-atom model † 188
7.5 Computational speedup through upscaling 193
7.6 Concluding remarks 196
References 197

7.1 Introduction

Peridynamics is a nonlocal continuum theory with similarities to molecular dynamics, and has sometimes been referred to as a continuum formulation of molecular dynamics [36]. In an atomistic system described by molecular dynamics, the force on an atom requires the computation of all the forces between that atom and its neighbors. Similarly, the force on a material point in peridynamics depends on the

*First published in P. Seleson, M. L. Parks, M. Gunzburger, and R. B. Lehoucq. Peridynamics as an upscaling of molecular dynamics, *Multiscale Modeling and Simulation*, 8(1): 204–227, 2009. Copyright ©2009 Society for Industrial and Applied Mathematics. Reprinted with permission. All rights reserved.

†First published in P. Seleson, M. L. Parks, and M. Gunzburger. Peridynamic state-based models and the embedded-atom model, *Communications in Computational Physics*, 15(1): 179–205, 2014, published by Global Science Press.

interactions between that point and all other points within its neighborhood [32, 34]. In both cases, the range of interaction is assumed to be finite and is determined by a cutoff radius in molecular dynamics, or by a horizon in peridynamics. An important feature shared between peridynamics and molecular dynamics is that both theories possess pairwise and multibody interaction models. Pairwise interactions restrict the class of feasible deformations in a material. In peridynamics, for instance, the assumption of pairwise interactions results in deformations with a fixed Poisson's ratio of $\nu = 1/4$ in 3D or $\nu = 1/3$ in 2D [32, 34]. The same restrictions have been observed in molecular dynamics [10, 15, 41]. Multibody interactions eliminate these restrictions and allow for a general material response. A main difference between peridynamics and molecular dynamics is that the former is a continuum theory, whereas the latter is a discrete theory. In peridynamics, a modeler can choose a discretization with a specified number of points. In contrast, the number of atoms in a molecular dynamics simulation is determined by the physical system under consideration. As a result, peridynamic simulations can employ a number of discretization points orders of magnitude smaller than the number of atoms in a corresponding fully atomistic simulation, while still preserving characteristics of the molecular dynamics model [30].

The ability to replace an atomistic system with a lower-resolution representation which retains important physical aspects, in a simplified form, of the original system has been the focus of coarse graining. Coarse-graining methods often derive effective potentials between a reduced number of coarse-grained particles to simulate systems at the mesoscale, where fully atomistic simulations are unfeasible and classical continuum models are not suitable. In computational biology, for instance, it is common to group atoms into coarse-grained sites or beads and describe the internal forces of a given system as effective interactions between beads. Examples of coarse-graining approaches in the literature include the elastic network model [38], the MARTINI force field [19], a force-matching approach [13], a reverse Monte Carlo approach [18], and the direct Boltzmann inversion approach [39], among many others. Reviews of coarse-grained models can be found in, e.g., [25, 4]. In contrast to those approaches, we attempt to replace atomistic models by a generalized nonlocal continuum formulation, rather than by a reduced discrete representation. We restrict our discussion to crystalline structures at zero temperature.

The similarities between peridynamics and molecular dynamics motivated various research avenues. Software development to implement peridynamic models within the LAMMPS molecular dynamics code is described in [21, 25]. Upscaling of molecular dynamics models to peridynamic models has been presented in [31, 30]. By upscaling we refer to a *continuous upscaling*, where a discrete model is replaced by a continuum model, which preserves certain characteristics of the discrete one. A statistical coarse-graining approach for molecular dynamics at finite temperatures into peridynamics was proposed in [17], and a statistical mechanical foundation of peridynamics based on ensemble averages was presented in [16]. Attempts to couple molecular dynamics to peridynamics through a hierarchical multiscale modeling framework appeared in [23, 22], where an exchange of information is established between different models in a hierarchy.

The remainder of this chapter is as follows. A brief overview of molecular dynamics is provided in Section 7.2. In Section 7.3, we review a meshfree discretization of peridynamic models, which leads to a particle-based description of peridynamic systems. The central part of this chapter is in Section 7.4, where we derive upscaled peridynamic models for pairwise and multibody atomistic interactions, following [31, 30], capable of preserving important features of molecular dynamics simulations at a greatly reduced computational cost. In Section 7.5, we present discussions about computational savings, limitations, and extensions of upscaled peridynamic models. Concluding remarks are given in Section 7.6.

7.2 Molecular dynamics

Molecular dynamics (MD) provides a means to achieve a detailed microscopic description of systems at the atomistic level. The underlying assumption is that the evolution of a system can be tracked by following the dynamics of its constituent atoms. Due to the fact that general analytical solutions are not feasible for $\mathcal{N}$-body problems, numerical solutions are the only alternative [24].

In MD, a discrete description of matter is assumed and the information about the positions and momenta of all the atoms in a given system is required at all times. The physics describing MD systems is classical. In such systems, relativistic effects are, in general, negligible, and quantum effects can be included as corrections if needed. Following Newton's second law, the governing equation for an atom i in a system is given by

$$m_i \frac{d^2\mathbf{y}_i}{dt^2} = \sum_{j(\neq i)} \mathbf{f}_{ij} + \mathbf{f}_i^{\text{ext}}, \tag{7.1}$$

with m_i the mass of atom i, $\mathbf{y}_i = \mathbf{y}_i(t)$ the position of atom i at some time t, $\mathbf{f}_{ij}$ the force exerted on atom i by atom j, and $\mathbf{f}_i^{\text{ext}}$ an external force applied on atom i. For computational expediency, it is usually assumed that an atom i only interacts with a surrounding atom j if $\|\mathbf{y}_j - \mathbf{y}_i\| \leqslant r_{\text{cut}}$. We refer to r_{cut} as the *cutoff radius*. An illustration of an MD system is given in Figure 7.1. To follow the dynamics of a given MD system, we need to integrate the equations of motion (7.1) for each atom in time, provided initial positions and velocities of all the atoms in the system. Different algorithms for time integration in MD, with varying degrees of accuracy, have been proposed in the literature. These algorithms include: Verlet, Leap Frog, and velocity Verlet, among others [9].

Forces in MD possess the following properties. By Newton's third law,

$$\mathbf{f}_{ji} = -\mathbf{f}_{ij}, \tag{7.2}$$

which represents the *law of action-reaction*. Equation (7.2) ensures conservation of linear momentum [2]. Conservation of angular momentum is satisfied when the force

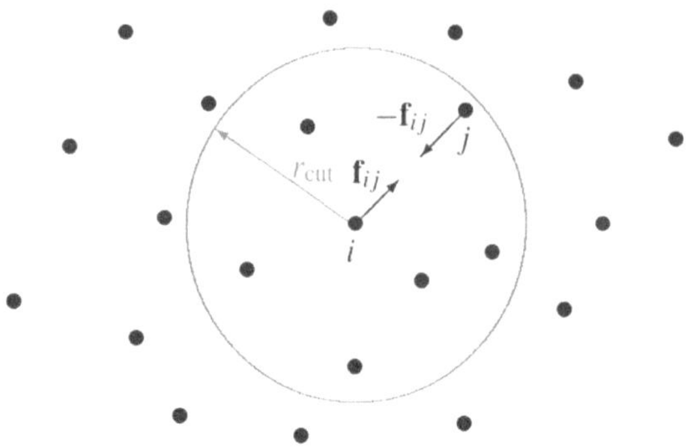

Figure 7.1: Illustration of a system of interacting atoms in molecular dynamics. An atom interacts with other surrounding atoms, closer to it than a given cutoff radius r_{cut}. The force that a neighbor atom j exerts on a given atom i is denoted by $\mathbf{f}_{ij}$. By Newton's third law, $\mathbf{f}_{ji} = -\mathbf{f}_{ij}$.

between each pair of atoms is aligned with their relative position vector in the current configuration [2], i.e.,

$$\mathbf{f}_{ij} \times (\mathbf{y}_j - \mathbf{y}_i) = \mathbf{0}. \tag{7.3}$$

Forces in MD often derive from potential functions. In such cases, the total force on an atom i can be written as

$$\mathbf{F}_i = \sum_{j(\neq i)} \mathbf{f}_{ij} + \mathbf{f}_i^{\text{ext}} = -\frac{\partial U}{\partial \mathbf{y}_i}, \tag{7.4}$$

with $U = U(\mathbf{y}_1, \mathbf{y}_2, \ldots, \mathbf{y}_{N_a})$ the potential energy of a system of N_a atoms. Forces which derive from potentials are conservative. For a more comprehensive review of MD and its implementation, the reader is referred to, e.g., [24, 9].

7.3 A meshfree discretization of peridynamic models

Numerical implementations of (continuum) models in peridynamics (PD) require the use of discretization schemes. Different discretization approaches can be used for this purpose [8]. In this section, we focus on the most commonly used discretization method in PD to date: the meshfree method proposed in [33]. This method has drawn interest in the PD community, due to the simplicity of its implementation and its lower computational cost, in comparison to other discretization approaches. Other more computationally expensive methods, however, may provide higher accuracy and better convergence rates of numerical results.

We begin with the state-based PD equation of motion [34] for a given point $\mathbf{x}$ in a bounded body $\mathcal{B}$ at time $t \geqslant 0$,

$$\rho(\mathbf{x})\frac{\partial^2\mathbf{u}}{\partial t^2}(\mathbf{x},t) = \int_{\mathcal{B}} \left\{\underline{\mathbf{T}}[\mathbf{x},t]\langle\mathbf{x}'-\mathbf{x}\rangle - \underline{\mathbf{T}}[\mathbf{x}',t]\langle\mathbf{x}-\mathbf{x}'\rangle\right\} dV_{\mathbf{x}'} + \mathbf{b}(\mathbf{x},t). \tag{7.5}$$

In (7.5), ρ is the mass density, $\mathbf{u}$ is the displacement field, $\underline{\mathbf{T}}$ is the force vector state field, and $\mathbf{b}$ is the body force density field. Conservation of linear momentum is satisfied in state-based PD models, due to the antisymmetry of the integrand in (7.5), similarly to (7.2). In addition, analogously to (7.3), ordinary materials in state-based PD [34] also satisfy conservation of angular momentum. Conditions to conserve angular momentum for general state-based PD models are provided in [34, Proposition 8.1]. In the state-based PD theory, materials characterized by force vector state fields which derive from a scalar-valued strain energy density function are called elastic materials [34]. We assume that $\underline{\mathbf{T}}[\mathbf{x},t]\langle\xi\rangle = \mathbf{0}$ for $\|\xi\| > \delta$, with δ the PD horizon.

In the meshfree method presented in [33], the body is discretized with a set of points and each point $\mathbf{x}_j$ is assigned a corresponding material region or *cell* with volume V_j. A collocation approach is then used to evaluate the state-based PD equation of motion (7.5) at each discretization point, and the integral is discretized with a composite midpoint quadrature rule. The resulting equation is

$$\rho_i\frac{d^2\mathbf{u}_i}{dt^2} = \sum_{j\in\mathcal{F}_i} \left\{\underline{\mathbf{T}}[\mathbf{x}_i,t]\langle\mathbf{x}_j-\mathbf{x}_i\rangle - \underline{\mathbf{T}}[\mathbf{x}_j,t]\langle\mathbf{x}_i-\mathbf{x}_j\rangle\right\} V_j + \mathbf{b}_i, \tag{7.6}$$

with $\rho_i := \rho(\mathbf{x}_i)$, $\mathbf{u}_i := \mathbf{u}(\mathbf{x}_i,t)$, and $\mathbf{b}_i := \mathbf{b}(\mathbf{x}_i,t)$. The *family of i* is defined as

$$\mathcal{F}_i := \{j \neq i : \mathbf{x}_j \in \mathcal{B}; \|\mathbf{x}_j - \mathbf{x}_i\| \leqslant \delta\}. \tag{7.7}$$

The quadrature weight V_j in (7.6) represents an approximation of the volume of the intersection between the neighborhood of $\mathbf{x}_i$ and the jth cell. In [20, 21], a volume correction factor for V_j was introduced to account for partial overlappings between the neighborhood of a point and neighbor cells. An approximation scheme to improve the above meshfree method with similar correction factors, but using an extension of the family of a point (cf. (7.7)) and a different choice of quadrature points was proposed in [3]. Furthermore, improved schemes based on analytical calculations of areas and centroids of intersections between the neighborhood of a point and neighbor cells for two-dimensional PD models, as well as on the use of influence functions, were presented in [27, 28]. Influence functions are used in PD to weight the contribution of bonds within the neighborhood of a point. Influence functions were introduced in [34] and their role in PD was studied in [29]. Alternative discretization methods for the strong form of the PD equation of motion have been considered in the literature. In [14], a collocation method based on Gaussian integration was proposed. An adaptive integration method based on the trapezoidal rule with error control was introduced in [40]. In [11], different meshing strategies comparing numerical results for cubic meshes and centroidal Voronoi tesellations were investigated.

Equation (7.6) has the same computational structure as (7.1). In fact, both equations represent discrete systems of interacting particles. This has allowed the implementation of PD within existing MD software [20, 21]. It can also be shown that the interaction between two arbitrary small volumes in a PD body resembles the classical interaction between two discrete particles [35]. In contrast to Equation (7.1), Equation (7.6) is based on density quantities because it is derived from a continuum equation.

7.4 Upscaling molccular dynamics to peridynamics

This section presents the upscaling of MD models to PD models. We derive in Section 7.4.1 the upscaling of a one-dimensional nonlocal linear springs MD model, and we explore connections between the MD and upscaled PD models, through a higher-order gradient theory. We demonstrate that upscaled PD models can reproduce nonlocal dispersion effects inherent to the MD model and observe that these effects are lost in the corresponding classical (local) wave equation. Then, in Section7.4.2, we apply the upscaling process to a three-dimensional multibody interaction model of the form of the embedded-atom model. We show that important quantities from the MD model are preserved throughout the upscaling process, and we demonstrate with two-dimensional simulations that upscaled PD models can be used to reproduce dynamics of MD models at a greatly reduced computational cost.

7.4.1 A one-dimensional nonlocal linear springs model *

Let the interactions between atoms in a one-dimensional MD system be given by nonlocal linear springs. Each atom is assumed to have mass m and be connected to N neighbors to the left and to the right. See Figure 7.2 for an illustration. We denote the lattice constant by a and the spring constant by K_j, for the interaction between a given atom and its jth nearest neighbors, given by

$$K_j := \frac{K(N)}{|ja|}, \tag{7.8}$$

with $K(N)$ depending on the number of nearest-neighbor interactions N. The larger the equilibrium distance between two atoms, the weaker their connecting spring. Neglecting external forces, i.e., $\mathbf{f}_i^{\text{ext}} \equiv \mathbf{0}$ in (7.1), the MD equation of motion is

$$m\frac{d^2y_i}{dt^2}(t) = \sum_{\substack{j=-N \\ j\neq 0}}^{N} \frac{K(N)}{|ja|}\Big[y_{i+j}(t) - y_i(t) - ja\Big], \quad i = 1,..,N_x, \tag{7.9}$$

*First published in P. Seleson, M. L. Parks, M. Gunzburger, and R. B. Lehoucq. Peridynamics as an upscaling of molecular dynamics, *Multiscale Modeling and Simulation*, 8(1): 204–227, 2009. Copyright ©2009 Society for Industrial and Applied Mathematics. Reprinted with permission. All rights reserved.

with N_x the number of atoms in the system, and $y_{i+j}(t)$ the current position at time t of an atom with reference position $x_i + ja$.

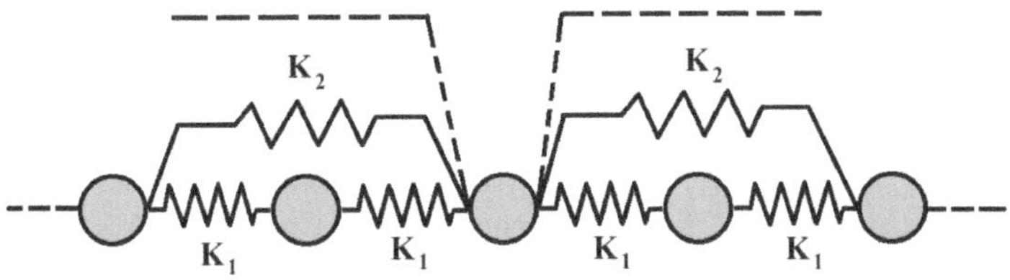

Figure 7.2: Illustration of a one-dimensional nonlocal linear springs molecular dynamics system. The spring constant of the interaction between an atom and its jth nearest neighbors is denoted by K_j.

We present below a higher-order gradient (HOG) partial differential equation (PDE) derived from (7.9), and a corresponding dispersion relation, for comparison with an upscaled PD model. The goal is to demonstrate that we can derive upscaled PD models which, under certain assumptions, possess a governing equation identical to that HOG PDE as well as the same dispersion relation of the MD model. Let $y = y(x,t)$ be a continuous and sufficiently function of x and t, and let the mass density be $\rho = m/a$. Using the *inner expansion* technique [1, 31], we get from (7.9) that

$$\rho \frac{\partial^2 y}{\partial t^2} = \frac{2K(N)}{a^2} \left[\left(\sum_{j=1}^{N} j \right) \frac{a^2}{2!} \frac{\partial^2 y}{\partial x^2} + \left(\sum_{j=1}^{N} j^3 \right) \frac{a^4}{4!} \frac{\partial^4 y}{\partial x^4} + \left(\sum_{j=1}^{N} j^5 \right) \frac{a^6}{6!} \frac{\partial^6 y}{\partial x^6} + \cdots \right],$$

where the dependence on x and t is omitted for brevity. Applying closed-form expressions for the summations and taking $N \gg 1$, keeping only the dominant terms, we obtain the HOG PDE

$$\frac{\partial^2 y}{\partial t^2} = \frac{K_1}{\rho} \left[\frac{\partial^2 y}{\partial x^2} + \frac{(Na)^2}{24} \frac{\partial^4 y}{\partial x^4} + \frac{(Na)^4}{1080} \frac{\partial^6 y}{\partial x^6} + \cdots \right], \tag{7.10}$$

where $K(N) = 2K_1/(N(N+1))$ (cf. (7.15)). The corresponding dispersion relation can be derived by assuming a displacement field of the form of a plane wave. Let $y(x,t) = x + e^{i(kx-\omega t)}$ in (7.9), with ω the angular frequency and k the wavenumber. Then, we obtain

$$\omega^2 = \sum_{j=1}^{N} 2\frac{K(N)}{m} \frac{1}{ja} \Big[1 - \cos(kja) \Big].$$

Using a Taylor expansion in each summation term, we get

$$\omega^2 = 2\frac{K(N)}{m} \left[\left(\sum_{j=1}^{N} j \right) \frac{ak^2}{2!} - \left(\sum_{j=1}^{N} j^3 \right) \frac{a^3k^4}{4!} + \left(\sum_{j=1}^{N} j^5 \right) \frac{a^5k^6}{6!} - \left(\sum_{j=1}^{N} j^7 \right) \frac{a^7k^8}{8!} + \cdots \right].$$

Using an additional Taylor expansion for the function $\sqrt{1+x}$, as well as closed-form expressions for the summations in the limit of $N \gg 1$ while keeping only the dominant terms, we obtain

$$\omega = \sqrt{\frac{K_1}{\rho}} k \left[1 - \frac{1}{48} (kNa)^2 + \frac{17}{69120} (kNa)^4 + \cdots \right], \tag{7.11}$$

where again $K(N) = 2K_1/(N(N+1))$ (cf. Eq. (7.15)). We observe that the limit $kNa \ll 1$ is the limit of long wavelengths relative to the MD interaction range. Therefore, the nonlinear terms in (7.11) reflect contributions of relatively short wavelengths. In fact, the leading terms in (7.10) and (7.11) correspond, respectively, to the classical local wave equation

$$\frac{\partial^2 y}{\partial t^2} = \frac{K_1}{\rho} \frac{\partial^2 y}{\partial x^2} \tag{7.12}$$

and its linear dispersion relation

$$\omega = \sqrt{\frac{K_1}{\rho}} k, \tag{7.13}$$

with $\sqrt{K_1/\rho}$ the wave speed.

We seek an upscaled PD model which can preserve the energy of the given MD model. For that purpose, we look at the energy density per particle in a nonlocal linear springs system with inter-particle distance Δx and with M nearest-neighbor interactions. Assuming continuous functions for the energy density per particle $\varphi(x,t)$ and the current position $y(x,t)$, we can write

$$\varphi(x,t) = \sum_{\substack{j=-M \\ j\neq 0}}^{M} \frac{1}{\Delta x} \frac{1}{2} \frac{K(M)}{|j\Delta x|} \left(y(x+j\Delta x,t) - y(x,t) - j\Delta x \right)^2.$$

For an isotropic deformation, i.e., $y(x,t) - x = \alpha x$ with α a constant, we obtain

$$\varphi(x,t) = \alpha^2 K(M) \frac{M(M+1)}{2}. \tag{7.14}$$

To derive an upscaled PD model, we determine the dependence of the constant $K(M)$ on the number of nearest-neighbor interactions, M. For that purpose, we compare the energy density per particle of a system with M nearest-neighbor interactions with the one of a nearest-neighbor system, under the same isotropic deformation. Using (7.14), we obtain

$$K(M) = \frac{2}{M(M+1)} K_1, \tag{7.15}$$

with $K_1 := K(1)$. This relation allows us to express the corresponding equation of motion (cf.(7.9)) as

$$\rho(x)\frac{\partial^2 y}{\partial t^2}(x,t) = \sum_{\substack{j=-M \\ j\neq 0}}^{M} \frac{2K_1}{|j\Delta x| M(M+1)(\Delta x)^2}\left(y(x+j\Delta x,t) - y(x,t) - j\Delta x\right)\Delta x, \tag{7.16}$$

with $\rho(x) = m/\Delta x$. Taking (7.16) as a discretization of an integral, based on a composite midpoint quadrature rule, we obtain, in the limit of $M \gg 1$ and $\Delta x \ll 1$ (with a fixed product $M\Delta x = \delta$), the upscaled PD model

$$\rho(x)\frac{\partial^2 y}{\partial t^2}(x,t) = \int_{-\delta}^{\delta} \frac{c}{|\xi|}\left(y(x+\xi,t) - y(x,t) - \xi\right)d\xi, \tag{7.17}$$

with $c = 2K_1/\delta^2$. Note that, in this case, the PD horizon δ is the same as the MD interaction range.

We now show that, under certain assumptions, the upscaled PD model (7.17) and the reference MD model (7.9) have the same evolution equation and dispersion relation. Assuming a smooth deformation and using a Taylor expansion for $y(x+\xi,t)$ in (7.17) (as in [7]), we obtain

$$\rho\frac{\partial^2 y}{\partial t^2} = \int_{-\delta}^{\delta} \frac{c}{|\xi|}\left(\frac{\partial y}{\partial x}\xi + \frac{1}{2!}\frac{\partial^2 y}{\partial x^2}\xi^2 + \frac{1}{3!}\frac{\partial^3 y}{\partial x^3}\xi^3 + \frac{1}{4!}\frac{\partial^4 y}{\partial x^4}\xi^4 + \cdots - \xi\right)d\xi,$$

which, after integration, gives

$$\frac{\partial^2 y}{\partial t^2} = \frac{K_1}{\rho}\left[\frac{\partial^2 y}{\partial x^2} + \frac{\delta^2}{24}\frac{\partial^4 y}{\partial x^4} + \frac{\delta^4}{1080}\frac{\partial^6 y}{\partial x^6} + \cdots\right]. \tag{7.18}$$

This equation recovers (7.10), the HOG PDE corresponding to the reference MD model, for $\delta = Na$. The PD dispersion relation is obtained by substituting $y(x,t) = x + e^{i(kx-\omega t)}$ into (7.17), giving

$$\omega^2 = \int_0^{\delta} 2\frac{c}{\rho\xi}(1-\cos(k\xi))d\xi =: \frac{M(k)}{\rho}.$$

Note that, as explained in [32, 36], propagation of real waves in a PD material requires wave speeds to be real for every wave number, i.e., $M(k) > 0, \forall k \neq 0$. Using a Taylor expansion for the function $\cos(x)$, we obtain after integration

$$\omega^2 = 2\frac{c}{\rho}\left(\frac{k^2\delta^2}{2!2} - \frac{k^4\delta^4}{4!4} + \frac{k^6\delta^6}{6!6} + \cdots\right).$$

Using an additional Taylor expansion for the function $\sqrt{1+x}$, we get

$$\omega = \sqrt{\frac{K_1}{\rho}}k\left[1 - \frac{1}{48}(k\delta)^2 + \frac{17}{69120}(k\delta)^4 + \cdots\right]. \tag{7.19}$$

This equation recovers (7.11), the dispersion relation for the reference MD model, for $\delta = Na$. Analogously to the MD model, here the limit $k\delta \ll 1$ corresponds to long wavelengths relative to the PD horizon. As such, the nonlinear terms in (7.19) represent contributions of relatively short wavelengths. The classical dispersion relation is recovered in the limit as $k\delta \to 0$.

We perform one-dimensional simulations of wave propagation to demonstrate that the upscaled PD model (7.17) closely reproduces the dynamics of the reference MD model (7.9). In the simulations below, we choose $\rho = 1$, $K_1 = 1$, and $\delta = 5$. We compare the nonlocal MD and PD models with the classical wave equation (7.12). We let (7.16) represent an exact MD equation of motion provided an atomistic resolution, a discretization of the PD equation of motion, and a suitable approximation for the classical wave equation provided a high spatial resolution with a small enough horizon. We implement a velocity Verlet algorithm for time integration. Let the domain be $\Omega = [0, 1000]$. We assume an initial profile given by $y(x,0) = x + p(x)$, with $p(x)$ a smooth 21st–order polynomial satisfying $p \equiv 0$ on $[0,490] \cup [510,1000]$, $p(500) = 1$, and $p'(x) = p''(x) = \ldots = p^{(10)}(x) = 0$ for $x = 490, 500, 510$ [1, 31]. The time step is chosen to depend on the spatial resolution, based on a stability analysis [31, 33], and is given by

$$\Delta t = \Delta t_0 \sqrt{\frac{\rho}{K(N) \sum_{i=1}^{N} \frac{1}{i}}} \Delta x, \tag{7.20}$$

with $\Delta t_0 = 0.85$ a "safety factor."

In Figure 7.3, we compare: (a) an MD simulation of 4,001 atoms with $N = 20$, (b) a coarse PD simulation with 2,001 discretization points and $N = 10$, and (c) a classical wave equation with 100,001 nodes and $N = 20$. The plots show the entire simulation, where the x-axis represents the reference configuration (the position of each atom/point/node in the original grid) and the y-axis represents time (from top to bottom). The colors * represent mass density, given by $(y')^{-1}$ and numerically approximated at x_i as $(x_{i+1} - x_{i-1})/(y_{i+1} - y_{i-1})$. The numerical results demonstrate that the PD simulation closely reproduces the MD simulation at a reduced computational cost. Estimating the computational cost of a simulation as the product of the number of atoms/points, the number of nearest-neighbor interactions, and the number of time steps, we conclude that the computational cost of the PD simulation, in this case, is about $1/5$ the cost of the MD simulation. Furthermore, the dynamics of the classical local wave equation significantly differ from the dynamics of the MD and PD nonlocal equations. In particular, the numerical dispersion exhibited in PD and MD is lost in the local model.

---

*Please see the electronic version of the handbook, or the original source, for the figures in color.

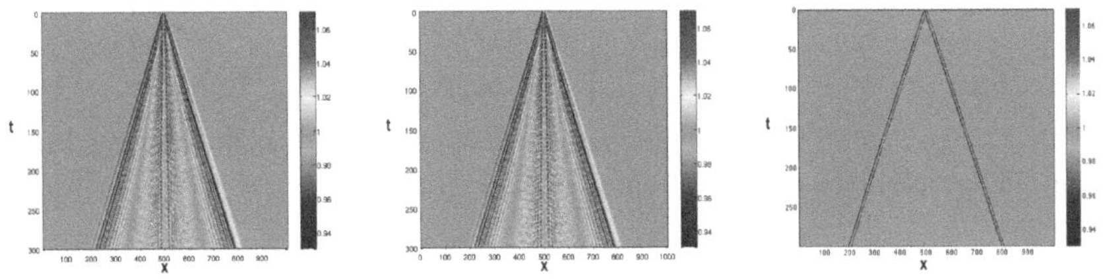

(a) Molecular Dynamics (b) Peridynamics (c) Classical Wave Equation

Figure 7.3: Density evolution for (a) the molecular dynamics model, (b) the peridynamic model, and (c) the classical wave equation.

7.4.2 A three-dimensional embedded-atom model *

The *embedded-atom model* (EAM) [5, 6] is a multibody interaction MD model suitable for the description of metallic systems. In the EAM, each atom is viewed as an impurity embedded in a host composed of other atoms. The potential energy of EAM systems is expressed as a combination of a volume-dependent and pairwise interaction terms, given by

$$E_{\text{pot}} = \sum_i F_i(\rho_{h,i}) + \frac{1}{2} \sum_{\substack{i,j \\ i \neq j}} \phi_{ij}(r_{ij}). \tag{7.21}$$

In (7.21), $\phi_{ij}(r_{ij})$ is a core-core repulsive pair potential between atoms i and j as a function of their distance r_{ij}, and $F_i(\rho_{h,i})$ is the energy to embed atom i in the host electron density $\rho_{h,i}$.

To demonstrate the upscaling process, we assume an embedding function of the form

$$F(\rho) = \frac{1}{2} c \left(\frac{\rho}{\rho_0} - 1 \right)^2, \tag{7.22}$$

with c a constant and ρ_0 a host electron density in a reference configuration. In addition, we take the host electron density at the position of atom i as

$$\rho_{h,i} = \sum_{j(\neq i)} f_{ij}(r_{ij}) \qquad \text{with} \qquad f_{ij}(r) = \chi_{ij}^{\delta} f_0 \frac{\mathrm{v}_a}{(r_{ij}^0)^{p-1}} r, \tag{7.23}$$

where $r_{ij} := \|\mathbf{y}_i - \mathbf{y}_j\|$ with $\mathbf{y}_i = \mathbf{y}_i(t)$ the position of atom i at time t, $r_{ij}^0 := \|\mathbf{x}_j - \mathbf{x}_i\|$ with $\mathbf{x}_i$ the reference position of atom i, f_0 is a constant, v_a is the Wigner–Seitz cell of a corresponding atomistic crystalline structure, p is a model parameter, and χ_{ij}^{δ} is

*First published in P. Seleson, M. L. Parks, and M. Gunzburger. Peridynamic state-based models and the embedded-atom model, *Communications in Computational Physics*, 15(1): 179–205, 2014, published by Global Science Press.

defined as

$$\chi_{ij}^{\delta} := \begin{cases} 1 & r_{ij}^{0} \leqslant \delta, \\ 0 & \text{otherwise,} \end{cases}$$

with $\delta > 0$ a "cutoff radius" in the reference configuration, analogous to a PD horizon. In the reference configuration $r_{ij} = r_{ij}^{0}$, so that

$$\rho_0 = \sum_{j\in\mathcal{F}_i} f_0 \frac{\mathrm{v}_a}{(r_{ij}^{0})^{p-2}}, \tag{7.24}$$

where the *family* of atom i is defined as

$$\mathcal{F}_i := \{ j \neq i \ : \ \mathbf{x}_j \in \mathcal{B} \, ; \, r_{ij}^{0} \leqslant \delta \}.$$

We also assume a pair potential of the form

$$\phi_{ij}(r) = \frac{1}{2}\chi_{ij}^{\delta}\frac{k}{(r_{ij}^{0})^{p}}\left(r - r_{ij}^{0}\right)^{2}, \tag{7.25}$$

with k a constant.

As in Section 7.4.1, we derive an upscaled PD model which preserves the energy of the MD system, under isotropic deformations, i.e., $r_{ij} - r_{ij}^{0} = \alpha r_{ij}^{0}$, $\forall i, j$ with α a constant. The energy density per particle of the EAM (7.21) for a homogeneous system is

$$e_i := \frac{1}{\mathrm{v}_a}F(\rho_{h,i}) + \frac{1}{\mathrm{v}_a}\frac{1}{2}\sum_{j\in\mathcal{F}_i}\phi_{ij}(r_{ij}). \tag{7.26}$$

Under isotropic deformations,

$$\frac{\rho_{h,i}}{\rho_0} = \frac{\displaystyle\sum_{j\in\mathcal{F}_i} f_0 \frac{\mathrm{v}_a}{(r_{ij}^{0})^{p-1}} r_{ij}}{\displaystyle\sum_{j\in\mathcal{F}_i} f_0 \frac{\mathrm{v}_a}{(r_{ij}^{0})^{p-2}}} = \frac{\displaystyle f_0 \mathrm{v}_a \sum_{j\in\mathcal{F}_i} \frac{1}{(r_{ij}^{0})^{p-2}}(1+\alpha)}{\displaystyle f_0 \mathrm{v}_a \sum_{j\in\mathcal{F}_i} \frac{1}{(r_{ij}^{0})^{p-2}}} = 1+\alpha.$$

Therefore, under isotropic deformations, the first term on the right-hand side of (7.26),

$$\frac{1}{\mathrm{v}_a}F(\rho_{h,i}) = \frac{1}{2}\frac{c}{\mathrm{v}_a}\left(\frac{\rho_{h,i}}{\rho_0} - 1\right)^{2}, \tag{7.27}$$

is preserved under a change in v_a (as done in the upscaling process) provided that $\bar{c} := c/\mathrm{v}_a$ is constant. The embedding energy function is then written as

$$F(\rho) = \frac{1}{2}\bar{c}\mathrm{v}_a\left(\frac{\rho}{\rho_0} - 1\right)^{2}.$$

Under the same isotropic deformation, the second term on the right-hand side of (7.26),

$$\frac{1}{\mathrm{v}_a}\frac{1}{2}\sum_{j\in\mathcal{F}_i}\phi_{ij}(r_{ij}) = \frac{1}{\mathrm{v}_a}\frac{1}{2}\sum_{j\in\mathcal{F}_i}\frac{1}{2}\frac{k}{(r_{ij}^{0})^{p}}\left(r_{ij} - r_{ij}^{0}\right)^{2} = \frac{1}{4}\frac{k}{\mathrm{v}_a}\sum_{j\in\mathcal{F}_i}\frac{1}{(r_{ij}^{0})^{p-2}}\alpha^{2} = \frac{1}{4}\bar{k}\alpha^{2},$$

is also preserved in the upscaling process provided that $\bar{k} := k\rho_0/f_0 \mathrm{v}_a^2$ is constant. The pairwise potential is then written as

$$\phi_{ij}(r) = \frac{1}{2}\chi_{ij}^{\delta}\frac{\bar{k}f_0}{\rho_0}\frac{\mathrm{v}_a^2}{(r_{ij}^0)^p}\left(r - r_{ij}^0\right)^2.$$

The equation of motion for this EAM can be expressed as (see [30])

$$m_i\frac{d^2\mathbf{y}_i}{dt^2} = \sum_{j\in\mathcal{F}_i}\left\{\left[\frac{\bar{c}}{\rho_0}\left(\frac{\rho_{h,i}}{\rho_0}-1\right)+\frac{\bar{c}}{\rho_0}\left(\frac{\rho_{h,j}}{\rho_0}-1\right)\right]\frac{f_0\mathrm{v}_a^2}{(r_{ij}^0)^{p-1}}+\frac{\bar{k}f_0}{\rho_0}\frac{\mathrm{v}_a^2}{(r_{ij}^0)^p}\left(r_{ij}-r_{ij}^0\right)\right\}$$
$$\times\frac{\mathbf{y}_j-\mathbf{y}_i}{\|\mathbf{y}_j-\mathbf{y}_i\|}+\mathbf{f}_i^{\text{ext}}, \tag{7.28}$$

with m_i the mass of atom i and $\mathbf{f}_i^{\text{ext}} = \mathbf{f}_i^{\text{ext}}(t)$ an external force on atom i. We divide (7.28) by v_a, and we define $\rho_i := m_i/\mathrm{v}_a$ and $\mathbf{b}_i := \mathbf{f}_i^{\text{ext}}/\mathrm{v}_a$. Taking the summation in (7.28) as a discretization of an integral, based on a composite midpoint quadrature rule, we obtain

$$\rho\frac{\partial^2\mathbf{y}}{\partial t^2} = \int_{\mathcal{H}(\mathbf{x},\delta)}\left\{\left[\frac{\bar{c}f_0}{\rho_0}\left(\frac{\rho_h}{\rho_0}-1\right)+\frac{\bar{c}f_0}{\rho_0}\left(\frac{\hat{\rho}_h}{\rho_0}-1\right)\right]\frac{1}{\|\hat{\mathbf{x}}-\mathbf{x}\|^{p-1}}\right.$$
$$\left.+\frac{\bar{k}f_0}{\rho_0}\frac{1}{\|\hat{\mathbf{x}}-\mathbf{x}\|^p}\left(\|\hat{\mathbf{y}}-\mathbf{y}\|-\|\hat{\mathbf{x}}-\mathbf{x}\|\right)\right\}\frac{\hat{\mathbf{y}}-\mathbf{y}}{\|\hat{\mathbf{y}}-\mathbf{y}\|}dV_{\hat{\mathbf{x}}}+\mathbf{b}, \tag{7.29}$$

with $\rho = \rho(\mathbf{x})$, $\mathbf{y} = \mathbf{y}(\mathbf{x},t)$, $\hat{\mathbf{y}} := \mathbf{y}(\hat{\mathbf{x}},t)$, $\rho_h = \rho_h(\mathbf{x},t)$, $\hat{\rho}_h := \rho_h(\hat{\mathbf{x}},t)$, $\mathbf{b} = \mathbf{b}(\mathbf{x},t)$, and

$$\rho_{h,i} = \sum_{j\in\mathcal{F}_i}\frac{f_0\mathrm{v}_a}{(r_{ij}^0)^{p-1}}r_{ij} \approx f_0\int_{\mathcal{H}(\mathbf{x},\delta)}\frac{1}{\|\check{\mathbf{x}}-\mathbf{x}\|^{p-1}}\|\check{\mathbf{y}}-\mathbf{y}\|dV_{\check{\mathbf{x}}} =: \rho_h(\mathbf{x},t), \tag{7.30}$$

where $\check{\mathbf{y}} := \mathbf{y}(\check{\mathbf{x}},t)$. The integrations in (7.29) and (7.30) are over the neighborhood of $\mathbf{x}$,

$$\mathcal{H}(\mathbf{x},\delta) := \{\hat{\mathbf{x}}\in\mathcal{B} : \|\hat{\mathbf{x}}-\mathbf{x}\| \leqslant \delta\}. \tag{7.31}$$

We demonstrate that the upscaled PD model (7.29) has the functional form of a *linear peridynamic solid* (LPS) model [34]. For simplicity, we demonstrate this connection for points in the bulk of a body. Let an influence function be $\underline{\omega}\langle\xi\rangle = 1/\|\xi\|^p$. Using the notation $\xi := \check{\mathbf{x}} - \mathbf{x}$ and $\eta := \mathbf{u}(\check{\mathbf{x}},t) - \mathbf{u}(\mathbf{x},t)$, we can write (cf. (7.30))

$$\rho_h(\mathbf{x},t) = f_0\int_{\mathcal{H}(\mathbf{0},\delta)}\underline{\omega}\langle\xi\rangle\|\xi\|\|\xi+\eta\|dV_\xi,$$

$$\rho_0 = f_0\int_{\mathcal{H}(\mathbf{0},\delta)}\underline{\omega}\langle\xi\rangle\|\xi\|^2dV_\xi = f_0m,$$

with m the PD weighted volume [34]. Using the definition of the dilatation $\theta[\mathbf{x},t]$ [34], we obtain

$$\left(\frac{\rho_h(\mathbf{x},t)}{\rho_0}-1\right) = \frac{\theta[\mathbf{x},t]}{3}.$$

As a consequence, we can express the upscaled PD equation of motion (7.29) as

$$\rho(\mathbf{x})\frac{\partial^2\mathbf{u}}{\partial t^2}(\mathbf{x},t) = \int_{\mathcal{H}(\mathbf{0},\delta)} \left\{ \frac{\bar{c}}{3m}\left(\theta[\mathbf{x},t]+\theta[\mathbf{x}+\xi,t]\right)\underline{\omega}\langle\xi\rangle\|\xi\| + \frac{\bar{k}}{m}\underline{\omega}\langle\xi\rangle\underline{\mathbf{e}}[\mathbf{x},t]\langle\xi\rangle \right\} \times \frac{\xi+\eta}{\|\xi+\eta\|}dV_{\xi} + \mathbf{b}(\mathbf{x},t), \tag{7.32}$$

with $\underline{\mathbf{e}}[\mathbf{x},t]$ the extension scalar state [34]. Equation (7.32) is identical to the LPS equation of motion. This can be observed by defining two elastic moduli:

$$G = \frac{\bar{k}}{30} \quad \text{and} \quad K = \frac{1}{9}\left(\bar{c}+\frac{1}{2}\bar{k}\right). \tag{7.33}$$

As in Section 7.4.1, we run wave propagation simulations to compare the upscaled PD model (7.29) with the reference MD EAM (7.28). The domain is chosen as a two-dimensional plate $\mathcal{B} = (-0.5,0.5)\times(-0.5,0.5)$. We assume atoms are distributed on a square lattice with lattice constant a, and the number of nearest-neighbor interactions in the MD model is chosen as $N = 12$. Accordingly, the upscaled PD model is discretized with square grids with grid spacing $h = a, 2a, 4a$ in PDLAMMPS [20, 21], taking advantage of the connection between the upscaled PD model and the LPS model, and the PD horizon is taken as $\delta = 12a$. The initial condition for both the MD and upscaled PD models is constructed in a similar way to the one in Section 7.4.1. We take $p(r)$ as a smooth 21st-order piecewise polynomial such that $p \equiv 0$ for $r \geqslant 0.2$, $p(0) = 1e-4$, and $p'(r) = p''(r) = ... = p^{(10)}(r) = 0$ for $r = 0, 0.2$. Given an atom/point with reference position $\mathbf{x}$, its initial position is given by $\mathbf{y} = \mathbf{x} + p(r)\hat{\mathbf{r}}$ with $\mathbf{r} = \mathbf{x} - \mathbf{x}_0$ and $\mathbf{x}_0$ the center of the plate, $r := \|\mathbf{r}\|$, and $\hat{\mathbf{r}} := \mathbf{r}/r$; see Figure 7.4. We evolve the simulations in time using the velocity-Verlet algorithm with final time T. The time step in the PD simulations is adapted with the spatial discretization, and taken as $\Delta t = \Delta t_0 h/a$, where Δt_0 is the time step in the MD simulation. We assume no external forces, i.e., $\mathbf{f}_i^{\text{ext}} \equiv \mathbf{0}$ in (7.28) and $\mathbf{b} \equiv \mathbf{0}$ in (7.29). In Table 7.1, we present the parameters used in the simulations.

Table 7.1: Parameters for the MD EAM and upscaled PD simulations.

ρ	$\bar{k}$	$\bar{c}$	f_0/ρ_0	a	Δt_0	T	p
8,960	$1.44e12$	$5.4e11$	1	$1.25e-3$	$1.25e-9$	$2.5e-5$	1

In Figure 7.5, we compare the MD potential energy (a) and the upscaled PD strain energy, for different grid spacings (b,c,d), at the final time T. In Figure 7.6(a), we compare the radial displacements relative to the center of the plate, at the final time T, for atoms/points with reference position along the x-axis. The dynamics of the MD system are preserved by upscaled PD simulations, up to discretization errors, at

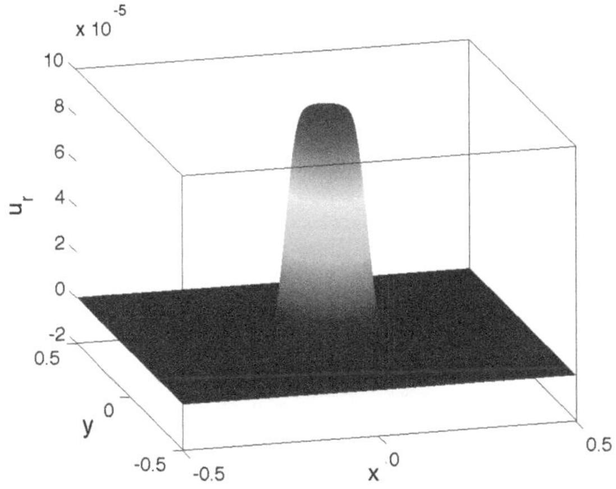

Figure 7.4: Radial displacement profile assigned to the initial configuration for the two-dimensional simulations in Figure 7.5.

a greatly reduced computational cost, as demonstrated in Table 7.2. An upscaled PD simulation with $h = 4a$ achieves a speedup of about 10^3 and a reduction of memory usage per processor of about 95%, relative to the MD simulation. All computations were run on 64 cores of Sandia's Red Sky supercomputer. To show that the energy is preserved throughout the upscaling process, we compare in Figure 7.6(b) the kinetic energy (black dotted lines), the potential/strain energy (blue dashed lines *), and the total energy (red solid lines) between the MD and coarse upscaled PD simulations.

Table 7.2: Comparison between the computational cost of the molecular dynamics (MD) simulation and the coarse upscaled peridynamic (PD) simulations presented in Figure 7.5.

	MD	PD ($h = 2a$)	PD ($h = 4a$)
Number of atoms/points	640,000	160,000	40,000
Number of bonds	296,960,000	17,920,000	1,120,000
Memory usage/processor (Mbytes)	151.63	19.92	7.33
Total time (seconds)	14,564.30	446.45	14.48
Speedup	1	32.62	1,005.82

*For the color figures please see the electronic version of the handbook or the original article where these appeared first.

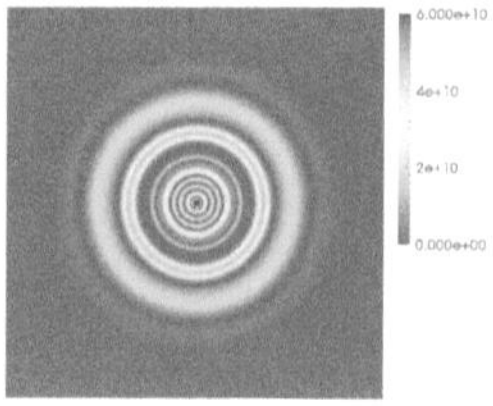

(a) MD

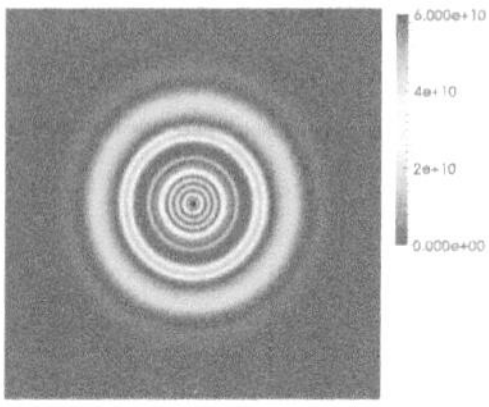

(b) Fine-grid PD: $h = a$

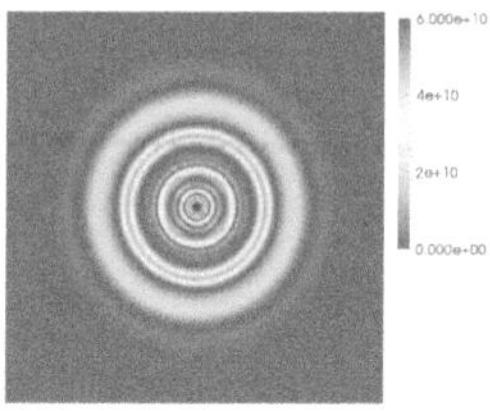

(c) Coarse-grid PD: $h = 2a$

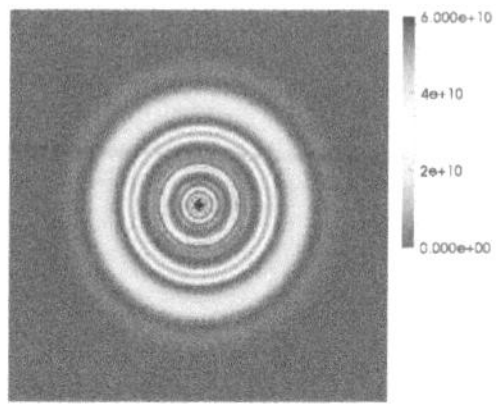

(d) Coarse-grid PD: $h = 4a$

Figure 7.5: Molecular dynamics (MD) potential energy (a) and upscaled peridynamic (PD) strain energy (b,c,d) profiles for two-dimensional wave propagation on a plate. The magnitude of the energy depends on the choices of $\bar{k}$ and $\bar{c}$ (cf. Table 7.1).

7.5 Computational speedup through upscaling

The dimension of the problem plays an important role in the computational savings of upscaled PD simulations, relative to MD simulations. To demonstrate this, let a body $\mathcal{B}$ be given by a d-dimensional unit cube. For an MD simulation, let the atoms be located along a uniform lattice with lattice constant a. Correspondingly, for an upscaled PD simulation, we discretize the body with a uniform grid with grid spacing h. Given the number of nearest-neighbor interactions in the MD model, N, we let the PD horizon be $\delta = Na$. For time integration, we take Δt_0 as the MD time step and assume an upscaled PD time step $\Delta t = \Delta t_0 h/a$. The computational cost of a simulation can be estimated by the product of the number of atoms/points in the body, the number of neighbor interactions per atom/point, and the number of time steps. We can thus estimate the speedup of an upscaled PD simulation with $h = na$, $n \in \mathbb{N}$, relative to the reference MD simulation, by

$$\frac{\text{MD computational cost}}{\text{PD computational cost}} = \frac{(1/a)^d \times C_d (Na)^d / a^d \times (T/\Delta t_0)}{(1/h)^d \times C_d \delta^d / h^d \times (T/\Delta t_0)(a/h)} = \left(\frac{h}{a}\right)^{2d+1} = n^{2d+1}, \tag{7.34}$$

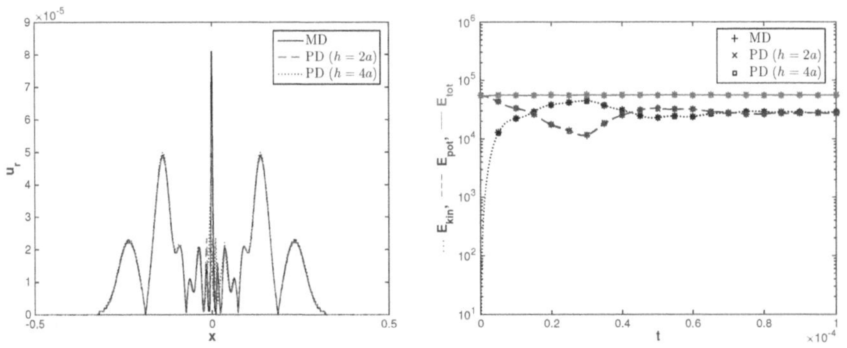

(a) Radial displacements (t = T)

(b) Energy evolution

Figure 7.6: (a) Radial displacements relative to the center of the plate, in the simulations presented in Figure 7.5 at the final time, for atoms/points with reference positions along the x-axis. We compare the results of a molecular dynamics (MD) simulation with lattice constant a and coarse upscaled peridynamic (PD) simulations with mesh sizes of $h = 2a, 4a$. (b) Comparison of the energy evolution between the MD and the coarse upscaled PD simulations. We plot the kinetic energy (black dotted lines), the potential/strain energy (blue dashed lines), and the total energy (red solid lines) for each of the simulations.

where T is the total simulation time and $C_d = 2, \pi$, or $4\pi/3$ for $d = 1, 2$, or 3, respectively. We observe that a one-dimensional simulation with $h = 2a$ would result in a speedup of about $8\times$ (same order as the $5\times$ speedup obtained in Section7.4.1), whereas a two-dimensional simulation with $h = 2a$ would result in a speedup of about $32\times$ (cf. Table 7.2). Moreover, a two-dimensional simulation with $h = 4a$ would result in a speedup of about $10^3\times$ (cf. Table 7.2), whereas a three-dimensional simulation with $h = 4a$ would result in a speedup of about $16{,}000\times$ (cf. [30]).

The upscaling of a one-dimensional nonlocal linear springs model in Section7.4.1 and a three-dimensional EAM in Section7.4.2 preserved the MD interaction range throughout the upscaling process, i.e., we choose $\delta = Na$. As the maximum grid spacing in meshfree discretizations of PD models is usually $h = \delta$, the maximum computational speedup achievable with such a discretization would be $\sim N^{2d+1}$ (cf. (7.34)). Note that in many MD models, it is common to assume a number of nearest-neighbor interactions of $N = 3$ or $N = 4$. Higher computational speedups, with this type of upscaled PD models, may be obtained by employing alternative discretization methods for PD models which do not restrict the choice of h by the horizon size; for example, finite elements with piecewise linear basis functions ([37]). Note, however, that there exists a limitation in the mesh size that one can employ in a PD simulation for upscaling purposes. The larger the ratio h/δ the closer a nonlocal discrete scheme to a local discrete scheme [26], leading to an eventual loss of the capability of representing nonlocal effects inherent to MD models (cf. Section7.4.1).

7.6 Concluding remarks

This chapter explored connections between molecular dynamics and peridynamics. Both theories are nonlocal and employ interactions between atoms/points at finite distances. Even though peridynamics is a continuum theory, certain discretizations of the peridynamic equation of motion produce particle-like descriptions of systems. Such discretizations possess the same computational structure as molecular dynamics equations. These resemblances motivated the study of deeper connections between these two theories. In particular, we demonstrated that since peridynamic equations have length scales, it is possible to derive upscaled peridynamic models from molecular dynamics ones. This upscaling process replaces a discrete model by a continuum one, capable of preserving characteristic nonlocal effects. By using coarse discretizations of upscaled peridynamic models we achieved great computational savings, demonstrated through wave propagation numerical examples.

The goal of this chapter was to present connections between peridynamics and molecular dynamics, through relatively simple mathematical and computational examples. Consequently, we chose to focus on structured molecular dynamics models, to upscale to peridynamics, for which the interaction between atoms depends on some reference configuration, as is the case in commonly used peridynamic models for engineering applications. Many molecular dynamics potentials in the literature are structureless, i.e., they only depend on a current configuration. The upscaling methods described in this chapter can be adapted to such models, as demonstrated in [31, 30]. The upscaling examples provided preserved the molecular dynamics interaction range in the upscaling process. However, this is not a necessary condition for upscaling in all molecular dynamics models. For example, in the upscaling of a structureless Lennard–Jones model in [31] and a structureless EAM in [30] the molecular dynamics interaction range is not preserved. In [30], for instance, the discretization of the upscaled structureless EAM preserved instead the number of nearest-neighbor interactions.

Upscaling studies in peridynamics to date are restricted to molecular dynamics models for crystalline structures at zero temperature. Related methods for amorphous or soft matter, and for finite-temperature systems, are yet to be developed. For such systems, techniques from statistical mechanics may be of use.

References

[1] M. Arndt and M. Griebel. Derivation of higher order gradient continuum models from atomistic models for crystalline solids. *Multiscale Modeling and Simulation*, 4(2):531–562, 2005.

[2] A. M. Bedford and W. Fowler. *Engineering Mechanics: Dynamics*, 5th ed. Upper Saddle River, NJ: Prentice–Hall, 2008.

[3] F. Bobaru and Y. D. Ha. Adaptive refinement and multiscale modeling in 2D peridynamics. *Journal for Multiscale Computational Engineering*, 9(6): 635–639 (2011) .

[4] S. Cranford and M. J. Buehler. Coarse-graining parametrization and multiscale simulation of hierarchical systems. Part I: theory and model formulation. In P. Derosa and T. Cagin, editors, *Multiscale Modeling: From Atoms to Devices*, chapter 2, pages 13–34. Boca Raton, FL: CRC Press, 2010.

[5] M. S. Daw and M. I. Baskes. Semiempirical, quantum mechanical calculation of hydrogen embrittlement in metals. *Physical Review Letters*, 50(17): 1285–1288, 1983.

[6] M. S. Daw and M. I. Baskes. Embedded-atom method: Derivation and application to impurities, surfaces, and other defects in metals. *Physical Review B*, 29(12): 6443–6453, 1984.

[7] E. Emmrich and O. Weckner. On the well-posedness of the linear peridynamic model and its convergence towards the Navier equation of linear elasticity. *Communications in Mathematical Sciences*, 5(4): 851–864, 2007.

[8] E. Emmrich and O. Weckner. The peridynamic equation and its spatial discretisation. *Mathematical Modelling and Analysis*, 12(1): 17–27, 2007.

[9] D. Frenkel and B. Smit. *Understanding Molecular Simulation: From Algorithms to Applications*, volume 1 of *Computational Science Series*, 2nd ed. Academic Press, 2002.

[10] G. N. Greaves. Poisson's ratio over two centuries: challenging hypotheses. *Notes and Records of the Royal Society*, 67(1): 37–58, 2013.

[11] S. F. Henke and S. Shanbhag. Mesh sensitivity in peridynamic simulations. *Computer Physics Communications*, 185(1): 181–193, 2014.

[12] F. Herman and S. Skillman. *Atomic Structure Calculations*. Englewood Cliffs, NJ: Prentice-Hall, Inc., 1963.

[13] S. Izvekov and G. A. Voth. A multiscale coarse-graining method for biomolecular systems. *Journal of Physical Chemistry B*, 109(7): 2469–2473, 2005.

[14] B. Kilic and E. Madenci. Structural stability and failure analysis using peridynamic theory. *International Journal of Non-Linear Mechanics*, 44(8): 845–854, 2009.

[15] R. Lakes. Deformation mechanisms in negative Poisson's ratio materials: Structural aspects. *Journal of Materials Science*, 26:2287–2292, 1991.

[16] R. B. Lehoucq and M. P. Sears. Statistical mechanical foundation of the peridynamic nonlocal continuum theory: Energy and momentum conservation laws. *Physical Review E*, 84: 031112, 2011.

[17] R. B. Lehoucq and S. A. Silling. Statistical coarse-graining of molecular dynamics into peridynamics. Sandia Report, SAND2007-6410, Sandia National Laboratories, October 2007.

[18] A. P. Lyubartsev and A. Laaksonen. Calculation of effective interaction potentials from radial distribution functions: A reverse Monte Carlo approach. *Physsical Review E*, 52(4): 3730–3737, 1995.

[19] S. J. Marrink, H. J. Risselada, S. Yefimov, D. P. Tieleman, and A. H. de Vries. The MARTINI force field: Coarse grained model for biomolecular simulations. *Journal of Physical Chemistry B*, 111(27): 7812–7824, 2007.

[20] M. L. Parks, R. B. Lehoucq, S. J. Plimpton, and S. A. Silling. Implementing peridynamics within a molecular dynamics code. *Computer Physics Communications*, 179(11): 777–783, 2008.

[21] M. L. Parks, P. Seleson, S. J. Plimpton, R. B. Lehoucq, and S. A. Silling. Peridynamics with LAMMPS: A user guide. Sandia Report SAND2010-5549, Sandia National Laboratories, August 2010.

[22] R. Rahman, J. T. Foster, and A. Haque. A multiscale modeling scheme based on peridynamic theory. *International Journal for Multiscale Computational Engineering*, 12:223–248, 2014.

[23] R. Rahman and A. Haque. A peridynamic formulation based hierarchical multiscale modeling approach between continuum scale and atomistic scale. *International Journal of Materials Science and Engineering*, 1(3): 1250029, 2012.

[24] D. C. Rapaport. *The Art of Molecular Dynamics Simulation*. Cambridge University Press, Cambridge, 1995.

[25] M. G. Saunders and G. A. Voth. Coarse-graining methods for computational biology. *Annual Review of Biophysics*, 42: 73–93, 2013.

[26] P. D. Seleson. *Peridynamic Multiscale Models for the Mechanics of Materials: Constitutive Relations, Upscaling from Atomistic Systems, and Interface Problems*. PhD thesis, Florida State University, 2010. Electronic Theses, Treatises and Dissertations. Paper 273.

[27] P. Seleson. Improved one-point quadrature algorithms for two-dimensional peridynamic models based on analytical calculations. *Computer Methods in Applied Mechanics and Engineering*, 282: 184–217, 2014.

[28] P. Seleson and D. J. Littlewood. Convergence studies in meshfree peridynamic simulations. *Computers and Mathematics with Applications*, 71(11): 2432–2448, 2016.

[29] P. Seleson and M. L. Parks. On the role of the influence function in the peridynamic theory. *International Journal for Multiscale Computational Engineering*, 9(6): 689–706, 2011.

[30] P. Seleson, M. L. Parks, and M. Gunzburger. Peridynamic state-based models and the embedded-atom model. *Communications in Computational Physics*, 15(1): 179–205, 2014.

[31] P. Seleson, M. L. Parks, M. Gunzburger, and R. B. Lehoucq. Peridynamics as an upscaling of molecular dynamics. *Multiscale Modelling and Simulation*, 8(1): 204–227, 2009.

[32] S. A. Silling. Reformulation of elasticity theory for discontinuities and long-range forces. *Journal of the Mechanics and Physics of Solids*, 48: 175–209, 2000.

[33] S. A. Silling and E. Askari. A meshfree method based on the peridynamic model of solid mechanics. *Computers and Structures*, 83: 1526–1535, 2005.

[34] S. A. Silling, M. Epton, O. Weckner, J. Xu, and E. Askari. Peridynamic states and constitutive modeling. *Journal of Elasticity*, 88: 151–184, 2007.

[35] S. A. Silling and R. B. Lehoucq. Convergence of peridynamics to classical elasticity theory. *Journal of Elasticity*, 93: 13–37, 2008.

[36] S. A. Silling, M. Zimmermann, and R. Abeyaratne. Deformation of a peridynamic bar. *Journal of Elasticity*, 73: 173–190, 2003.

[37] X. Tian and Q. Du. Analysis and comparison of different approximations to nonlocal diffusion and linear peridynamic equations. *SIAM Journal of Numerical Analysis*, 51(6): 3458–3482, 2013.

[38] M. M. Tirion. Large amplitude elastic motions in proteins from a single-parameter, atomic analysis. *Physical Review Letters*, 77(9): 1905–1908, 1996.

[39] W. Tschöp, K. Kremer, J. Batoulis, T. Bürger, and O. Hahn. Simulation of polymer melts. I. Coarse-graining procedure for polycarbonates. *Acta Polymerica*, 49: 61–74, 1998.

[40] K. Yu, X. J. Xin, and K. B. Lease. A new adaptive integration method for the peridynamic theory. *Modelling and Simulation in Materials Science and Engineering*, 19: 045003, 2011.

[41] G.-F. Zhao, J. Fang, and J. Zhao. A 3D distinct lattice spring model for elasticity and dynamic failure. *International Journal for Numerical and Analytical Methods in Geomechanics*, 35: 859–885, 2011.

Chapter 8

Absorbing Boundary Conditions with Verification

Raymond A. Wildman
US Army Research Laboratory

George A. Gazonas
US Army Research Laboratory

CONTENTS

8.1 Introduction 201
8.2 A PML for state-based peridynamics 202
8.2.1 Two-dimensional (2D), state-based peridynamics review 202
8.2.2 Auxiliary field formulation and PML application 204
8.2.3 Numerical examples 210
8.3 Verification of cone and center crack problems 213
8.3.1 Dimensional analysis of Hertzian cone crack development in brittle elastic solids 213
8.3.2 State-based verification of a cone crack 216
8.3.3 Bond-based verification of a center crack 218
8.4 Verification of an axisymmetric indentation problem 219
8.4.1 Formulation 220
8.4.2 Analytical verification 221
References 222

8.1 Introduction

In this chapter, a peridynamic method for modeling indentation experiments in fused silica and other chemically substituted glasses is described and hinges on the use of an absorbing boundary condition to facilitate smaller simulation regions [25]. As becomes clear later, state-based peridynamics [22] allows for an auxiliary field formulation, which is necessary to implement a perfectly matched layer (PML), the absorbing boundary layer used here. A PML is essentially an anisotropic, fictitious absorbing material, formulated in the frequency domain. In addition, several examples of verification are given, including exact analytical solutions to wave propagation problems and static indentation problems, and also a dimensional analysis solution to a crack length problem.

While most peridynamics work has focused on simulating problems with free or fixed boundary conditions, there are applications in which the simulation of an infinite medium may be useful, such as crack propagation in a half-space or indentation problems. Absorbing boundary conditions are one way of simulating an infinite medium as any impinging waves are suppressed so they do not reflect back into the simulation. A PML is such an absorbing boundary, and was originally introduced for electromagnetic simulations [4, 6]. PMLs differ from traditional absorbing boundary conditions in that they are an absorbing layer with a finite width, placed between the computation region of interest and the truncation of the grid or mesh. They can also be thought of as an anisotropic absorbing material, which is why the flexibility of state-based peridynamics is necessary [25]. A PML was also applied to one-dimensional peridynamics [24], which used the results of Du et al. [7] to formulate an auxiliary field equation. This approach required a banded matrix representation of the auxiliary field, which may be memory prohibitive in higher dimensions.

A low order discretization will be used—a forward Euler method for the temporal discretization, and standard one-point integration with point-matching in space—though the method presented does not rely on a specific discretization and could be extended. The PML is demonstrated on three types of problems: a wave propagation problem to demonstrate the effectiveness of the PML, two crack propagation problems (one state-based and one bond-based), and an axisymmetric indentation problem.

8.2 A PML for state-based peridynamics

In this section, the formulation of a linear elastic, state-based peridynamics is reviewed, and used in the application of a PML. The formulation is performed in the Laplace domain so that the temporal discretization is more straight forward. Several numerical examples show the efficacy of the absorbing layer.

8.2.1 Two-dimensional (2D), state-based peridynamics review

The continuum equation of motion in an elastic solid can be stated as

$$\rho \frac{\partial^2}{\partial t^2}\mathbf{u} = \nabla \cdot \overline{\sigma} + \mathbf{b} = \nabla \cdot \left(\overline{\overline{\mathbf{c}}} : \overline{\varepsilon}\right) + \mathbf{b}, \tag{8.1}$$

where $\rho(\mathbf{x})$ is the density; $\mathbf{u}(\mathbf{x},t)$ is the displacement; $\overline{\sigma}(\mathbf{x},t)$ is the stress tensor; $\overline{\varepsilon}(\mathbf{x},t)$ is the strain tensor; $\overline{\overline{\mathbf{c}}}$ is the stiffness tensor for plane strain, defined by Young's modulus E, Poisson's ratio ν, or Lamé parameters λ and μ; and $\mathbf{b}(\mathbf{x})$ is a body force

[16]. (Throughout, boldface type denotes a vector and a boldface variable with one overbar denotes a rank-two tensor and two overbars denote a rank-four tensor.) Peridynamics proposes replacing $\nabla \cdot \overline{\sigma}$ with a nonlocal operation that nonetheless also represents a force. Here, state-based peridynamics is used [22], rather than the original bond-based version [20]. A PML application requires an auxiliary field formulation, as it is essentially an anisotropic absorbing material, if a non-physical one. Consequently, a state-based peridynamic formulation is necessary to implement the required constitutive relations in the absorber. The state-based peridynamics uses a family of bonds to determine a given force rather than a single bond independently. This more general approach allows for inelastic behavior and more general elastic behavior, and is governed by

$$\rho \frac{\partial^2}{\partial t^2}\mathbf{u} = \int_{\mathcal{H}_\mathbf{x}} \left(\overline{\mathbf{T}}[\mathbf{x},t]\langle \mathbf{x}' - \mathbf{x}\rangle - \overline{\mathbf{T}}[\mathbf{x}',t]\langle \mathbf{x} - \mathbf{x}'\rangle\right) dV' + \mathbf{b}, \tag{8.2}$$

where $\mathcal{H}_\mathbf{x}$ is the horizon region, defined as a circle centered at $\mathbf{x}$ with radius δ, and $\overline{\mathbf{T}}[\mathbf{x},t]\langle \mathbf{x}' - \mathbf{x}\rangle$ is a peridynamic vector state, with parameters in the square brackets indicating variables that act as arguments to any functions that define the vector state and variables in the angle brackets acting as arguments to the vector state itself. In the state-based formulation, the deformation gradient, given by

$$\overline{\mathbf{F}} = \overline{\mathbf{I}} + \mathbf{u}\nabla, \tag{8.3}$$

can be approximated as a vector state as

$$\overline{\mathbf{F}}[\mathbf{x},t] = \left[\int_{\mathcal{H}_\mathbf{x}} C(|\xi|)\left(\mathbf{Y}[\mathbf{x},t]\langle \xi\rangle \otimes \xi\right) dV_{\mathbf{x}'}\right] \overline{\mathbf{K}}^{-1}, \tag{8.4}$$

where $C(|\xi|) = \exp(-|\xi|^2/\delta^2)$ is a shape function (taken as a Gaussian distribution here and with the horizon $\mathcal{H}_\mathbf{x}$ extended so that the shape function decays to an arbitrary, small value, 10^{-6}), $\overline{\mathbf{K}}$ is a shape tensor given by

$$\overline{\mathbf{K}}[\mathbf{x},t] = \int_{\mathcal{H}_\mathbf{x}} C(|\xi|)\left(\xi \otimes \xi\right) dV_{\mathbf{x}'}, \tag{8.5}$$

with the components of its inverse defined as

$$\overline{\mathbf{K}}^{-1} = \begin{bmatrix} k_{xx}^{\text{inv}} & k_{xy}^{\text{inv}} \\ k_{yx}^{\text{inv}} & k_{yy}^{\text{inv}} \end{bmatrix}, \tag{8.6}$$

and $\mathbf{Y}$ is a deformation vector state given by

$$\mathbf{Y}[\mathbf{x},t]\langle \xi\rangle = \eta + \xi, \tag{8.7}$$

with $\eta = \mathbf{u}[\mathbf{x}',t] - \mathbf{u}[\mathbf{x},t]$ and $\xi = \mathbf{x}' - \mathbf{x}$ [9].

The deformation gradient can now be substituted into Hooke's law and strain-displacement relations, giving the stress $\overline{\sigma}$ in terms of $\mathbf{u}$ in plane strain

$$\overline{\sigma} = \overline{\overline{\mathbf{c}}} : \overline{\varepsilon}, \tag{8.8}$$

where

$$\bar{\varepsilon}[\mathbf{x},t] = \frac{1}{2}(\nabla\mathbf{u} + \mathbf{u}\nabla) \approx \frac{1}{2}\left(\overline{\mathbf{F}}[\mathbf{x},t] + \overline{\mathbf{F}}[\mathbf{x},t]^{\mathrm{T}} - 2\bar{\mathbf{I}}\right), \tag{8.9}$$

and, unless otherwise noted, linear elastic, plane strain is assumed throughout, giving the constitutive tensor

$$\bar{\bar{\mathbf{c}}} = \begin{bmatrix} \lambda + 2\mu & \lambda & 0 \\ \lambda & \lambda + 2\mu & 0 \\ 0 & 0 & 2\mu \end{bmatrix}. \tag{8.10}$$

Ultimately, the peridynamic vector state $\mathbf{T}$ for plane strain elasticity is given by [22, 9]

$$\overline{\mathbf{T}}[\mathbf{x},t]\langle\xi\rangle = C(|\xi|)\bar{\sigma}[\mathbf{x},t]\overline{\mathbf{K}}^{-1}\xi, \tag{8.11}$$

combined with Equations (8.8–8.10).

8.2.2 Auxiliary field formulation and PML application

The first step in formulating a PML is to construct an analytic continuation to the complex plane, such as $\hat{x} = x + ig(x)$, where $g(x)$ is a given function describing some deviation from the real axis [13]. This mapping has the effect of transforming traveling waves of the form e^{ikx}, where $k = \omega/c$ is the wave number, into evanescent waves of the form $e^{ikx}e^{-kg(x)}$, thus attenuating such waves wherever $g(x) > 0$. It can be shown that applying a PML involves substituting for spatial derivatives using [13]

$$\frac{\partial}{\partial x} \to \frac{1}{1 + i\frac{\phi(x)}{\omega}}\frac{\partial}{\partial x}. \tag{8.12}$$

The function $\phi(x)$ defines the extent of the PML region, i.e., when $\phi(x) = 0$ the original wave equation is obtained, and when $\phi(x) > 0$, traveling waves decay exponentially. Typically, $\phi(x)$ transitions from 0 to a constant value using a smooth function to prevent numerical reflections at the boundary between the absorbing and computational regions. Before applying a PML directly to the peridynamic equation, Equation (8.1) is treated so that the PML application to peridynamics is clear. It is convenient to convert Equation (8.1) to the Laplace domain, assuming e^{-st} time dependence, and express the wave equation as two coupled first-order partial differential equations, the first in $\tilde{\mathbf{u}}$ and the second in $s\tilde{\bar{\psi}} = \tilde{\bar{\sigma}}$

$$\begin{aligned} \rho s\tilde{\mathbf{u}} &= \nabla \cdot \tilde{\bar{\psi}} \\ s\tilde{\bar{\psi}} &= \bar{\bar{\mathbf{c}}} : \tilde{\bar{\varepsilon}}, \end{aligned} \tag{8.13}$$

where the Laplace transform of a variable is indicated by $\mathcal{L}\{f(t)\} = \tilde{f}(s)$. Expanding Equation (8.13) into components gives five coupled equations

$$\begin{aligned}
\rho s\tilde{u}_x &= \frac{\partial}{\partial x}\tilde{\psi}_x + \frac{\partial}{\partial y}\tilde{\psi}_\tau \\
\rho s\tilde{u}_y &= \frac{\partial}{\partial x}\tilde{\psi}_\tau + \frac{\partial}{\partial y}\tilde{\psi}_y \\
s\tilde{\psi}_x &= (\lambda + 2\mu)\frac{\partial}{\partial x}\tilde{u}_x + \lambda\frac{\partial}{\partial y}\tilde{u}_y \\
s\tilde{\psi}_y &= \lambda\frac{\partial}{\partial x}\tilde{u}_x + (\lambda + 2\mu)\frac{\partial}{\partial y}\tilde{u}_y \\
s\tilde{\psi}_\tau &= \mu\left(\frac{\partial}{\partial y}\tilde{u}_x + \frac{\partial}{\partial x}\tilde{u}_y\right).
\end{aligned} \tag{8.14}$$

The state-based formulation can be expanded into components that can be then matched to terms in Equation (8.14). Following this approach yields a viable method for performing PML substitutions. First, the state-based peridynamic equations listed above in Equations (8.2–8.11) can be written explicitly as

$$\begin{aligned}
\rho s\tilde{u}_x[\mathbf{x},s] &= \int_{\mathcal{H}_\mathbf{x}} C(|\xi|)\left(\tilde{\psi}_x[\mathbf{x},s]k_{xx}^{\text{inv}} + \tilde{\psi}_\tau[\mathbf{x},s]k_{yx}^{\text{inv}}\right)\xi_x dV_{\mathbf{x}'} \\
&+ \int_{\mathcal{H}_\mathbf{x}} C(|\xi|)\left(\tilde{\psi}_x[\mathbf{x}',s]k_{xx}^{\prime\text{inv}} + \tilde{\psi}_\tau[\mathbf{x}',s]k_{yx}^{\prime\text{inv}}\right)\xi_x dV_{\mathbf{x}'} \\
&+ \int_{\mathcal{H}_\mathbf{x}} C(|\xi|)\left(\tilde{\psi}_x[\mathbf{x},s]k_{xy}^{\text{inv}} + \tilde{\psi}_\tau[\mathbf{x},s]k_{yy}^{\text{inv}}\right)\xi_y dV_{\mathbf{x}'} \\
&+ \int_{\mathcal{H}_\mathbf{x}} C(|\xi|)\left(\tilde{\psi}_x[\mathbf{x}',s]k_{xy}^{\prime\text{inv}} + \tilde{\psi}_\tau[\mathbf{x}',s]k_{yy}^{\prime\text{inv}}\right)\xi_y dV_{\mathbf{x}'}, \\
\rho s\tilde{u}_y[\mathbf{x},s] &= \int_{\mathcal{H}_\mathbf{x}} C(|\xi|)\left(\tilde{\psi}_\tau[\mathbf{x},s]k_{xx}^{\text{inv}} + \tilde{\psi}_y[\mathbf{x},s]k_{yx}^{\text{inv}}\right)\xi_x dV_{\mathbf{x}'} \\
&+ \int_{\mathcal{H}_\mathbf{x}} C(|\xi|)\left(\tilde{\psi}_\tau[\mathbf{x}',s]k_{xx}^{\prime\text{inv}} + \tilde{\psi}_y[\mathbf{x}',s]k_{yx}^{\prime\text{inv}}\right)\xi_x dV_{\mathbf{x}'} \\
&+ \int_{\mathcal{H}_\mathbf{x}} C(|\xi|)\left(\tilde{\psi}_\tau[\mathbf{x},s]k_{xy}^{\text{inv}} + \tilde{\psi}_y[\mathbf{x},s]k_{yy}^{\text{inv}}\right)\xi_y dV_{\mathbf{x}'} \\
&+ \int_{\mathcal{H}_\mathbf{x}} C(|\xi|)\left(\tilde{\psi}_\tau[\mathbf{x}',s]k_{xy}^{\prime\text{inv}} + \tilde{\psi}_y[\mathbf{x}',s]k_{yy}^{\prime\text{inv}}\right)\xi_y dV_{\mathbf{x}'},
\end{aligned} \tag{8.15}$$

$$
\begin{aligned}
s\tilde{\psi}_x[\mathbf{x},s] &= (\lambda+2\mu)\left[\int_{\mathcal{H}_\mathbf{x}} C(|\xi|)\left(\tilde{Y}_x[\mathbf{x},s]\xi_x k_{xx}^{\text{inv}} + \tilde{Y}_x[\mathbf{x},s]\xi_y k_{yx}^{\text{inv}}\right) dV_{\mathbf{x}'} - 1\right] \\
&\quad + \lambda\left[\int_{\mathcal{H}_\mathbf{x}} C(|\xi|)\left(\tilde{Y}_y[\mathbf{x},s]\xi_x k_{xy}^{\text{inv}} + \tilde{Y}_y[\mathbf{x},s]\xi_y k_{yy}^{\text{inv}}\right) dV_{\mathbf{x}'} - 1\right], \\
s\tilde{\psi}_y[\mathbf{x},s] &= \lambda\left[\int_{\mathcal{H}_\mathbf{x}} C(|\xi|)\left(\tilde{Y}_x[\mathbf{x},s]\xi_x k_{xx}^{\text{inv}} + \tilde{Y}_x[\mathbf{x},s]\xi_y k_{yx}^{\text{inv}}\right) dV_{\mathbf{x}'} - 1\right] \\
&\quad + (\lambda+2\mu)\left[\int_{\mathcal{H}_\mathbf{x}} C(|\xi|)\left(\tilde{Y}_y[\mathbf{x},s]\xi_x k_{xy}^{\text{inv}} + \tilde{Y}_y[\mathbf{x},s]\xi_y k_{yy}^{\text{inv}}\right) dV_{\mathbf{x}'} - 1\right], \\
s\tilde{\psi}_\tau[\mathbf{x},s] &= \mu\int_{\mathcal{H}_\mathbf{x}} C(|\xi|)\left(\tilde{Y}_x[\mathbf{x},s]\xi_x k_{xy}^{\text{inv}} + \tilde{Y}_x[\mathbf{x},s]\xi_y k_{yy}^{\text{inv}}\right) dV_{\mathbf{x}'} \\
&\quad + \mu\int_{\mathcal{H}_\mathbf{x}} C(|\xi|)\left(\tilde{Y}_y[\mathbf{x},s]\xi_x k_{xx}^{\text{inv}} + \tilde{Y}_y[\mathbf{x},s]\xi_y k_{yx}^{\text{inv}}\right) dV_{\mathbf{x}'}.
\end{aligned}
\tag{8.16}
$$

Though no spatial derivatives appear in Equations (8.15) or (8.16), the correspondence of each term involving a spatial integral to those in Equation (8.14) involving a spatial derivative is apparent and the PML substitutions can be made accordingly. Here, the substitution given in Equation (8.12) can be made for all spatial derivatives, written as

$$
\frac{\partial}{\partial x} \to \frac{s}{s+\phi_x(x)}\frac{\partial}{\partial x} \tag{8.17}
$$

in the Laplace domain, and similarly defined for the y-directed derivatives. Using the first of Equation (8.14) as an example, the PML substitutions give

$$
\begin{aligned}
&\rho s\tilde{u}_x = \frac{s}{s+\phi_x}\frac{\partial}{\partial x}\tilde{\psi}_x + \frac{s}{s+\phi_y}\frac{\partial}{\partial y}\tilde{\psi}_\tau \\
\Rightarrow &\rho\,(s+\phi_x)(s+\phi_y)\,\tilde{u}_x = (s+\phi_y)\frac{\partial}{\partial x}\tilde{\psi}_x + (s+\phi_x)\frac{\partial}{\partial y}\tilde{\psi}_\tau.
\end{aligned}
\tag{8.18}
$$

The remaining components of Equation (8.14) can be expanded in a similar way.

Wherever $\phi > 0$, $\mathbf{u}$ and $\overline{\sigma}$ will exponentially decay. Before discretization, any change in ϕ will not result in any reflections, so the region of interest would have $\phi = 0$, and the PML region could have a discontinuity in applying ϕ. In practice, however, numerical reflections can result from discontinuous material parameters after discretization, so it is better to use a smooth transition for ϕ. Here, the PML region can be divided into two parts, one in which ϕ is a constant value, and in the other ϕ ramps up to that constant value following a Gaussian distribution. An example is shown in Fig. 8.1, with the width of the constant region set to 0.1 m and the width of the Gaussian region 0.2 m. The variance of the distribution is set so that the minimum value in the Gaussian region is 10^{-6}.

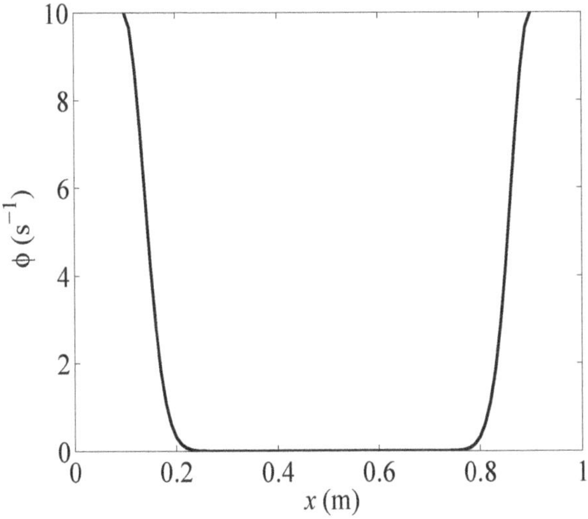

Figure 8.1: An example of a PML across one dimension using a Gaussian ramp.

For the peridynamic equations, the first equation in 8.15 can be rewritten as

$$\begin{aligned}\rho\,(s+\phi_x)\,(s+\phi_y)\,\tilde{u}_x = {} & (s+\phi_y)\int_{\mathcal{H}_{\mathbf{x}}} C(|\xi|)\left(\tilde{\psi}_x[\mathbf{x},s]k_{xx}^{\text{inv}}+\tilde{\psi}_\tau[\mathbf{x},s]k_{yx}^{\text{inv}}\right)\xi_x dV_{\mathbf{x}'} \\ & +(s+\phi_y)\int_{\mathcal{H}_{\mathbf{x}}} C(|\xi|)\left(\tilde{\psi}_x[\mathbf{x}',s]k_{xx}'^{\text{inv}}+\tilde{\psi}_\tau[\mathbf{x}',s]k_{yx}'^{\text{inv}}\right)\xi_x dV_{\mathbf{x}'} \\ & +(s+\phi_x)\int_{\mathcal{H}_{\mathbf{x}}} C(|\xi|)\left(\tilde{\psi}_x[\mathbf{x},s]k_{xy}^{\text{inv}}+\tilde{\psi}_\tau[\mathbf{x},s]k_{yy}^{\text{inv}}\right)\xi_y dV_{\mathbf{x}'} \\ & +(s+\phi_x)\int_{\mathcal{H}_{\mathbf{x}}} C(|\xi|)\left(\tilde{\psi}_x[\mathbf{x}',s]k_{xy}'^{\text{inv}}+\tilde{\psi}_\tau[\mathbf{x}',s]k_{yy}'^{\text{inv}}\right)\xi_y dV_{\mathbf{x}'},\end{aligned} \tag{8.19}$$

with the remaining equations following similarly [25]. Correspondence to spatial derivatives in Equation (8.14) can be identified by the component of ξ in each integrand for the displacement equations. For the stress equations, correspondence can be garnered by matching the constitutive parameters in each equation. The final step involves simply converting back to the time domain and implementing a forward Euler discretization scheme in time, and the standard one-point integration method [21] in space.

For the temporal discretization, forward Euler will be used to simplify the presentation and implementation. Higher-order temporal discretizations can be used, though they lead to more terms in what follows. The Laplace domain was used throughout to facilitate the temporal discretization of the final equations. Because differentiation in the Laplace domain is represented by s, approximations to s can be directly substituted in terms of appropriate differencing schemes represented in the z-transform domain. This technique is used in the design of digital filters, where it is known as "filter design by approximation of derivatives" or "the bilinear transformation" [17] and in integral equation methods where it is known as "convolution quadrature" [14, 15] or "finite difference delay modeling" [23]. A forward Euler approximation can be stated as the substitution

$$s \to \frac{z-1}{\Delta t}, \tag{8.20}$$

where z is the z-transform * variable representing a unit advance and Δt is the time step size. Substitution into Equation (8.19) gives

$$\begin{aligned}
\rho\,(z-1+\Delta t\phi_x)\,(z-1+\Delta t\phi_y)\,U_x
&= (z-1+\Delta t\phi_y)\int_{\mathcal{H}_\mathbf{x}} C(|\xi|)\left(\Psi_x[\mathbf{x},z]k_{xx}^{\mathrm{inv}} + \Psi_\tau[\mathbf{x},z]k_{yx}^{\mathrm{inv}}\right)\xi_x dV_{\mathbf{x}'}\\
&+ (z-1+\Delta t\phi_y)\int_{\mathcal{H}_\mathbf{x}} C(|\xi|)\left(\Psi_x[\mathbf{x}',z]k_{xx}^{\prime\mathrm{inv}} + \Psi_\tau[\mathbf{x}',z]k_{yx}^{\prime\mathrm{inv}}\right)\xi_x dV_{\mathbf{x}'}\\
&+ (z-1+\Delta t\phi_x)\int_{\mathcal{H}_\mathbf{x}} C(|\xi|)\left(\Psi_x[\mathbf{x},z]k_{xy}^{\mathrm{inv}} + \Psi_\tau[\mathbf{x},z]k_{yy}^{\mathrm{inv}}\right)\xi_y dV_{\mathbf{x}'}\\
&+ (z-1+\Delta t\phi_x)\int_{\mathcal{H}_\mathbf{x}} C(|\xi|)\left(\Psi_x[\mathbf{x}',z]k_{xy}^{\prime\mathrm{inv}} + \Psi_\tau[\mathbf{x}',z]k_{yy}^{\prime\mathrm{inv}}\right)\xi_y dV_{\mathbf{x}'},
\end{aligned} \tag{8.21}$$

where capital letters indicate the z-transform of a variable [25]. Expanding the quadratic term on the left hand side gives

$$\begin{aligned}
(z-1+\Delta t\phi_x)\,(z-1+\Delta t\phi_y) &= \Delta t^2\phi_x\phi_y + \Delta t\phi_x z - \Delta t\phi_x\\
&\quad + \Delta t\phi_y z + z^2 - z\\
&\quad - \Delta t\phi_y - z + 1\\
&= z^2 + (\Delta t\phi_x + \Delta t\phi_y - 2)\,z + \Delta t^2\phi_x\phi_y - \Delta t\phi_x - \Delta t\phi_y + 1.
\end{aligned} \tag{8.22}$$

Multiplying by z^{-2} and rearranging gives an update equation in terms of z, which can be converted to a time-stepping method via the inverse z-transform † (assuming

*The z-transform is defined here as $X(z) = Z\{x[n]\} = \sum_{n=0}^{\infty} x[n]z^{-n}$ for causal signals.
†The only necessary property is the delay: $x[n-k] \leftrightarrow z^{-k}X(z)$.

vanishing initial conditions and an appropriate region of convergence) as

$$\begin{aligned}
u_x[\mathbf{x},l] = &-(\gamma_x+\gamma_y)\,u_x[\mathbf{x},l-1] - \gamma_x\gamma_y u_x[\mathbf{x},l-2] \\
&+\frac{\Delta t}{\rho}\int_{\mathcal{H}_\mathbf{x}} C(|\xi|)\left(\psi_x[\mathbf{x},l-1]k_{xx}^{\text{inv}} + \psi_\tau[\mathbf{x},l-1]k_{yx}^{\text{inv}}\right)\xi_x dV_{\mathbf{x}'} \\
&+\frac{\Delta t}{\rho}\int_{\mathcal{H}_\mathbf{x}} C(|\xi|)\left(\psi_x[\mathbf{x}',l-1]k_{xx}'^{\text{inv}} + \psi_\tau[\mathbf{x}',l-1]k_{yx}'^{\text{inv}}\right)\xi_x dV_{\mathbf{x}'} \\
&+\frac{\Delta t}{\rho}\gamma_y\int_{\mathcal{H}_\mathbf{x}} C(|\xi|)\left(\psi_x[\mathbf{x},l-2]k_{xx}^{\text{inv}} + \psi_\tau[\mathbf{x},l-2]k_{yx}^{\text{inv}}\right)\xi_x dV_{\mathbf{x}'} \\
&+\frac{\Delta t}{\rho}\gamma_y\int_{\mathcal{H}_\mathbf{x}} C(|\xi|)\left(\psi_x[\mathbf{x}',l-2]k_{xx}'^{\text{inv}} + \psi_\tau[\mathbf{x}',l-2]k_{yx}'^{\text{inv}}\right)\xi_x dV_{\mathbf{x}'} \qquad (8.23)\\
&+\frac{\Delta t}{\rho}\int_{\mathcal{H}_\mathbf{x}} C(|\xi|)\left(\psi_x[\mathbf{x},l-1]k_{xy}^{\text{inv}} + \psi_\tau[\mathbf{x},l-1]k_{yy}^{\text{inv}}\right)\xi_y dV_{\mathbf{x}'} \\
&+\frac{\Delta t}{\rho}\int_{\mathcal{H}_\mathbf{x}} C(|\xi|)\left(\psi_x[\mathbf{x}',l-1]k_{xy}'^{\text{inv}} + \psi_\tau[\mathbf{x}',l-1]k_{yy}'^{\text{inv}}\right)\xi_y dV_{\mathbf{x}'} \\
&+\frac{\Delta t}{\rho}\gamma_x\int_{\mathcal{H}_\mathbf{x}} C(|\xi|)\left(\psi_x[\mathbf{x},l-2]k_{xy}^{\text{inv}} + \psi_\tau[\mathbf{x},l-2]k_{yy}^{\text{inv}}\right)\xi_y dV_{\mathbf{x}'} \\
&+\frac{\Delta t}{\rho}\gamma_x\int_{\mathcal{H}_\mathbf{x}} C(|\xi|)\left(\psi_x[\mathbf{x}',l-2]k_{xy}'^{\text{inv}} + \psi_\tau[\mathbf{x}',l-2]k_{yy}'^{\text{inv}}\right)\xi_y dV_{\mathbf{x}'},
\end{aligned}$$

where l is the time step number, $\gamma_x = \Delta t\phi_x - 1$, and $\gamma_y = \Delta t\phi_y - 1$ [25]. A stress component update equation becomes, e.g.,

$$\begin{aligned}
\psi_x[\mathbf{x},l] = &-(\gamma_x+\gamma_y)\,\psi_x[\mathbf{x},l-1] - \gamma_x\gamma_y\psi_x[\mathbf{x},l-2] \\
&+\Delta t\,(\lambda+2\mu)\left[\int_{\mathcal{H}_\mathbf{x}} C(|\xi|)\left(\tilde{Y}_x[\mathbf{x},l-1]\xi_x k_{xx}^{\text{inv}} + \tilde{Y}_x[\mathbf{x},l-1]\xi_y k_{yx}^{\text{inv}}\right)dV_{\mathbf{x}'} - 1\right] \\
&+\Delta t\gamma_y\,(\lambda+2\mu)\left[\int_{\mathcal{H}_\mathbf{x}} C(|\xi|)\left(\tilde{Y}_x[\mathbf{x},l-2]\xi_x k_{xx}^{\text{inv}} + \tilde{Y}_x[\mathbf{x},l-2]\xi_y k_{yx}^{\text{inv}}\right)dV_{\mathbf{x}'} - 1\right] \\
&+\Delta t\lambda\left[\int_{\mathcal{H}_\mathbf{x}} C(|\xi|)\left(\tilde{Y}_y[\mathbf{x},l-1]\xi_x k_{xy}^{\text{inv}} + \tilde{Y}_y[\mathbf{x},l-1]\xi_y k_{yy}^{\text{inv}}\right)dV_{\mathbf{x}'} - 1\right] \\
&+\Delta t\lambda\gamma_x\left[\int_{\mathcal{H}_\mathbf{x}} C(|\xi|)\left(\tilde{Y}_y[\mathbf{x},l-2]\xi_x k_{xy}^{\text{inv}} + \tilde{Y}_y[\mathbf{x},l-2]\xi_y k_{yy}^{\text{inv}}\right)dV_{\mathbf{x}'} - 1\right].
\end{aligned} \qquad (8.24)$$

Finally, Equation (8.23) can be discretized spatially using a simple one-point integration and point match testing, giving, for the x-component of displacement,

$$\begin{aligned}
u_x[\mathbf{x}_i,l] = &-(\gamma_x+\gamma_y)\,u_x[\mathbf{x}_i,l-1]-\gamma_x\gamma_y u_x[\mathbf{x}_i,l-2]\\
&+\frac{\Delta t}{\rho}\sum_{j=1}^{N_i}C(|\xi_{ij}|)\left(\psi_x[\mathbf{x}_i,l-1]k^{\text{inv}}_{i,xx}+\psi_\tau[\mathbf{x}_i,l-1]k^{\text{inv}}_{i,yx}\right)\xi_x V_j\\
&+\frac{\Delta t}{\rho}\sum_{j=1}^{N_i}C(|\xi_{ij}|)\left(\psi_x[\mathbf{x}_j,l-1]k^{\text{inv}}_{j,xx}+\psi_\tau[\mathbf{x}_j,l-1]k^{\text{inv}}_{j,yx}\right)\xi_x V_j\\
&+\frac{\Delta t}{\rho}\gamma_y\sum_{j=1}^{N_i}C(|\xi_{ij}|)\left(\psi_x[\mathbf{x}_i,l-2]k^{\text{inv}}_{i,xx}+\psi_\tau[\mathbf{x}_i,l-2]k^{\text{inv}}_{i,yx}\right)\xi_x V_j\\
&+\frac{\Delta t}{\rho}\gamma_y\sum_{j=1}^{N_i}C(|\xi_{ij}|)\left(\psi_x[\mathbf{x}_j,l-2]k^{\text{inv}}_{j,xx}+\psi_\tau[\mathbf{x}_j,l-2]k^{\text{inv}}_{j,yx}\right)\xi_x V_j\\
&+\frac{\Delta t}{\rho}\sum_{j=1}^{N_i}C(|\xi_{ij}|)\left(\psi_x[\mathbf{x}_i,l-1]k^{\text{inv}}_{i,xy}+\psi_\tau[\mathbf{x}_i,l-1]k^{\text{inv}}_{i,yy}\right)\xi_y V_j\\
&+\frac{\Delta t}{\rho}\sum_{j=1}^{N_i}C(|\xi_{ij}|)\left(\psi_x[\mathbf{x}_j,l-1]k^{\text{inv}}_{j,xy}+\psi_\tau[\mathbf{x}_j,l-1]k^{\text{inv}}_{j,yy}\right)\xi_y V_j\\
&+\frac{\Delta t}{\rho}\gamma_x\sum_{j=1}^{N_i}C(|\xi_{ij}|)\left(\psi_x[\mathbf{x}_i,l-2]k^{\text{inv}}_{i,xy}+\psi_\tau[\mathbf{x}_i,l-2]k^{\text{inv}}_{i,yy}\right)\xi_y V_j\\
&+\frac{\Delta t}{\rho}\gamma_x\sum_{j=1}^{N_i}C(|\xi_{ij}|)\left(\psi_x[\mathbf{x}_j,l-2]k^{\text{inv}}_{j,xy}+\psi_\tau[\mathbf{x}_j,l-2]k^{\text{inv}}_{j,yy}\right)\xi_y V_j,
\end{aligned} \tag{8.25}$$

where V_j is the volume of node j, N_i is the number of nodes in the neighborhood of node i, and $\xi_{ij}=\mathbf{x}_j-\mathbf{x}_i$ [25].

8.2.3 Numerical examples

As a first example, the PML formulated above is used to terminate a linear elastic full-space excited by a Gaussian distribution as an initial condition. Specifically, the x-directed displacement is set to

$$u_x(\mathbf{x},t=0)=e^{-200|\mathbf{x}-\mathbf{p}_{\text{mid}}|^2}, \tag{8.26}$$

where $\mathbf{p}_{\text{mid}}$ is the mid-point of the computational region, which in this example was defined as $0\le x,y\le 1$ and discretized with $\Delta x=\Delta y=0.01$ m. Young's modulus for the region was set to 1 Pa, Poisson's ratio was $1/4$, and the density was $1\ \text{kg}/\text{m}^3$. The PML region was defined as the 0.3-m border around the 1-m by 1-m region and used a Gaussian ramp with a width of 0.2 m, finally reaching a maximum of 50 s^{-1} for the remaining 0.1 m. For the Gaussian kernel, a horizon size of $\delta=1.1\Delta x$ was used, and for the Heaviside kernel, a horizon of $\delta=3.1\Delta x$ was used.

The simulation was run with both the Heaviside and Gaussian kernels, with the total strain energy shown in Fig. 8.2. The Gaussian kernel (dotted line) shows the largest drop in energy, reaching a minimum of 5.6×10^{-7}, and the Heaviside kernel (dashed line) decreases to $1. \times 10^{-4}$.

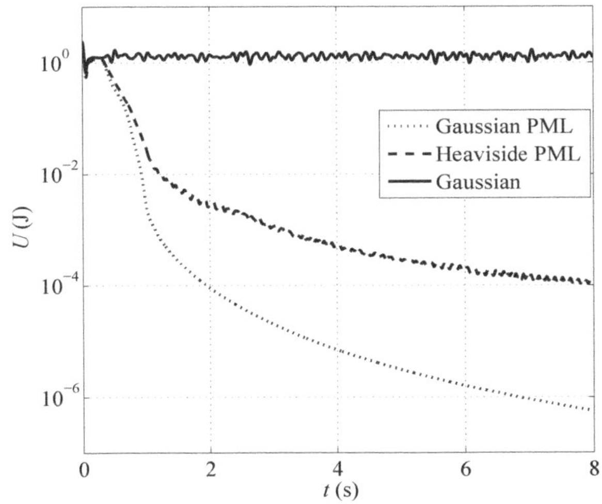

Figure 8.2: Total strain energy in a simulation terminated by a PML.

A bounded simulation is shown for reference as the solid line, which used a fixed displacement boundary condition and the Gaussian kernel. Fig. 8.3 shows a waterfall plot of the x-directed displacement along the $y = 0.5$ line for the Gaussian kernel with the PML function ϕ_x shown in gray on the far end of the plot (corresponding to $t = 1$ s). Fig. 8.4 shows the absolute value of the x-directed displacement at the edge of the PML region, in simulations terminated by a PML and with a fixed boundary condition. The wave is absorbed at the boundary with minimal reflections: As can be seen, the plots align for a time, and where they deviate (indicating a reflection from the hard boundary), the PML simulation remains in decay.

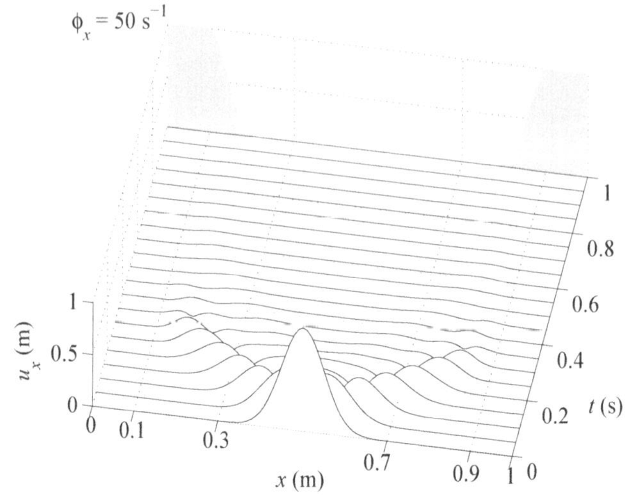

Figure 8.3: x-directed displacement at $y = 0.5$ m, terminated by a PML.

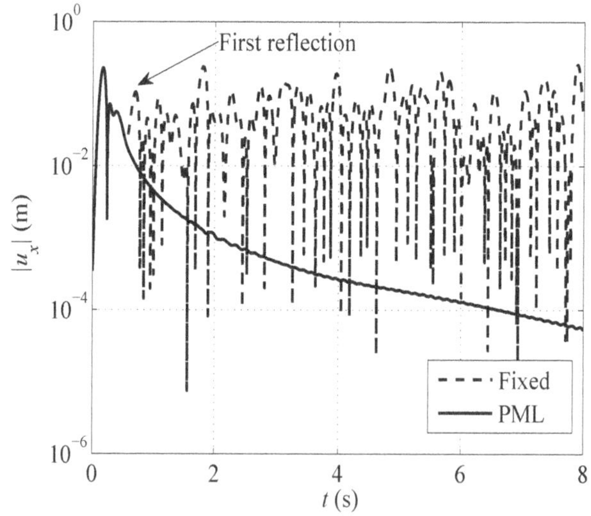

Figure 8.4: x-directed displacement at $x = 0.3$, $y = 0.5$ m. The solid line shows results terminated by a PML, and the dashed line used a fixed boundary condition.

For verification, the method was compared with an exact analytical solution. Consider a cylindrically symmetric wave propagating in an infinite elastic medium with the same constitutive parameters as the above example, and with an initial condition given by

$$u_r(r) = b\left(\frac{r}{a}\right)\left[1+\left(\frac{r}{a}\right)^2\right]^{-\frac{3}{2}} \tag{8.27}$$

where r is the radial coordinate and $b = 1$ and $a = 0.1$ are used in this example. The exact solution is given by Eringen and Suhubi [8]

$$\begin{aligned} u(r,t) &= \frac{br}{\sqrt{2}aR^6}\sqrt{R^2+\alpha}\left(2\alpha - R^2\right), \\ \alpha &= 1+\frac{r^2-c^2t^2}{a^2}, \\ R^2 &= \sqrt{\alpha^2+\frac{4c^2t^2}{a^2}}, \end{aligned} \tag{8.28}$$

where c is the longitudinal wave speed. This problem was simulated in a 2D region, 2 m by 2 m and $\Delta x = \Delta y = 0.01$ m, terminated by a PML with the same dimensions and magnitude as the above problem. The Gaussian kernel was used with a horizon size of $\delta = 0.75\Delta x$, with an actual cutoff of 0.028 m. The results are shown in Fig. 8.5, with the exact solution shown as the solid line and the peridynamic solution shown as the dashed line. The peridynamic solution shows good agreement with the exact solution.

8.3 Verification of cone and center crack problems

In this section, cone and center crack development in brittle, elastic half- and full-spaces are considered. First, dimensional analysis is used to give a relationship between load strength and crack length in a 2D half-space and 2D full-space. This result is used to further verify the above method and also a standard bond-based peridynamics formulation without any absorbing boundary conditions.

8.3.1 Dimensional analysis of Hertzian cone crack development in brittle elastic solids

In this section, a methodology (from [10]) is outlined for the derivation of similarity relationships for elastic solids exhibiting stable crack growth induced by rigid indenters; as will be shown in a subsequent section, such relationships are invaluable for verification of computational models that involve crack propagation. The application of dimensional analysis to the development of stable cone cracks appears to have

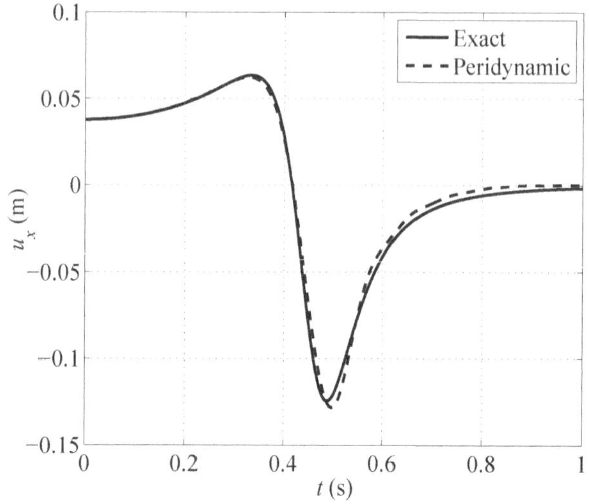

Figure 8.5: x-directed displacement at $x = 0.5$, $y = 0$ m. The solid line shows the exact solution and the dashed line results terminated by a PML.

been first addressed by Roesler [18] and Benbow [3], and later more fully analyzed by Barenblatt [1]. The universal dimensionless relationship,

$$D = \left(\frac{P}{K}\right)^{\frac{2}{3}}, \tag{8.29}$$

relates the width D of the base of a cone crack induced by a flat punch (cylindrical indenter) to the load P, and cohesive modulus K of the medium (Fig. 8.6). Cone crack development in glass confirms the universal scaling law described in Equation (8.29). In addition to the relevant physical parameters, Poisson's ratio, ν, and the modulus of cohesion, K, play an important role in describing the mechanics of the indentation problem. According to similarity methods described by Sedov [19] and others, since there are five relevant physical variables, and three fundamental dimensions of M, L, and T, this results in two possible dimensionless products, which can be formed from the physical variables.

The two dimensionless groups can be derived for this problem by first writing the relevant physical parameters in the following multiplicative form:

$$(D)^{k_1}(d_0)^{k_2}(P)^{k_3}(K)^{k_4}(\nu)^{k_5} = 1, \tag{8.30}$$

and then expressing Equation (8.30) in terms of the fundamental dimensions:

$$(L)^{k_1}(L)^{k_2}(MLT^{-2})^{k_3}(ML^{-1/2}T^{-2})^{k_4}(1)^{k_5} = 1. \tag{8.31}$$

The fundamental dimension terms in Equation (8.31) can be factored as follows:

$$(M)^{k_3+k_4}(L)^{k_1+k_2+k_3-k_4/2}(T)^{-2k_3-2k_4} = 1, \tag{8.32}$$

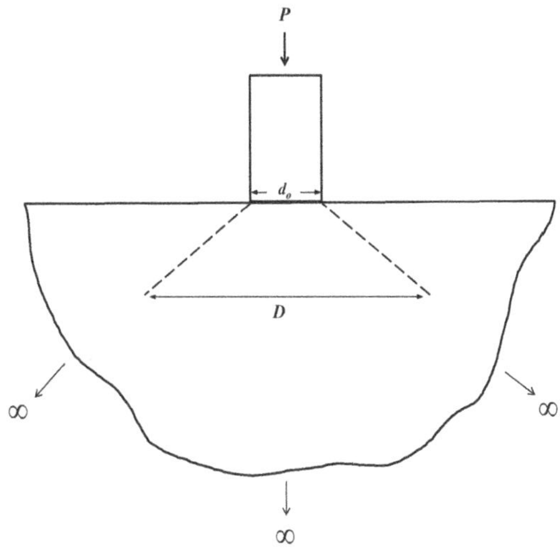

Figure 8.6: Dimensional analysis for the axisymmetric Hertzian cone crack problem.

resulting in the following system of equations:

$$\begin{aligned} k_3 + k_4 &= 0, \\ k_1 + k_2 + k_3 - k_4/2 &= 0, \\ -2k_3 - 2k_4 &= 0, \end{aligned} \tag{8.33}$$

that are solvable using the constants that appear in Table 8.1.

Table 8.1: Constants appearing in dimensional analysis.

	k_1	k_2	k_3	k_4	k_5
π_1	1	0	$-2/3$	$2/3$	1
π_2	0	1	$-2/3$	$2/3$	1

The dimensionless groups, also known as π-groups from the Buckingham π theorem [5], can be written as

$$\begin{aligned} \pi_1 &= DP^{-2/3}K^{2/3}\nu, \\ \pi_2 &= d_0P^{-2/3}K^{2/3}\nu. \end{aligned} \tag{8.34}$$

Since the π-groups in Equation (8.34) are dimensionless, they can be equated and solved for the width D of the base of the cone crack,

$$D = \left(\frac{P}{K}\right)^{2/3} \nu f\left(d_0\left(\frac{K}{P}\right)^{2/3}, \nu\right). \tag{8.35}$$

Thus, a log-log plot of D versus P, for $D >> d_0$, will have a slope of $2/3$, as it appears that the contribution from the third function in Equation (8.35) is negligible. The fracture scaling law can also be used to verify and validate finite element, peridynamic, material point, MD, and other computational methods currently in use for simulating fracture resulting from indentation experiments.

For fractures that develop under 2D states of stress such as the problem of the symmetric wedging of a thick plate [2], the wedging force can be approximated by two equal and opposite concentrated tractions P per unit length of plate thickness. Using the dimensional analysis methods outlined earlier in this section, a dimensionless relation can be derived, for 2D stress and deformation fields, which only depends on crack length l, load P, and cohesive modulus K [2]:

$$l = \left(\frac{P}{K}\right)^2. \tag{8.36}$$

This result is used to verify peridynamic simulations of fracture in the following sections.

8.3.2 State-based verification of a cone crack

Crack propagation in a half-space can be useful for modeling physical phenomenon such as indentation experiments. As an example, such a problem is modeled as a body force applied to a finite region with small pre-cracks in a region terminated on three sides with PMLs. One addition to the algorithm for this problem was a drag term, used to reduce noise. For crack problems with a sudden force application, noise and oscillations can cause hot spots and undesirable cracking. To remedy this, a drag term can be added to smooth oscillations, by adjusting the nodal velocity as

$$\mathbf{v}^*[\mathbf{x}_i, l] = (1-D)\,\mathbf{v}[\mathbf{x}_i, l] + \frac{D}{N_b}\sum_{j=1}^{N_b} \mathbf{v}[\mathbf{x}_j, l], \tag{8.37}$$

where D is the drag coefficient and N_b is the remaining number of bonds in the family of node n.

An absorbing boundary ensures that no reflections from the boundaries interfere with the crack propagation, possibly causing it to deviate. Fig. 8.7 gives a schematic of the problem: the extent of the computation region is designated by the solid line, the PML ramp begins at the dashed line, and the PML plateaus at the dotted line. The computational region was 70 mm wide and 35.25 mm high, the PML region began at 15 mm from each edge (except the top), and peaked at 5 mm to a value of $5 \times 10^6\ \mathrm{s}^{-1}$. The node spacing was 0.496 mm and the time step size was 1 ns. For material values, the density was 2235 kg/m^3, Young's modulus was 65 GPa, Poisson's ratio was 0.2, and the fracture criteria used a fracture energy of 204 J/m^2. The failure criteria used in this simulation was bond-based, i.e., a bond failed if it was stretched past a given limit, determined by the fracture energy [12]. The maximum relative bond stretch was then 2.971×10^{-3}. The load was applied across a 10-mm region, centered at the top surface, with pre-cracks on each edge with a length of two nodes or 0.993 mm. The simulation was run for a total of 10 μs, and the cracks were measured manually from the edge of the pre-cracks to the extent of the damaged area.

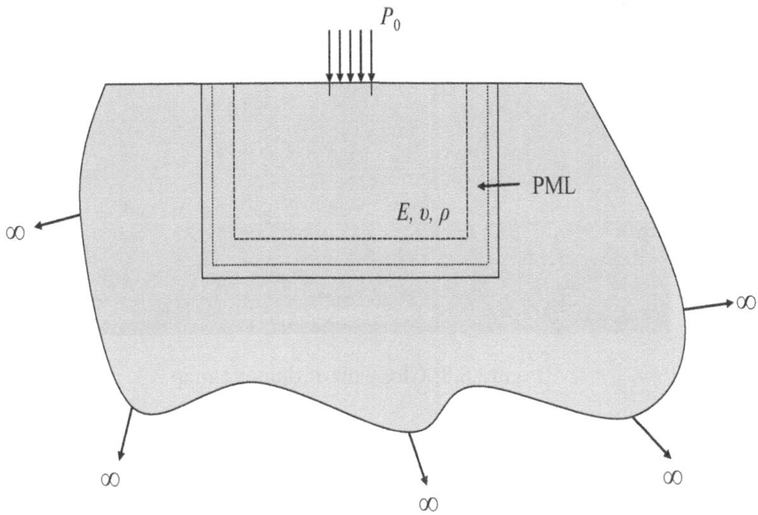

Figure 8.7: Schematic diagram of the problem for crack propagation in a half-space.

Fig. 8.8 shows a close-up of the damaged area from a result that had an applied load of 250×10^9 N, yielding a 2.11-mm crack. As can be seen, the crack extends three nodes down and three nodes across. Finally, the applied load was varied between 140×10^9 N and 500×10^9 N, with the crack length versus applied load shown in Fig. 8.9 as the dots. A curve, shown as the solid line in Fig. 8.9, was fit using the form motivated by the universal scaling law presented in Section 8.3.1:

$$\ell = Ap^s, \tag{8.38}$$

where ℓ is the crack length in meters and p is the applied load in newtons. A linear fit was computed in log-log space using least squares giving a power law of $s = 1.96$ and $A = 4.3 \times 10^{-8}$ m/N^s. This fit matches well with the predicted, dimensional analysis law shown in Equation (8.36). Though the loading is different here, because the problem is 2D the dimensional analysis still applies giving the correct power law form. In other words, for all 2D problems with stable crack growth in infinite regions, a load-crack length relation similar to that of Eq. 8.38 with a power of $s = 2$ should be expected, while in 3D, a power of $s = 2/3$ is expected.

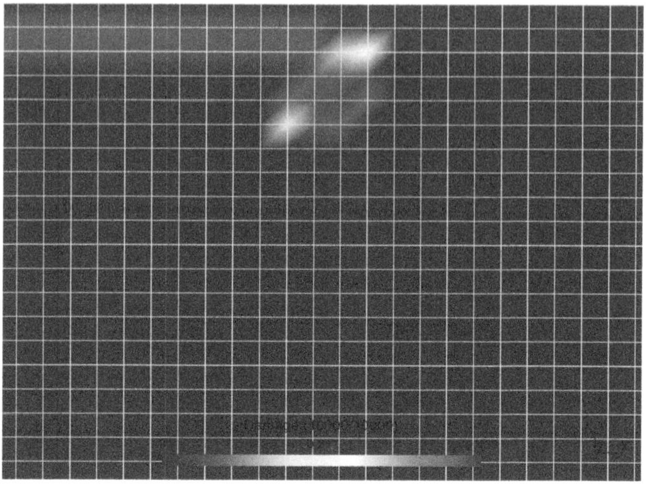

Figure 8.8: Close-up of damage map.

8.3.3 Bond-based verification of a center crack

A center crack opening problem was simulated with a bond-based, non-PML terminated peridynamics method. The problem involves applying equal but oppositely directed point loads in an infinite region. The stable crack length is related to the applied load as shown in Equation (8.36). Here, rather than applying the load at a single node (which may cause instability or large oscillations), the load was spread over three nodes on each side at the center of a square region. Bond-based peridynamics was used with a node spacing of 0.5 mm, a time step size of 1 ns, Young's modulus of 65 GPa, density 2235 kg/m^3, and horizon size of 1.5 mm with a Heaviside shape function. The load was applied slowly with a Gaussian ramp, spread over 500 time steps. The region was not terminated with a PML, though the region size (100 mm by 100 mm) was large enough to avoid issues associated with crack-boundary interactions. The criterion for bond damage was a critical relative stretch of 0.00185. A small pre-crack was added between the two point loads and the bonds surrounding the loads were set to a no-fail condition. This setup was run for varying loads, between 4×10^{11} and 9×10^{11} N, and the crack half length was measured between the center of the region and the crack tip, with a crack tip being a region with a broken bond ratio of more than 30%. The results are shown in Fig. 8.10 on

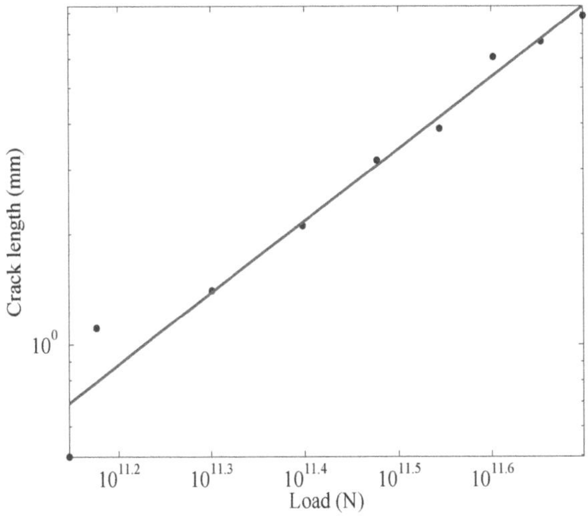

Figure 8.9: Crack length versus applied force for indentation into an elastic half-space using state-based peridynamics. Dots represent data points from the peridynamic simulation and the solid line is a least-squares fit.

a log-log scale. As in the above example (see Equation (8.38)), a linear curve fit in log-log space (shown in Fig. 8.10 as the solid line) was computed using least squares to determine the exponent of the power law, and found to be 1.94, very near 2, the value determined by dimensional analysis in Equation (8.36).

8.4 Verification of an axisymmetric indentation problem

In this final section, an axisymmetric peridynamic formulation is derived based on the formulation of Section 8.2. As will be shown, an axisymmetric formulation is straightforward, as the derivatives can be matched directly to those in the plane strain formulation and the undifferentiated terms carry over to the peridynamics formulation unaltered. Consequently, the application of a PML is also straightforward. The formulation is verified using exact solutions for an axisymmetric indentation problem of a rigid, circular punch into an elastic half-space as derived in [11].

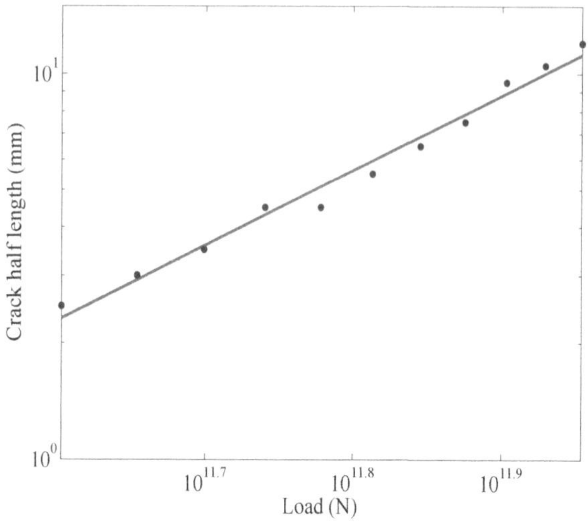

Figure 8.10: Crack half length versus load for a center crack problem using bond-based peridynamics.

8.4.1 Formulation

A state-based, axisymmetric formulation can be developed by again considering the form of the deformation gradient and matching the necessary derivatives. To review, an axisymmetric problem in cylindrical coordinates is considered such that there is no displacement variation or component in the azimuthal coordinate θ (i.e. $u_\theta = 0$ and $\partial u/\partial\theta = 0$). The gradient of the displacement in polar coordinates is given by

$$\nabla\mathbf{u} = \begin{bmatrix} \frac{\partial u_r}{\partial r} & 0 & \frac{\partial u_r}{\partial z} \\ 0 & \frac{1}{r}u_r & 0 \\ \frac{\partial u_z}{\partial r} & 0 & \frac{\partial u_z}{\partial z} \end{bmatrix}, \tag{8.39}$$

giving the strain as

$$\overline{\varepsilon} = \begin{bmatrix} \frac{\partial u_r}{\partial r} & 0 & \frac{1}{2}\left(\frac{\partial u_z}{\partial r} + \frac{\partial u_r}{\partial z}\right) \\ 0 & \frac{1}{r}u_r & 0 \\ \frac{1}{2}\left(\frac{\partial u_z}{\partial r} + \frac{\partial u_r}{\partial z}\right) & 0 & \frac{\partial u_z}{\partial z} \end{bmatrix}. \tag{8.40}$$

Next, assuming the stress tensor has the same components as the strain tensor, the divergence of the stress is expanded as:

$$\nabla\cdot\overline{\sigma} = \begin{bmatrix} \frac{\partial\sigma_{rr}}{\partial r} + \frac{\partial\tau_{rz}}{\partial z} + \frac{1}{r}\left(\sigma_{rr} - \sigma_{\theta\theta}\right) \\ 0 \\ \frac{\partial\sigma_{zz}}{\partial z} + \frac{\partial\tau_{rz}}{\partial r} + \frac{1}{r}\tau_{rz} \end{bmatrix}, \tag{8.41}$$

where

$$\overline{\sigma} = \begin{bmatrix} \sigma_{rr} & 0 & \tau_{rz} \\ 0 & \sigma_{\theta\theta} & 0 \\ \tau_{rz} & 0 & \sigma_{zz} \end{bmatrix}. \tag{8.42}$$

Hooke's law for axisymmetric elastic bodies can then be stated as

$$\overline{\sigma} = \overline{\overline{\mathbf{c}}} : \overline{\varepsilon} = \begin{bmatrix} \lambda+2\mu & \lambda & \lambda & 0 \\ \lambda & \lambda+2\mu & \lambda & 0 \\ \lambda & \lambda & \lambda+2\mu & 0 \\ 0 & 0 & 0 & 2\mu \end{bmatrix} \begin{bmatrix} \varepsilon_{rr} \\ \varepsilon_{\theta\theta} \\ \varepsilon_{zz} \\ \varepsilon_{rz} \end{bmatrix}, \tag{8.43}$$

where λ and μ are Lamé's parameters. Ultimately, the above can be given in terms of six coupled equations in the Laplace domain, following the form of Equation (8.14), as

$$\begin{aligned}
\rho s \tilde{u}_r &= \frac{\partial \psi_{rr}}{\partial r} + \frac{\partial \psi_{rz}}{\partial z} + \frac{1}{r}(\psi_{rr} - \psi_{\theta\theta}) \\
\rho s \tilde{u}_z &= \frac{\partial \psi_{zz}}{\partial z} + \frac{\partial \psi_{rz}}{\partial r} + \frac{1}{r}\psi_{rz} \\
s\tilde{\psi}_{rr} &= (\lambda + 2\mu)\frac{\partial}{\partial r}\tilde{u}_r + \lambda\frac{1}{r}u_r + \lambda\frac{\partial}{\partial z}\tilde{u}_z \\
s\tilde{\psi}_{\theta\theta} &= \lambda\frac{\partial}{\partial r}\tilde{u}_r + (\lambda + 2\mu)\frac{1}{r}u_r + \lambda\frac{\partial}{\partial z}\tilde{u}_z \\
s\tilde{\psi}_{zz} &= \lambda\frac{\partial}{\partial r}\tilde{u}_r + \lambda\frac{1}{r}u_r + (\lambda + 2\mu)\frac{\partial}{\partial z}\tilde{u}_z \\
s\tilde{\psi}_{rz} &= \mu\left(\frac{\partial}{\partial z}\tilde{u}_r + \frac{\partial}{\partial r}\tilde{u}_z\right).
\end{aligned} \tag{8.44}$$

Now, comparing Equations (8.14) and (8.44), all terms involving spatial derivatives have direct counterparts in each equation. In other words, the axisymmetric formulation only involves additional variables that are not differentiated in space. Thus, the state-based peridynamic formulation from Equations (8.15) and (8.16) and the PML formulation and discretization from Equation (8.25) can be directly applied and the appropriate (undifferentiated) terms are added where necessary. A computational implementation would then simply substitute all terms involving components and variations in r and z for those in x and y, respectively.

8.4.2 Analytical verification

In what follows, a 100 by 100 node grid was used for all axisymmetric peridynamic simulations. The PML for 2D peridynamics described above is employed in this axisymmetric solution at finite radial and vertical distances $\rho = \zeta \approx 8$ within the computational domain to simulate the zero vertical displacement condition that exists at infinity. A mean pressure $p_m = \frac{2\varepsilon}{\pi a}\frac{E}{1-\nu^2}$ is induced in the elastic half-space with rigid punch radius $a = 5$ mm, indentation depth $\varepsilon = 0.5$ mm, Young's modulus, $E =$ 65 GPa, and Poisson's ratio, $\nu = 0.20$ is 4.3 GPa.

The solution for the vertical displacements in an elastic half-space due to an axisymmetric indentor [11],

$$\frac{u_z(\rho,\zeta)}{\varepsilon} = \frac{1}{\pi}\left(2\arg\left(\frac{(\zeta-i)(1+H)}{\rho}\right) + \frac{\zeta}{\nu-1}\Im\left(\frac{1}{(\zeta-i)H}\right)\right) \tag{8.45}$$

where $H(\rho,\zeta) = \sqrt{1+\frac{\rho^2}{(\zeta-i)^2}}$ and $\Im(\cdot)$ is the imaginary part, is illustrated in Figure 8.11 (solid lines), where a numerical peridynamic solution (dashed lines) is verified at dimensionless depths, $\zeta = 0, \zeta = 1.7$, and $\zeta = 5.4$. Note that the vertical displacements on the surface are given by

$$\frac{u_z(\rho,\zeta=0)}{\varepsilon} = \begin{cases} \frac{2}{\pi}\sin^{-1}\left(\frac{a}{r}\right), & r > a \\ 1, & r \le a \end{cases}. \tag{8.46}$$

The solution for the radial displacements [11],

$$\frac{u_r(\rho,\zeta)}{\varepsilon} = \frac{1}{\pi(\nu-1)\rho}\left((1-2\nu)\Im((\zeta-i)(H-1)) + \zeta\Im\left(\frac{1}{H}\right)\right) \tag{8.47}$$

of the half-space is illustrated in Figure 8.12 (solid lines), where the numerical peridynamic solution (dashed lines) is verified at dimensionless depths, $\zeta = 0, \zeta = 1.7$, and $\zeta = 5.4$. The analytical radial surface displacements of the half-space are generally non-monotonic and the computed peridynamic results accurately reproduce the theoretical trends as a function of depth within the half-space; note, however, that the computed radial displacements undergo a small spatial wrinkle at $\rho = 1$ at the corner of the rigid punch, and it may be surmised that this is related to singularities present in the stress and strain fields that may influence the non-singular displacement fields. Since such behavior at $\rho = 1$ is not present in the computed vertical displacement field illustrated in Figure 8.11, it is also possible that this is a result of a solution that has not fully converged.

References

[1] G. I. Barenblatt. The mathematical theory of equilibrium cracks in brittle fracture. *Advances in Applied Mechanics*, 7: 55–129, 1962.

[2] G. I. Barenblatt. Scaling phenomena in fatigue and fracture. *International Journal of Fracture*, 138(1–4): 19–35, 2006. 11th International Conference on Fracture, Torino, ITALY, MAR 20–25, 2005.

[3] J. J. Benbow. Cone cracks in fused silica. *Proceedings of the Physical Society*, 75(5): 697–699, 1960.

[4] J.P. Berenger. A perfectly matched layer for the absorption of electromagnetic-waves. *Journal Computational Physics*, 114(2): 185–200, 1994.

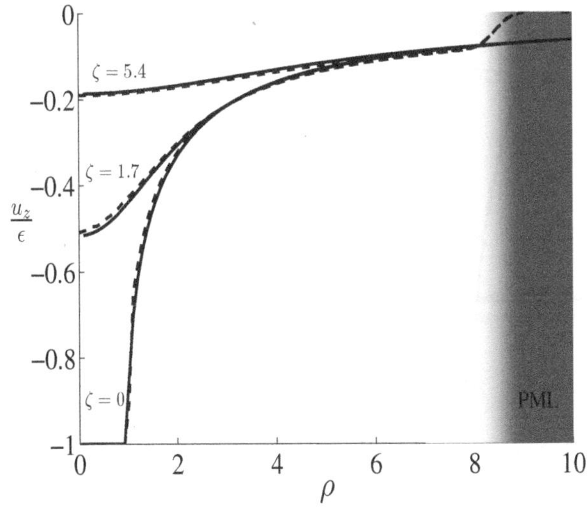

Figure 8.11: Verification of the peridynamics results (dashed lines) using the analytical solution (solid lines; Equation (8.45)) for normalized vertical displacements $\frac{u_z}{\varepsilon}$ in an isotropic elastic half-space, $\nu = 0.2$, due to a rigid flat-ended cylindrical punch; the mean pressure induced in the half-space is $p_m = 4.3$ GPa.

[5] E. Buckingham. On physically similar systems; illustrations of the use of dimensional equations. *Physical Review*, 4: 345–376, 1914.

[6] W.C. Chew and W.H. Weedon. A 3D perfectly matched medium from modified Maxwell's equations with stretched coordinates. *Microwave and Optical Technology Letters*, 7(13): 599–604, 1994.

[7] Q. Du, M. D. Gunzburger, R. B. Lehoucq, and K. Zhou. A nonlocal vector calculus, nonlocal volume-constrained problems, and nonlocal balance laws. Sandia Report SAND-2010-8353J, Sandia National Laboratories, 2010.

[8] A. C. Eringen and E. S. Suhubi. *Elastodynamics, Vol. II, Linear Theory*. Academic Press, New York, 1975.

[9] J. T. Foster, S. A. Silling, and W. Chen. Viscoplasticity using peridynamics. *International Journal for Numerical Methods in Engineering*, 81(10):1242–1258, 2010.

[10] G. A. Gazonas, J. W. McCauley, I. G. Batyrev, R. C. Becker, S. Izvekov, T. Jenkins, P. Patel, B. M. Rice, B. E. Schuster, N. S. Weingarten, and R. A. Wildman. Multiscale modeling of non-crystalline ceramics (glass): Final Report. Technical Report ARL-TR-6353, U.S. Army Research Laboratory: Aberdeen Proving Ground, MD, 2013.

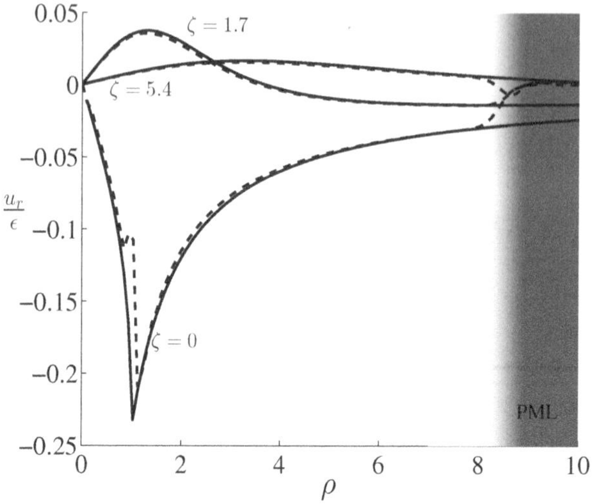

Figure 8.12: Verification of the peridynamics results (dashed lines) using the analytical solution (solid lines; Equation (8.47)) for normalized radial displacements $\frac{u_r}{\varepsilon}$ in an isotropic elastic half-space, $\nu = 0.2$, due to a rigid flat-ended cylindrical punch

[11] G. A. Gazonas and R. A. Wildman. The Mechanics of Axisymmetric Indentation Revisited. Technical Report ARL-TR-6528, U.S. Army Research Laboratory, 2013.

[12] Y. D. Ha and F. Bobaru. Studies of dynamic crack propagation and crack branching with peridynamics. *International Journal of Fracture*, 162(1-2): 229–244, 2010.

[13] S. G. Johnson. Notes on perfectly matched layers. http://math.mit.edu/~stevenj/18.369/pml.pdf, accessed 2010.

[14] C. Lubich. Convolution quadrature and discretized operational calculus 1. *Numerische Mathematik*, 52(2): 129–145, 1988.

[15] C. Lubich. Convolution quadrature and discretized operational calculus 2. *Numerische Mathematik*, 52(4):413–425, 1988.

[16] L.E. Malvern. *Introduction to the Mechanics of a Continuous Medium.* Upper Saddle River, NJ: Prentice-Hall, 1969.

[17] J.G. Proakis and D. G. Manolakis. *Digital Signal Processing: Principles, Algorithms, and Applications.* Upper Saddle River, NJ: Prentice-Hall, 1996.

[18] F. C. Roesler. Brittle fractures near equilibrium. *Proceedings of the Physical Society. Section B*, 69(10): 47–54, 1956.

[19] L. I. Sedov. *Similarity and Dimensional Methods in Mechanics*.10th ed. Boca Raton, FL: CRC Press, 1993.

[20] S. A. Silling. Reformulation of elasticity theory for discontinuities and long-range forces. *Journal of the Mechanics and Physics of Solids*, 48(1): 175–209, 2000.

[21] S. A. Silling and E. Askari. A meshfree method based on the peridynamic model of solid mechanics. *Computers & Structures*, 83(17-18): 1526–1535, 2005.

[22] S. A. Silling, M. Epton, O. Weckner, J. Xu, and E. Askari. Peridynamic states and constitutive modeling. *Journal of Elasticity*, 88(2):151–184, 2007.

[23] X.-B. Wang, R. A. Wildman, D. S. Weile, and P. Monk. A finite difference delay modeling approach to the discretization of the time domain integral equations of electromagnetics. *IEEE Transactions on Antennas and Propagation*, 56(8):2442–2452, 2008.

[24] R. A. Wildman and G. A. Gazonas. A perfectly matched layer for peridynamics in one dimension. Technical Report ARL-TR-5626, U.S. Army Research Laboratory, 2011.

[25] R. A. Wildman and G. A. Gazonas. A perfectly matched layer for peridynamics in two dimensions. *Journal of Mechanics of Materials and Structures*, 7 (8-9): 765–781, 2012.

Part IV

Modeling Material Failure and Damage

Chapter 9

Dynamic Brittle Fracture as an Upscaling of Unstable Mesoscopic Dynamics

Robert P. Lipton

Louisiana State University

CONTENTS

9.1 Introduction 229
9.2 The macroscopic evolution of brittle fracture as a small horizon limit of mesoscopic dynamics 234
9.3 Dynamic instability and fracture initiation 238
9.4 Localization of dynamic instability in the small horizon-macroscopic limit 240
9.5 Free crack propagation in the small horizon-macroscopic limit 241
9.6 Summary 242
References 243

9.1 Introduction

Dynamic brittle fracture is a multiscale phenomenon operating across a wide range of length and time scales. Contemporary approaches to brittle fracture modeling can be characterized broadly as bottom-up and top-down. Bottom-up approaches that take into account the discreteness of fracture process at the smallest length scales are often expressed through lattice models. This approach has provided penetrating insight into the dynamics of the fracture processes [6, 16, 17, 21]. On the other hand, numerical simulations of lattice models do not scale up to finite size samples with multiple freely propagating cracks. Complementary to the bottom-up approaches are top-down computational approaches using cohesive zone elements [14, 22]. Here the details of the process zone are collapsed onto an interfacial element with a force traction law given by the cohesive zone model [4]. More recently, cohesive zones have been applied within the extended finite element method [5, 18] to minimize the effects of mesh dependence on free crack paths. The current challenges facing these methods (indeed, all computational methods) include multiple growing cracks interacting in complex patterns.

In this chapter we depart from the aforementioned top-down and bottom up approaches to crack modeling and instead use peridynamics [19] to develop a nonlocal mesoscopic model for assessing the deformation state inside a cracking body. This model is used to connect the dynamics associated with unstable bonds at mesoscopic scales to dynamic free crack propagation inside a brittle material, as observed at macroscopic length scales. This strategy can be thought of as a bottom-up approach. However instead of starting at atomistic scales we pose the dynamics at the mesoscale and scale up to discover the macroscopic ramifications. A future goal would be to inform the mesoscopic model through atomistic or molecular dynamics at still smaller length scales.

The mesoscopic model is characterized through a nonlocal peridynamic potential

$$W^{\varepsilon}(\eta, x' - x).$$

The parameter ε indicates the relative length scale of the nonlocal interaction with respect to the characteristic dimensions of the material sample under consideration. Here $\eta = u(x') - u(x)$, where u is the displacement field inside the body evaluated at the points x and x' in the reference configuration. The range of nonlocal interaction is contained within the peridynamic neighborhood of x denoted by $\mathcal{H}_{\varepsilon}(x)$. The neighborhood is given by all points x' lying inside the sphere of radius ε centered at x. Points x' outside this neighborhood do not interact with x.

The mesomodel is developed within the context of an antiplane shear problem for a generic cross-section D of a long prismatic shaft. The antiplane displacement transverse to D is written $u(t,x)$. The relative displacement with respect to x is denoted by $\eta(t,x) = u(t,x') - u(t,x)$ and the antiplane shear strain with respect to x is written as $\mathcal{S}(x,t) = \eta(x)/|x' - x|$. For this problem the peridynamic neighborhood is denoted by $\mathcal{H}_{\varepsilon}(x)$ and is given by a disk of radius ε centered at x.

We investigate the behavior of the dynamics associated with the family of potentials $W^{\varepsilon}(\eta, x' - x)$ in the limit as $\varepsilon \to 0$. This limit represents the macroscopic limit for which the length scale of nonlocal interaction becomes infinitesimally small with respect to the shaft cross section D. We introduce a suitable class of peridynamic potentials associated with bonds that are initially elastic for small strains and exhibit softening after a critical strain. These potentials are of the generic form given by

$$W^{\varepsilon}(\eta, \varepsilon\xi) = \frac{1}{\varepsilon^3} J(|\xi|) f\left(\frac{|\eta|^2}{\varepsilon|\xi|}\right) = |\xi| (\mathcal{W}^{\varepsilon}(\mathcal{S}, \varepsilon|\xi|)), \tag{9.1}$$

where

$$\mathcal{W}^{\varepsilon}(\mathcal{S}, \varepsilon|\xi|) = \left(\frac{1}{\varepsilon^2} J(|\xi|) \frac{1}{\varepsilon|\xi|} f\left(\varepsilon|\xi|\mathcal{S}^2\right)\right); \tag{9.2}$$

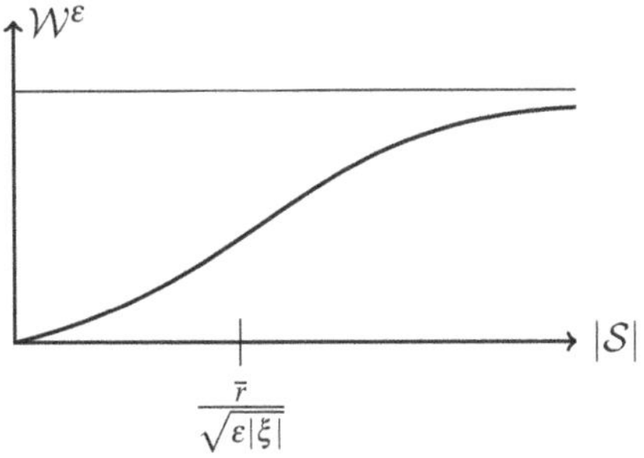

Figure 9.1: Convex-concave potential.

see Figure 9.1. Here we have made the change of variable $x' = x + \varepsilon\xi$, where ξ belongs to the unit disk $\mathcal{H}_1(0)$ centered at the origin. The quantity $\mathcal{W}^\varepsilon(\mathcal{S}, \varepsilon|\xi|)$ represents the energy per unit bond length per unit area due to a shear strain of magnitude $|\mathcal{S}|$. These potentials are of a general form and are associated with potential functions $f : [0,\infty) \to \mathbb{R}$ that are positive, smooth, and concave with the properties

$$\lim_{r\to 0^+} \frac{f(r)}{r} = f'(0) > 0, \qquad \lim_{r\to\infty} f(r) = f_\infty < \infty. \tag{9.3}$$

Here $J(|\xi|)$ is used to prescribe the influence of bonds of length $|\xi|$ on the bond force with $0 \le J(|\xi|) < M$ for $0 \le |\xi| \le 1$ and $J(|\xi|) = 0$ for $1 < |\xi|$. All potential energies given by (9.1) are convex-concave with respect to the magnitude of the shear strain $\mathcal{S}$ and have the infection point $\overline{r}/\sqrt{\varepsilon|\xi|}$. Here $\overline{r}$ is the inflection point for the function $r :\to f(r^2)$ (see Figure 9.1). This choice of potential delivers an initially elastic and then softening constitutive law for the bond force given by

$$\text{force} = \partial_\eta W^\varepsilon(\eta, \varepsilon\xi) = \varepsilon|\xi| \left(\frac{2}{\varepsilon^3} J(|\xi|) \frac{1}{\varepsilon|\xi|} f'\left(\varepsilon|\xi|\mathcal{S}^2\right) \mathcal{S} \right). \tag{9.4}$$

The bond force starts to fail when the antiplane shear strain $\mathcal{S}$ exceeds the critical shear strain,

$$|\mathcal{S}| > \frac{\overline{r}}{\sqrt{\varepsilon|\xi|}} = \frac{\overline{r}}{\sqrt{|x'-x|}} \tag{9.5}$$

(see Figure 9.2). Note that this is the same singularity strength associated with strain concentrations in the vicinity of a crack tip as in the classic theory of brittle fracture [11].

This class of potentials describes an ordinary state based elastic material with force state given by

$$\mathbf{T}^\varepsilon(x,t)\langle\varepsilon\xi\rangle = \partial_\eta W^\varepsilon(\eta(x), \varepsilon\xi). \tag{9.6}$$

The mesoscopic evolution $u^\varepsilon(x,t)$ is characterized by the peridynamic equation

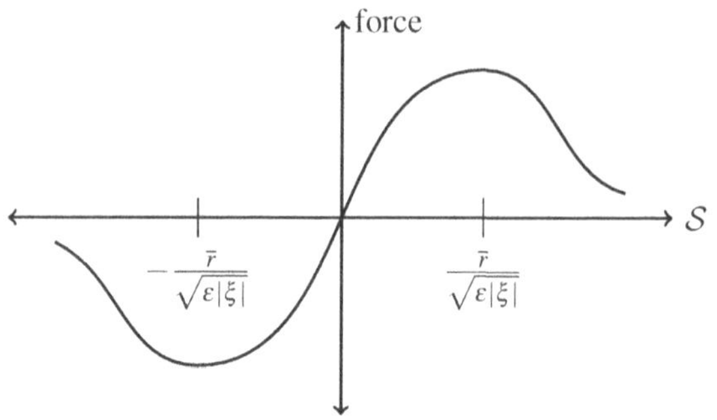

Figure 9.2: Unstable constitutive relation between force and antiplane shear strain.

$$\rho u_{tt}^{\varepsilon}(x,t) = \int_{\mathcal{H}_1(0)} (\mathbf{T}^{\varepsilon}(x,t)\langle\varepsilon\xi\rangle - \mathbf{T}^{\varepsilon}(x+\varepsilon\xi,t)\langle-\varepsilon\xi\rangle)\, d\xi + b(x,t) \quad (9.7)$$

where $b(x,t)$ is the prescribed body force density acting transverse to the cross section D in the reference configuration.

To define boundary conditions we introduce the slightly larger set D_α, with $D \subset D_\alpha$ given by the set of points x for which $dist(x,D) < \alpha$ (see Figure 9.3). For points x residing in $D_\alpha \sim D$ we set the displacement $u(t,x) = 0$ for all $0 \leq t \leq T$; this gives the nonlocal analogue of "Dirichlet" boundary conditions for peridynamic problems introduced in [9]. The initial conditions for the mesoscopic dynamics are given by the prescribed out of plane deformation and velocity fields given by

$$u(x,0) = u_0(x), \text{ and } u_t(x,0) = v_0(x). \quad (9.8)$$

In this formulation the initial conditions are chosen such that the initial vertical displacement $u_0(x)$ is smooth except across a system of preexisting cracks. Here the cracks are given by a system of curves of finite total length; the deformation suffers a finite jump discontinuity across each curve. Our choice of initial conditions is motivated by Linear Elastic Fracture Mechanics (LEFM). Here we investigate Mode III fracture in the context of antiplane shear. The initial condition is specified by a crack set K and displacement u_0. The gradient ∇u_0 is defined off the crack set and the displacement u_0 can suffer jumps across K. Griffith's theory of brittle fracture asserts that the energy necessary to produce a crack K is proportional to the crack length ℓ. For LEFM the total energy associated with bulk elastic and surface energy is given by

$$\int_D \mu|\nabla u_0|^2\, dx + \mathcal{G}_c\ell, \quad (9.9)$$

where μ is the shear modulus and $\mathcal{G}_c$ is the critical energy release rate for the material. In what follows we will assume that the bulk elastic energy and surface energy are

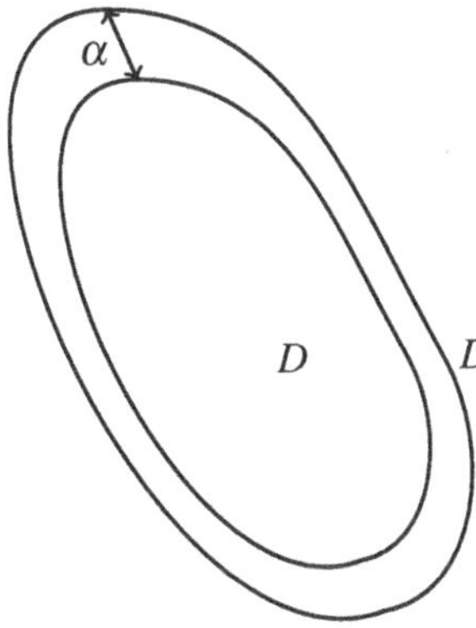

Figure 9.3: Domains D and D_α.

bounded as well as the initial velocity and displacement. For future reference we describe initial data u_0 and v_0 that satisfy these conditions as *LEFM initial data*.

This formulation of the mesoscopic dynamics is well posed for LEFM initial data [15], i.e., there is a unique mesoscopic evolution satisfying this initial boundary value problem and the mesoscopic evolution $u^\varepsilon(x,t)$ is uniformly bounded in the mean square norm over bounded time intervals $0 < t < T$, i.e.,

$$\max_{0<t<T} \left\{ \|u^\varepsilon(x,t)\|_{L^2(D)} \right\} < C. \tag{9.10}$$

Here $\|u^\varepsilon(x,t)\|_{L^2(D)} = \left(\int_D |u^\varepsilon(x,t)|^2\,dx\right)^{1/2}$ is the mean square norm over D and the upper bound C is independent of ε and depends only on the initial conditions and body force applied up to time T. (In this context we also point out the recent work of [10] verifying the well-posed nature nonlinear peridynamic problems with Lipschitz-continuous pairwise force functions.)

The mesoscopic evolution has several features to it. We introduce the total peridynamic elastic energy given by

$$\begin{aligned} PD^\varepsilon(u^\varepsilon(t)) &= \varepsilon^2 \int_D \int_{\mathcal{H}_1(0)} W^\varepsilon(\eta^\varepsilon(t), \varepsilon\xi)\,d\xi\,dx \\ &= \int_D \int_{\mathcal{H}_{\varepsilon(x)}} W^\varepsilon(\eta^\varepsilon(t), x' - x)\,dx'\,dx. \end{aligned} \tag{9.11}$$

Here we have suppressed the spatial variable and written $u^\varepsilon(x,t) = u^\varepsilon(t)$ and $u^\varepsilon(x + \varepsilon\xi, t) - u^\varepsilon(x,t) = \eta^\varepsilon(t)$. The mesoscopic evolution has uniformly bounded kinetic and elastic potential energy (see [15]).

Theorem 1 Bounds on kinetic and potential energy for the mesoscopic evolution *There exists a positive constant C depending only on T and independent of ε for* *which*

$$\sup_{0\leq t\leq T}\left\{PD^{\varepsilon}(u^{\varepsilon}(t))+\frac{\rho}{2}\|u_t^{\varepsilon}(t)\|_{L^2(D)}\right\}\leq C. \tag{9.12}$$

It is also uniformly continuous in time as measured by the mean square norm (see [15]).

Theorem 2 Continuous mesoscopic evolution in mean square norm
There is a positive constant K independent of $t_2 < t_1$ in $[0,T]$ and index ε for which

$$\|u^{\varepsilon}(t_1)-u^{\varepsilon}(t_2)\|_{L^2(D)}\leq K|t_1-t_2|. \tag{9.13}$$

The evolution also satisfies energy balance. The total energy of the mesoscopic evolution at time t is given by

$$\mathcal{EPD}^{\varepsilon}(t,u^{\varepsilon}(t))=\frac{\rho}{2}\|u_t^{\varepsilon}(t)\|^2_{L^2(D)}+PD^{\varepsilon}(u^{\varepsilon}(t))-\int_D b(t)u^{\varepsilon}(t)\,dx \tag{9.14}$$

and the total energy of the system at time $t=0$ is

$$\mathcal{EPD}^{\varepsilon}(0,u^{\varepsilon}(0))=\frac{\rho}{2}\|v_0\|^2_{L^2(D)}+PD^{\varepsilon}(u_0)-\int_D b(0)u_0\,dx. \tag{9.15}$$

The mesoscopic dynamics is seen to satisfy energy balance at every instant of the evolution (see [15]).

Theorem 3 Energy balance

$$\mathcal{EPD}^{\varepsilon}(t,u^{\varepsilon}(t))=\mathcal{EPD}^{\varepsilon}(0,u^{\varepsilon}(0))-\int_0^t\int_D b_t(\tau)u^{\varepsilon}(\tau)\,dx\,d\tau. \tag{9.16}$$

9.2 The macroscopic evolution of brittle fracture as a small horizon limit of mesoscopic dynamics

In this section we show how the dynamics of the mesoscopic model connects to dynamic free crack propagation inside brittle materials, as observed at macroscopic length scales. We begin by observing that the mesoscopic model is characterized through a nonlocal peridynamic potential $W^{\varepsilon}(\eta,x'-x)$ while at the macroscopic length scale the dynamics of the cracking body is characterized by the shear modulus μ and energy release rate $\mathcal{G}_c$. In order to make a connection between dynamics describing the state of the body as seen at two different length scales we need to discover the relationship between the potential $W^{\varepsilon}(\eta,x'-x)$ and the pair μ, $\mathcal{G}_c$. To reveal this connection we consider a family of dynamic mesoscopic evolutions u^{ε_k}, each associated with a fixed potential W^{ε_k} and horizon length ε_k, with $k = 1,2,\dots$ and $\varepsilon_k \to 0$. One can think of each $u^{\varepsilon_k}(x,t)$ as being the result of a perfectly accurate numerical simulation of a mesoscopic evolution associated with the potential W^{ε_k}. It is shown that the mesoscopic dynamics described by $u^{\varepsilon_k}(x,t)$ converges to a limit evolution $u^0(x,t)$ in the small horizon limit as $\varepsilon_k \to 0$ (see [15]). The limiting macroscopic evolution describes the dynamics of the cracked body when the peridynamic horizon is infinitesimally small with respect to the material specimen. Here the macroscopic free crack evolution is mediated through the pair μ and $\mathcal{G}_c$ given by the appropriate moments associated with the sequence of potentials W^{ε_k} [15] (see (9.17), (9.18), and Theorem 5 below).

It is of fundamental importance to emphasize that we do not impose relations between W^{ε_k} and the pair μ, $\mathcal{G}_c$; instead it is discovered that the mesoscopic dynamics $u^{\varepsilon_k}(x,t)$ associated with W^{ε_k} approaches the macroscopic dynamics $u^0(x,t)$ characterized by μ and $\mathcal{G}_c$ as given by the formulas (9.17) *and* (9.18) *when $\varepsilon_k \to 0$.*

In this way it is seen that the mesoscopic to macroscopic approach to free crack modeling described here is a *bottom-up* approach. The sequence of the mesoscopic evolutions u^{ε_k} is seen to converge to the macroscopic free crack evolution $u^0(x,t)$ in mean square, uniformly in time (see Theorem 4). The free crack evolution has bounded energy in the sense of classic brittle fracture mechanics described by (9.9) with μ and $\mathcal{G}_c$ related to the mesoscopic potentials via (9.17) and (9.18).

We now outline the conditions under which this convergence holds true. Suppose we are given a pair μ and $\mathcal{G}_c$ obtained from experiments performed on a material specimen. We calibrate the sequence of mesoscopic potentials W^{ε_k} so that the influence function J and potential function f are chosen to satisfy:

$$\mu = \pi f'(0) \int_0^1 r^2 J(r)dr, \tag{9.17}$$

and

$$\mathcal{G}_c = 2\pi f_\infty \int_0^1 r^2 J(r)dr, \tag{9.18}$$

where f_∞ is defined by (9.3). Here suitable choices of potential f and influence function J can always be found to satisfy (9.17) and (9.18) for any choice of μ and $\mathcal{G}_c$ provided that $\mu > 0$ and $\mathcal{G}_c > 0$.

For these choices of $\mu, \mathcal{G}, f$, and J, the mesoscopic dynamics is found to converge to a limiting macroscopic free crack evolution with the following properties (see [15]).

- The macroscopic evolution has uniformly bounded energy in the sense of linear elastic fracture mechanics for $0 \leq t \leq T$.
- The macroscopic evolution satisfies an energy inequality involving the kinetic energy of the motion together with the bulk elastic and surface energy associated with linear elastic fracture mechanics for $0 \leq t \leq T$.

- The macroscopic evolution is a solution of the linear elastic wave equation for points in space-time not on the crack set.

We now provide explicit conditions under which these properties are realized for the limit dynamics.

Hypothesis 1 *In order to assure convergence to a limit evolution we suppose that the magnitude of the deformations u^{ε_k} for the mesoscopic dynamics are bounded for $0 \leq t \leq T$ uniformly in ε_k. As of this writing, the uniform bound is assumed rather than proved purely for technical reasons.*

The convergence of the mesoscopic evolutions is described by the following theorem [15],

Theorem 4 Convergence of the mesoscopic dynamics to macroscopic dynamics *We prescribe LEFM initial data $u_0(x)$ and $v_0(x)$ and for each ε_k we solve for the mesoscopic evolution u^{ε_k}. Now consider a sequence of solutions u^{ε_k} associated with a vanishing peridynamic horizon $\varepsilon_k \to 0$. Suppose Hypothesis 1 holds true then (passing to a subsequence if necessary) the mesoscopic evolutions u^{ε_k} converge in mean square uniformly in time to a macroscopic evolution $u^0(x,t)$ with the same LEFM initial data, i.e.,*

$$\lim_{\varepsilon_k \to 0} \max_{0 \leq t \leq T} \left\{ \|u^{\varepsilon_k}(x,t) - u^0(x,t)\|_{L^2(D)} \right\} = 0 \tag{9.19}$$

and $u^0(x,0) = u_0(x)$ and $\partial_t u^0(x,0) = v_0(x)$.

To appropriately characterize the LEFM energy for the macroscopic dynamics, we generalize the notion of the gradient of a function and define an approximate gradient. The macroscopic evolution $u^0(x,t)$ has a generalized gradient $\nabla u^0(x,t)$ that satisfies [15],

$$\lim_{r \searrow 0} \frac{1}{\pi r^2} \int_{B(x,r)} \frac{|u^0(y,t) - u^0(x,t) - (y-x) \cdot \nabla u^0(x,t)|}{r} dy = 0. \tag{9.20}$$

The generalized gradient at each time is defined for points x that do not belong to the jump set of $u^0(x,t)$. To describe the jump set we recall a slightly generalized notion of the jump of a function across a plane with unit normal ν passing through a point x. Consider a disk $B(x,r)$ of radius r centered at x. Denote the set of points y inside the disk lying on one side of the plane by $B^-(x,r)$ and denote the set of points inside the disk belonging to the other side by $B^+(x,r)$. The "left" limit of u^0 at x, denoted by $u^0_-(x)$, satisfies

$$\lim_{r \searrow 0} \frac{1}{\pi r^2} \int_{B^-(x,r)} |u^0(y) - u^0_-(x)|\, dy = 0. \tag{9.21}$$

The "right" limit of u^0 denoted by $u^0_+(x)$ satisfies

$$\lim_{r \searrow 0} \frac{1}{\pi r^2} \int_{B^+(x,r)} |u^0(y) - u^0_+(x)|\, dy = 0. \tag{9.22}$$

Here the jump set of u^0 corresponds to the set of points for which $u^0_-(x) \neq u^0_+(x)$. The jump set at time t is written as $S_{u^0(t)}$ *. The limit deformation $u^0(x,t)$ is interpreted as the deformation state inside a cracked body. The crack set K is identified with the jump set $S_{u^0(t)}$ and ∇u^0 represents the usual shear strain in the elastic part of the body outside the crack (see, e.g., [2], [3]). The length of the jump set is given by its one-dimensional Hausdorff measure and is written as $\mathcal{H}^1(S_{u^0(t)})$. The Hausdorff measure corresponds to arc length for smooth curves. With these preliminaries in hand we can characterize the elastic energy of the macroscopic dynamics (see, [15]).

Theorem 5 Macroscopic dynamics has bounded LEFM energy
The macroscopic evolution has an approximate gradient $\nabla u^0(t,x)$ almost everywhere in D and the jump set $S_{u^0(t)}$ is the countable union of rectifiable arcs. Furthermore there exists a constant C depending only on T bounding the LEFM energy of the limit flow [15], i.e.,

$$\mu \int_D |\nabla u^0(t,x)|^2\,dx + \mathcal{G}_c\mathcal{H}^1(S_{u^0(t)}) \leq C \tag{9.23}$$

for $0 \leq t \leq T$.

We now present an energy inequality for the limit evolution. We denote the LEFM energy for the limit evolution $u^0(t)$ at time t as

$$LEFM(u^0(t),D) = \mu \int_D |\nabla u^0(t)|^2\,dx + \mathcal{G}_c\mathcal{H}^1(S_{u^0(t)}) \tag{9.24}$$

and the LEFM energy for the initial data is written

$$LEFM(u_0,D) = \mu \int_D |\nabla u_0|^2\,dx + \mathcal{G}_c\mathcal{H}^1(S_{u_0}). \tag{9.25}$$

The sum of energy and work for the deformation u^0 at time t is written

$$\mathcal{GF}(u^0(t),D) = \frac{\rho}{2}\|u^0_t(t)\|^2_{L^2(D)} + LEFM(u^0(t),D) - \int_D b(t)u^0(t)\,dx. \tag{9.26}$$

The sum of energy and work for the initial data u_0, v_0 is written

$$\mathcal{GF}(u_0,D) = \frac{\rho}{2}\|v_0\|^2_{L^2(D)} + LEFM(u_0,D) - \int_D b(0)u_0\,dx. \tag{9.27}$$

The energy inequality for the limit evolution u^0 is given by [15],

Theorem 6 Energy Inequality
For $0 \leq t \leq T$,

$$\mathcal{GF}(u^0(t),D) \leq \mathcal{GF}(u_0,D) - \int_0^t \int_D b_t(\tau)u^0(\tau)\,dx dt. \tag{9.28}$$

*The detailed analysis shows that at each time $u^0(x,t)$ belongs to the class of functions SBV or more generally GSBV to which these generalized notions of gradient and jump set apply [15]. For more on these spaces and their application to fracture see [1], [2], [3], [7].

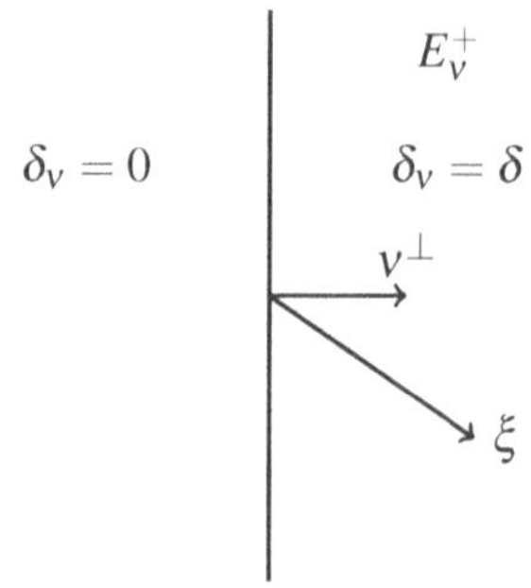

Figure 9.4: Jump discontinuity.

In order to claim that the macroscopic evolution is a solution of the wave equation for points (x,t) away from $S_{u^0(t)}$ we require a further analysis of the mesoscopic dynamics. This analysis is related dynamic instability and its localization as $\varepsilon_k \to 0$. This topic is taken up in the next two sections.

9.3 Dynamic instability and fracture initiation

In this section we present a fracture initiation condition that arises from the unstable force law (9.4). This initiation condition is manifested as a dynamic instability. In the following companion section we investigate the localization of dynamic instability as $\varepsilon_k \to 0$. Fracture nucleation conditions can be viewed as instabilities and have been identified for peridynamic evolutions in [20]. Here we investigate the nucleation criteria for the case at hand and postulate an additional condition for the most unstable direction along which the crack can nucleate.

First consider a time independent body force density b and a smooth equilibrium solution u of (9.7). Now perturb u in the neighborhood of a point x by adding a step function discontinuity denoted by the function δ_ν that takes the value zero on one side of a line passing through x parallel to the direction ν and takes the value δ on the other side. To fix ideas change coordinates so the point x becomes the origin and let $\nu^\perp$ denote the perpendicular to the line through 0. Then set $\delta_\nu(\xi) = 0$ for $\xi \cdot \nu^\perp < 0$ and $\delta_\nu(\xi) = \delta$ for $\xi \cdot \nu^\perp > 0$ (see Figure 9.4). The side of the line for which $\xi \cdot \nu^\perp > 0$ is denoted by E_ν^+.

Write $u^p = u + \delta_\nu$ and assume

$$\rho u_{tt}^p = -\nabla PD^\varepsilon(u^p) + b. \tag{9.29}$$

We regard δ as a small perturbation and expand the integrand of $\nabla PD^\varepsilon(u^p)$ in a Taylor series to recover the linearized evolution equation for the jump δ. The evolution equation is given by

$$\rho\delta_{tt} = \mathcal{A}_\nu \delta, \tag{9.30}$$

where

$$\mathcal{A}_\nu = -\left\{\int_{\mathcal{H}_1(0)\cap E_\nu^+} \varepsilon^2 \partial_\eta^2 W^\varepsilon(u(x+\varepsilon\xi)-u(x),\varepsilon\xi)d\xi\right\}. \tag{9.31}$$

Calculation shows that

$$\partial_\eta^2 W^\varepsilon(\eta,\varepsilon\xi) = \frac{2}{\varepsilon^4|\xi|}J(|\xi|)\left(f'\left(\varepsilon|\xi|\mathcal{S}^2\right)+2f''\left(\varepsilon|\xi|\mathcal{S}^2\right)\varepsilon|\xi|\mathcal{S}^2\right), \tag{9.32}$$

where $f'(\varepsilon|\xi|\mathcal{S}^2) > 0$ and $f''(\varepsilon|\xi|\mathcal{S}^2) < 0$. From this calculation we discover for $\mathcal{S}_c = \frac{\overline{r}}{\sqrt{\varepsilon|\xi|}}$ that

$$\partial_\eta^2 W^\varepsilon(\eta,\varepsilon\xi) > 0 \text{ for } |\mathcal{S}| < \mathcal{S}_c, \tag{9.33}$$

and

$$\partial_\eta^2 W^\varepsilon(\eta,\varepsilon\xi) < 0 \text{ for } |\mathcal{S}| > \mathcal{S}_c. \tag{9.34}$$

Here $\overline{r}$ is the inflection point for the function $r :\to f(r^2)$. Note that the critical strain $\mathcal{S}_c$ at which the dynamics becomes unstable is akin to the square root singularity seen at the crack tip in classical brittle fracture mechanics.

From (9.30) we see that for $\mathcal{A}_\nu > 0$ the jump can grow exponentially. It is evident that this can occur only if there are critically strained bonds, $|\mathcal{S}| > \mathcal{S}_c$, inside the peridynamic neighborhood $H_\varepsilon(x)$. In summary these results deliver the following proposition [15].

Proposition 9.1
Facture nucleation condition
Given a point x and a direction ν, a condition for crack nucleation at x along the direction ν is $\mathcal{A}_\nu > 0$.

Proposition 9.1 together with (9.34) provide the explicit link between dynamic instability and the critical shear strain where the constitutive law begins to soften.

More generally we see from the calculation that $\mathcal{A}_\nu$ depends on ν and we postulate that the directions ν^* along which cracks grow are the most unstable which satisfy the condition

$$\mathcal{A}_{\nu^*} = \max_\nu \mathcal{A}_\nu > 0. \tag{9.35}$$

9.4 Localization of dynamic instability in the small horizon-macroscopic limit

We present results that show that unstable peridynamic neighborhoods likely to nucleate discontinuities in the mesoscopic model concentrate on sets with zero area in the macroscopic limit. The discussion focuses on the basic unit of peridynamic interaction: the peridynamic neighborhoods $\mathcal{H}_\varepsilon(x)$ of diameter $\varepsilon > 0$ with centroids $x \in D$. Here we investigate the family of peridynamic evolutions $u^\varepsilon(t,x)$ at a fixed time t.

Consider a prototypical neighborhood $\mathcal{H}_\varepsilon(x)$ and the antiplane shear strain $\mathcal{S}^\varepsilon(x,y) = u^\varepsilon(t,y) - u^\varepsilon(t,x)|/|y-x|$ for y belonging to $\mathcal{H}_\varepsilon(x)$. The collection of points y inside $\mathcal{H}_\varepsilon(x)$ for which the shear strain $\mathcal{S}^\varepsilon(x,y)$ is beyond critical, i.e.,

$$|\mathcal{S}^\varepsilon(x,y)| > \mathcal{S}_c = \frac{\bar{r}}{\sqrt{|y-x|}}$$

is called the unstable subset of $\mathcal{H}_\varepsilon(x)$. This set is written as $\{y \text{ in } H_\varepsilon(x) : |\mathcal{S}^\varepsilon(x,y)| > \mathcal{S}_c\}$. We introduce the indicator function $\chi^{+,\varepsilon}(x,y)$ of this set taking the value 1 for y belonging to $\{y \text{ in } H_\varepsilon(x) : |\mathcal{S}^\varepsilon(x,y)| > \mathcal{S}_c\}$ and zero otherwise. The weighted area fraction of the neighborhood $\mathcal{H}_\varepsilon(x)$ occupied by the unstable set is written

$$P(\{y \text{ in } H_\varepsilon(x) : |\mathcal{S}^\varepsilon(x,y)| > \mathcal{S}_c\}) = \\ = \frac{1}{\varepsilon^2 m}\int_{\mathcal{H}_\varepsilon(x)} \chi^{+,\varepsilon}(x,y)\left|\frac{y-x}{\varepsilon}\right| J(\left|\frac{y-x}{\varepsilon}\right|)\,dy, \tag{9.36}$$

where the normalization constant $m = \int_{\mathcal{H}_1(0)} |\xi| J(|\xi|)\,d\xi$ is chosen so that $P(\{y \text{ in } \mathcal{H}_\varepsilon(x)\}) = 1$.

Definition 9.1 The neighborhood $\mathcal{H}_\varepsilon(x)$ is said to be unstable at time t if

$$P(\{y \text{ in } H_\varepsilon(x) : |\mathcal{S}^\varepsilon(x,y)| > \mathcal{S}_c\}) = 1. \tag{9.37}$$

To proceed choose $\delta > 0$ and consider a family of radii $\varepsilon_j = \frac{1}{2^k}$, $k = 1, \ldots$ and the collection of neighborhoods $\mathcal{H}_{\varepsilon_k}(x)$ with centroids x in the reference domain D. The set of centroids associated with unstable neighborhoods for some $\varepsilon_k < \delta$ at time t is denoted by $C_{\delta,t}$. Now define $C_{0,t}$ to be $C_{0,t} = \cap_{\delta>0} C_{\delta,t}$, i.e., the intersection of all centroids x which belong to unstable neighborhoods for every $\delta > 0$. Here $C_{0,t} \subset C_{\delta,t}$. An analysis shows that there is a localization of bond instability as the peridynamic horizon shrinks [15]. The localization is formulated in the following theorem.

Theorem 7 Localization of bond instability in the small horizon limit
The collection of centroids for unstable neighborhoods concentrates onto $C_{0,t}$. The set of centroids $C_{\delta,t}$ is decreasing with $\delta \to 0$ and there is a positive constant K

independent of t and δ for which

$$\begin{aligned} &Area\left(C_{\delta,t}\right) \leq K\delta, \text{ for } 0 \leq t \leq T \text{ with} \\ &Area\left(C_{0,t}\right) = \lim_{\delta \to 0} Area\left(C_{\delta,t}\right) = 0. \end{aligned} \tag{9.38}$$

This theorem shows that the nucleation sites concentrate on the centroids associated with the collection of unstable neighborhoods $C_{0,t}$ for which all bonds have become soft. For points x belonging to $C_{0,t}$ the dynamics of the surrounding motion no longer influences the dynamics at x. Intuitively this set of points provides an alternative description of the crack set in the reference configuration as seen in the small horizon-macroscopic limit. The concentration of instability is inevitable for this mesoscopic model and is directly linked to the constraint on the energy budget associated with the peridynamic motion as described by Theorem 1. This bound forces the localization of instability (see [15]). As it stands the fractal or Hausdorff dimension of the concentration set $C_{0,t}$ is yet to be determined. Future work will attempt to find the answer.

9.5 Free crack propagation in the small horizon-macroscopic limit

We make use of the localization of instability and identify conditions for which the limit evolution u^0 solves the wave equation. To begin we make a technical hypothesis on the regularity of the jump set of the limit dynamics $u^0(x,t)$.

Hypothesis 2 *We suppose that the crack set given by $S_{u^0(t)}$ is a closed set for $0 \leq t \leq T$.*

The next hypothesis applies to the concentration set associated with unstable neighborhoods and its relation to the crack set for the limit flow.

Hypothesis 3 *Theorem 7 shows that the centroids of neighborhoods for which all bonds have become soft concentrate on the lower dimensional set $C_{0,t}$. Recall that the dynamics associated with every point in the reference configuration belonging to the set $C_{0,t}$ is not affected by the motion of its surroundings. Motivated by this observation we assume that $S_{u^0(t)} = C_{0,t}$ for $0 \leq t \leq T$.*

The next hypothesis applies to neighborhoods $\mathcal{H}_{\varepsilon_k}(x)$ for which the shear strain is subcritical, i.e., $|\mathcal{S}|^{\varepsilon} < \overline{r}/\sqrt{|y-x|}$, for y in $\mathcal{H}_{\varepsilon_k}(x)$. These neighborhoods will be referred to as neutrally stable.

Hypothesis 4 *We suppose that $\varepsilon_k = \frac{1}{2^k} < \delta$ and $0 \leq t \leq T$. Consider the collection of centroids $C_{\delta,t}$. We fatten out $C_{\delta,t}$ and consider $\tilde{C}_{\delta,t} = \{x \in D : dist(x, C_{\delta,t}) < \delta\}$. We suppose that all neighborhoods $H_{\varepsilon_k}(x)$ that do not intersect the set $\tilde{C}_{\delta,t}$ are neutrally stable.*

With these conditions satisfied the macroscopic evolution u^0 is identified as a solution of the linear elastic wave equation [15].

Theorem 8 Wave equation

Suppose Hypotheses 2, 3, and 4 hold true. Then the limit evolution $u^0(t,x)$ is a solution of the wave equation

$$\rho u^0_{tt} = 2\mu \operatorname{div}(\nabla u^0) + b, on\ [0,T] \times D. \tag{9.39}$$

Here the second derivative u^0_{tt} is the time derivative in the sense of distributions of u^0_t and $\operatorname{div}(\nabla u^0)$ *is the divergence of the approximate gradient ∇u^0 in the distributional sense.*

Remark 9.1 For completeness we recall that the generalized gradient $\nabla u^0(x,t)$ and jump set $S_{u^0(t)}$ are defined over disjoint sets in $[0,T] \times D$ and the wave equation holds in the sense of distributions on the domain $[0,T] \times D$. ∎

Remark 9.2 The macroscopic limit of the mesoscopic dynamics model is given by the displacement–crack set pair $u_0(t,x)$, $S_{u^0(t)}$. The wave equation provides the dynamic coupling between elastic waves and the evolving fracture path inside the media. ∎

Remark 9.3 We conjecture that the Hypotheses 2, 3, and 4 hold true. ∎

Remark 9.4 We point out that the peridynamic constitutive model addressed in this work does not have an irreversibility constraint and the constitutive law (9.4) applies at all times in the peridynamic evolution. Because of this the crack set at each time is given by $S_{u^0(t)}$. Future work will investigate the effects of irreversibility (damage) in the peridynamic model. ∎

9.6 Summary

This model presented here is used to connect the dynamics associated with unstable bonds at mesoscopic scales to dynamic free crack propagation inside a brittle material, as observed at macroscopic length scales. This strategy can be thought of as a bottom up approach. The nonlocal peridynamic model for the mesoscopic dynamics put forth here encodes the effects of bond softening into the dynamics. Crack nucleation criteria are recovered here by viewing nucleation as a dynamic instability. This is similar in spirit to [20] and the work of [8] for phase transitions. The limit analysis shows that the macroscopic evolution has both bounded linear elastic energy and bounded Griffith surface energy and that these are expressed in terms of shear moduli μ and energy release rate $\mathcal{G}_c$. These parameters are determined explicitly by the peridynamic potential f and influence function J. The formulas for μ and $\mathcal{G}_c$ follow directly from the scaling limit without extra hypotheses. With these observations in hand we can turn this correspondence around and use measured values of μ and $\mathcal{G}_c$ to choose peridynamic potentials and influence functions appropriate for a particular material of interest.

It is pointed out that the constitutive model (9.4) does not include the effects of damage and bonds strained beyond critical can return to their subcritical constitutive behavior. However, with this caveat in mind the theory presented here strongly supports the viewpoint that this mesoscopic model can be used as a numerical tool for analyzing crack paths associated with brittle fracture by choosing the scaling parameter ε sufficiently small in (9.7) and running numerical simulations.

The mathematical tool set appropriate for extracting the limit behavior from this class of mesoscopic models is rooted in the theory of Γ-convergence and comes from the literature associated with the analysis of the Mumford–Shah functional and free discontinuity problems (see [12], and [13]). Further discussion including mathematical proofs of the results presented here have appeared in [15].

We conclude noting that the methods and ideas presented here can be extended to mode I fracture problems for bond-based models as well as to state-based formulations. Future work will attempt to address the effects of irreversibility and hysteresis in the mesoscopic model.

References

[1] L. Ambrosio. A compactness theorem for a special class of functions of bounded variation. *Bollettino dell'Unione Matematica Italiana*, 3-B: 857–881, 1989.

[2] L. Ambrosio and A. Brades. Energies in SBV and variational models in fracture mechanics. In P. Donnato D. Cioranescu, A. Damlamian, editors, *Homogenization and Applications to Materials Science*, volume 9, pages 1 – 22. Gakuto, Gakkotosho, Tokyo, Japan, 1997.

[3] H. Attouch, G. Buttazzo, and G. Michaille. *Analysis in Sobolev and BV Spaces: applications to PDEs and optimization.* SIAM, Philadelphia, PA, 2006.

[4] G.I. Barenblatt. The mathematical theory of equilibrium cracks in brittle fracture. *Advances in Applied Mechanics*, 7: 55–129, 1962.

[5] T. Belytschko and T. Black. Elastic crack growth in finite elements with minimal remeshing. *International Journal for Numerical Methods in Engineering*, 45: 601–620, 1999.

[6] M.J. Buehler, F.F. Abraham, and H. Gao. Hyperelasticity governs dynamic fracture at a critical length scale. *Nature*, 426: 141–146, 2003.

[7] G. Dal Maso and C.J. Larsen. Existence for wave equations on domains with arbitrary growing cracks. *Rendiconti Lincei–Matematica e Applicazioni*, 22: 387–408, 2011.

[8] K. Dayal and K. Bhattacharya. Kinetics of phase transformations in the perydanimic formulation of continuum mechanics. *Journal of the Mechanics and Physics of Solids*, 54: 1811–1842, 2006.

[9] Q. Du, M. Gunzburger, R. Lehoucq, and K. Zhou. A nonlocal vector calculus, nonlocal volume-constrained problems, and nonlocal balance laws. *Mathematical Models and Methods in Applied Sciences (M3AS)*, 23: 493–540, 2013.

[10] E. Emmrich and D. Puhst. Well-posedness of the peridynamic model with Lipschitz continuous pairwise force function. *Communications in Mathematical Sciences*, 11: 1039–1049, 2013.

[11] L.B. Freund. *Dynamic Fracture Mechanics*. Cambridge University Press, Cambridge, UK, 1998.

[12] M. Gobbino. Finite difference approximation of the Mumford-Shah functional. *Communications on Pure and Applied Mathematics*, 51: 197–228, 1998.

[13] M. Gobbino and M.G. Mora. Finite difference approximation of free discontinuity problems. *Proceedings of the Royal Society of Edinburgh: Section A Mathematics*, 131: 567–595, 2001.

[14] A. Hillerborg, M. Modeer, and P.E. Petersson. Analysis of crack formation and crack growth by means of fracture mechanics and finite elements. *Cement and Concrete Research*, 6: 731–781, 1976.

[15] R. Lipton. Dynamic brittle fracture as a small horizon limit of peridynamics. *Journal of Elasticity*, 117(1): 21–50, 2014.

[16] M. Marder. Supersonic rupture of rubber. *Journal of the Mechanics and Physics of Solids*, 54: 491–532, 2006.

[17] M. Marder and S. Gross. Origin of crack tip instabilities. *Journal of the Mechanics and Physics of Solids*, 43: 1–48, 1995.

[18] N. Moës, J. Dolbow, and T. Belytschko. A finite element method for crack growth without remeshing. *International Journal for Numerical Methods in Engineering*, 46: 131–150, 1999.

[19] S.A. Silling. Reformulation of elasticity theory for discontinuities and long-range forces. *Journal of the Mechanics and Physics of Solids*, 48: 175–209, 2000.

[20] S.A. Silling, O. Weckner, E. Askari, and F. Bobaru. Crack nucleation in a peridynamic solid. *International Journal of Fracture*, 162: 219–227, 2010.

[21] L.I. Slepyan. *Models and Phenomena in Fracture Mechanics*. Springer, Berlin, Germany, 2002.

[22] X.P. Xu and A. Needleman. Numerical simulations of fast crack growth in brittle solids. *Journal of the Mechanics and Physics of Solids*, 42: 1397–1434, 1994.

Chapter 10

Crack Branching in Dynamic Brittle Fracture

Florin Bobaru

University of Nebraska-Lincoln

Guanfeng Zhang

University of Nebraska-Lincoln

CONTENTS

10.1 Introduction 246
10.2 A brief review of literature on crack branching 249
10.2.1 Theoretical models and experimental results on dynamic brittle fracture and crack branching 249
10.2.2 Computations of dynamic brittle fracture based on FEM 251
10.2.3 Dynamic brittle fracture results based on atomistic modeling 253
10.2.4 Dynamic brittle fracture based on particle and lattice-based methods 253
10.2.5 Phase-field models in dynamic fracture 254
10.2.6 Results on dynamic brittle fracture from peridynamic models 255
10.3 Brief review of the bond-based peridynamic model 256
10.4 An accurate and efficient quadrature scheme 260
10.5 Peridynamic results for dynamic fracture and crack branching 262
10.5.1 Crack branching in soda-lime glass 265
10.5.1.1 Load case 1: stress on boundaries 265
10.5.1.2 Load case 2: stress on pre-crack surfaces 269
10.5.1.3 Load case 3: velocity boundary conditions 272
10.5.2 Crack branching in homalite 275
10.5.2.1 Load case 1: stress on boundaries 276
10.5.2.2 Load case 2: stress on pre-crack surfaces 278
10.5.2.3 Load case 3: velocity boundary conditions 282
10.5.3 Influence of sample geometry 285
10.5.3.1 Load case 1: stress on boundaries 285
10.5.3.2 Load case 2: stress on pre-crack surfaces 287
10.5.3.3 Load case 3: velocity boundary conditions 289

10.6 Discussion of crack branching results 291
10.7 Why do cracks branch? 296
10.8 The importance of nonlocal modeling in crack branching 304
10.9 Conclusions 307
Acknowledgments 308
References 309

10.1 Introduction

In this chapter we conduct a detailed peridynamic analysis of the dynamic brittle fracture phenomenon with a particular emphasis on crack branching. The content of the chapter is mostly based on the recent paper [13] which appeared in a special issue celebrating 50 years of the *International Journal of Fracture*. A number of corrections have been implemented, and another difference is that the color figures are now plotted with a bi-color scheme, to render them meaningful even when printed in black and white. Simulation movies that accompanied [13] as supplementary material, and are mentioned in the text here, can also be found at a dedicated link at http://engineering.unl.edu/mme/florin-bobaru/.

As mentioned in [28], at some scale, all fracture is dynamic. The dynamics of atomic bond rupture plays a role in crack propagation even when a crack at the macroscopic scale appears to be advancing quasi-statically. As the atomic bond rupture is a dynamic process, the authors of the review article [28] state that "the dynamic fracture problem is the most fundamental in the science of fracture." Dynamic fracture covers largely disparate spatial scales: on the smallest scale, dynamic fracture can help us understand how the introduction of nanostructures can affect the ultimate strength of a material, while at the geophysical scale we can understand the conditions required for a fault in the earth's crust to lose stability and produce a massive earthquake.

Dynamic fracture has been studied from several different angles by researchers from different communities. Two main lines of attack for understanding this problem have been atomistic modeling (classical and quantum) and continuum mechanics modeling.

Atomistic modeling has sometimes produced puzzling results (see [108]). For example, crack branching cannot be reproduced by 2D atomistic models [78], as branching attempts lead to one of the branches to eventually arrest. Three-dimensionality has been proposed as a potential solution for achieving crack branching in atomistic models of amorphous materials. Further studies that used realistic crystalline potentials have failed to reproduce experimental results (see [46], [20]). Somewhat surprisingly, atomistic simulations seem to require quantum mechanical calculations in order to enable an accurate quantitative description for mode I crack propagation (the simplest mode of fracture) at low speeds of a single crack in single crystal silicon (see [59]). The complexity of dynamic fracture in amorphous or polycrystalline materials is significantly enhanced by the physical interplay of multiple length scales due to the presence of microdefects or grain boundaries. For these types of problems, atomistic or quantum calculations cannot yet provide any answers.

In addition to the obstacles in predicting the observed behavior in dynamic brittle fracture with atomistic models mentioned above, such models have some obvious limitations [108]: the small size of the sample modeled (usually micrometer or sub-micrometer) and the short time scales that can be accessed (usually nanoseconds). Cracks in brittle solids are controlled by waves (see [87]) and the influence of elastic waves bouncing from the boundaries of the sample cannot be neglected if one is to predict the behavior of growing cracks in such solids. In trying to predict how cracks propagate in brittle solids it is, therefore, essential to model the entire sample, a feat not achievable by atomistic models. The alternative then remains modeling at a continuum level. Doubts about whether continuum approaches would *ever* be capable in successfully modeling the evolution of dynamic cracks, through fragmentation, have persisted for decades (see [100]). One source for these doubts has been attributed to the breaking down of Linear Fracture Mechanics (LEFM) for regions near the crack tip, as is reviewed in [20]. A number of authors have introduced models that modify the classical picture by introducing a certain region with nonlinear elastic behavior near the crack tip as well of a length-scale [20]. Such models have been successful in correctly representing the opening profile at the crack tip in soft brittle materials (aqueous gels). It is not yet known whether this approach is still valid in more standard brittle materials, in which there is no clear separation between nonlinear elastic scales and dissipative ones (see [20]).

The *branching of a crack* (splitting of a crack in two or more cracks) is one of the characteristics of dynamic brittle fracture (see [23], [83]). An increase in the fracture surface roughness is consistently observed in experiments prior to branching ([83]). The increase in the fracture surface roughness is tightly connected to the crack path instability. Crack path instabilities, in turn, have been proposed as being responsible for the observed limiting crack speed in amorphous brittle materials. This speed is well below the theoretical limiting speed, rarely reaching above 0.6 of the Rayleigh wave speed (see [84], page 191). The source of the observed instabilities has been assigned to a number of potential mechanisms happening in the process zone around the crack tip:

- Ravi-Chandar and Knauss [86] suggested that the process zone is significant even in nominally brittle materials, and in this zone nucleation, growth, and coalescence of *microcracks* occurs; the dynamics of evolution of these processes triggers microscopic path instabilities that eventually lead to crack branching. This mechanism has been observed in glassy polymers like PMMA [88] and polystyrene (see [55], page 141) where conical markings are seen on the crack surface.

- Hull [54] suggested that twisting and tilting of the stress vector on the crack front (thus mixed-mode loading around the crack tip), due to microscale variations in the symmetries, results in crack path instabilities, surface roughness, and eventual crack branching. This mechanism has been suggested for glasses or brittle epoxy resins in which the forming of microcracks around the crack tip has not been observed in experiments, and stresses required to activate the very small flaws near the crack tip would have to approach the theoretical strength of the material, thus making bond breakage at the tip the more likely scenario (see [6], [55], page 145).

Crack path instabilities have been addressed in a series of atomistic or theoretical investigations (see, e.g., [25], [62]). A vast literature on exposing various types of instabilities in dynamic fracture is reviewed in [20]. In our opinion, however, *the actual mechanisms that trigger these instabilities have not been clearly identified or satisfactorily explained.* Indeed, the authors of the review article [20] state: "is it quite likely that the dynamics of the near-tip zone could play an important role in unraveling the physical mechanisms driving instabilities of rapid cracks." What causes this instability and how is it generated? What is the sequence of events that lead, in one case, to a crack propagating straight, and in another case, instabilities spring up and induce micro-branching and roughening of the crack surface, and eventually macro-branching?

The crack propagation speed has been demonstrated not to be the main reason leading to crack branching. The experimental evidence points to the following likely scenario: when a critical stress intensity factor is reached, the crack splits into two or more branches, each propagating with about the same speed as the parent crack but with a smaller process zone (see [84], page 214).

In this chapter we aim to answer the following questions:

1. Is it possible to use a continuum model to predict dynamic brittle fracture, in particular crack branching?
2. What is the minimal set of input data that is able to reproduce the most important characteristics observed in crack branching?
3. Are quantum calculations the only option for explaining the dynamic instabilities that precede crack branching?
4. What are the causes of the dynamic instabilities and how do these lead to crack branching?
5. Why do cracks in amorphous materials branch well before reaching propagation speeds close to the Rayleigh wave speed, the theoretical limiting speed of mode I cracks?
6. How do stress waves and loading conditions affect crack branching?

As we shall see, a key factor in having a model that reproduces the essential features of crack branching is being able to capture, at least in some average sense, the evolution of the process zone around the crack tip. The peridynamic model satisfies these requirements. We will also see that micro-branching instability is not an intrinsically 3D pehnomenon, contrary to some recent conclusions that the micro-branching instability is intrinsically a 3D phenomenon and that it "cannot be modeled directly by a 2-dimensional" theory (see [19]). We will show that some essential features of crack instabilities are not controlled by the dimensionality, but by how the elastic wave energy flows around the crack tip and by the dynamic evolution of the process zone. Our results confirm previous statements made based on experimental evidence and theoretical considerations that "the process of branching is governed by the inner problem and not the outer problem that is treated by the (classical) continuum elastodynamics" (see [84], page 214) and this observation explains why cracks branch well before reaching speeds projected by elastodynamic theories to be required for a crack to branch ([107]). The role played by "structural inertia" (see [74], [73]) has been pointed out in branching of concrete. The results obtained in this chapter explain the mechanisms through which the dynamics of strain energy around the crack tip leads to branching.

We will explain the reasons for the angle of branching changing past the branching event (the branched cracks do not grow straight, but may curve after a while). We explain what controls the initial angle of branching, as well as what conditions favor microbranching. We shall see that the peridynamic model explains all these different characteristic features of crack branching without any conditions or extra criteria for crack growth. The fact that only elastic wave propagation and a proper damage model (in this case the nonlocal damage model in peridynamics, which only uses fracture energy information) are sufficient to capture all these different features of the dynamic brittle fracture phenomenon indicates that all other effects or criteria used in earlier models are secondary/not needed to explaining this phenomenon. The results contained here also demonstrate that atomistic resolution and quantum calculations are not necessary in order to understand dynamic brittle fracture and, in particular, the crack branching instability.

10.2 A brief review of literature on crack branching

In this section we provide a brief review of the vast literature dedicated to the subject of dynamic brittle fracture, and in particular to crack branching. Dynamic brittle crack growth remains a challenging problem after almost a century of research dedicated to it [28]. To investigate the mechanism of dynamic brittle fracture, theoretical models, numerical simulations, and experiments have been employed over the years (see [20]).

10.2.1 Theoretical models and experimental results on dynamic brittle fracture and crack branching

Various criteria have been proposed to explain crack curving and branching in dynamic brittle fracture. Early on, the crack propagation velocity was thought to control crack branching and a crack branching criterion was proposed based on elastodynamic theoretical results (see [107]) showing that, for a crack of constant length translating with a constant velocity in an infinite medium, the maximum circumferential stress shifts from the symmetry line to lines that make an angle of 60° with the direction of propagation of the crack when the crack speed exceeds the 0.73 fraction of the Rayleigh wave speed. The Rayleigh wave speed is the theoretical limiting velocity for propagating cracks in linear elastic materials (see [38]). The limiting crack speed is also discussed in [103]. In practice, cracks never reach such high speeds in isotropic amorphous brittle materials, losing stability and branching when they reach fractions of 0.35 to 0.65 of the Rayleigh wave speed (see [31, 86], and Table 11.1 in [84]). In crystalline materials, in which cracks may become trapped on cleavage planes, the propagation speed can reach close to the Rayleigh wave speed (see [35]). Another problem with the Yoffe solution is that experimentally observed branching angles are between 10° and 45°.

The constant length crack moving with constant velocity under mode II conditions is a physically possible process in seismic events where slip stop after the propagation of the disturbance. Broberg [24] gave the solution for this mode II Yoffe problem. Crack propagation for semi-infinite cracks has been studied by Freund [37] and Eshelby [34] in an infinite 2D linear elastic medium. These models show that cracks propagate smoothly and continuously accelerate until they achieve the limiting velocities. However, in reality, crack path instabilities are triggered well before such speeds are attained and crack surfaces present the "mirror, mist, hackle" patterns before branching takes place (see, e.g., [83], [6], and Chapter 5 in [55]). Noticing the discrepancies between experimental observation and analytical predictions, Gao [39] indicated that the linear elasticity might be inadequate because it misses possible large deformation near the crack tip. The resulting model obtains a local limiting speed that provides an explanation for the onset of mirror-mist transition observed in dynamic fracture experiments.

Crack path instabilities may lead to crack branching. Criteria for crack path curving and crack branching under single or mixed modes were proposed based on, for example, the maximum circumferential-stress or minimal strain energy criteria. The criterion that the crack growth direction is normal to the maximum circumferential-stress was introduced in [32]. Using this model, Streit and Finnie [102] proposed a directional stability criterion on the hypothesis that the crack instability is caused by an off-axis microcrack, ahead of the crack tip within a critical distance, connecting with the main crack. The crack path kinks or bifurcates once the critical distance is greater than the maximum radius for which the maximum circumferential stress lies on the symmetry axis. Extending Streit and Finnie's ideas, Ramulu and Kobayashi [82] put forward a *dynamic* crack curving criterion by adding the nonsingular stress or remote stress to the directional stability criterion. The dynamic crack curving criterion is capable of predicting the crack angle in both pure Mode I and mixed Mode I and II conditions. Results showed that positive nonsingular stress enhances the instability while the negative one reduces the fracture angle. This work also proposed a "long-range" interaction in dynamic fracture problems, where a sufficient criterion for crack branching is introduced and tested experimentally. The criterion involves a "characteristic distance" r_0 that depends on the dynamic state of stress near the crack tip. According to [82], crack branching occurs when $r_0 \leq r_c$, where r_c is a material property thought to be related to the distance between microvoids and the crack tip. When an off-axis microvoid gets within a critical distance r_c from the crack tip, it gets activated by the stress field and directional stability of the propagating crack is lost leading to a deflection of the straight crack path. The loss of directional stability, together with the necessary criterion for crack branching ($K_I \geq K_{Ib}$), leads to branching of a propagating straight crack. The role played by microcracks around the crack tip in crack branching has been discussed in, for example, [86] and [88].

Besides the microvoids and microcracks-induced crack path instabilities, the tilting and twisting of the stress vector at the crack front due to mixed-mode loading has been studied as another potential source that can lead to crack branching. The formation of "lances" in glass, as the result of superposition of mode I and mode III, was quantitatively investigated by Sommer [99]. The fracture plane, which is perpendicular to the loading axis in mode I, separates by lances as the result of torsion of tensile stress. A critical angle of rotation was proposed as a necessary condition for lance formation. Using a modified Sommers experiment, Hull [54] confirmed that a critical rotation angle is needed for "river lines" nucleation. Hull explained the progressive coarsening of river lines and the generation of helicoid surfaces in mixed mode I/III conditions. In inorganic glasses, no evidence of microcracks ahead of the crack tip is found and the source of roughness of the crack surface is placed on local changes in the path of the growing crack arising *"from some form of local instability at the tip of the crack related to the dynamical properties of the cracks"* (see [55], page 145). We remark that the specific "dynamic properties" creating these instabilities have not been specified. The picture proposed in [54], [6], and [55] to explain branching in inorganic glasses involves local tilting of the crack out of the main plane. The tilted cracks grow a short distance. With steps forming between the main crack and the tilted cracks, crack overlapping effects occur. When the dimensions of the tilted elements are comparable to the dimensions of the test sample, macroscopic bifurcation occurs (see [55]).

Reasons for crack surface roughening were also recently investigated by comparing responses of perturbation of growing cracks with shear pulse perturbations [16]. In this work it was determined that the mode III loading, introduced by the shear waves interacting with the crack front, is responsible for the observed roughening, via local twists that do not produce fragmenting. When a dynamically growing crack running at speeds below 0.4 C_R interacted with a localized heterogeneity (in the form of a groove scratched on the surface of the sample), undulations are produced on the crack surface as a result of the interactions between the crack front and the shear waves radiated from the groove, and not from the crack front waves (acoustic emissions) as suggested by others [68]. At speeds closer to 0.45 C_R, which may be a threshold for developing microbranches, whether emissions of crack front waves induce roughness or not is still unknown and needs to be investigated.

10.2.2 Computations of dynamic brittle fracture based on FEM

Computational modeling aimed at predicting dynamic brittle fracture has been performed to simulate the crack propagation and crack branching using finite element method (FEM) or various modified versions of the FEM. Existing FEM models are based on the classical continuum mechanics equations, which are described by partial differential equations. In order to solve problems with discontinuities, such as cracks, special techniques need to be devised. Most of these techniques essentially treat every new configuration as a problem in a new domain, since the crack creates new surfaces and the boundaries of the initial domain change. This *requires tracking of the crack surface* in some way. Other complications relate to describing crack initiation, determining the direction of propagation, and the propagation speed. All of these aspects require laws of propagation (or "kinetic relations") and most of the time they are set up in an ad hoc manner: they may work for one example, but not for another. These approaches also assume that dynamic crack growth is describable by a surface generally determined by the separation between the elements or tracked by a smooth level-set curve. This excludes capturing roughness on the crack surface that is associated with dynamic brittle fracture.

One of the simplest methods to modeling propagating cracks is the element-deletion (also known as "element-erosion") method. Such methods suffer from non-convergence in the limit of the mesh size going to zero. Recent improvements of these approaches have been produced using variational formulations (see [76]) or nonlocal averaging of displacement gradients schemes (see [69]) as those employed in the nonlocal models of [33]. The results in [76] match well quasi-static type of crack growth under mixed-mode conditions. While a length-scale is introduced in the element erosion model of [76, 75] in order to regularize the problem and ensure convergence in the limit of the mesh being refined, it is not yet known how this method performs in the crack branching problem.

Alternatives to element-erosion techniques are cohesive zone FEM models (see e.g. [105, 26]). Such models remove the need of pre-knowledge of the crack path. The crack path, however, is still forced to follow the particular mesh used, since cracks can only propagate along the element boundaries (see [106, 26, 70]). Since the correct, actual crack path (which minimizes the strain energy) of the propagation process may not be computed correctly, there are significant departures from the true energy released during the crack propagation event. In such cases, reliable prediction of strength of brittle ceramics under impact, for example, becomes difficult. The XFEM (see, e.g., [30]) allows cracks to pass through the finite elements leading to better approximations of the crack path. Interestingly, in dynamic brittle fracture problems, one may need to drastically modify the input fracture energy in the model in order to obtain crack propagation speeds similar to those seen in experiments (see [100]). Subdivision of cut elements for numerical integration purposes increases complexity and cost for this method. The method also requires phenomenological damage models and branching criteria as input, and tracking of the crack path using, for example, level sets. A recent review of methods that use tracking of the crack path has been recently provided in [79].

10.2.3 Dynamic brittle fracture results based on atomistic modeling

In atomistic models, like Molecular Dynamics (MD) simulations, under sufficiently high loading conditions that lead to instability of the crack path, cracks can branch without a specific criterion (see [108]). MD models, however, are not able to capture the crack propagation speed or the angle of crack branching correctly. For instance, MD simulations show instabilities that lead, shortly after the bifurcation of a crack, to the propagation of only one of the two branches, the other being arrested. The limited space and time scales that can be addressed by such models prevent investigating what could happen if this behavior of branching attempts (followed by the arrest of one of the branches) were monitored over macroscopic scales (perhaps, microns to millimeters and hundreds of microseconds) which are relevant to the crack branching phenomenon (see [83]). Moreover, the initial branching angle computed with MD in [108] is greater than 45°, where experiments show much smaller crack branching angles (see [83]). Reference [86] showed that the branching angle is influenced by the crack tip stress state, in particular the nonsingular term.

Paradoxically, simulations based on regular arrangement of atoms (crystalline solids) give solutions (see [1]) that are observed, at much larger scales, in noncrystalline materials! In crystals, cleavage fracture is observed in experiments and crack speed may reach a significant fraction of the Rayleigh wave speed (close to 0.9) without exhibiting branching (see [23]). Addition of hyperelasticity at the crack tip (see [2, 25]) does not improve on this, so far, unexplained behavior of atomistic models.

It is important to notice that for MD simulations to correctly predict dynamic brittle fracture, one may need to model the entire structure in order to capture, for example, stress wave reflections from the boundaries. The authors of [86] were the first to show that crack branching occurs in "infinite specimens" even in the absence of stress wave reflections from the boundaries, and also showed that elastic stress waves play a major role in the evolution of dynamic brittle cracks. An alternative to modeling the entire structure by MD could be using multiscale models capable of transferring waves between the scales correctly. Whether this is a viable approach is still an open question.

10.2.4 Dynamic brittle fracture based on particle and lattice-based methods

Particle and lattice-based modeling of brittle fracture has been researched for several decades (see, e.g., [29, 65]). A recent review is presented in [66]. Lattice dynamics models produce ranges of "forbidden velocities" [65] which are not observed in experiments. For a band of velocities beginning at zero and proceeding up to around 30% of the relevant wave speed, steady crack motion is impossible (see [65]), while in experiments, all such speeds are accessible for a crack (see [84]). Changing from a triangular to a square lattice changes the critical velocities substantially. A phenomenon called "intermittency" is thought to be at the origin of the instability of the crack path (see [65]).

In [14], rigid-body-spring networks were used to simulate brittle fracture in homogeneous, isotropic materials. In this model, the body is discretized using Voronoi diagrams, and the particles such created are connected via springs. Spring degradation or spring deletion is used to model the fracture initiation and propagation. The random geometry lattice created by the Voronoi structure greatly reduces the preferential direction for crack propagation seen in regular geometry lattice spring networks (see [56]). Results of a reinforced concrete beam show that fracture patterns obtained by simulation are close to experimental observations.

In the "particle cracking model" introduced in [80], a crack advances by breaking particles in a sequence. The crack propagation speed in crack branching reaches values of about $0.6C_R$, which is close to what is observed in experiments. However, the crack patterns obtained after branching appear non-physical, with some unexplained features, like sub-branching in regions which are expected to be unloading. Dynamic failure and fragmentation has also been studied with a semi-Lagrangian reproducing kernel particle method (RKPM) in [42], where a stabilized non-conforming nodal is introduced. The method is able to show effectiveness for examples of a projectile penetration into a concrete block and bullet into a steel plate. For the brittle damage of the concrete block, not all of the failure features seen in experiments are reproduced by the computations.

10.2.5 Phase-field models in dynamic fracture

In order to circumvent the problems associated with numerically tracking the propagating discontinuity representing a crack, phase field models (see [4, 7, 21, 77, 101, 22, 17]) approximate the fracture surface by a phase-field: a function of position (and time for dynamic problems) with values in the interval $[0, 1]$, with value 1 away from the crack and 0 inside the crack. By introducing a limiting parameter (in effect a length-scale) into the model, phase-field approaches eliminate the problem with the lack of convergence under grid refinement that classical methods suffer from (see [21]). Note that the material-point erosion model (see [75]) mentioned above uses a similar approach to regularize the problem and achieve convergence.

In phase-field models, one minimizes the free energy functional comprised of the strain energy density and the regularized crack surface density. These models are capable of representing with some success many characteristic of fracture, like crack initiation, propagation, branching, acoustic emissions, and fracture instabilities. Two approaches have been favored in this area: dynamic phase-field fracture models based on Landau–Ginzburg-type phase-field evolution equations (e.g., [58, 101]) and phase-field formulations based on Griffith's theory (see [21, 22]). Some authors (see [17]) prefer the phase-field formulation of the Bourdin-type since the physical properties of Griffith's theory are well understood and have proven useful in engineering applications. The regularization parameter that defines the width of the failure zone in phase-field models permits (see [17]) obtaining the "crack thickening" before branching that may be a representation of roughening of the crack surface before branching observed in experiments.

Some issues in the Landau–Ginzburg-type formulations related to the evolution of the phase-field away from the crack tip, even when the sample is brought back to equilibrium, are pointed out in [22]. The results in [22], however, are not intended to be compared yet with experiments in terms of crack propagation speed and other measures. The phase-field model for dynamic fracture [17] used in [47] shows branching when one would not expect it (in the Kalthoff–Winkler test), branching patterns that do not resemble those seen in experiments, and branching angles that are much smaller than those observed in experiments.

In a recent review article (see [101]) the authors note that it "is not yet clear how to relate quantitatively phase field model predictions of dynamical branching instabilities to experimental observations," and that "phase field models that incorporate a more realistic description of the process zone are presently needed to make predictions more broadly applicable." It is not yet clear how one can introduce in phase-field models effects related to the discrete nature of the material, including lattice trapping effects that affect macroscopically observable properties (see [90]), nonlinear elastic effects, post-failure contact between interacting fragments, or severe plastic deformation.

10.2.6 Results on dynamic brittle fracture from peridynamic models

Peridynamics is a reformulation of the classical continuum mechanics equations that allows a natural treatment of discontinuities in the solution of equation of motion for continua by employing the concept of nonlocal interactions (see [95]). Integration, rather than differentiation, is used to compute the total force density acting on a certain material volume, and deformation gradients are not used in the formulation. Peridynamics differs from other nonlocal methods such as those described in [33], [60], and [89], or those reviewed in [5], for at least two fundamental reasons:

1. The deformation gradient and strains (spatial derivatives of displacements) are not used in peridynamics. Other nonlocal methods average strains over the nonlocal region. Spatial derivatives of the displacement field become undefined when discontinuities, like cracks, emerge and this requires special treatment and algorithms in models that employ such derivatives in their formulation.

2. Damage is introduced in the peridynamic method at the microlevel in the constitutive model for the peridynamic bonds between material points. When the relative elongation of the bond reaches a failure criterion related to the material's fracture energy, the bond breaks. Fracture surfaces or diffuse damage happen autonomously as a consequence of this definition. Dealing with multiple interacting cracks of arbitrary shapes in complex geometries becomes as easy as dealing with a single straight crack. In this way, peridynamics integrates damage and fracture under a single model for material failure.

In references [43, 44, 45], the crack branching problem has been treated using the bond-based version of peridynamics. Convergence results have also been provided there. Peridynamics correctly predicts important elements of dynamic crack propagation: the shape of the crack paths, the general profile of the crack propagation speed (similar to the experimental one reported in [35]), attempted and successful branching events (similar to those observed in [83]), secondary cracks caused by wave reflections from the boundaries, and the relation between the way strain energy is delivered into the fracture zone and the evolution of the fracture process (as reported in the experiments in [87, 86]). Note that in [43, 44, 45] the fracture energy for glass used was that reported in the literature at branching. The fracture energy at crack initiation is significantly lower than that measured at branching. This happens because in measurements, one measures the straight advancement of the crack and cannot account for all of the micro-cracks and roughness of the advancing crack. Therefore, the fracture energy measured at around the branching point appears to be much larger than the energy measured at crack initiation. The results shown in the present contribution use the fracture energy at initiation, and if we were to measure the "apparent" fracture energy in the branching region we would likely obtain values similar to those measured in experiments, because the model allows for the expenditure of fracture energy by autonomously creating damage away from the original crack surface in the form of roughness/microcracks. We notice that one way in which results change when the input fracture energy is changed is the amplitude of the applied loading required to generate crack propagation.

The peridynamic results, in terms of the crack propagation speed and the crack path, converged (see [43]) once the horizon (the nonlocal parameter) reached submillimeter values. These values may be related to the characteristic length discussed in [83, 82], and further clarifications for this connections are made later in the chapter (see Section 10.8).

Peridynamics' success with modeling the fundamental feature of dynamic fracture, namely, crack branching, is perhaps the reason for the excellent results obtained with the peridynamic (PD) model in impact on brittle targets problems (see, e.g., [10, 52]).

10.3 Brief review of the bond-based peridynamic model

The peridynamic equations of motion at a point $\mathbf{x}$ and time t are ([95]):

$$\begin{aligned}\rho(\mathbf{x})\ddot{\mathbf{u}}(\mathbf{x},\mathbf{t}) = &\int_{H_{\mathbf{x}}} \mathbf{f}(\mathbf{u}(\hat{\mathbf{x}},t) - \mathbf{u}(\mathbf{x},t), \hat{\mathbf{x}} - \mathbf{x})\, \mathrm{d}V_{\hat{\mathbf{x}}} \\ &+ \mathbf{b}(\mathbf{x},t) \text{ for } \mathbf{x} \in \Omega \text{ and } t \in [t_0, \infty)\end{aligned} \tag{10.1}$$

where Ω is the domain occupied by the body, t_0 is some initial time, $\mathbf{u}$ is the displacement vector field, $\mathbf{b}$ is the body force vector, and $\mathbf{f}$ is the pairwise force function in the peridynamic bond that connects material points $\mathbf{x}$ and $\hat{\mathbf{x}}$. The integral is defined over a region $H_{\mathbf{x}}$ called the *horizon region*, or simply the *horizon*. It is natural to consider the horizon as a small sphere (disk in 2D, interval in 1D) centered at the current point. We will call the radius of the sphere δ and refer to it also as the *horizon*. From the context it will always be clear whether we refer to the region or its radius when we mention the word "horizon." A discussion on the meaning, selection, and use of the peridynamic horizon and its relation to crack branching in brittle materials has appeared in [11].

For a microelastic material [95], a pairwise potential exists such that

$$\mathbf{f}(\xi,\eta) = \frac{\partial\omega(\xi,\eta)}{\partial\eta}, \tag{10.2}$$

where $\xi = \hat{\mathbf{x}} - \mathbf{x}$ is the relative position and $\eta = \mathbf{u}(\hat{\mathbf{x}},t) - \mathbf{u}(\mathbf{x},t)$ is the relative displacement between points $\hat{\mathbf{x}}$ and $\mathbf{x}$. A *linear microelastic material* is defined by a micropotential ω as:

$$\partial\omega(\eta,\xi) = \frac{c(\xi)s^2\|\xi\|}{2}, \tag{10.3}$$

where the $c(\xi)$ is the micromodulus function and

$$s = \frac{\|\eta+\xi\| - \|\xi\|}{\|\xi\|} \tag{10.4}$$

is the *relative elongation* of the bond connecting $\hat{\mathbf{x}}$ and $\mathbf{x}$. For a horizon region with spherical symmetry, the corresponding pairwise force becomes:

$$\mathbf{f}(\xi,\eta) = \begin{cases} \frac{\eta+\xi}{\|\eta+\xi\|}c(\xi)s & ,\|\xi\| \leq \delta \\ 0 & ,\|\xi\| > \delta. \end{cases} \tag{10.5}$$

Assuming a specific form of the isotropic micromodulus function $c(\xi) = c(\|\xi\|)$, for example constant over the horizon region or varying linearly with $\|\xi\|$ (see Fig. 10.1), one finds the parameters in these representations by calibrating the strain energy density computed with peridynamics to the classical strain energy density for a homogeneous deformation, such as equal bi-axial strain. The calibration is made for a point in the bulk, at least a distance δ away from the boundaries so that the node has a complete horizon region. Material points close to or on the boundaries of domain Ω have horizon regions that are partial disks. If one performs the calibration for these points, the resulting micromodulus value for the bonds connected to such points will be higher than that obtained for the points in the bulk, since the integration area/volume is smaller here. When this special calibration for points within a distance δ from a boundary is not employed, and the bond stiffness for points near the boundary is assigned to be the same as that for nodes in the bulk, the effective material behavior near the boundaries is slightly softer than in the bulk. As the horizon δ decrease to zero, the thickness of this softer zone decreases and becomes negligible. More on this "skin effect" is discussed in [44]. To reduce, or in some cases eliminate, the surface effect, for any horizon size, several options have been proposed in the literature:

- introduce ghost nodes ([40]) (also called fictitious nodes, see [72]) outside of the domain. This approach may become cumbersome for bodies with complex geometries.

- compute approximate corrections of the micromoduli for nodes on or near the surface (see, for example, [63], and Section 4.2 in [64]). These approaches are exact only for regular boundaries and homogeneous deformations.

A detailed discussion and comparison of the performance of surface-correction methods is given in [61].

In anisotropic materials, the micromodulus function $c(\xi)$ is defined by at least two parameters, and various ways to calibrate the models have been proposed (see [105, 49, 50, 41, 53, 71]).

For examples of micromodulus functions in 3D see [97], in 2D see [43], and in 1D see [12]. Because of its faster convergence properties to the classical elasticity solutions (as the horizon goes to zero), in this work we use the 2D *conical* micromodulus (see Fig. 10.1) with the plane stress assumptions (since all of the computational tests are done, as in experiments, on thin plates):

$$c(\xi) = C_1 \left(1 - \frac{\|\xi\|}{\delta}\right) = \frac{24E}{\pi\delta^3(1-\nu)} \left(1 - \frac{\|\xi\|}{\delta}\right) \tag{10.6}$$

where E is Young's modulus and ν is the Poisson ratio (fixed to 1/3 in this 2D plane stress case).

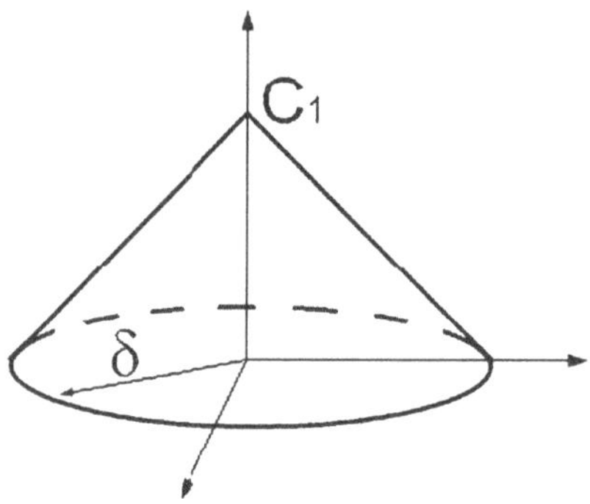

Figure 10.1: The conical micromodulus function

To introduce damage, a law for peridynamic bond-failure has to be established. In PD, bonds can break irreversibly (see [98]), or reversibly (see [8]) when they are meant to represent van der Waals-like interactions, for example. Here we assume irreversibility of bond breaking:

$$f(\xi,\eta,\mathbf{x},t) = \begin{cases} f(\xi,\eta,\mathbf{x}) & \text{if } s(\xi,\mathbf{x},\tilde{t}) < s_0, \text{ for all } 0 \le \tilde{t} \le t, \\ 0 & \text{otherwise} \end{cases} \tag{10.7}$$

where s_0 is the critical value of bond relative elongation for breakage. When a bond reaches this critical value, the break is irreversible and the bond no longer sustains a force. This critical value, which could be made to depend on the bond length $\xi = \|\xi\|$ or on the state of local damage (see [44]), for example, is defined by matching the fracture energy G_0 of the material to the energy required by the peridynamic model at a point in the bulk to completely separate a body in two at that point with a fracture surface. This separation requires breaking all bonds that initially connected points on opposite sides of the fracture surface. The value for s_0 in 3D is given in [97]. In 2D, the connection between the critical relative elongation and the material fracture energy is given by (see, e.g. [43]):

$$G_0 = 2\int_0^{\delta}\int_z^{\delta}\int_0^{\cos^{-1}(z/\xi)}\left[\frac{cs_0^2}{2}\right]\xi\,\mathrm{d}\theta\,\mathrm{d}\xi\,\mathrm{d}z. \tag{10.8}$$

For the conical micromodulus functions, the critical relative elongation is obtained as:

$$s_0 = \sqrt{\frac{5\pi G_0}{9E\delta}}. \tag{10.9}$$

Similar to the "skin effect" discussed above for the computation of the micromodulus function, the critical relative elongation s_0 value is affected near the boundary of the domain, or in a region where damage is already present. This happens because the above calculation is based on points in the bulk. For points on or near the boundary, the domain of integration is smaller, and when matched to the same G_0 value, the resulting critical relative elongation would be higher than in the bulk. Therefore, when a constant s_0, computed for a point in the bulk, is used in computations, the peridynamic material model will effectively have slightly weaker bonds for points near the surface than for those in the bulk. The same happens in regions where damage is present, such as on the surface of a propagating crack. As the horizon goes to zero, the weaker "skin" region goes to zero and the effect is minimized. Some attempts to correct this for regions where damage is present have been implemented in the code EMU ([96]) and showed better results in fragmentation problems. The correction, also discussed and used in [44], consists in increasing, in an approximate way, the value of the s_0 for points with a certain amount of damage, proportional to the damage amount. This model is referred to as the "*damage-dependent peridynamic model*." The constant s_0 model is used in the present chapter. For models that use corrections of this value depending on the damage present at a node, or on the node being near a material surface, or on the amount of compression in bonds that share a node with the bond in question, please see Chapter 13. For the crack branching problem, these corrections have a minimal effect and they do not change in any way the conclusions reached.

10.4 An accurate and efficient quadrature scheme

With the peridynamic formulation introduced in the previous section, we are ready to compute solutions to dynamic fracture problems. Here we will only briefly review on such possible discretization. In principle, the peridynamic equations can be discretized using the finite element method, or any other method appropriate to compute solutions to integro-differential equations. This approach, however, soon hits well-known obstacles and difficulties for problems with evolving topologies, like those in dynamic fracture and fragmentation. Instead, meshfree-types discretizations are preferred in peridynamics simulations of dynamic failure of materials. The discretization proposed in [97] uses the mid-point integration scheme (equivalent to a one-point Gaussian integration) for the domain integral.

$$\begin{aligned}&\int_{H_{\mathbf{x}}} \mathbf{f}(\mathbf{u}(\hat{\mathbf{x}},t)-\mathbf{u}(\mathbf{x},t),\hat{\mathbf{x}}-\mathbf{x})\,\mathrm{d}V_{\hat{\mathbf{x}}}\\ &\simeq \sum_{j\in \mathrm{Fam}(i)} c(\xi_{ij})s_{ij}V_{ij}\end{aligned} \tag{10.10}$$

where $\mathrm{Fam}(i)$ is the family of nodes j with their area (volume) covered, fully or partially, by the horizon region of nodes i, ξ_{ij} is the bond length between nodes i and j, s_{ij} is the relative elongation for the same bond, and V_{ij} is the area (volume) of node j estimated to be covered by the horizon of node i.

Simple and regular discretization grids are preferable for many reasons and such grids will be used here (see also Chapter 5). When nonuniform grids are used with this type of meshfree discretization approach, special care needs to be taken (see, e.g., [9]). Here, we give a simple algorithm for the numerical integration of the domain integral in Eq. (10.1), first introduced in [48] and also shown in [9] (where the "−" sign on line 7 should be a "+" sign). The main advantage of this algorithm compared with one that simply checks whether a node is inside (in which case it adds the contribution to the integral as if its entire volume is inside the horizon) or outside the horizon region (in which case it does not add any contribution to the integral, as if its entire volume is outside the horizon) is that as the ratio $m = \delta/\Delta x$ increases (for a fixed horizon value), the numerical convergence (in terms of strain energy density, for example) is monotonic ([48]).

In the one-point Gauss quadrature method, each node has a certain volume (area in 2D, length in 1D) associated with it. Figure 10.2(a) shows nodes covered by horizon δ of node $\mathbf{p}$. The square indicates the area of an arbitrary node $\mathbf{q}$. The area of the horizon region used in the spatial integration of the peridynamic equations is composed of the nodal areas that are covered, fully or partially, by the horizon region (see Fig. 10.2(b)).

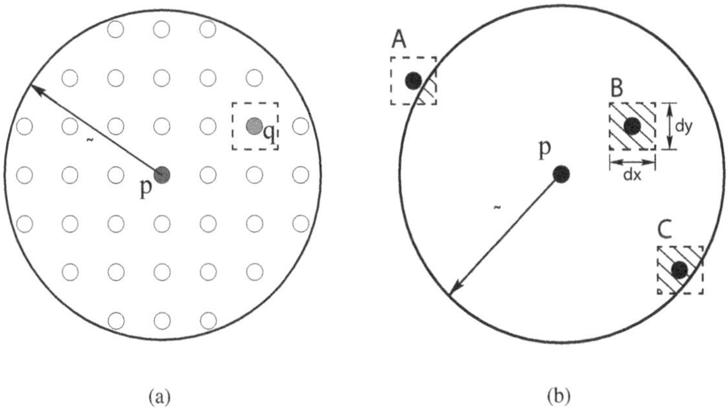

Figure 10.2: (a) One-point Gauss quadrature for the spatial integration of the peridynamic equations; (b) exact areas of various nodes that participate in the quadrature or node **p**.

In [48] several different algorithms have been proposed and tested for approximating the exact nodal areas covered by the horizon. Algorithm 6 adjusts the areas for nodes that are partially covered by the horizon by a certain factor. Note that with this algorithm, the search for the family of nodes needs to consider points that are *outside of the horizon*, up to a distance $\delta + \Delta x/2$. This algorithm gives results which do not fluctuate much with respect to variations in $m = \delta/\Delta x$. Note that when the node's location is exactly δ away from the current node, about half of its area is covered by the horizon region, and the algorithm uses half the node's area.

Algorithm 6 Approximate calculation for area of a node covered by the horizon of the current node

---

if $\|\xi\| < \delta - \frac{\Delta x}{2}$ **then**
  factor = 1.0
else if $\|\xi\| \leq \delta + \frac{\Delta x}{2}$ **then**
  factor = $\frac{\delta + \frac{\Delta x}{2} - \|\xi\|}{\Delta x}$
else
  factor = 0.0
end if
area covered = $(\Delta x)^2 \times$ factor

In computing the quantities in Eq.(10.10) for the case in which the micromodulus function c depends on the bond length, some adjustments are useful to increase the accuracy of the mid-point quadrature scheme. We distinguish three cases for nodes in the family of nodes of some current node i, shown in Algorithm 7: a node's area is fully covered by the horizon, the node is inside the horizon but not all of its area is covered, the node is outside the horizon but has some of its area covered by the horizon. For each case, we compute the micromodulus at the node's actual location or at a slightly modified one (see Algorithm 7).

This algorithm, for the conical micromodulus, produces a close-to-monotonous convergence to the exact classical value in terms of strain energy density for a homogeneous deformation as m increases (see [48]). We note that [91] proposed an algorithm based on analytical calculation of the covered areas to perform the numerical spatial integration. This algorithm shows asymptotically monotonic convergence and also demonstrates slight improvements compared with the one shown in Algorithm 7. However, this more "exact" algorithm may be more expensive to use than other approximating algorithms, like Algorithm 7 below, introduced in [48].

Algorithm 7 Computing the micromodulus function for the conical case

if $\|\xi\| < \delta - \frac{\Delta x}{2}$ **then**
  factor = 1.0
  Compute $c(\xi)$ value with $\xi = \|\xi\|$
else if $\|\xi\| \leq \delta + \frac{\Delta x}{2}$ **then**
  factor $= \frac{\delta + \frac{\Delta x}{2} - \|\xi\|}{\Delta x}$
  Compute $c(\xi)$ value with $\xi = \delta - \Delta x \times \frac{\text{factor}}{2}$
else
  factor = 0.0
end if

10.5 Peridynamic results for dynamic fracture and crack branching

To obtain a more complete understanding of how crack branching is generated and of the factors that influence it, we perform a series of peridynamic computational tests that consider:

1. two different material systems, to observe the influence of Young's modulus, density, and fracture energy on dynamic crack propagation and branching;

2. three different loading types, to investigate the role of stress waves interactions with the tip of the advancing crack;

3. two different geometries, to analyze how reflected stress waves from the boundaries affect the branching process.

Stress waves have a strong influence on dynamic crack propagation (see e.g. [87], [84]). Our focus in the computations below is to investigate how stress waves control crack branching. We utilize two material systems, soda-lime glass and homalite, for which substantial experimental results are available in the published literature (e.g. [23, 35, 31, 83, 87, 84]). Homalite is a material representative of many other glassy polymers, like PMMA or polycarbonate, on which experimental results have appeared in, for example, [93, 15]. One reason more recent experiments have used glassy polymers instead of glass for studies on dynamic brittle fracture is the reduced wave speed (and therefore slower crack propagation speeds) in these materials due to their lower elastic modulus. This allows experimentalists to record the fracture events easier than in the case of glass.

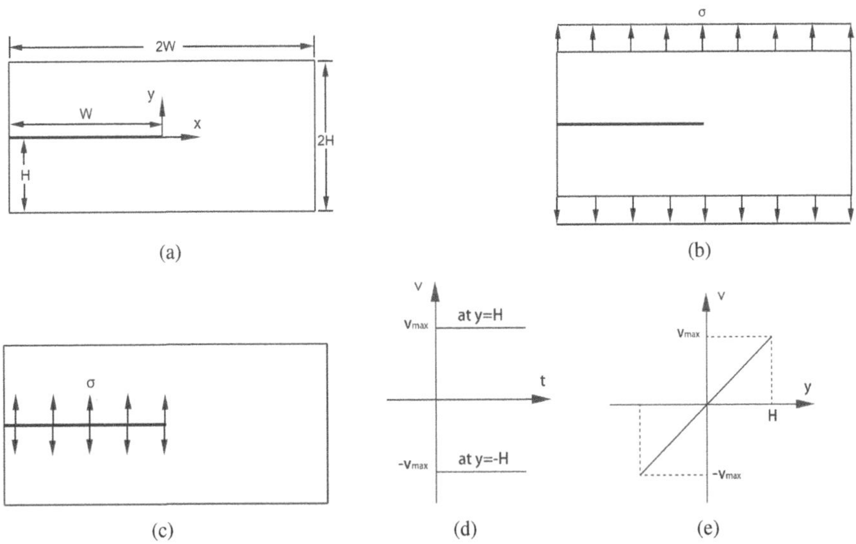

Figure 10.3: (a) Sample geometry; three loading conditions: stresses suddenly applied on top and bottom boundaries (b), stress applied on crack surfaces (c), and velocity boundary conditions (d). In (e): the initial conditions for the case shown in (d).

Different loading conditions have been used in experiments on crack branching. A dynamic loading on the boundaries of the sample has been used in [83], while in [84], in order to avoid the interaction between the propagating crack and waves reflected from the boundaries of the sample, a sudden loading was applied to the crack surfaces themselves to induce crack propagation and eventual crack branching. The simplest, from the point of view of experimental setup, is one of quasi-static loading of samples with sharper or blunter notches used in [23]. The crack propagates without branching from the sharp notch, but can branch once or multiple times from the blunt notch because of the much higher stress intensity factor reached before propagation. The first two types of loading conditions may, however, be preferred since they do not require changing the sample geometry (the bluntness of the crack tip) as the quasi-static loading conditions do in order to observe cases in which the crack propagates straight and does not branch, and cases when the crack branches. In these types of loading, branching or no branching is produced by higher or lower amplitude applied loadings. We implement three types of loading conditions for a pre-notched sample as follows:

- Load case 1: apply, suddenly, stresses on upper and lower boundaries of the sample (Fig. 10.3(b)),
- Load case 2: apply stresses on pre-crack surfaces (Fig. 10.3(c)); and
- Load case 3: apply velocity boundary conditions as in Fig. 10.3(d) and initial velocity conditions as in Fig. 10.3(e) to prevent a strong shock and its reflection from the boundaries. This mimics, to some extent, the quasi-static loading conditions.

We remark that for some quasi-static loading conditions used in experiments, a corresponding explicit-type simulation would be very costly, and solutions for this type of loading require a combination between an implicit solver (up to the point crack growth initiates) and an explicit solver (for the dynamic crack propagation phase). We note that cracks under quasi-static loading can run dynamically, and can branch. On the other hand, quasi-statically growing cracks cannot branch in an isotropic and homogeneous material. Quasi-statically growing cracks do branch in materials with microstructure, as is the case for fatigue-crack growth in materials in which cracks split when encountering tougher inclusions.

Geometry also has an effect on the crack branching problem since the brittle fracture process is influenced/controlled by stress waves, and wave reflections from the sample's boundaries, for example, will differ for different geometries. We will consider two sample geometries: one with twice the width but same length as the other. To remove the influence of asymmetry-induced rounding errors, we place the coordinate system at the centroid of each sample (see Fig. 10.3(a)).

We perform convergence studies for each loading case and material type to justify the selection for the peridynamic horizon size. The aim is to get a crack propagation speed that no longer changes when we change the horizon size. This means that we anticipate that there are no constant material length-scales (induced by the microstructure) that manifest themselves in this problem. This does not preclude existence of length-scales induced by the dynamics of the problem. The meaning of the horizon as a material length-scale is commented in [11]. In peridynamics, one can consider three types of convergence, as introduced in [12]. The δ-convergence, which holds the ratio $m = \delta/\Delta x$ fixed and decreases the magnitude of the horizon, the m-convergence, in which the spatial integration accuracy is increased (for a fixed horizon size), and the combined $\delta - m$ convergence. For a discussion on the form of the kernel under the integral in Eq. (10.1) that leads, for the 1-point Gaussian quadrature, to a consistent scheme, please see [27].

Because the m-convergence for dynamic fracture problems has been discussed before (see, e.g. [43, 44]) and a value of at least $m = 4$ is suggested for crack path independence on the grid, here we focus on the δ-convergence. We select a horizon size, and using a value of $m = 4$ we set the nodal spacing Δx for a uniform grid.

The time step for all computations on soda-lime glass samples is the same, $5\times10^{-3}\mu$s, while for homalite samples it is $5\times10^{-2}\mu$s. These time step sizes ensure stability when used with the finest grids (smallest horizons) used in this work.

10.5.1 Crack branching in soda-lime glass

A soda-lime glass sample with dimensions 10 cm by 4 cm is used to investigate the role of loading conditions. The sample has Young's modulus E = 72 GPa, density ρ = 2440 Kg/m^3, Poisson ratio $\nu = 0.22$ (the corresponding 2D plane stress peridynamic model has a Poisson ratio of 1/3), and energy release rate at crack initiation G_0 = 3.8 J/m^2 (see [104, 36]). Note that in references [43, 44], a much larger value for the fracture energy G_0 was used, namely the measured energy release rate *at branching*, instead of the fracture energy at crack initiation. The main difference between the results shown here and those from [43, 44] is the amplitude of the loading required to generate propagating and branching cracks. A higher fracture energy value requires a larger amplitude loading being applied. The crack will propagate with roughly the same speed, as the propagation is mostly controlled by elastic properties (wave speeds) of the material under consideration.

10.5.1.1 Load case 1: stress on boundaries

To perform the δ-convergence study, we use horizon sizes of 3.0 mm, 1.5 mm, and 0.8 mm. With $m = \delta/\Delta x$, the resulting discretizations have the following number of nodes, respectively: 7236; 28,670; 100,200. The crack propagation speeds for these three different horizon sizes are shown in Fig. 10.4. The loading amplitude is $\sigma = 2$ MPa, the crack grows straight for a while and branches at some point. The damage profile for this loading and with a horizon of 1 mm is shown in Fig. 10.5(b). The damage profiles, including the time of initiation of branching (which, for this loading amplitude, happens around 20.5 μs), are not sensitive to the horizon size (see also [43, 44]).

To calculate the crack velocity, crack tips are tracked by searching the most advanced (right-most) node with a damage index higher than a given threshold (0.3, for example). For each simulation we store outputs from about 200 time steps, and since total simulation times vary, the data-dump period varies between 0.15 to 0.75 μs for soda-lime glass. Crack speed is calculated at each time of the data dump using forward differences approximation. Data points on crack velocity plots are averaged velocities over a time period (usually 1/10 of total simulation time for that particular test), and the abscissa of a data point on these plots is the final time of this time period. For example, the total simulation time for results shown in Fig. 10.6(a) is 150μs; each data point shown represents the average velocity over a time interval of 15μs. We normalize velocities by the Rayleigh wave speed c_R of the material. For soda-lime glass, $c_R \approx 3102$ m/s based on the following equations (see [81]):

$$c_R \equiv (2-M_2^2)^2 - 4\sqrt{(1-M_1^2)(1-M_2^2)} = 0$$

in which

$$M_1 = \frac{c_R}{c_1}, M_2 = \frac{c_R}{c_2}$$

and $c_1 = \sqrt{(\lambda+2\mu)/\rho}, c_2 = \sqrt{\mu/\rho}$, where λ, and μ are Lamé's constants. The Rayleigh wave speed is the real root of this equation in the interval $(0, c_2)$.

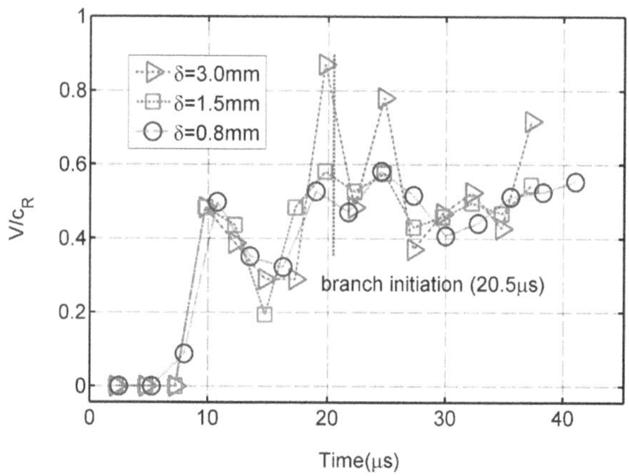

Figure 10.4: δ-convergence study in glass for the crack propagation speed using δ = 3, 1.5, and 0.8 mm, for a suddenly applied tensile stress on the top and bottom boundaries of magnitude σ = 2 MPa. Load is maintained constant in time for the duration of the simulation.

In Fig. 10.4 and similar ones later on, we draw a vertical line to show the branch initiation moment for each computation. Fig. 10.4 shows that this quantity, the time when branching takes place, is independent of the horizon size. Moreover, the difference between crack propagation speed results obtained with the smallest two horizon sizes is minimal. For this reason, the horizon size used in the remaining computations in this section is 1.0 mm, nodal spacing is $\Delta x = 0.25$ mm, and the total number of nodes is 64,160 for the grid of 401×160 nodes.

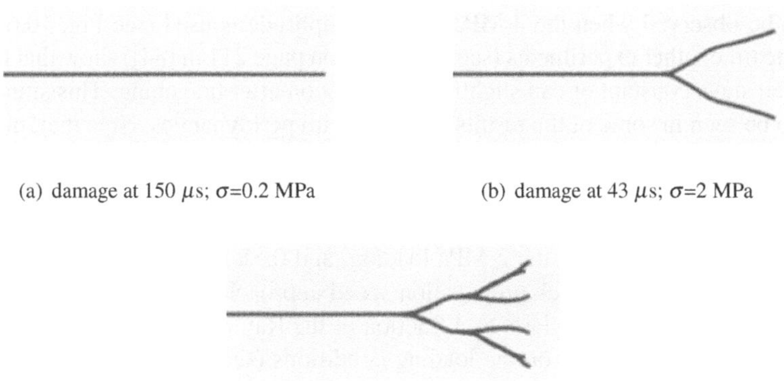

(a) damage at 150 μs; σ=0.2 MPa

(b) damage at 43 μs; σ=2 MPa

(c) damage at 33 μs; σ=4 MPa

Figure 10.5: Damage maps in glass under suddenly applied tensile stress on the boundaries with different amplitudes.

Damage maps for this type of boundary conditions (tensile stress suddenly applied on the boundaries) are shown in Fig. 10.5. We recall that the damage index $d(t)$ at a node at time t is computed as:

$$d(t) = \frac{n_{\text{broken}}(t)}{n_{\text{original}}}$$

where n_{original} is the total number of bonds originally connected at the current node, and n_{broken} is the number of broken bonds at this node at time t. In damage map figures in this chapter, we select a color scheme in which the darker color means that $d > 40\%$. If such damage levels organize themselves along a localized curve, that is an indication of a crack line that formed. The reason we choose 40% instead of 50% is that in the discrete system, to separate a body in two parts requires less than half of bonds to break at a point. As the horizon factor $m = \frac{\delta}{\Delta x}$ grows, the damage index for separation into two parts converges to 50%.

For a load of amplitude $\sigma = 0.2$ MPa applied, we obtain a straight crack. Larger applied load ($\sigma = 2$ MPa) generates one branching event, while multiple branching events develop under stress of 4 MPa magnitude applied suddenly on the top and bottom boundaries. Continued or cascading branching is one cause of fragmentation (see, for example, Chapter 16 in [67]). If the energy dissipated by the formation of the first branching event is overcome by the energy stored in the material, the branching process repeats, eventually leading to fragmentation.

We also observe that the crack branching angle becomes smaller as the magnitude of the applied stress increases. Fig. 10.6 shows the crack propagation speed profiles, normalized by the Rayleigh wave speed c_R, for these three different loading amplitudes. Some experimental works show that the crack speed decreases slightly after branching ([31], [83]), and some of our results confirm these observations. This can be observed when the 4 MPa loading amplitude is used (see Fig. 10.6). At the same time, other experiments (see Fig. 11.12 on page 211 in [84]) show that the crack speed stays constant or can slightly increase soon after branching. This situation can also be seen in some of the results obtained with peridynamics. Note that, in general, the computations show a crack propagation velocity magnitude that increases as the applied stress increases.

At branching, the crack propagation speed as a fraction of the Rayleigh wave speed, V/c_R, is 0.51 for the 2 MPa loading, and 0.55 for the 4 MPa loading amplitude. The maximum crack propagation speed approaches $0.6c_R$. In experiments, the limiting crack speed in glass, as a fraction of the Rayleigh wave speed, ranges from 0.47 to 0.66, depending on the loading conditions (see Table 11.1 in [84]). The location of branching moves closer to the starting location (tip of the pre-notch) when the amplitude of the applied stress increases. This is likely due to the increased stress intensity reached when the crack starts to propagate. In experiments conducted using quasi-static loading of glass slides that have blunter and blunter notches, the same behavior is observed: with the location of branching moving closer to the starting point the blunter the notch is (and therefore the higher the stress intensity at which the crack starts to propagate, see [23]).

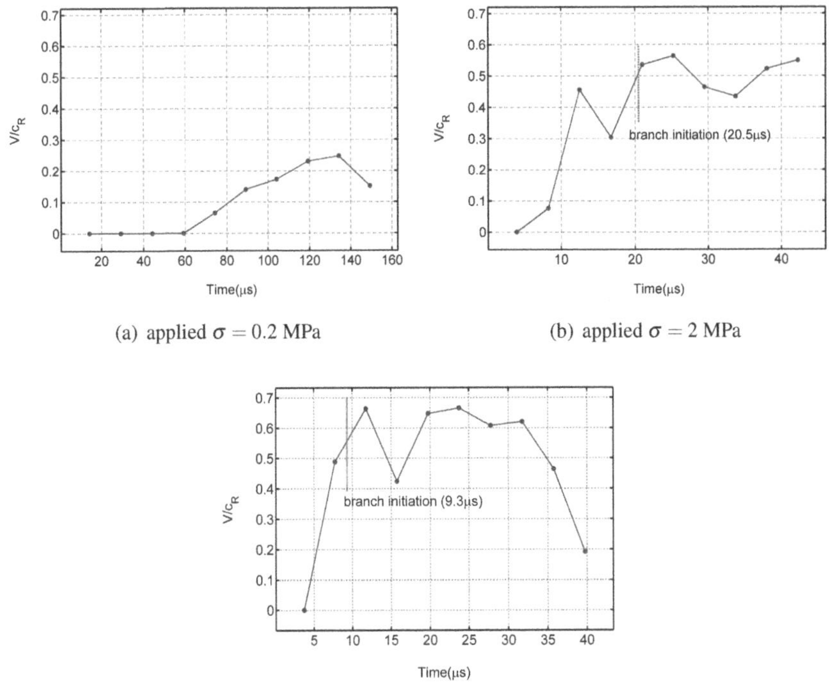

(a) applied $\sigma = 0.2$ MPa

(b) applied $\sigma = 2$ MPa

(c) applied $\sigma = 4$ MPa

Figure 10.6: Crack propagation speed profiles for glass under stress boundary conditions applied on the boundaries of the sample at different amplitudes.

10.5.1.2 Load case 2: stress on pre-crack surfaces

Electromagnetic loading of crack surfaces has been used in [84] (see also [85]) to generate sharp pulses with a rise to peak amplitude of about 25 μs and total duration of about 150 μs. Such pulses were able to generate surface pressure waves in the range of 1–20 MPa. Until waves reflected from the edges of the plate return to interact with the crack tip, this type of loading can be considered as taking place in an infinite plate. In this section we apply, suddenly, body forces on the nodes located on the crack surfaces to mimic the stress generated by electromagnetic loading conditions and investigate crack branching. These loads, once applied, are maintained constant throughout the simulation. We apply several loading magnitudes and notice that a loading equivalent to 0.5 MPa generates a straight propagating crack in this material, while a load of 3 MPa or higher leads to branching. For this material and sample size, branching still happens after reflected stress waves interact with the tip of the advancing crack. However, as we shall see in Section 10.5.2.2, for a material in which waves speeds are slower than in glass, e.g. homalite, branching can happen in the absence of interactions between the crack tip and waves reflected from the boundaries of the sample. In order to produce this behavior in a soda-lime glass sample, we would have had to consider a much larger sample, resulting in a much larger computation. Nevertheless, we perform a δ-convergence study for the crack propagation velocity with a loading magnitude of 3 MPa and the results are shown in Fig. 10.7. The damage maps corresponding to these different horizon sizes are identical, and they are given in Fig. 10.8(b). For the remainder of the results shown in this section, we use a horizon size $\delta = 1$ mm and nodal spacing of 0.25 mm, with a grid of 401×160 nodes.

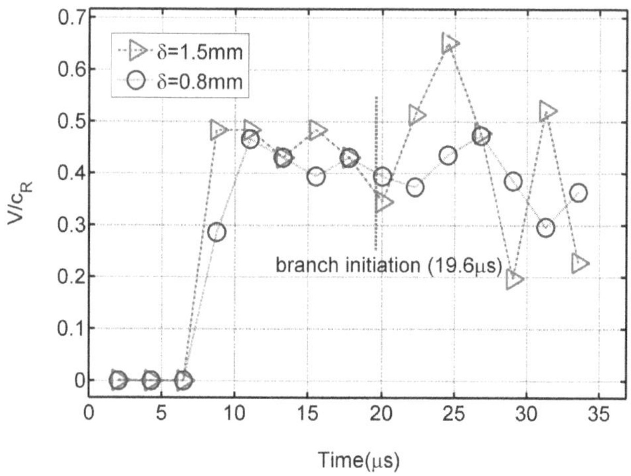

Figure 10.7: δ-convergence study in glass for the crack propagation speed using δ = 1.5 mm and 0.8 mm, with a suddenly applied tensile stress on crack surfaces of magnitude σ = 3 MPa. The load is maintained constant in time for the duration of the simulation.

Damage maps obtained under different stress levels (0.5, 3, and 6 MPa, respectively) applied on crack surfaces are shown in Fig. 10.8. The stress waves reflected from the boundaries return and meet the crack tip about 7.36 μs after the moment the loading is applied on the crack surfaces, therefore crack propagation and branching is, in these cases, influenced by reflected stress waves. Continued reflections from the boundaries and *reinforcement of stress waves* interact with the propagating crack, but branching occurs only when the dynamic conditions around the tip of the advancing crack are favorable. This point is discussed further in Section 10.7.

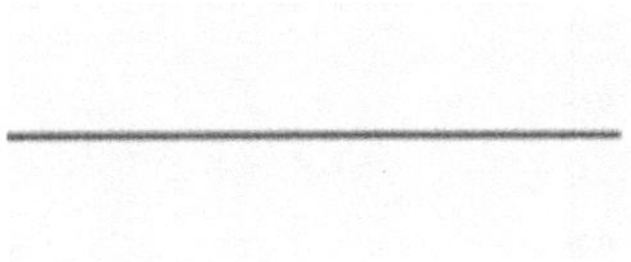

(a) damage at 113 μs; $\sigma = 0.5$ MPa

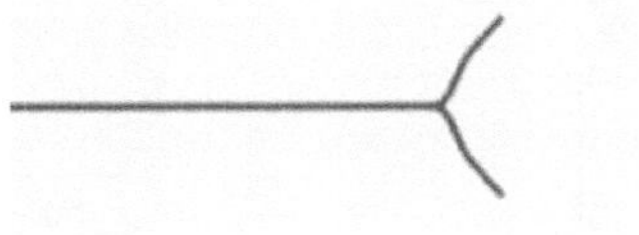

(b) damage at 35 μs; $\sigma = 3$ MPa

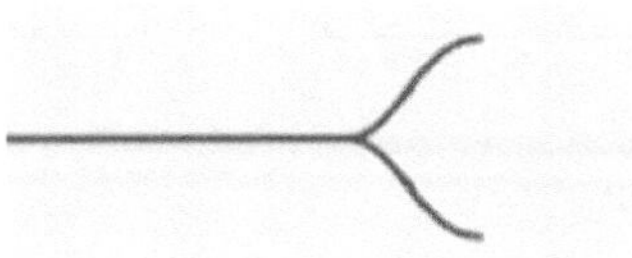

(c) damage at 30 μs; $\sigma = 6$ MPa

Figure 10.8: Damage maps for glass under stress on pre-crack surface with different stress amplitudes applied.

Increasing the loading magnitude further did not cause multiple branching events, but rather created fracture at the corners due to reinforcement of the waves bouncing from the boundaries. We note that under these loading conditions, the crack propagation velocity at branching is $V = 0.40c_R, 0.44c_R$ for the 3 MPa and 6 MPa applied stresses, respectively. The maximum crack speed is around $0.6c_R$, even with the larger loading amplitude. It is interesting to observe the significant change in the crack branching angle between the 3 MPa and 6 MPa loading conditions. These differences are caused by the particular conditions at which the dynamic interaction between the advancing crack tip and the reflected stress waves (longitudinal, shear, as well as the Rayleigh waves propagation on the crack surfaces) takes place (see Figs. 10.9(b) and 10.9(c)). Simulation movies that allow us to visualize these interactions are included in the tests on homalite (see Section 10.5.2).

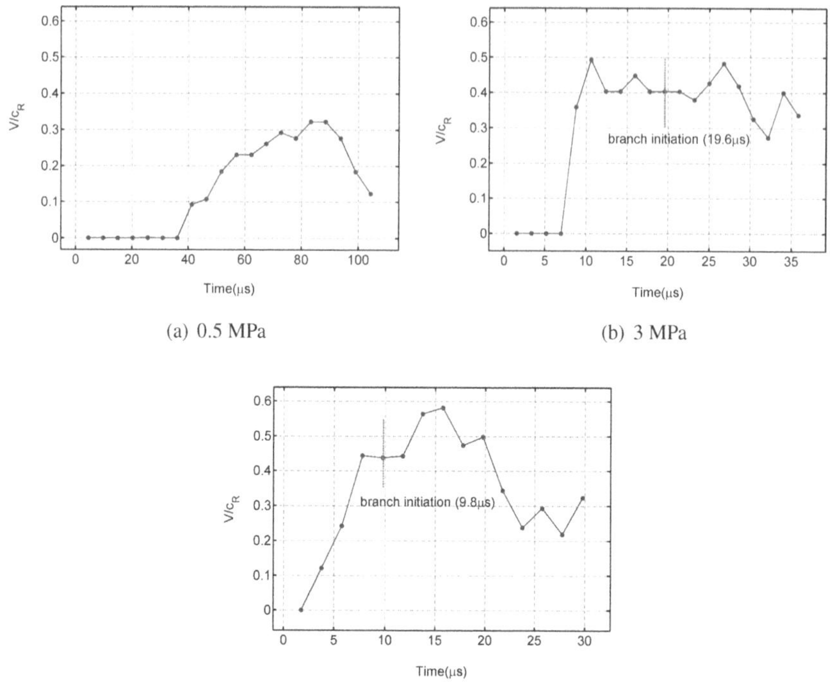

(a) 0.5 MPa

(b) 3 MPa

(c) 6 MPa

Figure 10.9: Crack propagation speed profiles for glass under stress boundary conditions applied on the pre-crack surfaces: (a) σ = 0.5 MPa, (b) σ = 3 MPa, (c) σ = 6 MPa.

10.5.1.3 Load case 3: velocity boundary conditions

In many experiments on crack branching (e.g. [23]), the sample is slowly loaded. When a critical stress intensity is reached around the crack tip, a pre-crack starts propagating. The sharpness of the pre-notch determines the elastic energy stored before the propagation starts. At higher elastic strain energy, the crack propagates and can branch once or multiple times (see [23]). Simulating the slow loading is costly with an explicit time-integration model like the one used in this work. A close simulacrum of this type of loading is one in which the sides of the plate are moving away from each other with constant velocity, with the initial velocity at all points in the plate being a linear interpolation of the values imposed on the sides (see Fig. 10.3(d) and 10.3(e)). While different from a quasi-static loading, this dynamic loading avoids, however, the strong shocks that the previous two loading cases generate.

We conduct a δ-convergence study in terms of the crack propagation speed for the case when the sides of the plate are moving apart from each other with constant velocity of 0.02 m/s. The results are shown in Fig. 10.10 and for this loading condition, for the three different horizon sizes used, the crack propagation speed is the same. The damage map is given in Fig. 10.11(a), the crack propagating straight. We notice that the crack initiation time and the overall trend for the crack propagation speed are essentially the same for the three different horizon sizes used, and especially so for the smallest two horizon sizes. Because of this, for the remaining results in this section a horizon size $\delta = 1$ mm is used, along with grid spacing $\Delta x = 0.25$ mm.

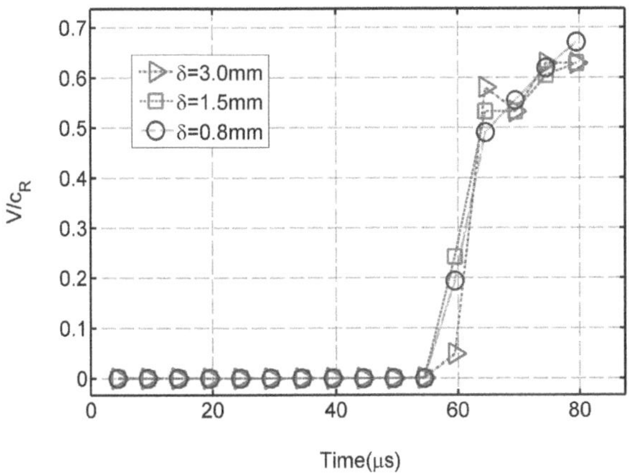

Figure 10.10: δ-convergence in terms of crack propagation speed in glass for velocity boundary conditions with applied velocity on top and bottom boundaries of $V_{max} =$ 0.02 m/s, using $\delta = 3$, 1.5, and 0.8 mm.

Damage maps presented in Fig. 10.11 demonstrate that as the intensity of loading increases, one branching event can take place, or a series of branchings followed by crack arrest can develop into a *fish-bone like pattern*. The reason for this extreme behavior can be understood given the type of loading used in which the top and bottom sides of the plate are forced apart at a constant rate. One can imagine that when one increases the loading rate, the material eventually exhibits multiple failure points and fragmentation in trying to accommodate the intense strain rates. The fish-bone failure patterns have been observed before in brittle fracture (see [94]). We also observe that, as the loading intensity increases, the location of the initiation of branching or branching attempts moves closer to the location of the pre-crack tip, similar to what is reported from experiments.

(a) damage at 83μs; applied $V_{max} = 0.02$ m/s

(b) damage at 45μs; applied $V_{max} = 0.06$ m/s

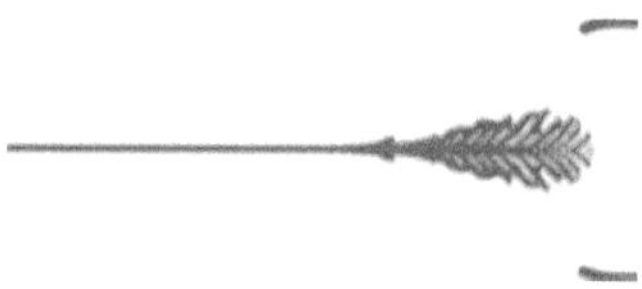

(c) damage at 24μs; applied $V_{max} = 0.2$ m/s

Figure 10.11: Damage maps in glass under velocity boundary conditions with different velocity amplitudes applied.

Corresponding to the first two plots in Fig. 10.11, we give the crack propagation velocity in Fig. 10.12. For the case in which the fish-bone profile is obtained it is difficult to decide which of the micro-cracks to focus on and we did not compute the crack propagation speed. We observe that, unlike in the previous two loading cases, in this imposed displacement conditions case, the crack velocity in the region of branching reaches beyond $0.6c_R$, and even higher soon after branching (see Fig. 10.11(b)). Such high crack propagation speeds in glass have been experimentally observed, for example, in [3]. The crack slows down slightly immediately after branching, but it then quickly recovers as the two branches propagate with even higher velocities than before the branching event. These results confirm the theories (see [84], page 214) that state that branching does not happen because the crack propagation speed reaches a certain threshold value, but rather that branching occurs when the strain energy in the region around the crack tip reaches a critical state. A region with "thicker" damage extends over a significant length and can be easily observed to occur *before branching* (see Fig. 10.11(b)). This could be the signature of roughening of the crack surface reported in all experiments to occur before branching ([83]). After branching, the thicker damage zones subside and the distinct branches propagate as "thin" damage zones, indicating that the roughness on the crack surfaces has disappeared. If the strain energy delivered into the crack tip areas of the two branches is sufficiently high, the process repeats itself and a second branching event can ensue. In Fig. 10.11(b)) we observe this phenomenon just about to happen as the branches approach the boundary of the sample.

These results support the ideas proposed by Ravi-Chandar (see [86] and [84], page 214) and based on experimental evidence: "*when a crack reaches a critical stage identified macroscopically by its stress intensity factor, it splits into two or more branches, each propagating with the same speed as the parent crack, but with a much reduced process zone.*" A more detailed discussion of the meaning of this thickening of damage ahead of the branching point is given in Section 10.7.

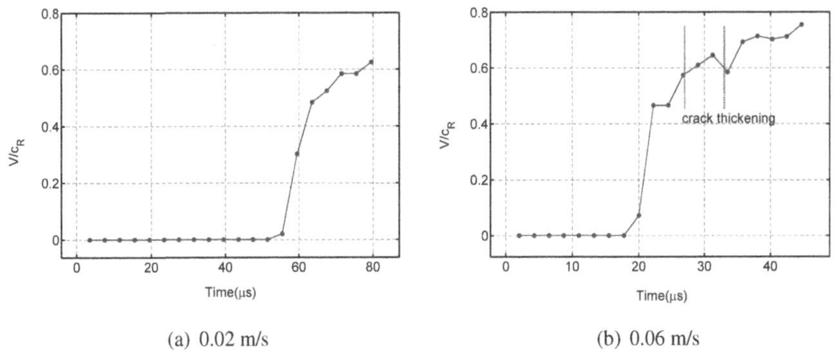

Figure 10.12: Crack propagation speed profiles for velocity boundary conditions: (a) V_{max} = 0.02 m/s, (b) V_{max} = 0.06 m/s

10.5.2 Crack branching in homalite

With elastic wave speeds and crack propagation speeds being so fast in glasses, a significant number of experiments on brittle fracture have switched to glassy polymers that allow easier observation of the crack branching phenomenon ([83, 84]). For example, in [84], a homalite sample is used with loadings on the crack surfaces (loading case 2) in order to observe crack branching without waves that reflect from the boundaries interfering with the tip of the advancing crack. In glass, such an experiment would require a much larger sample size, which is not practical. In this section we use a two-dimensional pre-cracked homalite sample with dimensions of 20 by 40 cm. Homalite has significantly different stiffness, density, and fracture energy compared with soda-lime glass. This investigation will also give us insight into the role and influence material properties have on crack branching.

The homalite's Young's modulus is $E = 4.55$ GPa, density is $\rho = 1230$ kg/m^3, fracture energy G_0 =38.46 J/m^2. We perform δ-convergence studies (with a fixed $m = \frac{\delta}{\Delta x} = 4$) for each of the loading cases mentioned before, and once the crack propagation speed shows little changes when using a smaller horizon size, we select that horizon size to compute the rest of the results in this section. The time step in all of the computations in this section is the same, 0.05 μs, which is stable for the finest grid used here.

10.5.2.1 Load case 1: stress on boundaries

For a tensile stress amplitude of 1 MPa suddenly applied on the top and bottom boundaries of the sample as in Fig. 10.3(b), the crack propagation speed for three different horizon sizes varies as shown in Fig. 10.13. The crack branches at around 260 μs and damage maps are identical (except for the "thickness" of the crack line, see [44]) for the three different horizon sizes, looking like the one shown in Fig. 10.14(b). Notice that the crack speed profiles are almost identical for δ = 4.0 mm and 3.0 mm. For the remainder of the computations in this section the horizon size is 4 mm, and, with a ratio $m = 4$, this leads to a grid spacing of 1 mm and a total number of nodes of 80,200 (401×200 nodes) for the sample dimensions selected above.

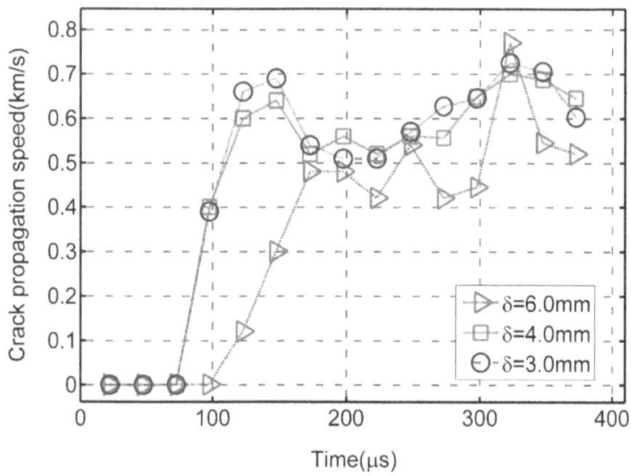

Figure 10.13: δ-convergence study in homalite for crack propagation speed using δ = 6.0, 4.0, and 3.0 mm, for a suddenly applied tensile stress (σ = 1 MPa) on the top and bottom boundaries. Load is maintained constant in time for the duration of the simulation.

Using this horizon size and grid spacing, we now perform calculations for the damage index under different loading magnitudes (see Fig. 10.14). With a loading of 0.2 MPa the crack propagates straight, while at 1 MPa the crack branches after growing straight for a while. With further increase of the loading amplitude, at 2 MPa, for example, there is an attempt of branching that is arrested almost immediately, after which a successful branching takes place, followed by secondary branching. The particular shape of the crack branches as well as the location of the branching events are all controlled by the wave interactions with the advancing crack. Given the different elastic wave speeds in homalite compared with those in soda-lime glass, these interactions happen at different times resulting in the observed differences in the shape of cracks (compare Figs. 10.5 and 10.14).

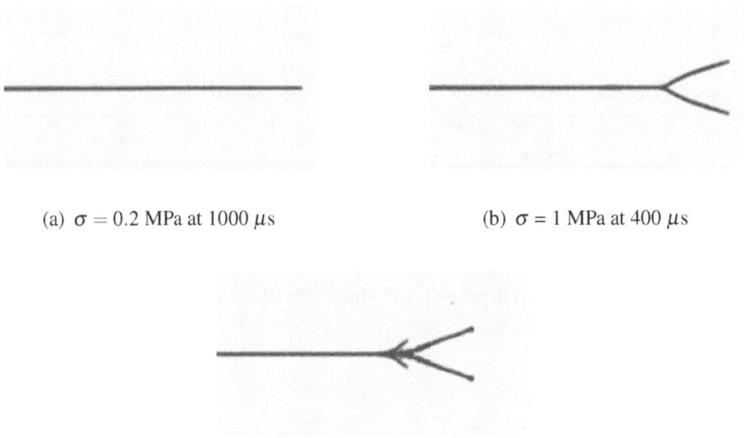

(a) $\sigma = 0.2$ MPa at 1000 μs (b) $\sigma = 1$ MPa at 400 μs

(c) $\sigma = 2$ MPa at 300μs

Figure 10.14: Damage maps for homalite under tensile stresses of various amplitudes applied on the top and bottom boundaries. Load is maintained constant in time.

The crack propagation speed profiles for the lower two loading cases are given in Fig. 10.15. The profile for the 2 MPa loading case is not provided because the unsuccessful branching attempt can make interpreting the results more difficult. We note that for the 1 MPa loading amplitude, the crack speed at branching is about $0.5c_R$, in line with experimental observations. The crack advances straight at this speed (or at even slightly higher speed) for a significant length before branching, again confirming results obtained in experiments (see [83, 84]).

Simulation movies for crack propagation and branching in homalite for this type of boundary conditions are shown in Movies 1, 2, and 3 (see supplementary materials for [13] or visit http://engineering.unl.edu/mme/florin-bobaru/). In these movies, the color indicates the vertical component of the velocity vector, allowing us to observe the propagation of stress waves and their interaction with the advancing crack. In Section 10.7 we offer a more detailed discussion of what the movies demonstrate.

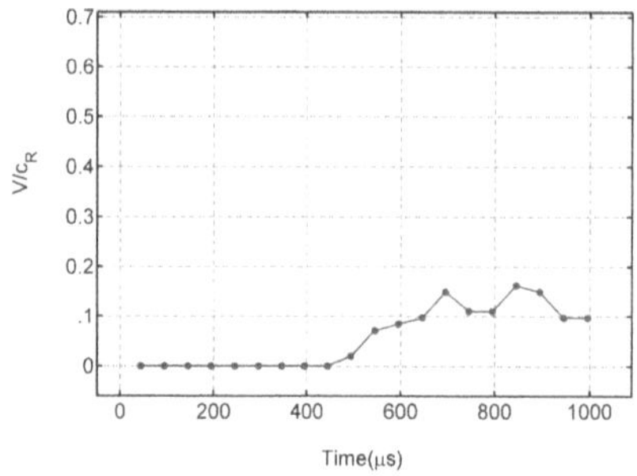

(a) applied $\sigma = 0.2$ MPa

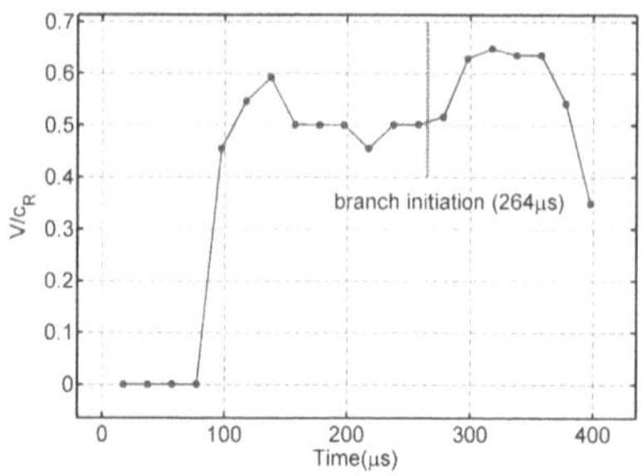

(b) applied $\sigma = 1$ MPa

Figure 10.15: Crack propagation speed profiles in homalite with tensile stresses of various amplitudes applied on the top and bottom boundaries: (a) σ = 0.2 MPa, (b) σ = 1 MPa

10.5.2.2 Load case 2: stress on pre-crack surfaces

With loads applied directly on the crack surface, for homalite material and the sample size chosen here it will be possible to observe crack branching without the interaction between the advancing crack and waves reflected from the boundaries.

We first perform a δ-convergence study in terms of crack propagation speed for a loading amplitude that leads to a single branching event. Results are shown in Fig. 10.16. The damage maps for the three different horizon sizes used are identical (except for the thickness of damage) to the one shown in Fig. 10.17(b). Because differences in crack propagation speed are negligible between the smallest two horizon sizes used, for the rest of the computations performed for this section we use the horizon size $\delta = 4$ mm, and with a ratio $m = 4$ this leads to a grid spacing of 1 mm and a total number of nodes of 80,200 (401×200 nodes) for the sample dimensions mentioned at the beginning of Section 10.5.2.

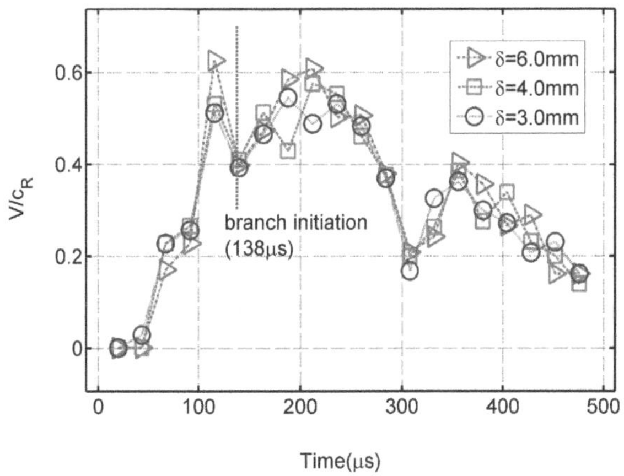

Figure 10.16: δ-convergence study for crack propagation speed in homalite, using δ = 6, 4, and 3 mm, and a loading magnitude of σ = 2 MPa applied on the pre-crack surfaces.

The damage maps for several different loading amplitudes are shown in Fig. 10.17, at the times listed in the figure captions. With the lower 0.3 MPa loading amplitude, the crack propagates straight without branching. From a sudden load of 2 MPa (and maintained constant in time) applied on the crack surfaces, the crack branches at around 135 μs, which is after the stress waves return and interact with the propagating crack. When the load is increased to 4 MPa, the crack branches at around 124 μs (see Fig. 10.18), which is still after the reflected waves meet the crack, this event happening at around 107 μs. We observe an interesting phenomenon: the loading amplitude is higher (4 MPa versus 2 MPa), but the point of branching is further away from the tip of the pre-crack (center of plate). This appears to contradict the results seen in the previous sections in which the branching location moved closer to the pre-crack tip when the loading amplitude was increased. The *paradox is explained* as follows: with the higher loading amplitude of 4 MPa, the straight crack moves faster and the reflected waves meet the tip further down the path compared with the 2 MPa loading case. In both of these cases, branching happens after the reflected waves interact with the advancing straight crack. In the 4 MPa loading case, subsequent interactions between the initial branches and stress waves result in a number of attempted branching and a successful secondary branching event over a time span of 400 μs (see Fig. 10.17(c)).

We now verify that the branching point moves closer to the pre-crack tip by applying higher loading amplitudes, as long as branching is not influenced by stress waves coming in from the boundaries (incident or reflected). With a loading amplitudes of 6 MPa and 8 MPa, respectively, we indeed see that branching moves closer to the pre-crack tip (see Figs. 10.17(d), 10.17(e)). In both of these cases the crack reaches sufficient conditions to branch *before* the waves return from the boundaries at around 72 μs and 50 μs, for the 6 MPa and 8 MPa loading amplitudes, respectively (see Fig. 10.18). The waves return to meet the crack after more than 100 μs.

We have thus confirmed that, as long as the branching mechanism is *not influenced by incoming stress waves interacting with the propagating crack*, higher loadings lead to branching taking place closer (as well as sooner) to the pre-crack tip location. This is further discussed in Section 10.7. When branching is *caused by waves that meet the advancing crack*, two things can happen under applying higher loading amplitudes:

(a) the location of branching moves closer to the pre-crack tip when conditions for branching are not met at the first interaction between the advancing crack and the incoming stress waves: this was the case in all of the loading cases of the glass sample, and loading cases 1 and 3 (see below); or

(b) the location of branching is farther from the pre-crack tip when conditions for branching are met at the first interaction between the advancing crack and the incoming stress waves: this was the case for the homalite sample and loading case 2, when the loading amplitude was high enough to cause branching but lower than 6 MPa, since at those values or higher, branching happens before incoming waves interfere with the crack.

Obviously, the differences between the glass sample and the homalite sample are only due to the different wave propagation speeds in these materials. Every type of branching behavior that we see in homalite (like the variation of branching location with the amplitude of applied loading mentioned above), we can see in glass as well for an appropriately sized sample geometry.

When compared with the results for soda-lime glass with the same type of loading, it is interesting to see that the patterns are similar when the branching mechanisms is the same between the two types of materials (see Figs. 10.9(b), 10.9(c), and 10.17(b)). The damage pattern changes significantly (in homalite, see Fig. 10.17(c)) when branching happens before the reflected wave return to meet the advancing crack paths.

The crack propagation speed profiles in homalite are shown in Fig. 10.18. The crack speed at branching reaches $0.42c_R$ when reflected waves induce it (see the 2 MPa loading case), and approaches $0.55c_R$ when branching is the direct result of the loading on the crack surfaces (see the 4 MPa, and 8 MPa loading cases).

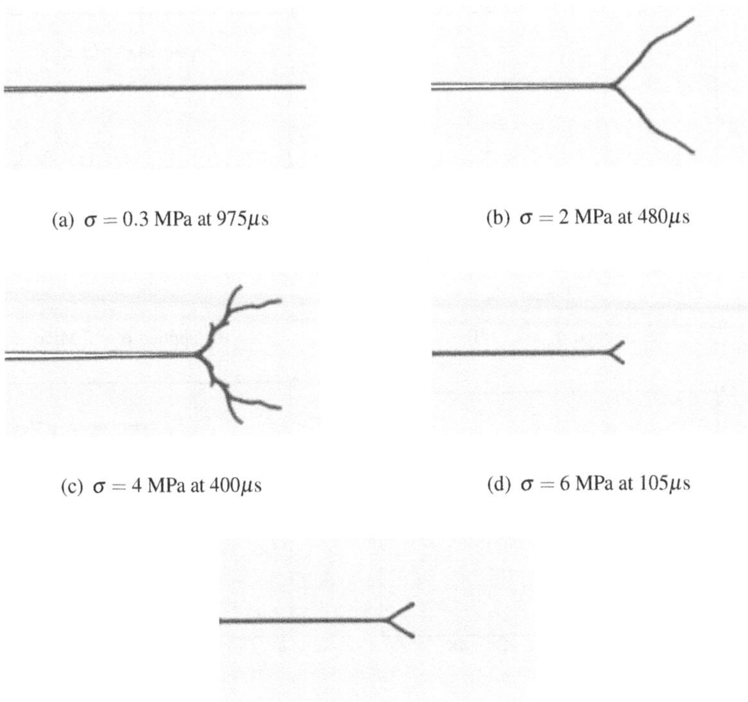

(a) $\sigma = 0.3$ MPa at 975μs

(b) $\sigma = 2$ MPa at 480μs

(c) $\sigma = 4$ MPa at 400μs

(d) $\sigma = 6$ MPa at 105μs

(e) $\sigma = 8$ MPa at 105μs

Figure 10.17: Damage maps for homalite under stress on pre-crack surface with different amplitudes.

Simulations of crack propagation and branching in homalite when the loading is applied on the crack faces are shown in Movies 4–7 (see supplementary materials for [13] or visit http://engineering.unl.edu/mme/florin-bobaru/) for various amplitudes of loading. As before, the color in these movies indicate the vertical velocity component. The simulations help us understand the reasons crack branching happens, and a detailed discussion is provided in Section 10.7.

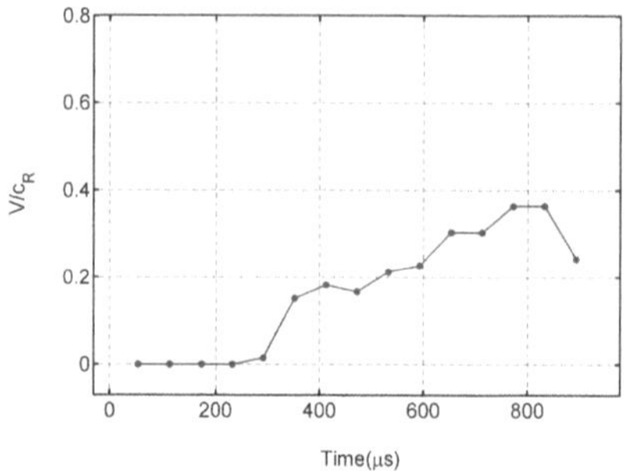

(a) applied $\sigma = 0.3$ MPa

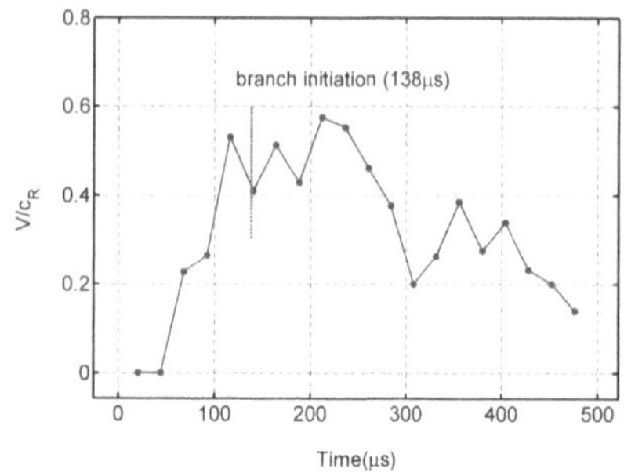

(b) applied $\sigma = 2$ MPa

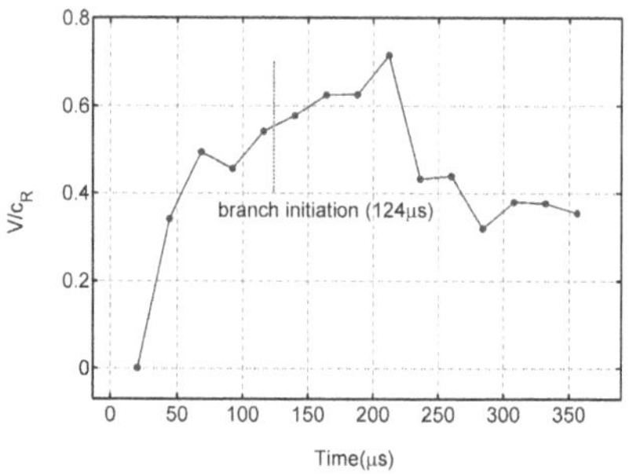

(c) applied $\sigma = 4$ MPa

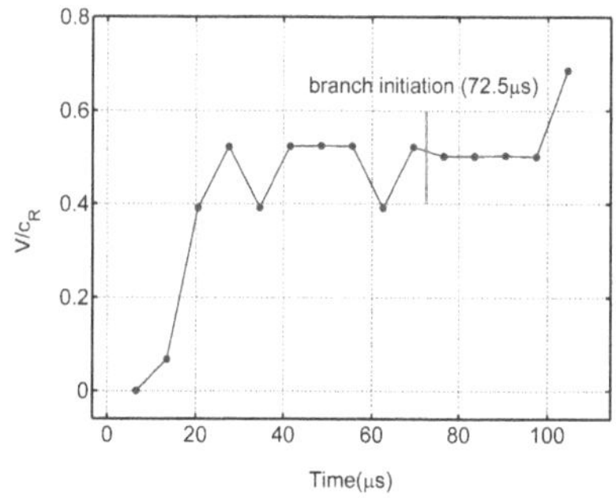

(d) applied $\sigma = 6$ MPa

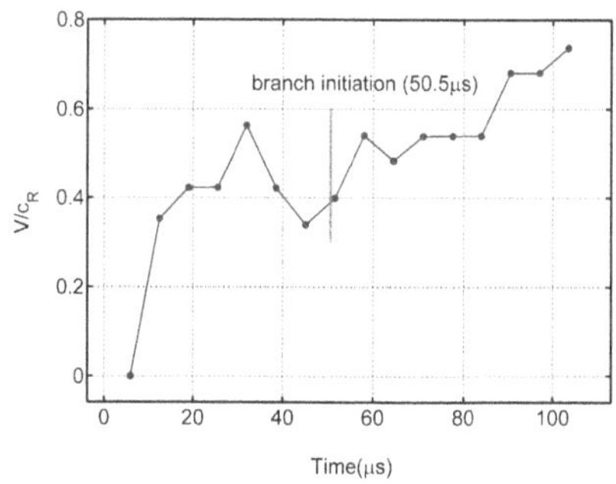

(e) applied $\sigma = 8$ MPa

Figure 10.18: Crack propagation speed profiles for homalite with stress applied on pre-crack surfaces at various amplitudes.

10.5.2.3 Load case 3: velocity boundary conditions

In the dynamic loading type that reduces the shocks compared with the previous two loading cases, the sides of the plate are moving away from each other with constant velocity enforced at the top and bottom boundaries, with the initial velocity at all points in the plate being a linear interpolation of the values imposed on the sides (see Figs. 10.3(d) and 10.3(e)). While different from a quasi-static loading, this dynamic loading is the closest one to it. We conduct a δ-convergence study in terms of the crack propagation speed for the case when the sides of the plate are separated by a constant velocity applied of 0.02 m/s on each of the top and bottom boundaries. The crack propagation speed for horizon sizes equal to 4 mm and 3 mm are shown in Fig. 10.19. The damage patterns obtained with both these horizon values are given in Fig. 10.20(b). With a horizon size $\delta = 6.0$ mm, the damage map, for the same loading amplitude, results in a fish-bone pattern, for which it is difficult to decide the crack propagation speed. The results shown justify using the $\delta = 4.0$ mm as the horizon size for the remaining computations in this section.

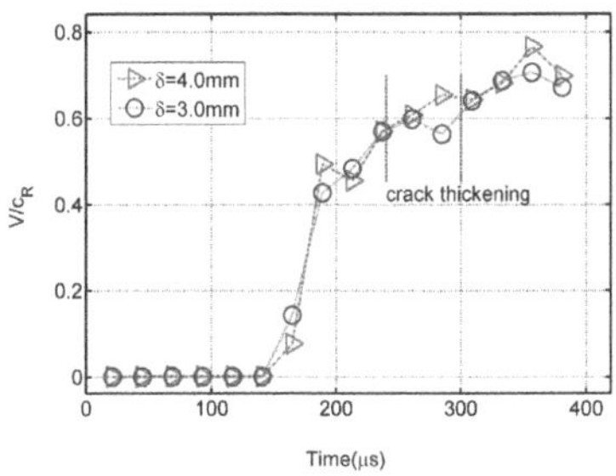

Figure 10.19: δ-convergence study for velocity boundary conditions using $\delta = 4.0$, and 3.0 mm for a loading magnitude of $V_{max} = 0.02$ m/s.

With this horizon size and $m = 4$, we test different loading amplitudes that lead to the pre-crack propagating straight (see Fig. 10.20(a)), branching once (see Fig. 10.20(b)), or splitting into a multitude of microbranches (see Fig. 10.20(c)).The results are similar to those seen in glass, with the only difference being that in homalite the applied velocities need to be higher to reach the similar patterns. This can be easily understood based on the differences in stiffness and fracture energy between the two materials.

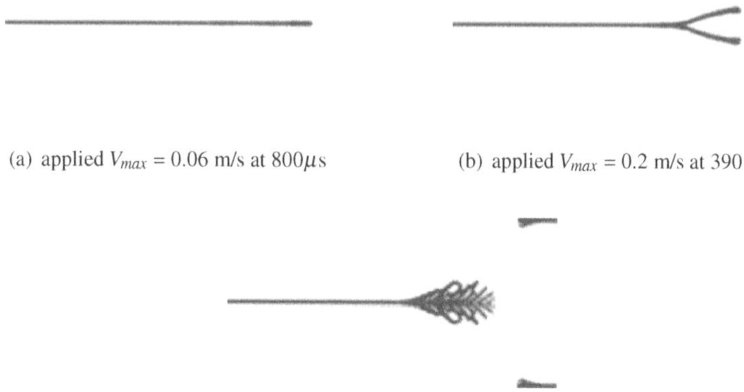

(a) applied V_{max} = 0.06 m/s at 800μs

(b) applied V_{max} = 0.2 m/s at 390 μs

(c) applied V_{max}= 1 m/s at 180 μs

Figure 10.20: Damage maps in homalite under velocity boundary conditions with different velocity amplitudes applied.

We notice again the thickening of the damage area just before the branching point. Once branching happens, the process zones are reduced significantly, only for them to start thickening again for a potential secondary branching event. As in the glass sample, once the material is ripped apart with higher intensity, multiple failure points can initiate simultaneously to accommodate the enforced displacements, leading to the familiar fish-bone failure pattern.

The crack speed profiles under the first two loading conditions are given in Fig. 10.21. For the loading in which the crack branches, the crack speed at branching is around $0.6c_R$. The slight drop in crack speed immediately after branching reported in experiments is seen here as well.

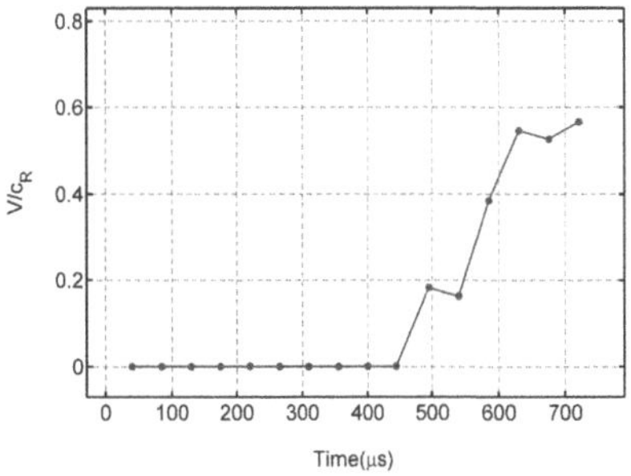

(a) applied $V_{max} = 0.06$ m/s

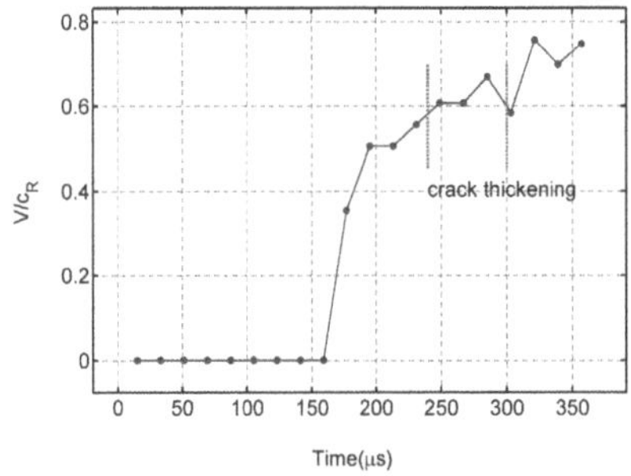

(b) applied $V_{max} = 0.2$ m/s

Figure 10.21: Crack propagation speed profiles for homalite under velocity boundary conditions for different applied amplitudes.

The evolution of crack propagation and branching in homalite is contained Movies 8–10 * , for the different magnitude of applied velocity boundary conditions. The color indicate the vertical velocity component. The results contained in these movies are commented on further in Section 10.7.

10.5.3 Influence of sample geometry

In this section, the influence of sample geometry is investigated for the soda-line glass material. The model is the same as in Section 10.5.1, except that we now use a *narrower geometry* with dimensions of 2 × 10 cm, instead of 4 × 10 cm. We monitor the crack paths/patterns, the crack propagation speed, and the angle of branching for the three different types of loadings used before. Calculations are performed using the same horizon size and discretization as in the case of the wider sample: $\delta = 1$ mm, $\Delta x = 0.25$ mm.

10.5.3.1 Load case 1: stress on boundaries

For this loading case (see Fig. 10.3(b)), a tensile stress with amplitude $\sigma = 0.2$ MPa applied on the top and bottom boundaries leads to a straight propagating crack, while increasing this amplitude to 2 MPa results in one branching happening at 14 μ; see Figs. 10.22(b) and 10.23(b). This is sooner than what happened in the wider sample case (see Fig. 10.6(b)), because stress waves travel shorter distances and return sooner from the boundaries in the narrower sample to create conditions favorable to branching. A similar behavior is seen in case of loading the sample with a 4 MPa stress amplitude (compare Figs. 10.23(c) and 10.6(c)). The branching path in the 2 MPa loading amplitude is similar between the wide and narrow samples, but for the 4 MPa amplitude, the second branching attempt arrests one of the branches in the

*See supplementary materials for [13] or visit http://engineering.unl.edu/mme/florin-bobaru/

narrow sample, whereas in the wider sample both branches propagate. The cracks arrest because of wave interactions with the advancing crack. The crack opening angles are only slightly modified by the narrower geometry, while the crack speeds attained at branching are lower than in the wider sample. This result confirms that the crack propagation speed cannot be a criterion for branching. The more intense wave reinforcement taking place in the narrower sample (because of wave reflections from the boundaries that are closer to one another than in the wider sample) leads to conditions sufficient for branching when the crack speed is in the $0.4c_R$–$0.5c_R$ range.

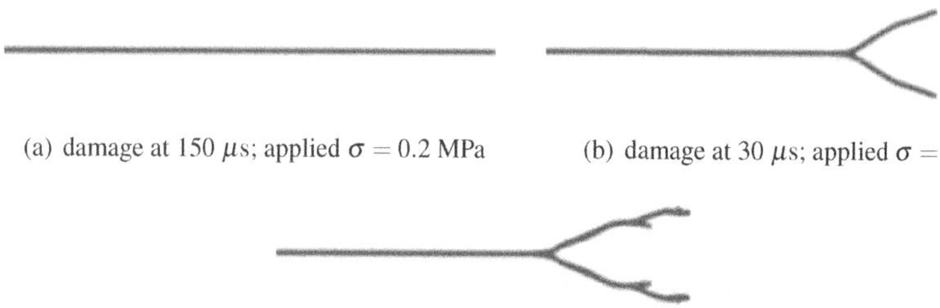

(a) damage at 150 μs; applied $\sigma = 0.2$ MPa

(b) damage at 30 μs; applied $\sigma = 2$ MPa

(c) damage at 25 μs; applied $\sigma = 4$ MPa

Figure 10.22: Damage maps for narrow sample of glass under suddenly applied tensile stresses, of various magnitudes, on the top and bottom boundaries.

After branching, we observe large variations in the crack propagation speed, likely induced by the elastic energy flow to the crack tip regions transported by the bouncing of stress waves from the boundaries and from the newly created crack surfaces.

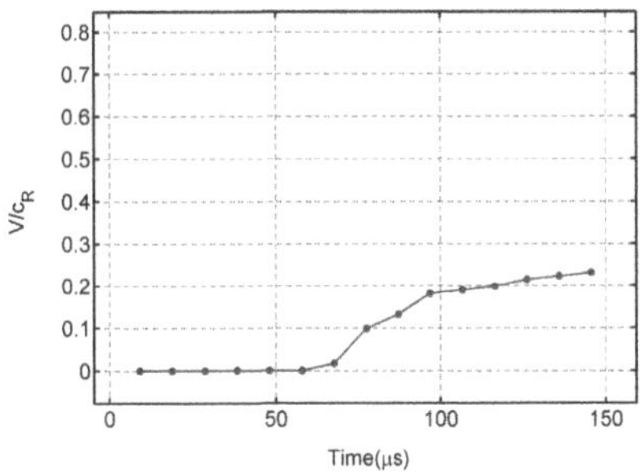

(a) applied $\sigma = 0.2$ MPa

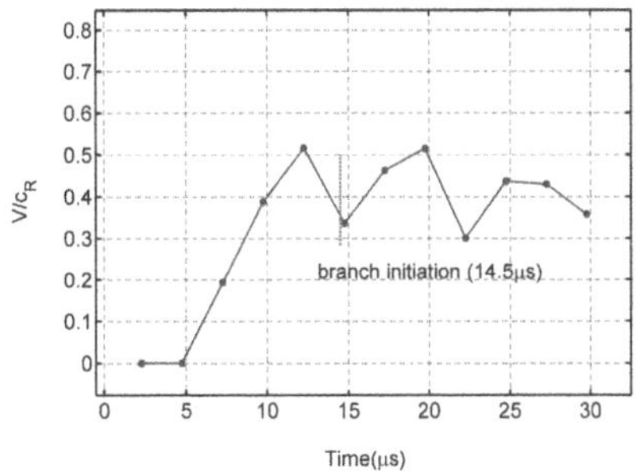

(b) applied $\sigma = 2$ MPa

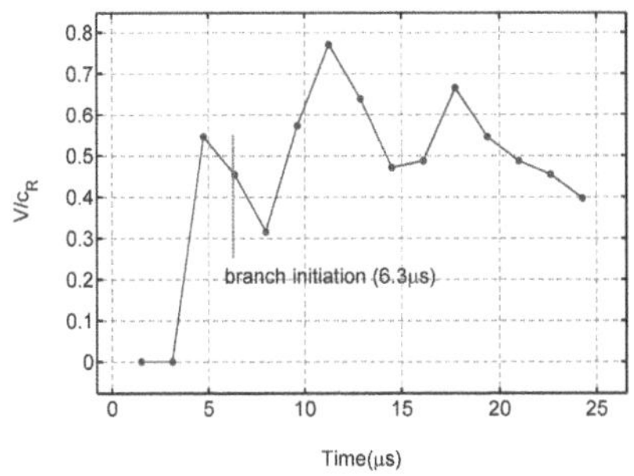

(c) applied $\sigma = 4$ MPa

Figure 10.23: Crack propagation speed profiles for glass under sudden tensile stress applied on the boundaries of the narrower sample at different amplitudes.

The evolution of crack propagation and branching in glass for this geometry are contained in Movies 11–13 *, for the stress amplitudes applied on the outer boundaries. The color scheme used in these movies represents the vertical velocity component. The results contained in these movies are commented further on in Section 10.7.

10.5.3.2 Load case 2: stress on pre-crack surfaces

When load is applied to the crack surfaces (see Fig. 10.3(c)) in the glass sample with a narrower profile, branching is again induced by stress waves bouncing back and forth from the boundaries. Compared with the wider sample, this time we need to apply smaller amplitude loads to get the crack to branch, consistent with the fact that wave reinforcement of reflected waves happens more frequently in the narrower sample. To obtain a straight crack, the previously used loading amplitude of 0.5 MPa (see Fig. 10.8(a)) needs to be reduced to around 0.2 MPa (see Fig. 10.24(a)); otherwise,

* See supplementary materials for [13] or visit http://engineering.unl.edu/mme/florin-bobaru/

the crack eventually branches. The stress wave influence on the crack opening angle is clearly visible when the loading amplitude is increased to 1 MPa, as the branching angle becomes very wide (see Fig. 10.24(b)). Further increasing the loading amplitude to 3 MPa leads to the branching point moving closer to the pre-crack tip, as expected (see Fig. 10.24(c)). With the 6 MPa load amplitude used in the results in Fig. 10.8(c), the narrower sample gets damaged in multiple places near the boundaries, confirming the expectation that more frequent stress waves reflections that are taking place in the smaller sample have an amplifying effect on stress reinforcements that lead to damage.

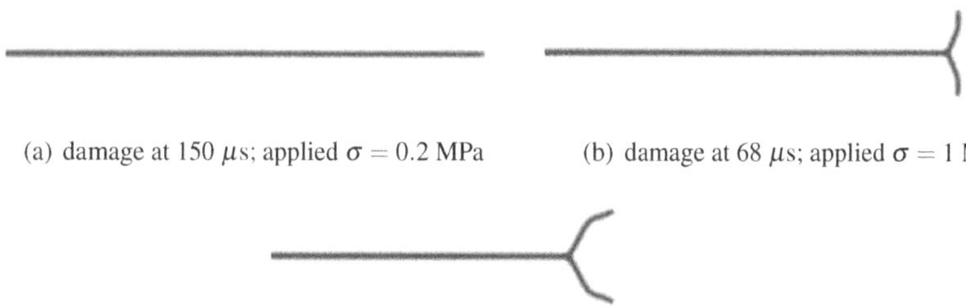

(a) damage at 150 μs; applied $\sigma = 0.2$ MPa

(b) damage at 68 μs; applied $\sigma = 1$ MPa

(c) damage at 39 μs; applied $\sigma = 3$ MPa

Figure 10.24: Damage maps for narrow sample of glass under stresses of different magnitudes applied on pre-crack surfaces.

We observe that in this narrow sample, branching can happen at surprisingly low crack propagation speeds. As seen from Figs. 10.25(b) and 10.25(c), crack propagation speed at branching only reaches around $0.3c_R$. These results provide further evidence that crack branching is not controlled by the crack propagation speed, but by how strain energy flows around the tip of the propagating crack. It would be interesting to confirm experimentally the possibility of crack branching in glass at such low crack propagation speeds. On different geometries and loading conditions, experimental results obtained from 1938 to 1970 by different authors list the limiting crack speed in glass (just before branching) in the range $0.47c_R - 0.66c_R$ (see Table 11.1 in [84]).

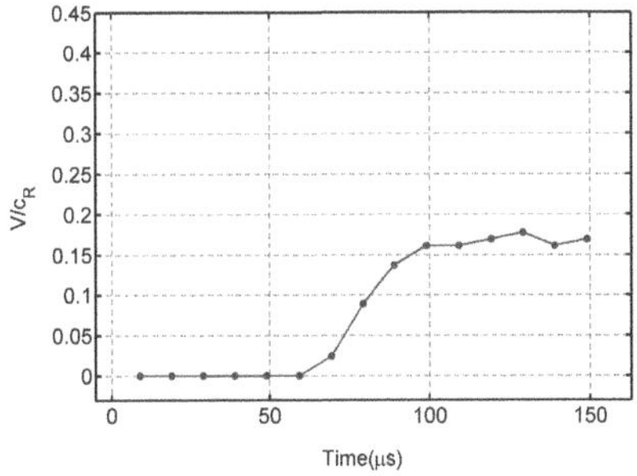

(a) applied $\sigma = 0.2$ MPa

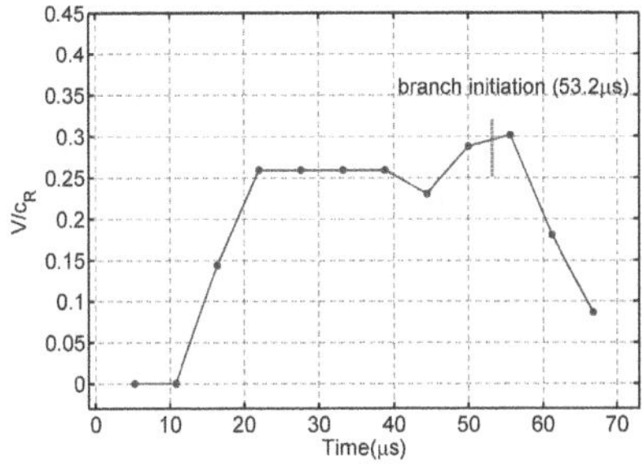

(b) applied $\sigma = 1$ MPa

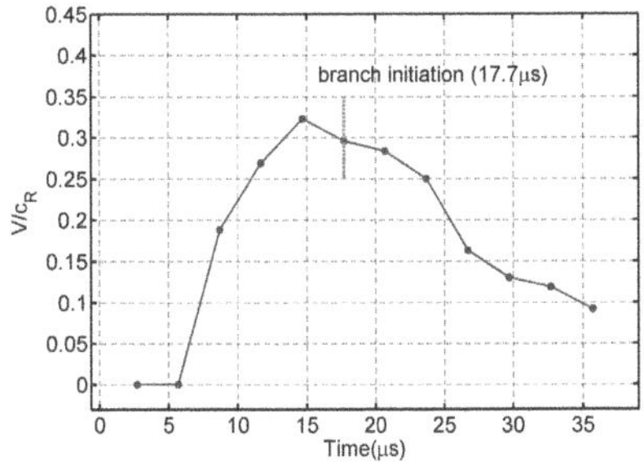

(c) applied $\sigma = 3$ MPa

Figure 10.25: Crack propagation speed in narrow sample of glass under tensile stresses of different magnitudes applied on the pre-crack surfaces.

The reasons for the large angles of branching observed in this case of geometry and loading conditions are clarified by investigating crack propagation and branching in simulation Movies 14–16 *. The color indicate the vertical velocity component and the stress waves interactions explain the particular branching patters. More comments are given in Section 10.7.

10.5.3.3 Load case 3: velocity boundary conditions

Applying the displacement-controlled boundary conditions as in Fig. 10.3(d) and initial conditions that attenuate a strong shock as in Fig. 10.3(e) on the narrower glass sample, we obtain the results shown in Fig. 10.26. Two of the applied boundary velocities are identical to those in the wider glass sample case and the corresponding damage profiles are similar (compare Fig. 10.11(a) with Fig. 10.26(a), and Fig. 10.11(b) with Fig. 10.26(b)). However, the corresponding crack propagation speeds are different (compare Fig. 10.27 with Fig. 10.12), with the crack speed in

* See supplementary materials for [13] or visit http://engineering.unl.edu/mme/florin-bobaru/

the narrower sample being faster than in the wider sample. Because of this, when the crack branches once (see Fig. 10.26(b)), branching takes place sooner in the narrow sample than in the wider sample. This is because the applied initial acceleration (thus force) is higher in the narrower sample than in the wider sample. For the same reason, using in the narrow sample the 0.2 m/s velocity boundary conditions employed earlier in the wider sample leads to cracks forming early near the boundaries of the sample. Lowering the loading amplitude to 0.1 m/s results in the fish-bone failure pattern observed in the wider sample loaded at 0.2 m/s (see Figs. 10.26(c) and 10.11(c)).

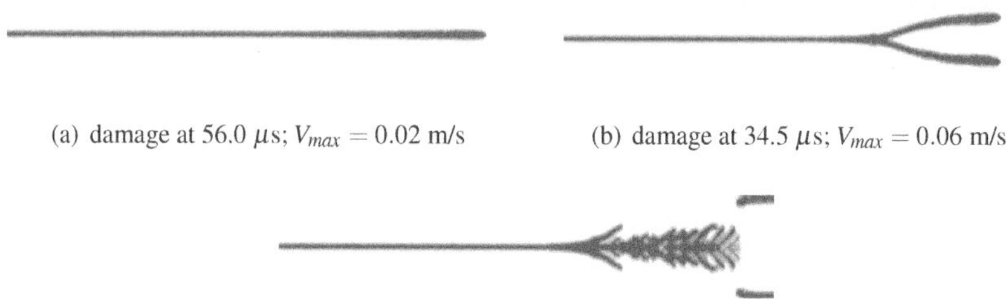

(a) damage at 56.0 μs; $V_{max} = 0.02$ m/s

(b) damage at 34.5 μs; $V_{max} = 0.06$ m/s

(c) damage at 27.0 μs; $V_{max} = 0.1$ m/s

Figure 10.26: Damage maps for narrow glass sample under velocity boundary conditions with different applied velocity amplitudes.

The thicker damage zone is easily noticeable before branching in Fig. 10.26(b). From the same figure, after branching, the crack paths display a reduced process zone that increases again as the cracks advance with the potential of a secondary branching event. Once the loading rate is sufficiently high (see Fig. 10.26(c)), a multitude of crack initiation points develop simultaneously in the material as the loading conditions are ripping the system apart. For this case it is not easy to define a crack propagation speed. While both the wide and narrow glass samples under these boundary conditions branch at around $0.6c_R$, we observe that in the case when the crack does not branch, the crack speed reaches close to $0.7c_R$, close to the measurement data provided by [3].

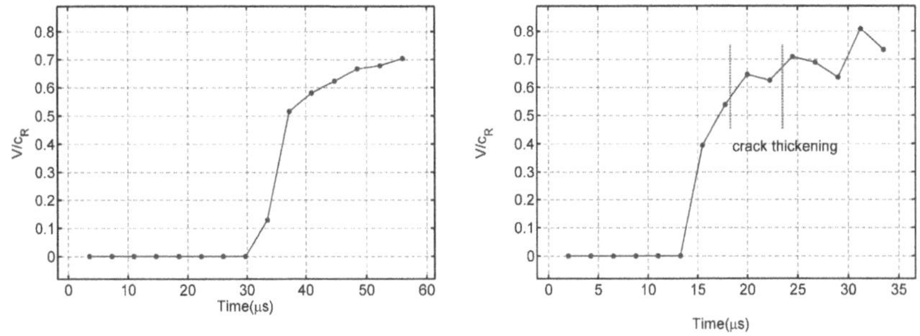

(a) 0.02 m/s

(b) 0.06 m/s

Figure 10.27: Crack propagation speed in narrow glass sample under velocity boundary conditions applied with different magnitudes for applied velocity: (a) $V_{max} = 0.02$ m/s, (b) $V_{max} = 0.06$ m/s

With the material being forced to move apart with a constant velocity imposed at its boundaries, the stress wave interactions are considerably different from the load-controlled cases. The simulations contained in Movies 17–19 * further explain the observed behavior. The vertical velocity component is used to observe the waves in the system. The enlarged crack-tip process zones near the branching points are very prominent here, and when the loading reaches some extreme values, the only way the material can accommodate the imposed deformations is via creating a multitude of microcracks. More on this issue is discussed in Section 10.7.

10.6 Discussion of crack branching results

In this section we summarize the results obtained so far in terms of crack propagation speed at branching and the angle of branching. In Table 10.1 we show values of crack propagation speed, relative to the Rayleigh wave speed, at the moment of branching. The only cases for which the data is not included are the loading case 3 under high strain rates (when fish-bone like damage is obtained) and for the attempted but unsuccessful branching event in Load case 1 in homalite under the higher loading amplitude (see Fig. 10.14c). The reason for these omissions is that tracking of branches becomes ambiguous when the main crack continues to propagate along with the branches that eventually arrest. The range of speeds at branching compare well with the limiting crack propagation speeds measured in glass and homalite by various experimentalists and summarized in reference [84] on page 191. For glass, the exper-

* See supplementary materials for [13] or visit http://engineering.unl.edu/mme/florin-bobaru/

imentally measured limiting velocity (that is, before a single crack branches) ranges from $0.47c_R$ to $0.66c_R$, while for homalite the reported spread is between $0.33c_R$ and $0.45c_R$.

From the crack propagation speed plots for load cases 1 and 2 (like the ones in Fig. 10.23(b)) we observe that branching tends to initiate after a peak in speed, likely due to the energy dissipated by the migration of damage away from the crack line (the expansion of the process zone). A slight drop in propagation speed is observed after branching, but soon after that, individual branches tend to recover the propagation speed from before the expansion of the process zone. In contrast, for loading case 3 the crack propagation speed profile is more monotonous, with some smaller variations in the region where the process zone enlarges prior to branching.

It is not surprising that when the narrower geometry sample is used, for load cases 1 and 2 the crack propagation speed at branching is lower than in the wider geometry: this only confirms the fact that stress waves reflected from the boundaries (which are now closer to the crack tip) can significantly alter the crack propagation speed by slowing it down in this case, and speeding it up in other cases, depending on the angle at which they meet with the propagating crack. Interestingly, in load case 3 the effect is opposite: the crack propagates slightly faster in the narrower geometry than in the wider one. That is because the imposed acceleration inside the sample is larger in the narrower sample than in the wider one, if the constant velocity applied on the boundaries is the same.

Table 10.1: Crack propagation speed at branching, computed by peridynamics, as a fraction of the Rayleigh wave speed (V/c_R).

Load case	Soda-lime glass		Homalite		Narrower geometry	
	Loading amplitude	V/c_R	Loading amplitude	V/c_R	Loading amplitude	V/c_R
1	2 MPa	0.51	1 MPa	0.51	2 MPa	0.35
	4 MPa	0.55	2 MPa	-	4 MPa	0.47
2	3 MPa	0.40	2 MPa	0.42	1 MPa	0.29
	6 MPa	0.44	4 MPa	0.55	3 MPa	0.30
			6 MPa	0.51		
			8 MPa	0.39		
3	0.06 m/s	0.63	0.2 m/s	0.59	0.06 m/s	0.66
	0.2 m/s	-	1 m/s	-	0.1 m/s	-

Crack branching angles, measured between the symmetry line (the original direction of propagation of the crack before branching) and one of the branches, are shown in Table 10.2. Another name used for these angles is "bifurcation half angles" (see [38]). Some authors refer to the branching angle as the angle between the branches (see e.g. [87]), which is twice the angle we use here. As crack branches propagate and interact with stress waves, their path can change direction and the angle of branching changes. This is why, in producing the data in Table 10.2, we use the damage maps close to the branching point. We observe that in the majority of tests presented here, the crack branching angles are between 15° and 35°. The only cases when the branching angle is around 60° is when stress waves reflected from the boundaries interact with the advancing crack. This confirms the experimental observations in [87]. We note that smallest branching angles are for loading conditions that mimic quasi-static loading, and these angles are similar to those seen from experiments in [23]. For loading case 2, in which the applied loading magnitude is sufficiently high to produce crack branching *before* stress waves reflected from the boundaries return to interact with the advancing crack, crack branching angles are in the 30° to 35° range (homalite with 6 MPa and 8 MPa loading amplitudes). These cases are equivalent to modeling branching in an unbounded medium. Note that the analytical result provided by the Yoffe solution for a crack of constant length that translates with a constant velocity in an unbounded medium (therefore, no wave reflections from boundaries) gives an angle of branching of 60° (see [107]), almost double the value from our predictions, as well as what is reported from experiments.

Table 10.2: Crack branching angles (between the symmetry line and a branch) computed by peridynamics, estimated at the branching location from damage maps.

Load case	Soda-lime glass		Homalite		Narrower geometry	
	Loading amplitude	Angle	Loading amplitude	Angle	Loading amplitude	Angle
1	2 MPa	28.2°	1 MPa	26.9°	2 MPa	32.0°
	4 MPa	21.8°	2 MPa	20.0°*	4 MPa	21.9°
2	3 MPa	63.0°	2 MPa	43.9°	1 MPa	64.0°
	6 MPa	35.1°	4 MPa	38.2°	3 MPa	60.0°
			6 MPa	35.8°		
			8 MPa	31.4°		
3	0.06 m/s	15.6°	0.2 m/s	16.4°	0.02 m/s	13.4°
	0.2 m/s	-	1 m/s	-	0.06 m/s	-

*crack angle of the successful branching in figure 10.14, not the one whose branches arrest.

10.7 Why do cracks branch?

The experimental investigations in [16] and [92] discuss the role of crack front waves in dynamic brittle fracture. The peridynamic simulations in this work show that crack surface and crack front waves may be responsible for the "migration" of damage away from the original crack line (roughening of the crack surface) when the dynamic loading conditions are met for this to happen. The reason, as described below, appears to be a *pile up of stress waves* traveling on the crack surfaces near the crack tip.

Loading case 2 (sudden loading on the crack surfaces), with an amplitude large enough that produces crack branching before the waves return from the boundaries to interact with the advancing crack, makes things easier to monitor and understand. The scenario we observe from the velocity vector plots (see Movie 21 * and Fig. 10.28) is as follows:

- **Phase I: dynamic crack propagation.** Material in front of the advancing crack is pulled towards it, and the larger the loading amplitude is, the faster the material moves *against* the advancing crack.

- **Phase II: stress waves "pile up" against crack tip.** Stress waves generated by the propagating crack are emitted and travel into the body and along the crack surface as Rayleigh waves; the Rayleigh waves on the crack surfaces are responsible for the rotation induced in the velocity vectors for nodes on or near these surfaces. When loading amplitude (strain energy flow into the process zone) is sufficiently high, these waves are generated at a faster rate and they "pile up" against the material in front of the crack tip, which is pulled even stronger against the advancing crack. This causes waves to be reflected/ deflected at an angle, leading to having bonds that are at different orientation than those perpendicular to the crack line to be stretched beyond their critical stretch.

- **Phase III: migration of damage and branching.** Bonds that are stretched to their maximum are away from the symmetry line, and damage (captured by peridynamics via nonlocal bond-breaking) "spreads" into the crack surfaces. This would induce significant crack surface roughening, observed in experiments. With the perturbed stress profile around the crack tip, bonds that, until now, were breaking due to the mode I loading (bonds that crossed the symmetry line and are perpendicular on it) start to relax as the waves pile up and reflections stretch other bonds beyond their critical limit. The branching phase is completed when two independent crack branches start advancing at an angle from the symmetry line.

*See supplementary materials for [13] or visit http://engineering.unl.edu/mme/florin-bobaru/

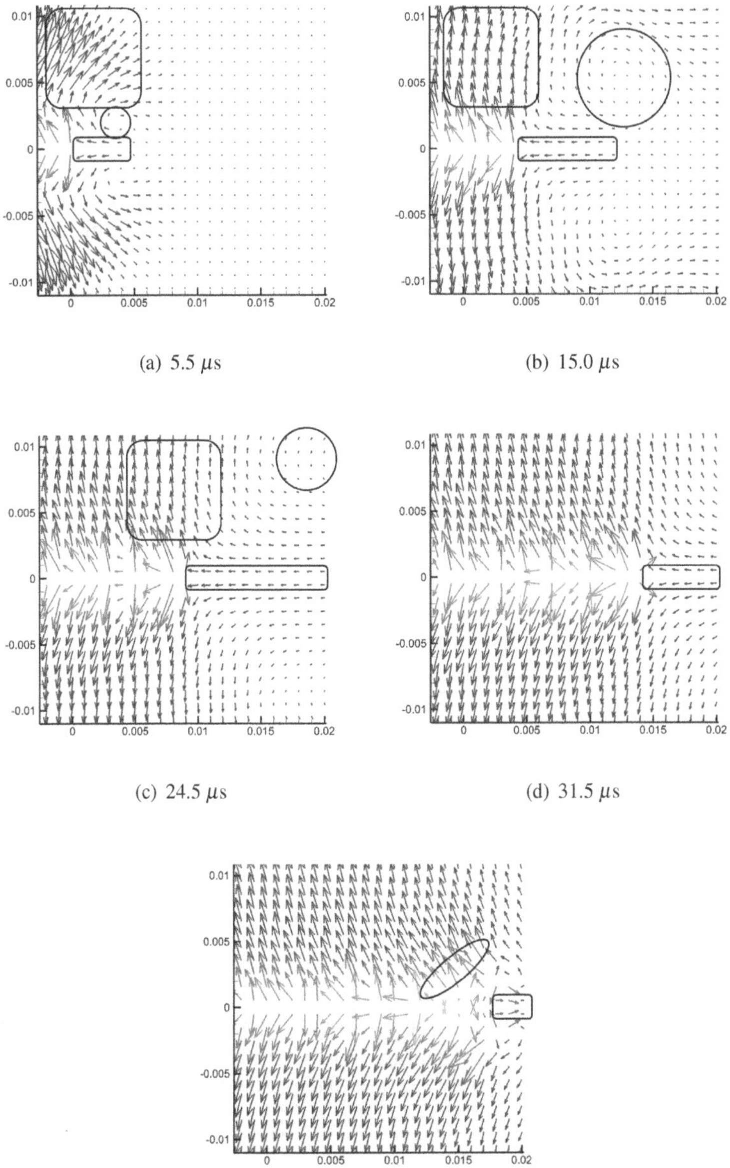

(a) 5.5 μs

(b) 15.0 μs

(c) 24.5 μs

(d) 31.5 μs

(e) 36.0 μs

Figure 10.28: Velocity vector plot at: (a) 5.5 μs, (b) 15.0 μs, (c) 24.5 μs, (d) 31.5 μs, (e) 36.0 μs for the loading on the crack surfaces' case with 8 MPa loading amplitude. The color (for color figures please see the electronic version of the handbook, or the online version of reference [13]). of the nodal velocity vectors represents the damage index at that node.

We explain this scenario based on interpreting the plots in Fig. 10.28, which show the velocity vector for nodes near the pre-crack tip under 8 MPa loading applied on the crack surfaces. The reason for using results with this loading amplitude is because the differences between the nodal velocities are more pronounced here and easier to follow. The length of the vectors shown in Fig. 10.28 are proportional to their corresponding nodal velocity magnitudes. The color * of the arrows indicates the damage index: darker arrows mean corresponding nodes have zero damage, while brighter arrows means that at least 40% bonds of the node have been broken. In the discrete model, a crack line leads to nodes on its surface to exhibit damage index values approaching, but slightly lower than, 50%. Regions of interests are marked by circles, squares, and rectangles. These regions change location as the crack propagates. Because of symmetry, we discuss only the nodes on the upper half of the model.

The propagation initiates soon after the sudden loading is applied (shown in Figure 10.28(a)) and round wave fronts (see square region) are moving away from the crack surface (see also Movie 20 †). The velocity of nodes ahead of the crack tip (inside the rectangular region) is opposite the crack propagation direction, so material ahead of the crack tip moves towards or against the propagating crack. A small vortex (inside circular region) with close-to-zero material points velocity exists between the two regions mentioned above, as a transition zone.

As the crack propagates (see Fig. 10.28(b), 10.28(c)), the regions mentioned above change location attempting to reach a steady state. The nodes directly in front of the propagating line move faster against it. As bonds break, waves are emitted from them and of particular interest are crack surface waves identified by the spinning of the velocity vectors' direction. We recall that Rayleigh waves induce elliptical motion of material points near a surface. In the early stages of crack propagation, nodes which have 40% damage are observed only on the closest rows to the symmetry line. The pile up of surface waves against the crack tip and the reflections induced by their interaction with the material ahead of the crack tip which moves against the direction of the advancing crack, leads to having a few rows of nodes away from the symmetry line that have that type of damage levels (see (Fig. 10.28(d)). This is the beginning of damage migration or thickening of the process zone. A similar "spread of damage" has been identified by [57] using the cell model for damage.

If the strain energy levels are not too large, a steady state settles in and the crack propagates straight, without branching. When, however, the flow of strain energy is too high, the waves created by the fracturing (breaking of bonds) process reach higher amplitudes and their pile up near the crack tip leads to bonds along directions other than those perpendicular to the original crack line being stretched beyond their critical level. This makes nodes in the elliptical region in Fig. 10.28(e) to start having damage index higher than those on the same "column" and lying on the symmetry line. Once this happens, material along the symmetry line relaxes, and the branching

---

*For color figures please see the electronic version of the handbook, or the online version of reference [13].

†See supplementary materials for [13] or visit http://engineering.unl.edu/mme/florin-bobaru/

of the crack process is complete, as the two branches continue to grow (see also Movie 20 *).

The argument that a sufficiently high level of strain energy needs to be flowing into the process zone for crack branching to take place is based on the following analysis of the results for the loading on the crack surface case with different loading magnitudes. We monitor the vertical component of the nodal velocity for three loading amplitudes: 0.3, 2, and 6 MPa applied suddenly on the crack surfaces (loading case 2). This allows us to see wave interactions and crack propagation clearly. In Figs. 10.29, 10.30, and 10.31, we show snapshot of the crack propagation process (movies containing the entire process are 4, 5, and 21 †). In these figures, darker color ‡ indicates a higher positive v_y-component, brighter color is for more negative v_y values, and in between represents regions with vertical velocities close to zero. In the beginning, waves travel from the pre-crack surface to the boundaries (see Figs. 10.29(a), 10.30(a), 10.31(a)). Under the 0.3 MPa applied stress, the crack does not start to propagate until after waves reflect from the boundaries and meet the pre-crack tip twice (see Fig. 10.29(b)). In this case, the crack propagates in a straight line, without branching. We observe the wavelets created by the propagation itself, but these are not strong enough to create branching (see Fig. 10.29(c)). For the 2 MPa loading amplitude, crack propagation starts before reflected waves return to meet the crack tip (see Fig. 10.30(b)). Crack branching, however, takes place only after waves reflected from the boundaries interact with the advancing crack (see Fig. 10.30(c)), specifically after these reflected waves bounce off the propagating crack surfaces. To obtain branching without the interference of waves reflected from the boundaries, one has to increase the applied loading amplitude to, for example, 6 MPa or higher. In this case, the crack starts propagating and branching takes place *before waves return* from the boundaries (see Figs. 10.31(b), 10.31(c)). The wavelets produced by the crack propagation itself are small relative to the large amplitude waves created by the applied loads and a little harder to see now. In the previous cases these were easier to see due to the lower amplitude waves generated by the lower amplitude of applied loading.

We can conclude that two essential factors are needed in dynamic brittle fracture for crack branching:

(a) a sufficient loading magnitude for crack propagation, and

(b) pile up of waves (created by the crack propagation process) in the process zone, or wave reinforcement between waves reflected from the boundaries and those created by the propagating crack.

The certain instability mentioned, in many references before but never fully explained, as being responsible for crack branching now has a clear explanation. We find that crack branching is not necessarily a three-dimensional phenomenon, as some recent papers have suggested. We see that the dynamic flow of strain energy

* See supplementary materials for [13] or visit http://engineering.unl.edu/mme/florin-bobaru/
† See supplementary materials for [13] or visit http://engineering.unl.edu/mme/florin-bobaru/
‡ For color figures please see the electronic version of the handbook.

around the crack tip is responsible for crack branching and that the elastic modulus and the fracture energy (energy release rate) control how waves propagate and how material damage occurs in the sample prior to branching. Based on observation of simulated results, the *macroscopic mechanism of dynamic brittle branching*, when there is no interaction between the advancing crack and incoming waves, could be summarized as follows:

- Applied loading on the pre-crack surfaces generates stress waves that travel on the crack surfaces and in the bulk;
- At sufficiently high loading amplitudes, the pre-crack starts propagating;
- When stress intensity is sufficiently high, waves generated by the propagation itself pile up against the crack-tip region and are deflected away by material just ahead of the crack tip which moves against the direction of crack propagation;
- The resulting wave interactions and reinforcements lead to loading conditions that create damage away from the crack line (surface), thus damage migrates into the crack surfaces;
- When the wave pile up is sufficiently strong, the amount of damage migration becomes sufficiently large, most of it happening off of the original direction of the crack. Thus the material directly in front of the original crack tip becomes relaxed; at this point crack branching ensues.

We note that the macroscopic mechanism of dynamic brittle branching described here is independent of the particular microscopic conditions of a particular microstructure. Such microscopic mechanisms proposed in the literature (microcracks ahead of the crack tip, or tilting of the stress vector, etc.) may exist and can influence the behavior in a given material system, but our results indicate that the larger picture, the macroscopic one described above, is *the "driving mechanism" for crack branching*.

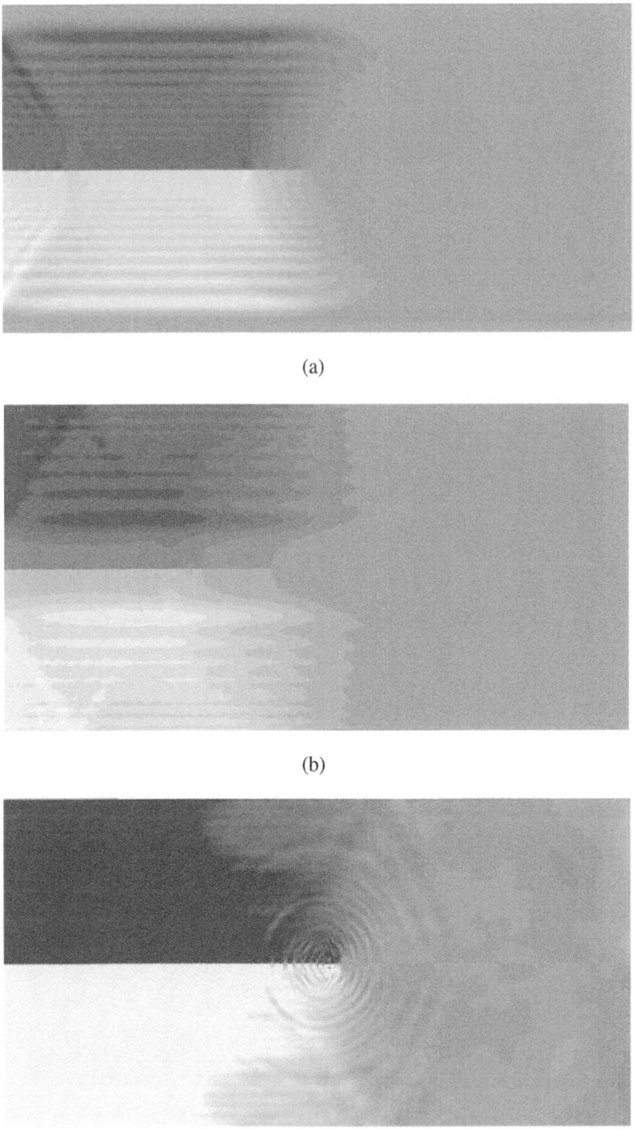

(a)

(b)

(c)

Figure 10.29: Vertical component of nodal velocity in homalite sample under 0.3 MPa stress applied on pre-crack surfaces: (a) waves created by the applied load approach the boundaries, (b) after several reflections and wave reinforcements, the crack starts propagating, (c) crack continues to propagate in a straight line without branching.

(a)

(b)

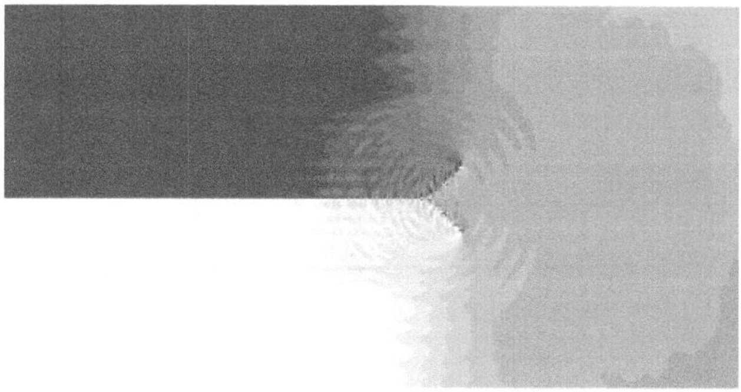

(c)

Figure 10.30: Vertical component of nodal velocity in homalite sample under 2 MPa stress applied on pre-crack surfaces: (a) waves propagate towards the boundaries, (b) crack propagation starts before reflected waves interact with the tip, (c) interaction of waves reflected from the boundaries with those produced by the crack propagation process leads to crack branching.

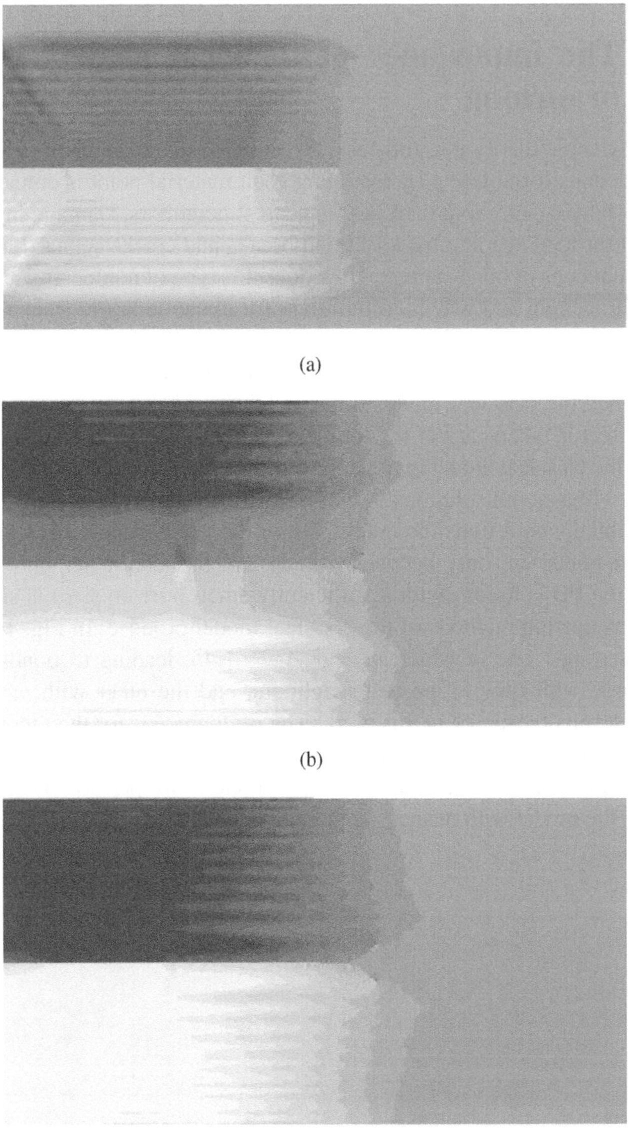

(a)

(b)

(c)

Figure 10.31: Vertical component of nodal velocity in homalite sample under 6 MPa stress applied on pre-crack surfaces (a) waves propagate from pre-crack surface, crack propagation starts almost as soon as the loads are applied, (b) branching takes place before reflected waves meet the propagating crack, (c) waves reflected from the sample's boundaries meet with the two propagating branches and influence their future path growth.

10.8 The importance of nonlocal modeling in crack branching

The peridynamic theory is a non-local formulation that extends the classical continuum mechanics formulation by assuming each material point is connected via peridynamics bonds with more than just its nearest neighbors. Damage in peridynamics is defined through keeping track of bond breaks and cracks appear/form as a necessary consequence of bond rupture. How does this type of nonlocality affect modeling of crack branching and why is it important for dynamic brittle fracture? We answer these questions in this section.

A peridynamic model of a linear elastic material under homogeneous deformation matches the classical model solution for any horizon size (assuming that the surface effect is taken care of by, for example, fictitious nodes method), and it converges to the classical model in the limit of the horizon size going to zero, for general problems with smooth solutions or with static cracks (see [51]).

In problems with dynamic cracks, when the peridynamic discrete model is set up so that bonds are only between a node and its nearest neighbor nodes (small m-ratio), the PD solution (with a sufficiently small horizon size) is similar to what some discretization methods of the classical model produce. In Fig. 10.32 we show two PD settings: one in which $m = \delta/\Delta x = 1.45$, leading to bonds between the central node with only its nearest neighbors, and the other with $m = 4$, used for all of the results presented in this work. The preference is for the second type, as m-convergence results demonstrated in [43] and [44], in order to obtain crack paths that are not influenced by the grid density. Nevertheless, we perform a dynamic fracture test using the model with nearest-neighbor only bonds.

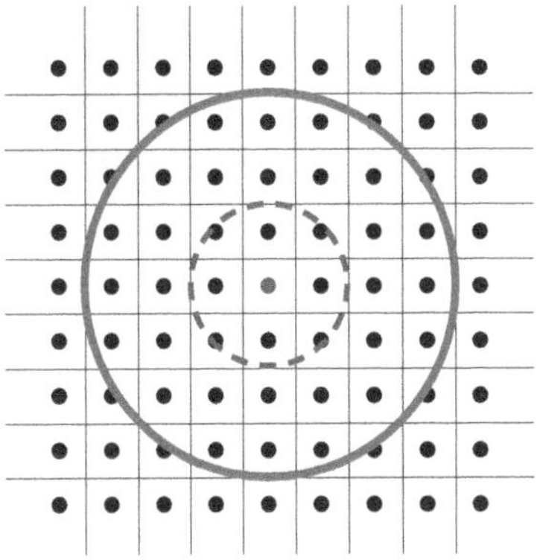

Figure 10.32: Two peridynamic models with two different horizons and the same discretization grid: one with horizon factor $m = \frac{\delta}{\Delta x} = 4.01$, shown by a solid circle, and the other with $m = 1.45$, shown by a dashed-line circle.

With the sudden tensile loading applied on the top and bottom edges of the homalite sample (load case 1) and a loading magnitude of 3 MPa, we obtain the results shown in Figure 10.33 for $m = 1.45$. For comparison, we also include the results obtained in [80] that use the "cracking particle model" for discretizing the classical (local) elasticity equations. The results with the PD model with $m = 4$ is similar to that shown in Fig. 10.14(c). Here we used the slightly higher loading magnitude because with 2 MPa loading the crack does not branch in the model that uses $m = 1.45$. In both the cracking particle model and the peridynamic model with *nearest-neighbor-only interactions* (NNOI), a large number of spurious microcracks form, which are not seen in experiments in homalite. The authors of [80] commented on the presence of these unexpected extra cracks: "*This is a rather spurious artefact which is undesirable but we do not know its origin or how to eliminate it at this time.*" We have tried velocity loading conditions (load case 3) with our peridynamic model that uses the NNOI and similar spurious cracks are obtained. We also observe spurious cracks in the NNOI peridynamic model for glass.

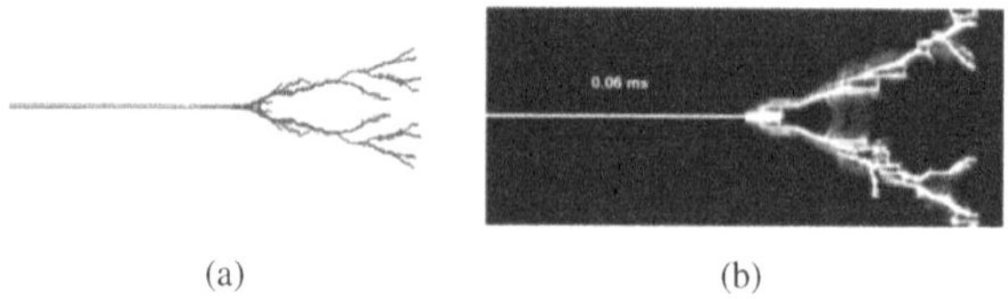

Figure 10.33: Damage maps for load case 1 (sudden loading on the top and bottom boundaries) with 3 MPa for a homalite sample: (a) results from the peridynamic model that only uses nearest-neighbor bonds (see dashed-line circle in Fig. 10.32), and (b) results from the cracking particle method of Rabczuk and Belytschko [80].

The main reason for the formation of spurious cracks in the peridynamic model that uses NNOI is that there are too few directions for bond breaking under a given applied load. The exiting bonds in this model would not likely break for loading that is not perpendicular to them, had we had bonds in all possible directions. With only a small number of different bond-directions available in a NNOI PD model, the crack paths that evolve depart from the real crack path. The strain energy landscape around the crack tip is also significantly different in the NNOI model than in a regular peridynamic model (one that uses a sufficiently large m value). Waves in the NNOI model propagate like in lattice-type models. We note that lattice models are known to produce zones of "forbidden velocities" for crack propagation (see [65]). Such zones are not observed in experiments (see [84] pp. 218-220).

The ability of generating damage in arbitrary directions (by the breaking of a bond stretched in a given, but arbitrary direction) is, perhaps, the most important characteristic that allows for the evolution of crack branches from the damage zone (increased process zone) caused by stress wave pile ups and wave reinforcements of rapidly growing cracks. We recall that the increased process zone was the mechanism postulated by [86] for crack branching. Methods that are, for example, employing surfaces for describing (and tracking) crack growth will likely fail to correctly capture the crack branching behavior. As we have seen in the previous sections, crack branching does not evolve as a crack surface that suddenly becomes two crack surfaces, but rather as a damage process-zone that widens (mirror-mist-hackles transition) and splits into two damage process-zones that continue to propagate with the narrower profile that the original crack was showing before its thickening prior to branching. This is what is seen in experiments and this is how branching happens in a PD model (which uses a sufficiently large m-ratio).

Remark: Several studies (see e.g. [25], [18]) have proposed the existence of a nonlinear elastic region near the crack tip in order to obtain crack-tip instabilities and branching. While for soft materials, the existence of a hyperelastic region around the crack tip is a reasonable assumption (see [18]), for stiff, glass-like materials it is not clear that this is true anymore (see [20]). With the results presented here we have shown that the nonlocality introduced by the simplest peridynamic model of a linear micro-elastic brittle material does not require nonlinear elastic behavior near the crack tip to predict crack branching and *explain the origin of instabilities that trigger crack branching*. In other words, while hyperelastic behavior near the crack tip may be a sufficient condition to produce crack branching, it is not a necessary condition.

10.9 Conclusions

In this chapter we presented some detailed studies of modeling crack branching in dynamic brittle fracture using peridynamics. Without special crack propagation and branching criteria, results show that the peridynamic model is capable of capturing all the details of dynamic crack branching in brittle materials reported in the experimental literature. We were able to propose an explanation for the origin of instabilities that lead to crack branching in brittle materials.

We verified the peridynamic models by comparing crack propagation velocities and crack branching angles with experimental results. The overall trend of speed profiles match the experimental values under similar types of loading conditions. The values obtained from the model for the crack propagation speed at branching ranged between $0.3c_R$ and $0.66c_R$ depending on the material (elastic modulus, density, and fracture energy), loading conditions, and sample geometry. A similar range of values is obtained in experiments, with the higher values observed when the loading amplitudes are increased. We revealed how stress waves can trigger crack branching even when the crack propagation speed is relatively low. Stress waves can also modify the crack opening angles. The values we obtained with the peridynamic model are close to those reported in experiments.

We performed convergence studies in terms of the size of the horizon going to zero and chose to use the horizon size under which the crack propagation speed no longer changes significantly when the horizon continues to decrease. From that point of view, the sufficiently small nonlocal region for glass was 1 mm while for homalite it was 4 mm. These values are in the same range with the *characteristic distance* found for crack branching in [83].

The peridynamic simulations allowed us to uncover a possible mechanism (macroscopic) for crack branching in brittle isotropic and homogeneous materials: under mode-I loading conditions in which waves coming from the boundaries do not interfere with the propagating crack, mixed-mode loading conditions near the crack tip are induced by stress waves (generated by the fracturing process) pile up around the crack tip due to deflection by the material ahead of the crack tip that moves against the direction of propagation of the crack. The wave pile up leads to "migration" of damage away from the symmetry line, indicating an enlargement of the process zone (possible roughening of the crack surface). When this migration is sufficiently large, material ahead of the crack tip is unloaded, and damage growth no longer happens along the original crack direction, but splits into two separate process zones that continue to propagate with a reduced thickness compared to that just before the branching event. When crack branching is a result of waves traveling towards the propagating crack, the mechanism responsible for branching is wave reinforcement.

We also explained why the peridynamic model is able to capture the crack branching phenomenon correctly. One significant feature of dynamic brittle branching is crack thickening, which is observed as roughing of the crack surface in experiments (the famous mirror-mist-hackle transition). Because the damage model in peridynamics is based on bond breaking, not on crack surface advancement, and since bonds cover virtually every possible direction (as long as the grid is fine relative to the horizon size), we are able to mimic with this model the evolution of the process zone observed by experimentalists in crack branching. The peridynamic model captures the correct amount of energy dissipated via increasing the process zone, reported in experiments to happen just before branching takes place. Without the right amount of energy (determined by the flow of strain energy around the crack tip) dissipated near that critical point, one gets spurious damage paths or the wrong crack propagation speed. In any dynamic simulation, results that follow a "slightly" wrong move depart further and further from the actual result.

It is noteworthy that the nonlocality introduced by this simplest peridynamic model of a linear micro-elastic brittle material does not require nonlinear elastic behavior near the crack tip to predict crack branching. Such an idea was used in [25] to provide an explanation for crack branching via atomistic modeling. We conclude that, at least to first order, a continuum-level linear elastic model and a capable damage model (the peridynamic model) are the only ingredients necessary to correctly predict crack branching in brittle, homogeneous and isotropic linear elastic materials. The input data for this model (material density, elastic constants, and fracture energy) appears to be the *minimal set* of input data possible. While more complex models may provide more accurate descriptions of the branching process for particular material types, the principles of crack branching in linear-elastic brittle materials can be reproduced with the proposed peridynamic model. The results answer the question we promised to answer at the beginning in Section 10.1, whether quantum or atomistic, or coupled across-scale modeling, is necessary for explaining the crack branching phenomenon. They are not, as long as the continuum-level model used is one of the type employed here (peridynamic).

Acknowledgments

This research has been supported by ARO/ARL (grant number W911NF1010431), program manager Dr. Larry Russell (ARO) and Dr. Chian-Fong Yen (ARL), and by the AFOSR's MURI Center for Material Failure Prediction Through Peridynamics, program managers Dr. David Stargel, Dr. Ali Sayir, and Dr. Fariba Fahroo.

References

[1] F. F. Abraham, D. Brodbeck, W.E. Rudge, and X. Xu. A molecular dynamics investigation of rapid fracture mechanics. *Journal of the Mechanics and Physics of Solids*, 45(9): 1595–1619, 1997.

[2] F.F. Abraham. Unstable crack motion is predictable. *Journal of the Mechanics and Physics of Solids*, 53:1071–1078, 2005.

[3] S. R. Anthony, J. P. Chubb, and J. Congleton. The crack-branching velocity. *Philosophical Magazine*, 22(180): 1201–1216, 1970.

[4] I. S. Aranson, V. A. Kalatsky, and V. M. Vinokur. Continuum field description of crack propagation. *Physical Review Letters*, 85: 118–121, 2000.

[5] Z. P. Bažant and M. Jirásek. Nonlocal integral formulations of plasticity and damage: Survey of progress. *Journal of Engineering Mechanics*, 128(11): 1119–1149, 2002.

[6] E. K. Beauchamp. Mechanisms for hackle formation and crack branching. In *Fractography of Glasses and Ceramics III*, volume 64 of *Ceramic Transactions*, pages 409–445. American Ceramic Society, 1996.

[7] D. N. Bhate, A. Kumar, and A. F. Bower. Diffuse interface model for electromigration and stress voiding. *Journal of Applied Physics*, 87(4): 1712–1721, 2000.

[8] F. Bobaru. Influence of van der Waals forces on increasing the strength and toughness in dynamic fracture of nanofibre networks: A peridynamic approach. *Modelling and Simulation in Materials Science and Engineering*, 15(5): 397–417, 2007.

[9] F. Bobaru and Y. D. Ha. Adaptive refinement and multiscale modeling in 2d peridynamics. *International Journal for Multiscale Computational Engineering*, 9(6): 635–660, 2011.

[10] F. Bobaru, Y. D. Ha, and W. Hu. Damage progression from impact in layered glass modeled with peridynamics. *Open Engineering*, 2(4): 551–561, 2012.

[11] F. Bobaru and W. Hu. The meaning, selection, and use of the peridynamic horizon and its relation to crack branching in brittle materials. *International Journal of Fracture*, 176(2): 215–222, 2012.

[12] F. Bobaru, M. Yang, L. F. Alves, S. A. Silling, E. Askari, and J. Xu. Convergence, adaptive refinement, and scaling in 1d peridynamics. *International Journal for Numerical Methods in Engineering*, 77(6): 852–877, 2009.

[13] F. Bobaru and G. Zhang Why do cracks branch? A peridynamic investigation of dynamic brittle fracture. *International Journal of Fracture*, 196(1): 59–98, 2015.

[14] J. E. Bolander and S. Saito. Fracture analyses using spring networks with random geometry. *Engineering Fracture Mechanics*, 61(5): 569–591, 1998.

[15] D. Bonamy and K. Ravi-Chandar. Interaction of shear waves and propagating cracks. *Physical Review Letters*, 91(23): 235502, 2003.

[16] D. Bonamy and K. Ravi-Chandar. Dynamic crack response to a localized shear pulse perturbation in brittle amorphous materials: On crack surface roughening. *International Journal of Fracture*, 134(1): 1–22, 2005.

[17] M. J. Borden, C. V. Verhoosel, M. A. Scott, T. J. R. Hughes, and C. M. Landis. A phase-field description of dynamic brittle fracture. *Computer Methods in Applied Mechanics and Engineering*, 217: 77–95, 2012.

[18] E. Bouchbinder, A. Livne, and J. Fineberg. Weakly nonlinear fracture mechanics: Experiments and theory. *International Journal of Fracture*, 162(1-2): 3–20, 2010.

[19] E. Bouchbinder, J. Mathiesen, and I. Procaccia. Branching instabilities in rapid fracture: Dynamics and geometry. *Physical Review E*, 71: 056118, 2005.

[20] E. Bouchbinder, T. Tamar Goldman, and J. Fineberg. The dynamics of rapid fracture: Instabilities, nonlinearities and length scales. *Reports on Progress in Physics*, 77(4): 046501, 2014.

[21] B. Bourdin, G. A. Francfort, and J. J. Marigo. The variational approach to fracture. *Journal of Elasticity*, 91: 5–148, 2008.

[22] B. Bourdin, C. J. Larsen, and C. L. Richardson. A time-discrete model for dynamic fracture based on crack regularization. *International Journal of Fracture*, 168: 133–143, 2011.

[23] F. P. Bowden, J. H. Brunton, J. E. Field, and A. D. Heyes. Controlled fracture of brittle solids and interruption of electrical current. *Nature*, 216: 38–42, 1967.

[24] K.B. Broberg. *Cracks and Fracture*. Academic Press, 1999.

[25] M. J. Buehler and H. Gao. Dynamical fracture instabilities due to local hyperelasticity at crack tips. *Nature*, 439(7074): 307–310, 2006.

[26] G. T. Camacho and M. Ortiz. Computational modelling of impact damage in brittle materials. *International Journal of Solids and Structures*, 33(20): 2899–2938, 1996.

[27] Z. Chen and F. Bobaru. Selecting the kernel in a peridynamic formulation: A study for transient heat diffusion. *Computer Physics Communications*, 197: 51–60, 2015.

[28] B. N. Cox, H. Gao, D. Gross, and D. Rittel. Modern topics and challenges in dynamic fracture. *Journal of the Mechanics and Physics of Solids*, 53(3): 565–596, 2005.

[29] P.A. Cundall and O.D.L. Strack. A discrete numerical model for granular assemblies. *Geotechnique*, 29: 47–65, 1979.

[30] J. Dolbow and T. Belytschko. A finite element method for crack growth without remeshing. *International Journal for Numerical Methods in Engineering*, 46(1): 131–150, 1999.

[31] W. Döll. Investigations of the crack branching energy. *International Journal of Fracture*, 11(1): 184–186, 1975.

[32] F. Erdogan and G. C. Sih. On the crack extension in plates under plane loading and transverse shear. *Journal of Basic Engineering*, 85(4): 519–525, 1963.

[33] A. C. Eringen. Nonlocal continuum field theories. Springer, 2002.

[34] J.D. Eshelby. The elastic field of a crack extending non-uniformly under general anti-plane loading. *Journal of the Mechanics and Physics of Solids*, 17(3): 177–199, 1969.

[35] J. E. Field. Brittle fracture: Its study and application. *Contemporary Physics*, 12(1): 1–31, 1971.

[36] A. C. Fischer-Cripps and I. Mustafaev. *Introduction to contact mechanics*. Springer, 2000.

[37] L. B. Freund. Crack propagation in an elastic solid subjected to general loadingi. constant rate of extension. *Journal of the Mechanics and Physics of Solids*, 20(3): 129–140, 1972.

[38] L. B. Freund. *Dynamic Fracture Mechanics*. Cambridge University Press, 1990.

[39] H. Gao. A theory of local limiting speed in dynamic fracture. *Journal of the Mechanics and Physics of Solids*, 44(9): 1453–1474, 1996.

[40] W. Gerstle, N. Sau, and S. A. Silling. Peridynamic modeling of plain and reinforced concrete structures. In *18th International Conference on Structural Mechanics in Reactor Technology (SMiRT 18)*, pages SMiRT18–B01–2, Beijing, China, August 7-12 2005.

[41] M. Ghajari, L. Iannucci, and P. Curtis. A peridynamic material model for the analysis of dynamic crack propagation in orthotropic media. *Computer Methods in Applied Mechanics and Engineering*, 276: 431–452, 2014.

[42] P. C. Guan, S. W. Chi, J. S. Chen, T. R. Slawson, and M. J. Roth. Semi-Lagrangian reproducing kernel particle method for fragment-impact problems. *International Journal of Impact Engineering*, 38(12): 1033–1047, 2011.

[43] Y. D. Ha and F. Bobaru. Studies of dynamic crack propagation and crack branching with peridynamics. *International Journal of Fracture*, 162(1-2): 229–244, 2010.

[44] Y. D. Ha and F. Bobaru. Characteristics of dynamic brittle fracture captured with peridynamics. *Engineering Fracture Mechanics*, 78(6): 1156–1168, 2011.

[45] Y. D. Ha and F. Bobaru. Dynamic brittle fracture captured with peridynamics. In *Proceedings of the IMECE 2011: The 2011 International Mechanical Engineering Congress & Exposition*, pages IMECE2011–65515. ASME, 2011.

[46] J. A. Hauch, D. Holland, M. P. Marder, and H. L. Swinney. Dynamic fracture in single crystal silicon. *Physical Review Letters*, 82: 3823–3826, 1999.

[47] M. Hofacker and C. Miehe. A phase field model of dynamic fracture: Robust field updates for the analysis of complex crack patterns. *International Journal for Numerical Methods in Engineering*, 93(3): 276–301, 2013.

[48] W. Hu, Y. D. Ha, and F. Bobaru. Numerical integration in peridynamics. Technical report, University of Nebraska-Lincoln, Department of Mechanical and Materials Engineering, 2010.

[49] W. Hu, Y. D. Ha, and F. Bobaru. Modeling dynamic fracture and damage in fiber-reinforced composites with peridynamics. *International Journal for Multiscale Computational Engineering*, 9(6): 707–726, 2011.

[50] W. Hu, Y. D. Ha, and F. Bobaru. Peridynamic model for dynamic fracture in unidirectional fiber-reinforced composites. *Computer Methods in Applied Mechanics and Engineering*, 217: 247–261, 2012.

[51] W. Hu, Y.D. Ha, F. Bobaru, and S.A. Silling. The formulation and computation of the nonlocal j-integral in bond-based peridynamics. *International Journal of Fracture*, 176(2): 195–206, 2012.

[52] W. Hu, Y. Wang, J. Yu, C.-F. Yen, and F. Bobaru. Impact damage on a thin glass plate with a thin polycarbonate backing. *International Journal of Impact Engineering*, 62: 152–165, 2013.

[53] Y. Hu, Y. Yu, and H. Wang. Peridynamic analytical method for progressive damage in notched composite laminates. *Composite Structures*, 108: 801–810, 2014.

[54] D. Hull. The effect of mixed mode I/III on crack evolution in brittle solids. *International Journal of Fracture*, 70(1): 59–79, 1994.

[55] D. Hull. *Fractography: Observing, measuring and interpreting fracture structure topography*. Cambridge University Press, 1999.

[56] M. Jirásek and Z. P. Bažănt. Particle model for quasibrittle fracture and application to sea ice. *Journal of Engineering Mechanics*, 121(9): 1016–1025, 1995.

[57] E. Johnson. Process region changes for rapidly propagating cracks. *International Journal of Fracture*, 55(1): 47–63, 1992.

[58] A. Karma, D. A. Kessler, and H. Levine. Phase-field model of mode III dynamic fracture. *Physical Review Letters*, 87: 045501, 2001.

[59] J. R. Kermode, T. Albaret, D. Sherman, N. Bernstein, P. Gumbsch, M. C. Payne, G. Csanyi, and A. De Vita. Low-speed fracture instabilities in a brittle crystal. *Nature*, 455: 1224–1227, 2008.

[60] I. A. Kunin. *Elastic Media with Microstructure. I - One-dimensional models.* Springer Verlag. Springer Series on Solid State Sciences, 1982.

[61] Q. Le and F. Bobaru. Surface corrections in peridynamic models for elasticity and fracture. *In review*, 2016.

[62] A. Livne, O. Ben-David, and J. Fineberg. Oscillations in rapid fracture. *Physical Review Letters*, 98: 124301, 2007.

[63] R. W. Macek and S.A. Silling. Peridynamics via finite element analysis. *Finite Elements in Analysis and Design*, 43(15): 1169–1178, 2007.

[64] E. Madenci and E. Oterkus. *Peridynamic Theory and Its Applications.* Springer, New York, 2014.

[65] M. Marder and S. Gross. Origin of crack tip instabilities. *Journal of the Mechanics and Physics of Solids*, 43(1): 1–48, 1995.

[66] M. Marder Particle methods in the study of fracture. *International Journal of Fracture*, 196(1): 169–188, 2015.

[67] M.A. Meyers. *Dynamic behavior of materials.* John Wiley & Sons, New York, 1994.

[68] J.W. Morrissey and J.R. Rice. Perturbative simulations of crack front waves. *Journal of the Mechanics and Physics of Solids*, 48(6-7): 1229–1251, 2000.

[69] M. Negri. A non-local approximation of free discontinuity problems in sbv and sbd. *Calculus of Variations and Partial Differential Equations*, 25(1): 33–62, 2006.

[70] M. Ortiz and A. Pandolfi. Finite-deformation irreversible cohesive elements for three-dimensional crack-propagation analysis. *International Journal for Numerical Methods in Engineering*, 44(9): 1267–1282, 1999.

[71] S. Oterkus and E. Madenci. Peridynamic analysis of fiber-reinforced composite materials. *Journal of Mechanics of Materials and Structures*, 7(1): 45–84, 2012.

[72] S. Oterkus, E. Madenci, and A. Agwai. Peridynamic thermal diffusion. *Journal of Computational Physics*, 265: 71–96, 2014.

[73] J. Ožbolt, J. Bošnjak, and E. Sola. Dynamic fracture of concrete compact tension specimen: Experimental and numerical study. *International Journal of Solids and Structures*, 50: 4270–4278,

[74] J. Ožbolt, A. Sharma, and H-W. Reinhardt. Dynamic fracture of concrete compact tension specimen. *International Journal of Solids and Structures*, 48: 1534–1543, 2011.

[75] A. Pandolfi, B. Li, and M. Ortiz. Modeling fracture by material-point erosion. *International Journal of Fracture*, 184: 3–16, 2013.

[76] A. Pandolfi and M. Ortiz. An eigenerosion approach to brittle fracture. *International Journal for Numerical Methods in Engineering*, 92(8): 694–714, 2012.

[77] A. J. Pons and A. Karma. Helical crack-front instability in mixed-mode fracture. *Nature*, 464(7285): 85–89, 2010.

[78] I. Procaccia and J. Zylberg. Propagation mechanism of brittle cracks. *Physical Review E*, 87: 012801, 2013.

[79] T. Rabczuk. Computational methods for fracture in brittle and quasi-brittle solids: State-of-the-art review and future perspectives. *ISRN Applied Mathematics*, 2013(849231): 38, 2013.

[80] T. Rabczuk and T. Belytschko. Cracking particles: A simplified meshfree method for arbitrary evolving cracks. *International Journal for Numerical Methods in Engineering*, 61(13): 2316–2343, 2004.

[81] M. Rahman and T. Michelitsch. A note on the formula for the Rayleigh wave speed. *Wave Motion*, 43(3): 272–276, 2006.

[82] M. Ramulu and A. S. Kobayashi. Dynamic crack curving - a photoelastic evaluation. *Experimental Mechanics*, 23(1): 1–9, 1983.

[83] M. Ramulu and A. S. Kobayashi. Mechanics of crack curving and branching — a dynamic fracture analysis. *International Journal of Fracture*, 27: 187–201, 1985.

[84] K. Ravi-Chandar. *Dynamic Fracture*. Elsevier, 2004.

[85] K. Ravi-Chandar and W. G. Knauss. Dynamic crack tip stresses under stress wave loading: A comparison of theory and experiment. *International Journal of Fracture*, 20: 209–222, 1982.

[86] K. Ravi-Chandar and W. G. Knauss. An experimental investigation into dynamic fracture: III. On steady-state crack propagation and crack branching. *International Journal of Fracture*, 26(2): 141–154, 1984.

[87] K. Ravi-Chandar and W. G. Knauss. An experimental investigation into dynamic fracture: IV. On the interaction of stress waves with propagating cracks. *International Journal of Fracture*, 26(3): 189–200, 1984.

[88] K. Ravi-Chandar and B. Yang. On the role of microcracks in the dynamic fracture of brittle materials. *Journal of the Mechanics and Physics of Solids*, 45(4): 535–563, 1997.

[89] D. Rogula. *Nonlocal Theory of Material Media.* Springer, 1982.

[90] F. Rösch and H.-R. Trebin. Brittle fracture in a complex metallic compound from an atomistic viewpoint: Nbcr 2 , a case study. *Europhysics Letters*, 85(5): 56002, 2009.

[91] P. Seleson. Improved one-point quadrature algorithms for two-dimensional peridynamic models based on analytical calculations. *Computer Methods in Applied Mechanics and Engineering*, 282: 184–217, 2014.

[92] E. Sharon, G. Cohen, and J. Fineberg. Crack front waves and the dynamics of a rapidly moving crack. *Physical Review Letters*, 88: 085503, 2002.

[93] E. Sharon and J. Fineberg. Microbranching instability and the dynamic fracture of brittle materials. *Physical Review B*, 54(10): 7128, 1996.

[94] E. Sharon, S. P. Gross, J. Fineberg. Local crack branching as a mechanism for instability in dynamic fracture. *Physical Review Letters*, 74(25): 5096, 1995.

[95] S. A. Silling. Reformulation of elasticity theory for discontinuities and long-range forces. *Journal of the Mechanics and Physics of Solids*, 48(1): 175–209, 2000.

[96] S. A. Silling. Emu: The first code based on the peridynamic theory of solid mechanics. Sandia National Laboratories, 2003.

[97] S. A. Silling and E. Askari. A meshfree method based on the peridynamic model of solid mechanics. *Computers & Structures*, 83(17): 1526–1535, 2005.

[98] S. A. Silling and F. Bobaru. Peridynamic modeling of membranes and fibers. *International Journal of Non-Linear Mechanics*, 40(2-3): 395–409, 2005.

[99] E. Sommer. Formation of fracture 'lances' in glass. *Engineering Fracture Mechanics*, 1(3): 539–546, 1969.

[100] J. H. Song, H. Wang, and T. Belytschko. A comparative study on finite element methods for dynamic fracture. *Computational Mechanics*, 42(2): 239–250, 2008.

[101] R. Spatschek, E. Brener, and A. Karma. Phase field modeling of crack propagation. *Philosophical Magazine*, 91(1): 75–95, 2011.

[102] R. Streit and I. Finnie. An experimental investigation of crack-path directional stability. *Experimental Mechanics*, 20(1): 17–23, 1980.

[103] A. N. Stroh. A theory of the fracture of metals. *Advances in Physics*, 6(24): 418–465, 1957.

[104] S. M. Wiederhorn. Fracture surface energy of glass. *Journal of the American Ceramic Society*, 52(2): 99–105, 1969.

[105] J. Xu, E. Askari, O. Weckner, and S. A. Silling. Peridynamic analysis of impact damage in composite laminates. *Journal of Aerospace Engineering*, 21(3): 187–194, 2008.

[106] X. P. Xu and A. Needleman. Numerical simulations of fast crack growth in brittle solids. *Journal of the Mechanics and Physics of Solids*, 42(9): 1397–1434, 1994.

[107] E. H. Yoffe. The moving Griffith crack. *Philosophical Magazine*, 42(330): 739–750, 1951.

[108] S. J. Zhou, P. S. Lomdahl, R. Thomson, and B. L. Holian. Dynamic crack processes via molecular dynamics. *Physical Review Letters*, 76(13): 2318, 1996.

Chapter 11

Relations between Peridynamic and Classical Cohesive Models

Scot M. Breitenfeld
University of Illinois at Urbana-Champaign

Philippe H. Geubelle
University of Illinois at Urbana-Champaign

Olaf Weckner
The Boeing Company

Stewart A. Silling
Sandia National Laboratories

CONTENTS

11.1 Introduction 317
11.2 Analytical PD-based normal cohesive law 319
11.2.1 Case 1 — No bonds have reached critical stretch 322
11.2.2 Case 2 — Bonds have exceeded the critical stretch 323
11.2.3 Numerical approximation of PD-based cohesive law 325
11.3 PD-based tangential cohesive law 329
11.3.1 Case 1 — No bonds have reached critical stretch 331
11.3.2 Case 2 — Bonds have exceeded the critical stretch 331
11.4 PD-based mixed-mode cohesive law 333
11.5 Conclusions 336
Acknowledgments 337
References 337

11.1 Introduction

Cohesive zone modeling (CZM) is a well established and researched method for modeling fracture. Independently introduced in the early sixties by Dugdale [3] and

Barrenblatt [1] to address the issue of the singularity of the crack tip field, CZM places ahead of the crack a nonlinear region governed by a cohesive failure law between the tractions resisting crack opening and the resulting displacement jump across the fracture surface. The approach regained a lot of attention two decades ago with the development of the cohesive finite element method, which incorporates the cohesive failure law in the constitutive response of interfacial (cohesive) elements placed between conventional (volumetric) finite elements [9, 16, 2, 4]. The cohesive finite element method has achieved remarkable success in the simulation of spontaneous crack propagation when the crack path is known a priori, such as in the failure of interfaces [15, 14, 8]. However, when the numerical method is used to model arbitrary crack growth, care must be exercised to address lattice dependency effects [12, 10, 17]. Nevertheless, CZM continues to be a valuable analysis tool in the study of fracture.

In PD, the criterion used to break the bonds between particles is a key component because it is the fundamental way in which discontinuities, such as cracks, are defined. A critical component of any bond failure criteria is the ability to relate the PD quantities to key classical measurable fracture parameters, i.e., the fracture toughness G_c. The first bond failure criterion introduced in [13] is the "bond stretch" criterion which is based on a critical value s_c of the bond stretch s

$$s = \frac{|\xi + \eta| - |\xi|}{|\xi|} = s_c, \tag{11.1}$$

where $\xi = \mathbf{x}' - \mathbf{x}$ and $\eta = \mathbf{u}' - \mathbf{u}$ as have been defined in earlier chapters. This value can be related to the fracture toughness G_c, i.e., the total energy per unit of fracture surface needed to completely separate two halves of the horizon. In terms of the critical bond stretch (11.1), Silling *et al.* [13] considered the work required to break a single bond

$$w_0(\xi) = \int_0^{s_c} g(s)d\eta = \int_0^{s_c} g(s)(\xi ds), \quad \xi = \|\xi\|. \tag{11.2}$$

Assuming a stiffness c of the bond and a linear function for $g(s) = cs$, the integration of (11.2) simplifies to

$$w_0(\xi) = \int_0^{s_c} cs(\xi ds) = \frac{cs_c^2\xi}{2}. \tag{11.3}$$

Therefore, the work G_c required to break all the bonds per unit fracture area is (Fig. 11.1)

$$G_c = \int_0^{\delta}\int_0^{2\pi}\int_z^{\delta}\int_0^{cos^{-1}z/\xi} \left(\frac{cs_0^2\xi}{2}\right)\xi^2 sin\phi d\phi d\xi dz, \tag{11.4}$$

leading to

$$G_c = \frac{\pi c s_c^2 \delta^5}{10}. \tag{11.5}$$

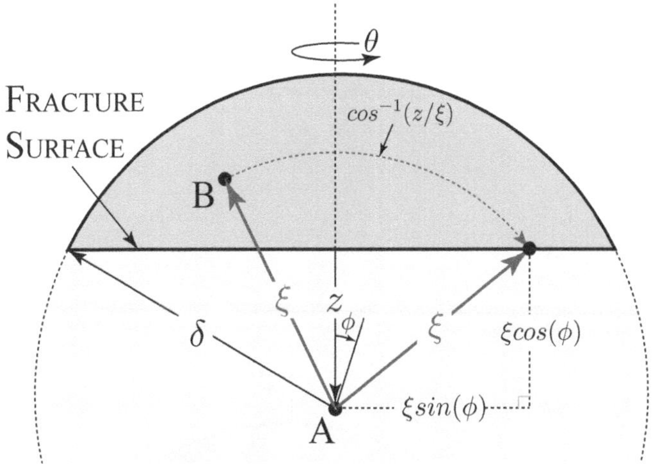

Figure 11.1: Geometrical representation of the integration bounds (11.4) for a horizon containing a bisecting fracture plane used in the "bond stretch" and "energy density" criteria. The total work to break all the bonds connecting point A to all points B in the spherical cap is the integration over all these connections for each point A along the line z.

Thus far, the key premise for the relation between the critical stretch s_c and the fracture toughness G_c is the disregard for the *failure process* taking place in breaking the bonds; only the final state of complete bond failure is considered. However, for cohesive modeling, this process is modeled by considering the stretch s of every bond, which is now a function of the crack face separation. Therefore, each bond will reach the critical stretch, s_c, at different crack opening displacement levels, resulting in a traction-separation law. The cohesive traction-separation laws in PD are investigated in the next sections for normal opening (Sections 11.2 and 11.2.3), tangential opening (Section 11.3), and mixed-mode loading (Section 11.4).

11.2 Analytical PD-based normal cohesive law

To derive analytically the normal traction-separation law described by PD, let us assume that a horizon is traversed by a fracture plane, which is then displaced normal to the crack surface by an opening distance Δ_n (Fig. 11.2). Defining $\xi = \|\xi\|$ and $\eta = \|\eta\|$, the length of the stretched bond, $\overline{AC}$, of segment $\overline{AB}$ is

$$\overline{AC} = \sqrt{\xi^2 + \Delta_n^2 + 2\xi\Delta_n \cos\phi}. \tag{11.6}$$

The stretch $s_{\overline{AC}}$ of segment $\overline{AC}$ is thus given by

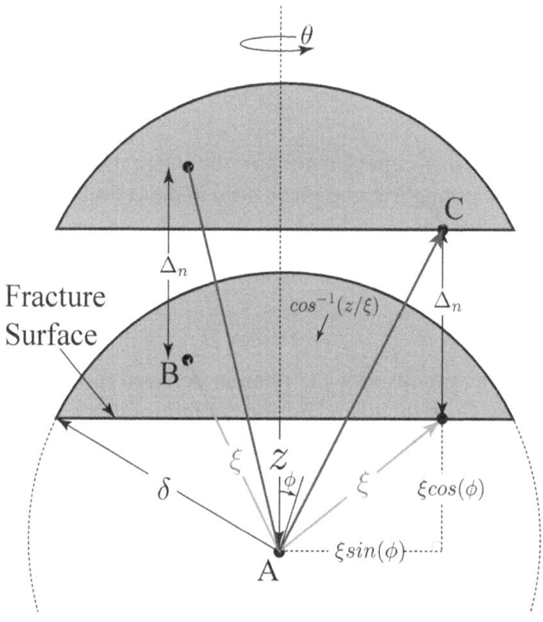

Figure 11.2: Normal opening, Δ_n, of the fracture surface. The dark shaded area corresponds to the portion of the horizon of point A that contains the end points of the bonds crossing the fracture surface.

$$s_{\bar{AC}} = \frac{\overline{AC}}{\xi} - 1 = \sqrt{1 + 2\left(\frac{\Delta_n}{\xi}\right)\cos(\phi) + \left(\frac{\Delta_n}{\xi}\right)^2} - 1. \tag{11.7}$$

Bonds which have exceeded the critical stretch, s_c, are found by setting $s_{\bar{AC}}$ (11.7) equal to s_c, i.e,

$$\sqrt{1 + 2\left(\frac{\Delta_n}{\xi}\right) cos(\phi) + \left(\frac{\Delta_n}{\xi}\right)^2} - 1 = s_c, \tag{11.8}$$

and solving (11.8), for ξ,

$$\xi_c = \frac{\Delta_n\left(\cos(\phi) + \sqrt{\cos^2(\phi) + 2s_c + s_c^2}\right)}{s_c\,(2 + s_c)}. \tag{11.9}$$

Thus, all bonds with a length less than ξ_c described by (11.9) have failed. Assuming a bond stiffness c, the force magnitude in segment $\overline{AC}$ is

$$\|\mathbf{T}_{AC}\| = c\frac{\eta}{\xi} = cs = c\left[\sqrt{1 + 2\left(\frac{\Delta_n}{\xi}\right) cos(\phi) + \left(\frac{\Delta_n}{\xi}\right)^2} - 1\right]. \tag{11.10}$$

The components of the total force acting on the fracture surface per unit of fracture area due to all the bonds connecting points A located below the fracture plane ($0 \leq z \leq \delta$) to their counterpart points B in the spherical cap above the fracture plane are then

$$\mathbf{T}_x = \int_0^{\delta}\int_{\theta_1}^{\theta_2}\int_{\phi_1}^{\phi_2}\int_{\xi_1}^{\xi_2} (\|\mathbf{T}_{AC}\| \sin\phi\cos\theta)\,\xi^2 \sin\phi\, d\xi\, d\phi\, d\theta\, dz, \tag{11.11}$$

$$\mathbf{T}_y = \int_0^{\delta}\int_{\theta_1}^{\theta_2}\int_{\phi_1}^{\phi_2}\int_{\xi_1}^{\xi_2} (\|\mathbf{T}_{AC}\| \sin\phi\sin\theta)\,\xi^2 \sin\phi\, d\xi\, d\phi\, d\theta\, dz, \tag{11.12}$$

$$\mathbf{T}_z = \int_0^{\delta}\int_{\theta_1}^{\theta_2}\int_{\phi_1}^{\phi_2}\int_{\xi_1}^{\xi_2} (\|\mathbf{T}_{AC}\| \cos\phi)\,\xi^2 \sin\phi\, d\xi\, d\phi\, d\theta\, dz, \tag{11.13}$$

where the fracture surface is in the xy-plane and z is normal to the crack surface. The force magnitude $\|\mathbf{T}_{AC}\|$ in each bond is independent of θ, and evaluating the $d\theta$ integral, with the bounds $\theta_1 = 0$ and $\theta_2 = 2\pi$, results in the $\mathbf{T}_z$, referred to hereafter as the normal traction $\mathbf{T}_n$, being the only non-zero component:

$$\mathbf{T}_x = (\sin 2\pi - \sin 0)\int_0^{\delta}\int_{\phi_1}^{\phi_2}\int_{\xi_1}^{\xi_2} (\|\mathbf{T}_{AC}\| \sin\phi)\,\xi^2 \sin\phi\, d\xi\, d\phi\; dz = 0, \tag{11.14}$$

$$\mathbf{T}_y = (\cos 2\pi - \cos 0)\int_0^{\delta}\int_{\phi_1}^{\phi_2}\int_{\xi_1}^{\xi_2} (\|\mathbf{T}_{AC}\| \sin\phi)\, \xi^2 \sin\phi \, d\xi \, d\phi \, dz = 0, \quad (11.15)$$

$$\mathbf{T}_z = \mathbf{T}_n = 2\pi \int_0^{\delta}\int_{\phi_1}^{\phi_2}\int_{\xi_1}^{\xi_2} (\|\mathbf{T}_{AC}\| \cos\phi)\, \xi^2 \sin\phi \, d\xi \, d\phi \, dz. \quad (11.16)$$

The integration bounds in (11.14)–(11.16) change depending on the number of broken bonds, which is itself dependent on the crack opening displacement Δ_n and critical stretch s_c.

11.2.1 Case 1 — No bonds have reached critical stretch

Assuming no bonds have reached the critical stretch s_c, from Fig. 11.2, the bounds of integration are

$$\begin{array}{ll} \phi_2 = cos^{-1}(z/\delta) & \xi_2 = \delta \\ \phi_1 = -cos^{-1}(z/\delta) & \xi_1 = z/cos\phi. \end{array} \quad (11.17)$$

To get the components of the force in the deformed configuration, i.e., the force components of vector AC, the angle ϕ with respect to AC is

$$\phi = \sin^{-1}\left(\frac{\xi sin\phi}{\sqrt{\xi^2 + \Delta_n^2 + 2\xi\Delta_n \cos\phi}}\right). \quad (11.18)$$

Using the property $\cos(\sin^{-1}\phi) = \sqrt{1-\phi^2}$, we get

$$\cos\phi = \sqrt{1 - \left(\frac{\xi sin\phi}{\sqrt{\xi^2 + \Delta_n^2 + 2\xi\Delta_n \cos\phi}}\right)^2}. \quad (11.19)$$

Substituting (11.19) into (11.16) gives the final form of the traction component normal to the crack surface,

$$\mathbf{T}_n = 2\pi \int_0^{\delta}\int_{\phi_1}^{\phi_2}\int_{\xi_1}^{\xi_2} \left(\|\mathbf{T}_{AC}\| \sqrt{1 - \left(\frac{\xi sin\phi}{\sqrt{\xi^2 + \Delta_n^2 + 2\xi\Delta_n \cos\phi}}\right)^2}\right) \xi^2 \sin\phi \, d\xi \, d\phi \, dz, \quad (11.20)$$

where $\|\mathbf{T}_{AC}\|$ is described by (11.10). In the sections that follow, the integrals are evaluated numerically using the Romberg integration function, a successively higher-order generalization of Simpson's rule [11].

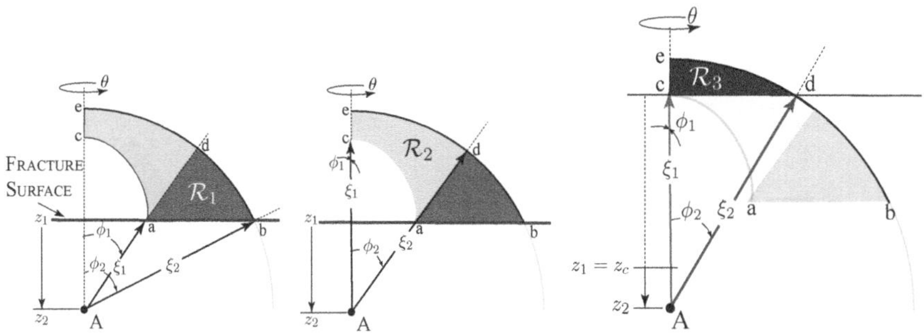

(a) Regions $\mathcal{R}_1$ and $\mathcal{R}_2$. (b) Region $\mathcal{R}_3$.

Figure 11.3: Case 2A integration bounds when $\xi_c < \delta$.

11.2.2 Case 2 — Bonds have exceeded the critical stretch

Case 2 handles the situation when any of the bonds crossing the fracture plane have exceeded the critical stretch. The integration bounds have to be split into two different cases, CASE 2A and CASE 2B, due to the drastic changes in the integration regions.

CASE 2A — ξ_c IS LESS THAN THE HORIZON SIZE δ

Case 2A corresponds to the situation when ξ_c (11.9) is less then the horizon size δ at $\phi = 0$ and $z = 0$ (Fig. 11.3(a))

$$\xi_c\,(\phi = 0, z = 0) = z_c = \frac{\Delta_n}{s_c}. \tag{11.21}$$

Using spherical coordinates centered at point A, the integration needs to be split into three different regions $\mathcal{R}_1$, $\mathcal{R}_2$, and $\mathcal{R}_3$ (Fig. 11.3). The colored regions are those where bonds have not failed. The bounds of integration for region $\mathcal{R}_1$ (Fig. 11.3(a)) for the traction forces $(T_n)^{\mathcal{R}_1}$ (11.20) are

$$\begin{array}{lll} z_2 = z_c & \phi_2 = \cos^{-1}(z/\delta) & \xi_2 = \delta \\ z_1 = 0 & \phi_1 = \frac{z\sqrt{\Delta_n(\Delta_n+2z)s_c(2+s_c)}}{\Delta_n(\Delta_n+2z)} & \xi_1 = \frac{z}{\cos\phi}, \end{array} \tag{11.22}$$

where the angle ϕ_1 is associated with point a, which is the intersection of the curves ab and ac (11.9), i.e., solving for ϕ in

$$\frac{\Delta_n\left(\cos\phi + \sqrt{\cos^2\phi + 2s_c + s_c^2}\right)}{s_c\,(2+s_c)} = \frac{z}{\cos\phi}. \tag{11.23}$$

The bounds of integration for region $\mathcal{R}_2$ (Fig. 11.3(a)) for the traction forces $(T_n)^{\mathcal{R}_2}$ (11.20) are

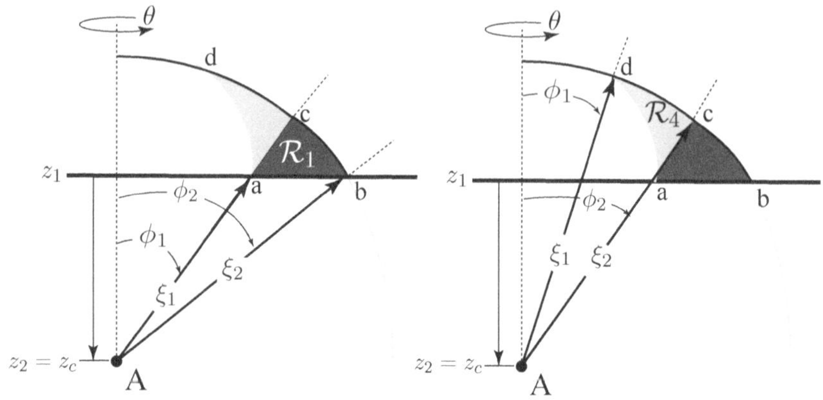

Figure 11.4: Case 2B integration bounds when $\xi_c > \delta$.

$$\begin{array}{lll} z_2 = z_c & \phi_2 = \frac{z\sqrt{\Delta_n(\Delta_n+2z)s_c(2+s_c)}}{\Delta_n(\Delta_n+2z)} & \xi_2 = \delta \\ z_1 = 0 & \phi_1 = 0 & \xi_1 = \frac{\Delta_n\left[\cos\phi+\sqrt{\cos^2\phi+s_c(2+s_c)}\right]}{s_c(2+s_c)}. \end{array} \tag{11.24}$$

Lastly, the bounds of integration for region $\mathcal{R}_3$ (Fig. 11.3(b)) for the traction forces $(T_n)^{\mathcal{R}_3}$ (11.20) are

$$\begin{array}{lll} z_2 = \delta & \phi_2 = \cos^{-1}(z/\delta) & \xi_2 = \delta \\ z_1 = z_c & \phi_1 = 0 & \xi_1 = \frac{z}{\cos\phi}. \end{array} \tag{11.25}$$

CASE 2B — ξ_c IS GREATER THAN THE HORIZON SIZE δ

Case 2B corresponds to larger values of Δ_n for which ξ_c given by (11.9) is larger than the horizon size δ at $\phi = 0$ and $z = 0$ (Fig. 11.4). Again using spherical coordinates, the integration needs to be split into two different regions, $\mathcal{R}_1$ (Fig. 11.4) and $\mathcal{R}_4$ (Fig. 11.4), where the bounds of integration for R_1 are the same as in the previous case (11.22). For region $\mathcal{R}_4$, the bounds of integration for the traction forces $(T_n)^{\mathcal{R}_4}$ (11.20) are

$$\begin{aligned} z_2 &= z_c \\ z_1 &= \frac{2\delta^2 s_c - \Delta_n^2 + \delta^2 s_c^2}{2\Delta_n\delta} \\ \phi_2 &= \frac{z\sqrt{\Delta_n(\Delta_n+2z)s_c(2+s_c)}}{\Delta_n(\Delta_n+2z)} \\ \phi_1 &= \pi - cos^{-1}\left[\frac{-2\delta^2 s_c + \Delta_n^2 - \delta^2 s_c^2}{2\Delta_n\delta}\right] \\ \xi_2 &= \delta \\ \xi_1 &= \frac{\Delta_n\left[cos\phi+\sqrt{cos^2\phi+s_c(2+s_c)}\right]}{s_c(2+s_c)}, \end{aligned} \tag{11.26}$$

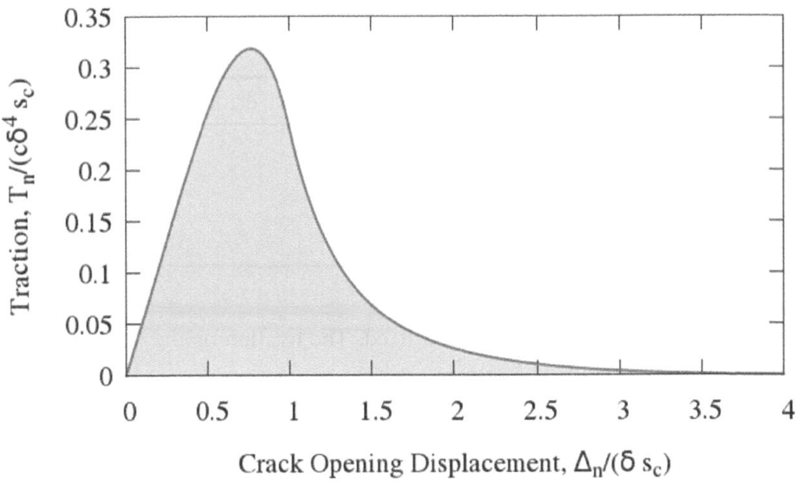

Figure 11.5: Analytical traction-separation law for normal opening. The area under the curve is the fracture toughness G_c (11.5).

where ϕ_1 is the angle associated with intersection point d.

The resulting evolution of the normal traction T_n as a function of the Δ_n is presented in Fig. 11.5. The area under the traction-separation curve is the fracture toughness G_c, which is found to be in excellent agreement (within 0.006%) with the expression (11.5) found earlier. The PD-based cohesive law has a profile typical of some of the polynomial and exponential relations adopted in the literature [9, 16]. Note that the critical value of the displacement jump corresponding to the maximum value of the normal cohesive traction is approximately $0.76/\delta s_c$. Using the semi-analytical method described above, it would be possible to solve the inverse problem and derive the bond failure relation that leads to a desired cohesive failure law, thereby allowing for a direct link between some of the cohesive failure laws available in the literature and the PD CZM.

11.2.3 Numerical approximation of PD-based cohesive law

Whereas Section 11.2 derived the 'analytical' cohesive law by essentially considering an infinite number of points A and points B, this section explores the effects of using a finite number of particles on approximating the analytical traction-separation law in normal opening (Fig. 11.5). The first particle arrangement is a cubic lattice (Fig. 11.6(a)(b)) using grid spacings dq/δ equal to 0.5, 0.25, 0.125 and .0625. The second particle arrangement is a non-uniform grid (Fig. 11.6(c)) created by using the tetrahedron's centroid in an unstructured lattice as the particle coordinates, and the particle's volume is set equal to the tetrahedron's volume. To model the progressive opening of the crack, the hemispherical cap is displaced rigidly by Δ_n and the bot-

Table 11.1: The relative error of the critical strain energy release rate, $(G_c/G_c^{exact} - 1)\,100\%$, as a function of grid spacing for a cubic arrangements of particles.

dq/δ	Degrees of freedom	T_n^{max} relative error	G_c relative error
0.5	39	30.8%	11.3%
0.25	633	3.5%	8.3%
0.125	2895	1.9%	2.8%
0.0625	24471	0.31%	0.9%

tom half's line of points are all held fixed. The traction on the fracture surface dq^2 highlighted in red (Fig. 11.6(d)) is

$$\mathbf{T}_n = \frac{\sum_{j=1}^{M}\sum_{i=1}^{N_j} cs_i (dq)_i^3 (dq)_j^3}{(dq)^2}, \tag{11.27}$$

where M is the total number of particles below the spherical cap and N_j is the number of particles in the horizon of particle j. For the non-uniform arrangement of particles, dq was assumed to be the average distance between all the particles, which is a reasonable approximation as the global edge lengths of the tetrahedron are fairly uniform throughout the lattice.

Looking first at the cubic particle arrangement, the traction-separation law is presented in Fig. 11.7 as a function of particle refinement. The cohesive traction converges to the analytical solution described in Section 11.2 as the grid is refined. The discretization effects are clearly evident for the coarser arrangements, leading to an overestimation of the maximum traction T_n^{max} by roughly 30% and the fracture toughness by roughly 10% (Table 11.1). As the grid spacing is reduced, the error quickly drops. A typical peridynamic simulation uses a discretization of the order $dq/\delta \gtrsim 0.33$ since simulations would be prohibitively expensive for lower values of dq/δ.

Next is the cohesive traction using a non-uniform particle arrangement (Fig. 11.8). The effects of the coarser arrangement are again evident. However, the variation of each bond's length throughout the domain appears to help the convergence of G_c as evident in Table 11.2 where, for fewer degrees of freedom, the relative error in G_c is less than for the cubic arrangement. On the other hand, the relative error in T_n^{max} is larger than that found in the cubic arrangement. Additionally, the dissimilitude of each bond's length causes the bonds to fail at different crack openings, smoothing out the net resistive force in comparison to the uniform particle arrangement.

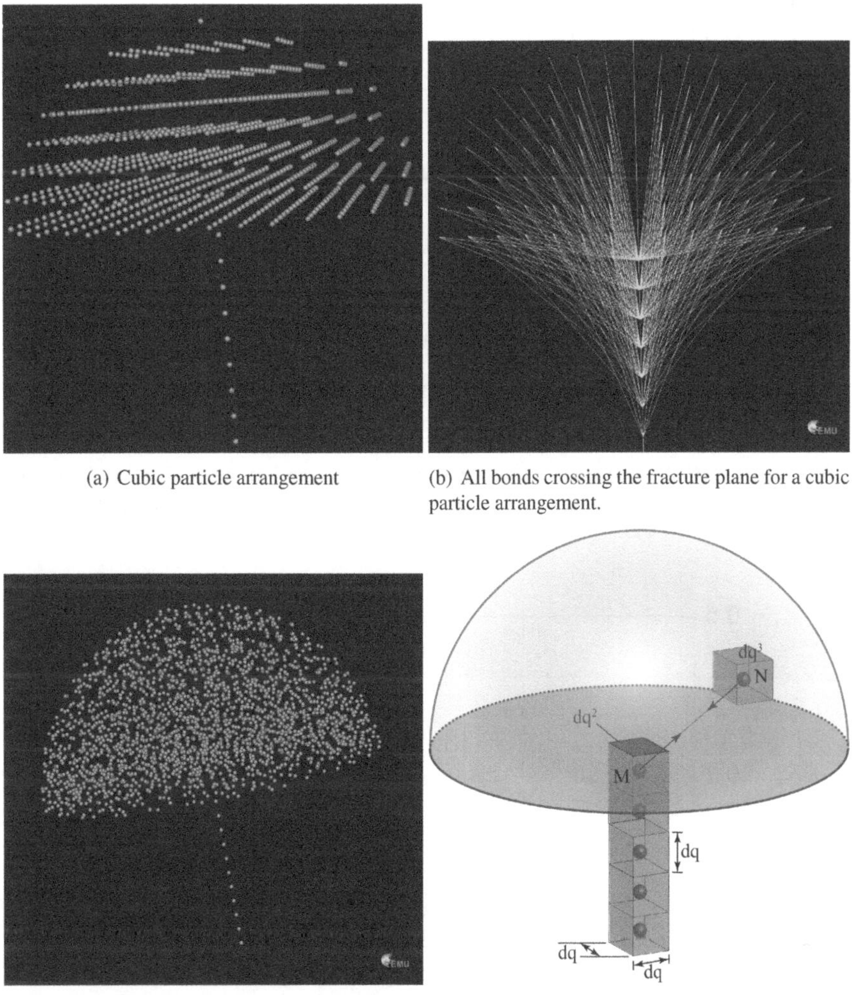

(a) Cubic particle arrangement

(b) All bonds crossing the fracture plane for a cubic particle arrangement.

(c) Non-uniform particle arrangement.

(d) Schematic for finding the traction on the highlighted area dq^2 in red.

Figure 11.6: Lattice arrangements for the study of normal opening cohesive traction laws using a discretized domain.

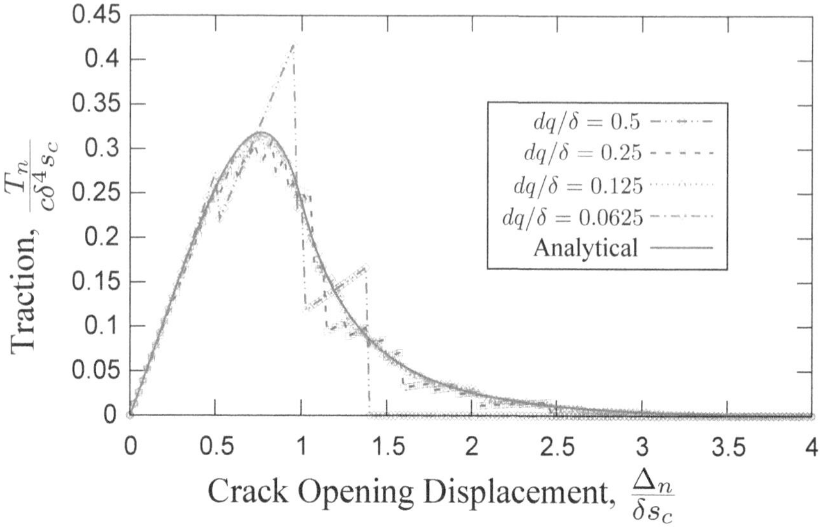

Figure 11.7: Cohesive traction-separation law obtained numerically for different particle spacings in a cubic arrangement compared to the exact solution (11.20).

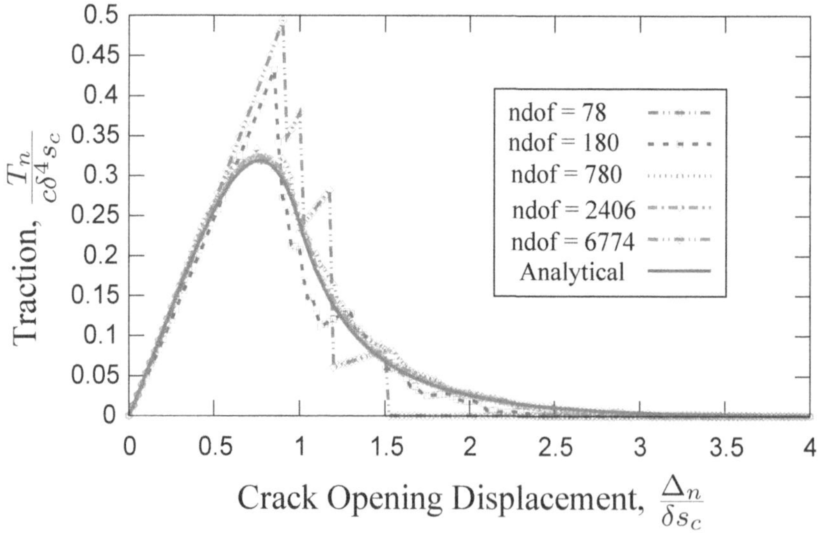

Figure 11.8: The cohesive traction versus the crack opening displacement using different particle spacings in a non-uniform arrangement compared to the exact solution (11.20).

Table 11.2: The relative error of the critical strain energy release rate, as a function of the degrees of freedom, for a non-uniform arrangement of particles.

Degrees of freedom	T_n^{max} relative error	G_c relative error
78	55.3%	4.3%
180	35.8%	5.1%
780	5.0%	4.4%
2406	2.2%	1.6%
6774	1.6%	0.31%

11.3 PD-based tangential cohesive law

In this section, the traction law for tangential opening is investigated. Assume a horizon is traversed by a fracture plane which is then displaced tangential to the crack surface an opening distance Δ_t (Fig. 11.9). The length of the stretched bond, $\overline{AC}$, for segment $\overline{AB}$ is

$$\overline{AC} = \sqrt{\Delta_t^2 + \xi^2 + 2\Delta_t \xi \sin(\phi)\sin(\theta)},$$

and the stretch s of segment $\overline{AC}$ is

$$s_{\overline{AC}} = \frac{\overline{AC}}{\xi} - 1 = \sqrt{1 + 2\left(\frac{\Delta_t}{\xi}\right)\sin(\phi)\sin(\theta) + \left(\frac{\Delta_t}{\xi}\right)^2} - 1, \tag{11.28}$$

which can be either tensile or compressive (Fig. 11.10). The bond, ξ, at which the critical stretch, s_c, is exceeded is obtained by setting (11.28) equal to s_c,

$$\sqrt{1 + 2\left(\frac{\Delta_t}{\xi}\right)\sin(\phi)\sin(\theta) + \left(\frac{\Delta_t}{\xi}\right)^2} - 1 = s_c, \tag{11.29}$$

and solving (11.29) for ξ

$$\xi_c = \frac{\Delta_t\left(\sin(\phi)\sin(\theta) + \sqrt{\sin^2(\phi)\sin^2(\theta) + 2s_c + s_c^2}\right)}{s_c(2 + s_c)}. \tag{11.30}$$

Depending on s_c and Δ_t (11.9), all bonds with a length less than ξ_c have failed. Additionally, in contrast to the normal opening analysis (Section 11.2), ξ_c is also dependent on θ.

The magnitude of the force in segment $\overline{AC}$ is

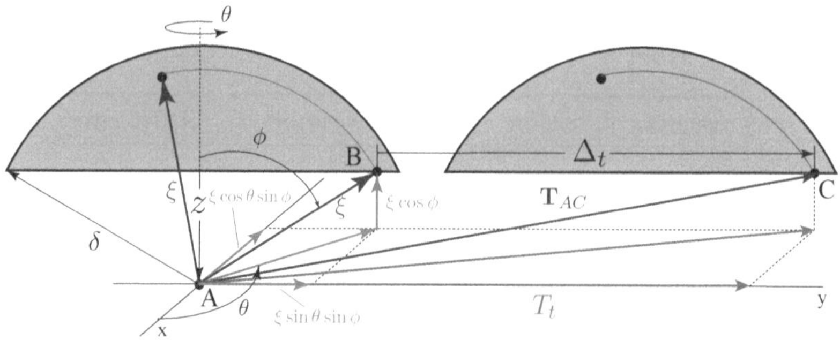

Figure 11.9: Tangential opening, Δ_t, of the fracture surface.

$$\|\mathbf{T}_{AC}\| = c\frac{\eta}{\xi} = cs = c\left[\sqrt{1+2\left(\frac{\Delta_t}{\xi}\right)\sin(\phi)\sin(\theta)+\left(\frac{\Delta_t}{\xi}\right)^2}-1\right]. \quad (11.31)$$

The components of the total force acting on the fracture surface per unit of fracture area due to all the bonds connecting points A to points B in the spherical cap are

$$\mathbf{T}_x = \int_0^{\delta}\int_{\theta_1}^{\theta_2}\int_{\phi_1}^{\phi_2}\int_{\xi_1}^{\xi_2}(\|\mathbf{T}_{AC}\|\sin\phi\cos\theta)\,\xi^2\sin\phi\,d\xi\,d\phi\,d\theta\,dz = 0, \quad (11.32)$$

$$\mathbf{T}_y = \mathbf{T}_t = \int_0^{\delta}\int_{\theta_1}^{\theta_2}\int_{\phi_1}^{\phi_2}\int_{\xi_1}^{\xi_2}(\|\mathbf{T}_{AC}\|\sin\phi\sin\theta)\,\xi^2\sin\phi\,d\xi\,d\phi\,d\theta\,dz, \quad (11.33)$$

$$\mathbf{T}_z = \int_0^{\delta}\int_{\theta_1}^{\theta_2}\int_{\phi_1}^{\phi_2}\int_{\xi_1}^{\xi_2}(\|\mathbf{T}_{AC}\|\cos\phi)\,\xi^2\sin\phi\,d\xi\,d\phi\,d\theta\,dz. \quad (11.34)$$

Note that in the tensile case the forces perpendicular to the opening displacement were zero. However, the forces in the tangential opening case are non-zero, making the shear opening case more complex.

The final expression of the shear traction T_t tangential to the crack face, in the direction of the displacement Δ_t, is thus

$$\mathbf{T}_t = \int_{\theta_1}^{\theta_2}\int_{z_1}^{z_2}\int_{\phi_1}^{\phi_2}\int_{\xi_1}^{\xi_2}\left\{\|\mathbf{T}_{AC}\|\frac{\Delta_t+\xi\sin\phi\sin\theta}{\sqrt{\xi^2\cos^2\phi+(\xi\sin\theta\sin\phi+\Delta_t)^2-z^2}}\right.$$
$$\left.\sqrt{1-\frac{\xi^2\cos^2\phi}{\xi^2\cos^2\phi+(\xi\sin\theta\sin\phi+\Delta_t)^2}}\right\}\xi^2\sin\phi\,d\xi\,d\phi\,dz\,d\theta, \quad (11.35)$$

where $\|\mathbf{T}_{AC}\|$ is given by (11.31).

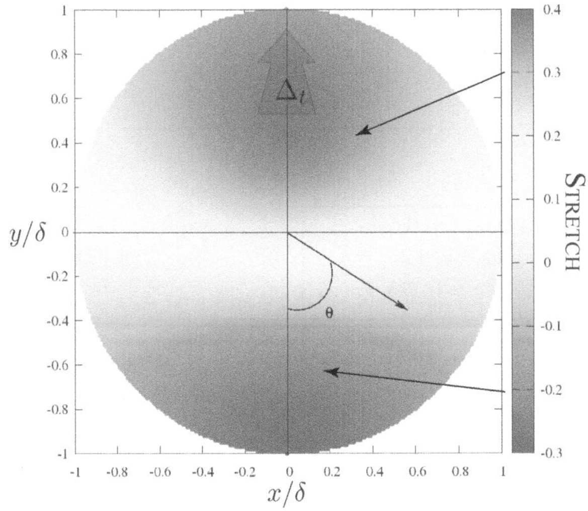

Figure 11.10: Shear opening stretch between points on plane $z/\delta \simeq 0$ and a point located at $z/\delta = 0.75$ for an opening displacement $\Delta_t/\delta = 0.5$ showing both tensile and compressive stretches.

11.3.1 Case 1 — No bonds have reached critical stretch

Assuming no bonds have reached the critical stretch s_c, from Fig. 11.9, the bounds of integration are

$$\begin{array}{cccc} \theta_2 = 2\pi & z_2 = \delta & \phi_2 = cos^{-1}(\min(z,\delta)/\delta) & \xi_2 = \delta \\ \theta_1 = 0 & z_1 = 0 & \phi_1 = 0 & \xi_1 = \min(\delta, z/cos\phi). \end{array} \tag{11.36}$$

The integration of (11.36) was performed analytically and the resulting nonlinear cohesive traction-separation law is plotted in Fig. 11.11 as the solid line. As was done in Section 11.2.3, an analysis of the traction-separation law for a cubic particle arrangement, using the same parameters, was performed (Fig. 11.11), showing convergence as the number of particles is refined. The area under the curve for $0 \le \Delta_t/\delta s_c \le 25$ can be used as a measure of the relative error with respect to the analytical solution and is summarized in Table 11.3.

11.3.2 Case 2 — Bonds have exceeded the critical stretch

Due to the added complexity associated with the θ dependence of the solution, extraction of the closed-form expression of the integration bounds in 11.35 in the presence of bond breakage has eluded our efforts. However, the mode II cohesive failure law can be extracted numerically using the same cubic lattice model adopted earlier in the normal opening case (Section 11.2.3). As mentioned earlier, the failure process is substantially more complex in the shear case since some of the bonds experience

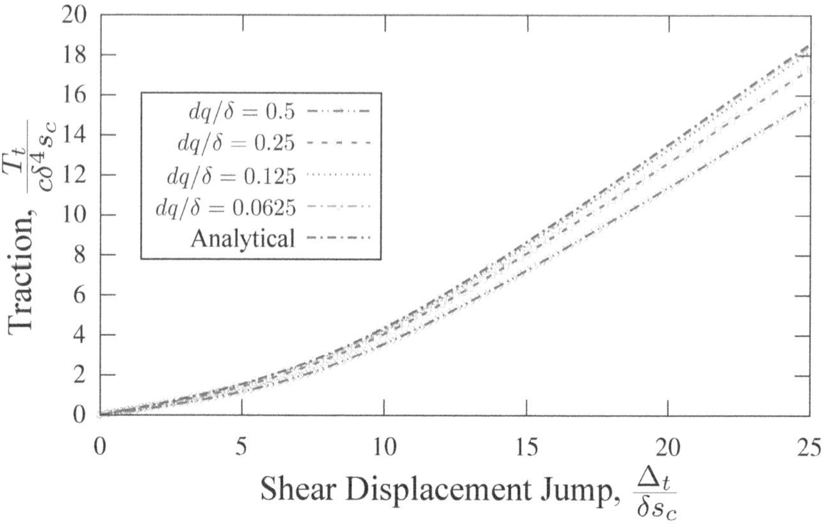

Figure 11.11: Tangential traction-separation law obtained analytically (11.35) and by using a series of refined cubic lattices in the absence of bond failure. A very good agreement is observed for $dq/\delta \leq 0.125$.

dq/δ	Degrees of freedom	Relative error
0.5	39	16.4%
0.25	633	8.9%
0.125	2895	3.2%
0.0625	24471	1.7%

Table 11.3: The relative error of the area under the curve for $0 \leq \Delta_t/\delta s_c \leq 25$ (Fig. 11.11) as a function of the cubic lattice grid spacing, dq, for shear opening Δ_t and assuming no bonds have broken.

compression while the "upper dome" of the horizon (i.e., the set of points located above the failure surface) is subjected to shear failure. Various scenarios can then be considered, depending on whether and at what level compressive failure of the bonds is allowed. These scenarios can be characterized by the parameter η defined as the (absolute value of the) ratio of critical stretch in compression to that in tension.

Fig. 11.12 presents the cohesive failure law in shear for the case $\eta = \infty$, i.e., when bonds are not allowed to break in compression. Curves corresponding to four grid spacing values ($dq/\delta = 0.5, 0.25, 0.125$, and 0.0625) are presented. As apparent in that figure, the absence of failure in compression leads to a substantial reversal of the shear cohesive traction after most of the bonds under tension have failed. This fact is illustrated by the two insets, which depict the evolution of the region (shaded in gray) corresponding to the unbroken bonds. For larger shear displacement jump values (such as $\Delta_t/\delta s_c = 10$), most of the bonds under tension have failed, and those still under compression create a shear cohesive traction in a direction opposite the applied displacement jump Δ_t. The cycling between tension and compression in a bond is geometrically related to the bond's rotation. For instance, a bond under compression reaches maximum compression when the bond is vertical. Further opening displacement causes the bond to rotate beyond vertical and to be loaded in tension, leading to eventual failure of the bond in tension at a critical stretch. Fig. 11.12 also illustrates a grid convergence of the cohesive law similar to that obtained in the tensile case: the traction-separation law converges for $dq/\delta \leq 0.125$. This convergence is quantified in Table 11.4 in terms of the relative error on the area under the traction-separation curve, i.e., the mode II fracture toughness. The relative error is with respect to the fracture toughness presented in Section 11.1, Eq. (11.5), which remains valid also under a mode II loading.

If we allow for failure under compression, we obtain the mode II cohesive laws shown in Fig. 11.13, which correspond to a lattice spacing $dq/\delta = 0.0625$. Nine values of η are considered, ranging from 0.25 to ∞. As apparent in that figure, no negative shear traction is observed for $\eta \lesssim 2$). As expected, as η decreases, i.e., as more bonds fail under compression, the failure process is "steeper." As shown in Fig. 11.14, which present the dependence of the fracture toughness on η, the mode II fracture toughness when the critical stretch values in tension and compression are equal ($\eta = 1$) is equal to that obtained in the absence of compressive failure of the bonds ($\eta = \infty$), i.e. $G_c(\eta = 1) \simeq G_c(\eta = \infty)$.

11.4 PD-based mixed-mode cohesive law

In this final section, we obtain the cohesive envelopes predicted by the PD modeling of cohesive failure, i.e., the dependence of the normal and tangential cohesive traction on the combined normal and shear displacement jumps. The results corresponding to $\eta = \infty$ and dq/δ=0.0625 are presented in Fig. 11.15. Since the failure process is

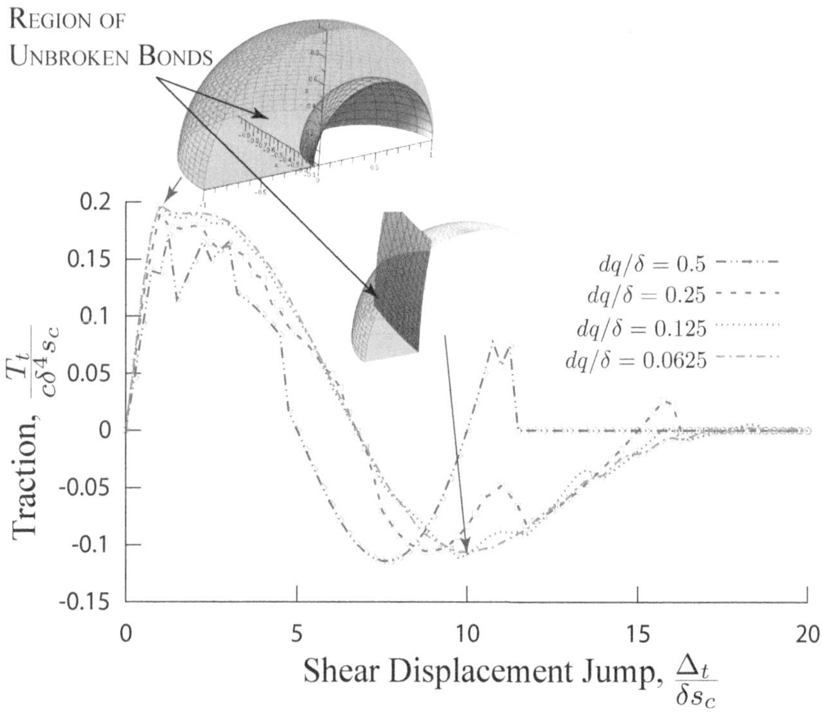

Figure 11.12: Traction-separation law shear when bonds break only in tension. The insets show the bond intact regions (shaded in gray) as the shear opening increases.

Table 11.4: Relative error of the fracture toughness, with respect to the fracture toughness G_c presented in Section 11.1, Eq. (11.5), as a function of the cubic lattice grid spacing, dq, for shear opening Δ_t and assuming no bonds break in compression ($\eta = \infty$).

dq/δ	Degrees of freedom	G_c relative error
0.5	39	16.4%
0.25	633	8.9%
0.125	2895	3.2%
0.0625	24471	1.7%

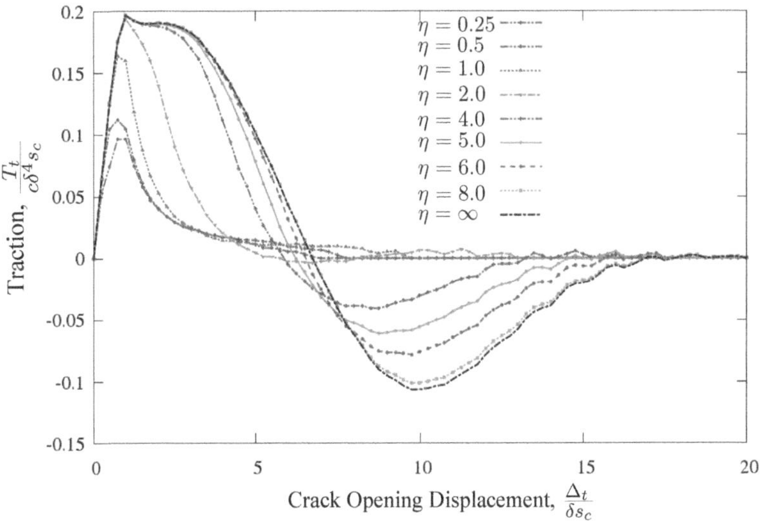

Figure 11.13: Traction-separation law obtained numerically with dq/δ=0.0625 when bonds are allowed to fail in compression. η is the ratio of critical stretch in compression to that in tension.

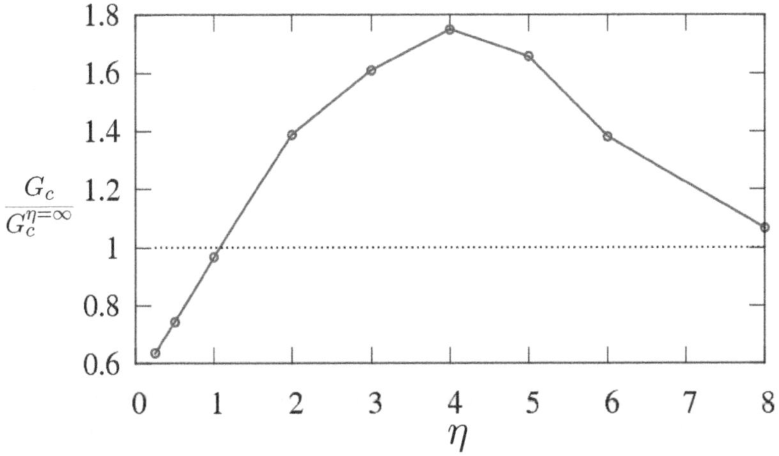

Figure 11.14: Variation of the fracture toughness G_c with respect to the ratio of critical stretch in compression to critical stretch in tension.

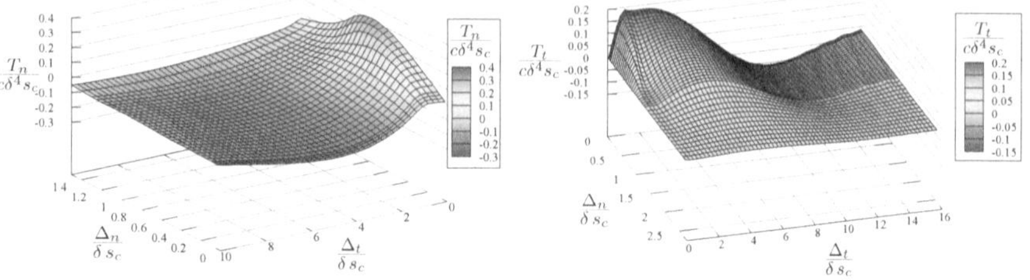

(a) Normal cohesive traction. (b) Shear cohesive traction.

Figure 11.15: Normal T_n and tangential T_t cohesive traction versus the normal Δ_n and tangential Δ_t displacement jumps for the case $\eta = \infty$.

path dependent, these results are obtained by holding Δ_n fixed and then varying Δ_t in the (Δ_n, Δ_t) plane. The coupling between normal and shear failure is clearly visible.

11.5 Conclusions

In this chapter, we investigated the analytical and numerical link between PD and cohesive zone modeling under mode I, mode II, and mixed-mode loading. It was observed that the mode I, mode II, and mixed-mode PD-based cohesive laws resulted in a profile typically found in some of the polynomial and exponential relationships found in the literature. By using the cohesive analysis outlined in this chapter, one should be able to derive the bond failure relation that leads to a desired failure law, thereby forming a link between PD CZM and cohesive laws available in the literature.

In general, there is a clear relation between the PD and classical cohesive law for when the crack opening is normal to the crack surface. However, when the opening is tangential to the crack surface, care must be taken to avoid nonphysical results, such as when the net resistive force switches sign contingent on the opening displacement with respect to the bond's alignment (Section 11.3). Additionally, in a PD simulation there are multiple sets of particles on either side of the fracture surface, and not just the one set considered in this work. The additional bonds caused by including these additional particles would help to further smooth the abrupt changes seen in the traction-separation law for the larger grid spacing.

An interesting addition to the PD cohesive modeling, as studied in this chapter, would be the use of a damage law as the fundamental mechanism for bond failure. Thus, instead of a bond failing catastrophically at a critical stretch, the strength of a bond would vary according to a damage law. Healing of the material (e.g. [5]) could be introduced into the cohesive law by allowing the damage within a bond to be reduced (i.e. allowing the bond to heal) in accordance with the phenomenological behavior of the material. Additionally, fatigue crack growth can be modeled by adopting a bond damage model which would capture the progressive degradation of the bonds due to the cyclic loading and unloading events [7, 6].

Acknowledgments

This work is supported by the Boeing Company, with Dr. O. Weckner as program manager, through Contract #556628 at the National Center for Supercomputing Applications located at the University of Illinois. The original version of the peridynamic code EMU and computational resources have been provided by Sandia National Laboratories.

References

[1] G.I. Barenblatt. The mathematical theory of equilibrium cracks in brittle fracture. *Advances in Applied Mechanics*, 7(C): 55–129, 1962.

[2] G. T. Camacho and M. Ortiz. Computational modelling of impact damage in brittle materials. *International Journal of Solids and Structures*, 33(20-22): 2899–2938, 1996.

[3] D. Dugdale. Yielding of steel sheets containing slits. *Journal of the Mechanics and Physics of Solids*, 8(2): 100–104, 1960.

[4] P.H. Geubelle and J.S. Baylor. Impact-induced delamination of composites: A 2D simulation. *Composites Part B: Engineering*, 29(5): 589–602, 1998.

[5] M.W. Keller, S.R. White, and N.R. Sottos. A self-healing poly(dimethyl siloxane) elastomer. *Advanced Functional Materials*, 17(14): 2399–2404, 2007.

[6] S. Maiti and P. H. Geubelle. Cohesive modeling of fatigue crack retardation in polymers: Crack closure effect. *Engineering Fracture Mechanics*, 73(1): 22–41, 2006.

[7] S. Maiti, C. Shankar, P. H. Geubelle, and J. Kieffer. Continuum and molecular-level modeling of fatigue crack retardation in self-healing polymers. *Journal of Engineering Materials and Technology*, 128(4): 595, 2006.

[8] M. Meo and E. Thieulot. Delamination modelling in a double cantilever beam. *Composite Structures*, 71(3-4): 429–434, 2005.

[9] A. Needleman. A continuum model for void nucleation by inclusion debonding. *Transactions of the ASME. Journal of Applied Mechanics*, 54(3): 525–31, 1987.

[10] K.D. Papoulia, S.A. Vavasis, C. Sam, and P. Ganguly. Analysis of mesh dependence in rigid cohesive interface finite element models. *11th International Conference on Fracture 2005, ICF11*, 2: 1235–1240, 2005.

[11] W.H. Press, B.P. Flannery, S.A. Teukolsky, and W.T. Vetterling. *Numerical Recipes in Fortran 77: The Art of Scientific Computing*. Cambridge University Press, 1992.

[12] J. Rimoli, J. Rojas, and F. Khemani. On the mesh dependency of cohesive zone models for crack propagation analysis. In *53rd AIAA/ASME/ASCE/AHS/ASC Structures, Structural Dynamics and Materials Conference 20th AIAA/ASME/AHS Adaptive Structures Conference 14th AIAA*, pages 1–9, Reston, Virigina, 23–26 April, 2012. American Institute of Aeronautics and Astronautics.

[13] S.A. Silling and E. Askari. A meshfree method based on the peridynamic model of solid mechanics. *Computers & Structures*, 83(17-18): 1526–1535, 2005.

[14] L. Sorensen, J. Botsis, T. Gmür, and L. Humbert. Bridging tractions in mode I delamination: Measurements and simulations. *Composites Science and Technology*, 68(12): 2350–2358, 2008.

[15] A. Turon, P.P. Camanho, J. Costa, and J. Renart. Accurate simulation of delamination growth under mixed-mode loading using cohesive elements: Definition of interlaminar strengths and elastic stiffness. *Composite Structures*, 92(8): 1857–1864, 2010.

[16] X.-P. Xu and A. Needleman. Numerical simulations of fast crack growth in brittle solids. *Journal of the Mechanics and Physics of Solids*, 42(9): 1397–1434, 1994.

[17] F. Zhou and J.F. Molinari. Dynamic crack propagation with cohesive elements: A methodology to address mesh dependency. *International Journal for Numerical Methods in Engineering*, 59(1): 1–24, 2004.

Chapter 12

Peridynamic Modeling of Fiber-reinforced Composites

Erdogan Madenci
University of Arizona

Erkan Oterkus
University of Strathclyde

CONTENTS

12.1 Introduction 339
12.2 Peridynamic analysis of a lamina 341
12.3 Peridynamic analysis of a laminate 347
12.4 Numerical results 349
12.5 Conclusions 354
12.6 Appendix A: PD material constants of a lamina 354
12.6.1 Simple shear 355
12.6.2 Uniaxial stretch in the fiber direction 356
12.6.3 Uniaxial stretch in the transverse direction 358
12.6.4 Biaxial stretch 359
12.7 Appendix B: Surface correction factors for a composite lamina 363
12.8 Appendix C: PD interlayer and shear bond constants of a laminate 368
12.9 Appendix D: Critical Stretch Values for Bond Constants 374
References 376

12.1 Introduction

Utilization of composite materials is increasing in many different industries, including aerospace, automotive, and marine industries, due to their superiority of specific strength and stiffness properties over those of metals. However, damage initiation and its subsequent propagation in composites are not understood as clearly as they are, for example, for metals because of the presence of stiff fibers embedded into a soft matrix material, causing inhomogeneity. In order to better understand failure mechanisms, Hallett and Wisnom [9] conducted experiments on double-edge-notched composite specimens made from E-glass. They reported the occurrence of matrix cracking before the ultimate failure for all specimens representing four different lay-ups when loaded in tension. Furthermore, it was reported that fiber failure initiated at the notch tip. Later, Green et al. [8] investigated the effect of scaling on the tensile strength of notched composites made from unidirectional carbon-fiber/epoxy pre-preg by considering hole diameter and laminate thickness as independent variables. These experiments showed that failure mechanisms in composites are very complex due to matrix cracking, fiber breakage, and delamination.

In order to investigate the behavior of cracks, Wu [19] considered unidirectional fiberglass-reinforced Scotch-plies with center cracks oriented in the direction of the fibers. The plies had fibers in the 0° and 45° directions and were loaded in tension, pure shear, and combined tension and shear. In all three types of loading, it was observed that the crack propagated in a direction co-linear with the initial crack.

It is, therefore, evident that the inhomogeneous nature of the composites must be retained in the analysis to predict the correct failure modes. Peridynamics offers the ability to predict complex failure patterns in composite structures. The peridynamic (PD) theory was applied successfully by Colavito et al. [5, 6] to predict damage in laminated composites subjected to low-velocity impact and static indentation. Askari et al. [2] and Xu et al. [20] also used peridynamic simulations to predict damage in laminates subjected to low-velocity impact and for notched laminated composites under biaxial loads. Kilic et al. [13] predicted the basic failure modes of fiber breakage, matrix cracking, and delamination in various laminates with a pre-existing central crack under tension by considering the matrix and fiber materials independently. As an application study, Oterkus et al. [15] demonstrated that PD analysis is capable of capturing bearing and shear-out failure modes in bolted composite lap-joints. More recently, Askari et al. [3] considered the effect of both high- and low-energy hail impacts against a toughened-epoxy, intermediate-modulus, carbon-fiber composite. Also, Hu et al. [10] predicted the basic failure modes of fiber breakage, matrix cracking, and delamination in laminates with a pre-existing central crack under tension. A multi-scale approach for modeling of composites by linking of micro- and macroscales was described by Alali and Lipton [1]. The analytical derivation of the PD material parameters, including thermal loading conditions, was recently given by Oterkus and Madenci [16]. They also demonstrated the constraints on material constants due to the assumption of pairwise interactions in bond-based peridynamics. An ordinary state-based formulation that overcomes these constraints was presented by Madenci and Oterkus [14].

This chapter presents a peridynamic formulation of fiber-reinforced composite materials subjected to mechanical and thermal loading conditions. The PD approach to model a lamina is first validated against analytical solutions by considering uniaxial tension and uniform temperature change. Then, damage growth patterns from a pre-existing crack in a lamina for different fiber orientations are computed and compared against experimental observations. This approach is further extended to analyze composite laminates and to predict damage growth patterns from a pre-existing crack in two distinct laminate constructions under tension. In the absence of a crack, the PD displacement predictions are compared against those of the classical laminate theory. In the presence of a crack, damage patterns are qualitatively compared against experimental observations.

12.2 Peridynamic analysis of a lamina

Silling [18] reformulated the continuum mechanics equations and replaced the divergence term in the classical equation of motion with an integral term, which makes the new form of the equation of motion applicable whether or not a discontinuity exists in the structure:

$$\rho(\mathbf{x})\,\ddot{\mathbf{u}}(\mathbf{x},t) = \int_H \mathbf{f}\left(\mathbf{x}' - \mathbf{x}, \mathbf{u}' - \mathbf{u}\right) dH + \mathbf{b}(\mathbf{x},t)\,. \tag{12.1}$$

In Eq. (12.1), the horizon, H, includes all the material points that the material point $\mathbf{x}$ can interact with inside the body. The interaction force or peridynamic force between material points $\mathbf{x}$ and $\mathbf{x}'$can be expressed as $\mathbf{f}(\mathbf{x}' - \mathbf{x}, \mathbf{u}' - \mathbf{u})$, and it is a function of the relative position vector, $\mathbf{x}' - \mathbf{x}$, and relative displacement vector, $\mathbf{u}' - \mathbf{u}$. The peridynamic force is along the same direction of the relative position of these material points in the deformed configuration, i.e., $\mathbf{y}' - \mathbf{y} = (\mathbf{x}' + \mathbf{u}') - (\mathbf{x} + \mathbf{u})$. For a fiber-reinforced composite lamina, the peridynamic force can be expressed as

$$\mathbf{f} = c\,(s - s^*)\,\frac{\mathbf{y}' - \mathbf{y}}{|\mathbf{y}' - \mathbf{y}|} = c\,\bar{s}\,\frac{\mathbf{y}' - \mathbf{y}}{|\mathbf{y}' - \mathbf{y}|}, \tag{12.2}$$

where c, s, $\bar{s}$, and s^* represent the peridynamic material parameter, total stretch, mechanical stretch, and thermal stretch between material points $\mathbf{x}$ and $\mathbf{x}'$, respectively. The total stretch, s, and the thermal stretch due to thermal loading, s^*, are defined as

$$s = \frac{|\mathbf{y}' - \mathbf{y}| - |\mathbf{x}' - \mathbf{x}|}{|\mathbf{x}' - \mathbf{x}|} \tag{12.3}$$

and

$$s^* = \alpha_{PD}\,\Delta T, \tag{12.4}$$

where α_{PD} and ΔT represent the peridynamic coefficient of thermal expansion and the temperature change, respectively. By using Eqs. (12.3) and (12.4), the mechanical stretch, $\bar{s}$, can be computed as

$$\bar{s} = s - s^* = \frac{|\mathbf{y}' - \mathbf{y}| - |\mathbf{x}' - \mathbf{x}|}{|\mathbf{x}' - \mathbf{x}|} - \alpha_{PD}\,\Delta T. \tag{12.5}$$

Since the fiber-reinforced composite lamina is an anisotropic material, the directional dependency must be included in the PD formulation. In order to represent the directional dependency, two different PD material parameters can be introduced for a fiber-reinforced composite lamina with a fiber orientation of θ, as shown in Fig. 12.1. In this figure, the material point q, which is a family member of the material point of interest i, represents material points that interact with material point i only along the fiber direction. However, the material point p, which is also a family member of the material point i, represents material points that interact with material point i in any arbitrary direction, including the fiber direction.

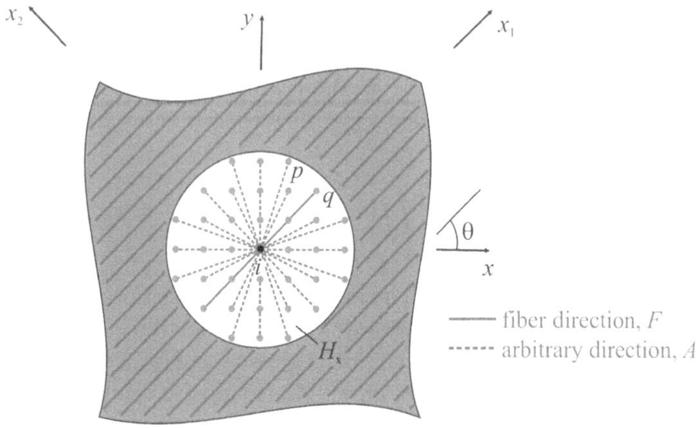

Figure 12.1: PD horizon (interaction domain) of material point i for a lamina with a fiber orientation of θ and PD bonds between material point i and other material points within its horizon.

The peridynamic parameters concerning the interaction of material points in the fiber direction only and in any arbitrary direction can be defined as c_F and c_A, respectively. Extending the procedure introduced by Gerstle et al. [7] for isotropic materials, the PD parameters c_F and c_A can be expressed in terms of the engineering material constants E_1, E_2, G_{12}, and ν_{12} by equating strain energy densities of a material point based on the classical continuum mechanics and PD theory for simple loading conditions.

Based on the classical continuum mechanics, the strain energy density of a material point, W^{CM}, for a two-dimensional composite lamina is expressed as

$$W^{CM} = \frac{1}{2}\sigma^T(\varepsilon - \varepsilon^*), \tag{12.6}$$

in which the stress, σ, total strain, ε, and thermal strain vectors, ε^*, are defined as

$$\sigma^T = \{ \sigma_{11} \quad \sigma_{22} \quad \sigma_{12} \}, \tag{12.7}$$

$$\varepsilon^T = \{ \varepsilon_{11} \quad \varepsilon_{22} \quad \gamma_{12} \}, \tag{12.8}$$

and

$$\varepsilon^{*T} = \{\ \varepsilon_{11}^{*} \quad \varepsilon_{22}^{*} \quad 0\ \}. \tag{12.9}$$

The stress and strain components given in Eq. (12.7)–(12.9) are defined with respect to the principal (material) coordinate system, (x_1, x_2). For a composite lamina with a fiber orientation of θ, the stress and strain components are related through the constitutive relation as

$$\sigma = \mathbf{Q}\,(\varepsilon - \varepsilon^{*}), \tag{12.10}$$

in which the stiffness matrix $\mathbf{Q}$ is defined as

$$\mathbf{Q} = \begin{bmatrix} Q_{11} & Q_{12} & 0 \\ Q_{12} & Q_{22} & 0 \\ 0 & 0 & Q_{66} \end{bmatrix} \tag{12.11}$$

where

$$Q_{11} = \frac{E_{11}}{1-\nu_{12}\nu_{21}}, \quad Q_{12} = \frac{\nu_{12}E_{22}}{1-\nu_{12}\nu_{21}}, \quad Q_{22} = \frac{E_{22}}{1-\nu_{12}\nu_{21}}, \quad Q_{66} = G_{12} \tag{12.12}$$

with $\nu_{12}/E_{11} = \nu_{21}/E_{22}$. Four independent material constants in Eq. (12.12) correspond to the elastic modulus in the fiber direction, E_{11}, elastic modulus in the transverse direction, E_{22}, the in-plane shear modulus, G_{12}, and the in-plane Poisson's ratio, ν_{12}. Furthermore, the thermal strain vector can be expressed in terms of thermal expansion coefficients in the fiber and transverse directions, α_{11} and α_{22}, respectively, for a temperature change, ΔT, as

$$\varepsilon^{*T} = \{\ \alpha_{11}\Delta T \quad \alpha_{22}\Delta T \quad 0\ \}. \tag{12.13}$$

Alternatively, the strain energy density of the same material point in PD theory, W^{PD}, can be obtained by integrating the micropotential of each PD interaction,

$$W^{PD} = \frac{1}{2}\sum_{q=1}^{Q} w_{qi} V_q + \frac{1}{2}\int_{H} w\, dH, \tag{12.14}$$

in which Q is the number of fiber bonds within the horizon of material point i. The micropotential of each PD interaction can be expressed as

$$w = \frac{1}{2} c(\phi)\, \bar{s}^2(\phi)\, \xi. \tag{12.15}$$

Note that the stretch of the bond between the material points i and p, $\bar{s}(\phi)$, is referenced to a polar coordinate system (ξ, ϕ) (see Fig. 12.2). Furthermore, for each peridynamic interaction, the peridynamic parameter $c(\phi)$ can be defined as

$$c(\phi) = \begin{cases} c_F + c_A & \phi = 0 \\ c_A & \phi \neq 0 \end{cases}. \tag{12.16}$$

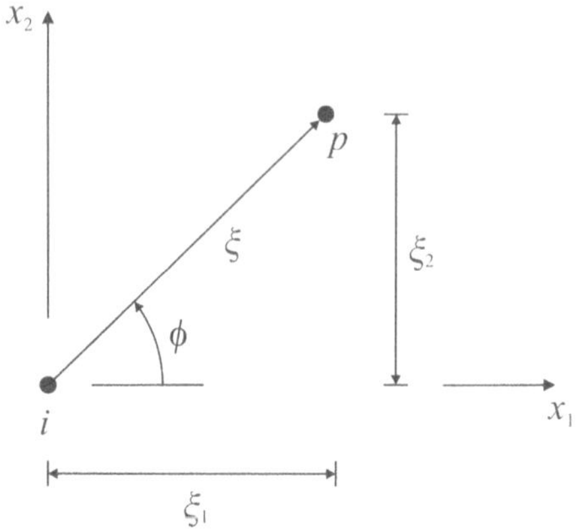

Figure 12.2: Components of the initial bond length between material points i and p.

The peridynamic coefficient of thermal expansion, $\alpha_{PD}(\phi)$, which is used to calculate the thermal stretch term for a lamina, is dependent on the bond orientation between material points i and p. It can be expressed in terms of the coefficients of thermal expansion, α_{11} and α_{22}, in the form

$$\alpha_{PD}(\phi) = \alpha_{11}\cos^2(\phi) + \alpha_{22}\sin^2(\phi). \tag{12.17}$$

After substituting Eq. (12.15) into Eq. (12.14), Eq. (12.14) can be rewritten as

$$W^{PD} = \frac{1}{2}\sum_{q=1}^{Q}\frac{c_F\bar{s}_{qi}^2\,\xi_{qi}}{2}V_q + \frac{1}{2}\int_H \frac{c_A\bar{s}^2\,\xi}{2}\,dH, \tag{12.18}$$

in which Q is the number of fiber bonds within the horizon of material point i. The initial length of the bond in the fiber direction and its mechanical stretch after deformation between material points q and i are denoted by ξ_{qi} and $\bar{s}_{qi}$, respectively. The volume of the material point q that interacts with material point i is denoted by V_q and can be approximately calculated as

$$V_q = \frac{\pi h \delta^2}{N}, \tag{12.19}$$

in which N is the number of material points within the horizon of material point i, and h is the thickness of the lamina.

In order to determine the bond constants c_F and c_A in terms of the engineering constants E_{11}, E_{22}, G_{12}, and ν_{12}, four different simple loading conditions can be considered, as explained in Appendix A. Equating the strain energy densities from PD theory and classical continuum mechanics for these loading conditions results in explicit expressions for c_F and c_A as

$$c_F = \frac{2E_1(E_1 - E_2)}{\left(E_1 - \frac{1}{9}E_2\right)\left(\sum\limits_{q=1}^{Q}\xi_{qi}V_q\right)} \tag{12.20}$$

and

$$c_A = \frac{8 E_1 E_2}{\left(E_1 - \frac{1}{9}E_2\right) \pi h \delta^3}, \tag{12.21}$$

along with constraints on material constants G_{12} and ν_{12} as

$$G_{12} = \frac{\nu_{12} E_2}{1 - \nu_{21} \nu_{12}} = \frac{E_1 E_2}{3\left(E_1 - \frac{1}{9}E_2\right)} \tag{12.22}$$

and

$$\nu_{12} = \frac{1}{3}. \tag{12.23}$$

As presented in Oterkus et al. [17], the constitutive or force-stretch relationships for the in-plane interactions (bonds) of two material points are shown in Fig. 12.3. The critical parameters that define the failure of these bonds under tension and compression are (s_{Ft}, s_{At}) and (s_{Fc}, s_{Ac}), respectively, which can be determined based on the experimental measurements. At least four different uniaxial tension test cases of an unnotched laminate should be performed by considering different lay-up configurations. As shown in Fig. 12.4, these four lay-up configurations can be chosen as (25/0/75), (50/0/50), (75/0/25), and (100/0/0), where each value in the lay-up configuration indicates the percentage of 0°, ±45°, and 90° plies in the laminate, respectively. Tensile failure strain of the (100/0/0) laminate corresponds to the critical stretch of a bond in fiber direction under tension, s_{Ft}. Linear extrapolation of the tensile failure strains for different lay-up configurations leads to the failure strain for the (0/0/100) laminate, i.e., the critical stretch of a bond in arbitrary direction under tension, s_{At}.

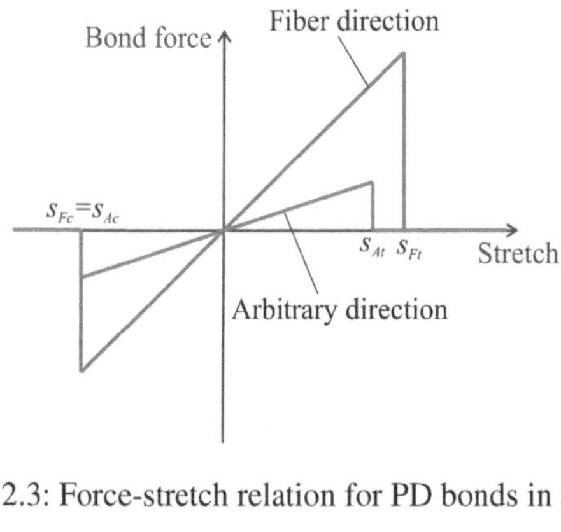

Figure 12.3: Force-stretch relation for PD bonds in a lamina.

The dominant failure mechanism for the fiber-reinforced composites under compression loading is microbuckling. Therefore, a simultaneous failure of bonds in fiber and arbitrary directions can be assumed under compression loading. Hence, a uniaxial compression test for an unnotched quasi-isotropic laminate with a (25/50/25) lay-up can be performed to determine the critical stretch of bonds in fiber and arbitrary directions under compression, s_{Fc} and s_{Ac}, respectively.

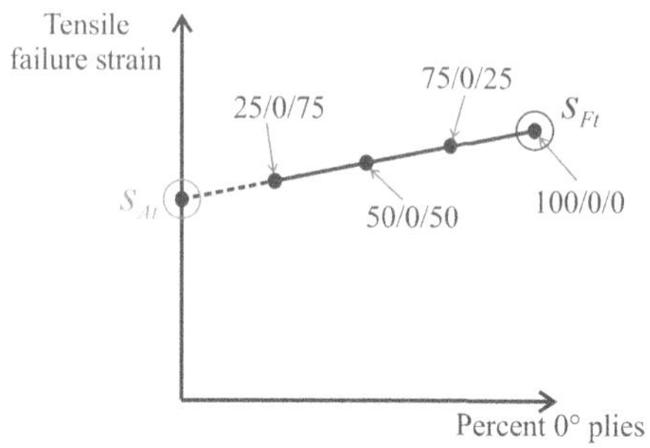

Figure 12.4: Determination of critical stretch values s_{Ft} and s_{At}.

Termination of the interaction between material points can be associated with the failure of material by modifying the peridynamic force relation given in Eq. (12.2) by introducing the failure parameter $\mu\left(\mathbf{x}'-\mathbf{x},t\right)$:

$$\mathbf{f}=\mu\left(\mathbf{x}'-\mathbf{x},t\right)c\left(s-s^*\right)\frac{\mathbf{y}'-\mathbf{y}}{\left|\mathbf{y}'-\mathbf{y}\right|}, \tag{12.24}$$

where the failure parameter can be defined as

$$\mu\left(\mathbf{x}'-\mathbf{x},t\right)=\begin{cases}1 & \text{if } s\left(\mathbf{x}'-\mathbf{x},t'\right)-s^*<s_c \text{ for all } 0<t'<t\\ 0 & \text{otherwise}\end{cases} \tag{12.25}$$

The inexplicit nature of local damage at a material point, $\mathbf{x}$, arising from the introduction of failure in the constitutive model, is removed by defining the local damage as

$$D(\mathbf{x},t)=1-\frac{\int_H \mu(\mathbf{x}'-\mathbf{x},t)dH}{\int_H dH}. \tag{12.26}$$

Surface correction is an important concept in PD theory. The peridynamic parameters given in Eqs. (12.20) and (12.21) are derived under the assumption that the material point located at $\mathbf{x}$ is in a single material with its complete neighborhood entirely embedded within its horizon. However, this assumption becomes invalid when the material point is close to free surfaces (Fig. 12.5). It results in a reduction in material stiffness near the free surfaces, and this stiffness reduction must be corrected. After determining the surface correction factor for each bond, the PD force in that bond is modified based on the associated surface correction factor. A detailed derivation of the surface correction factors is given in Appendix B.

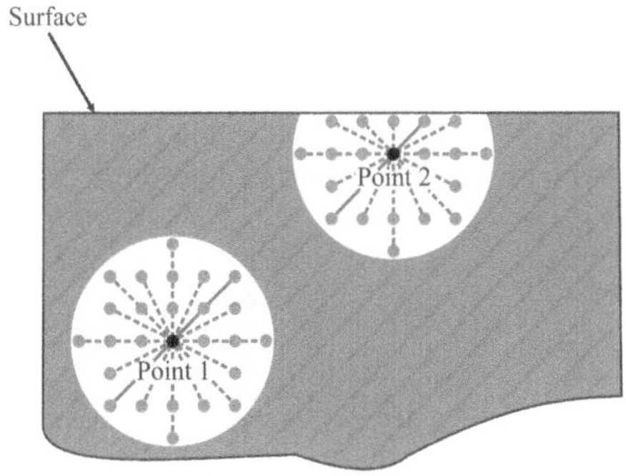

Figure 12.5: Surface effects in the domain of interest.

12.3 Peridynamic analysis of a laminate

The PD formulation for a composite lamina can be extended to consider a composite laminate. In order to capture the deformation behavior of a laminate in the thickness direction and define the interaction between neighboring plies, two additional bond constants between neighboring plies are introduced, as shown in Fig. 12.6.

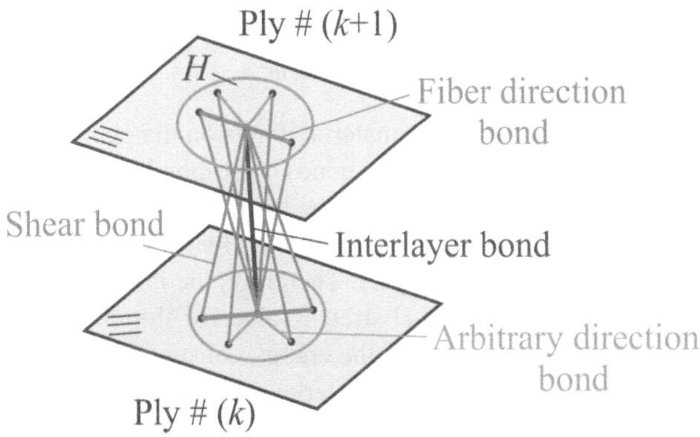

Figure 12.6: Four different bond constants for a fiber-reinforced composite laminate.

Transverse normal and shear deformations between material points located on adjacent (neighboring) layers are related through the bond constants c_N and c_S, respectively. As shown in Fig. 12.6, interlayer bonds only exist in the normal direction, whereas shear bonds exist in all directions between the neighboring plies. Hence, a material point can interact with two other material points via interlayer bonds that have the same in-plane coordinates.

As in the case of in-plane deformation in the case of a lamina, the peridynamic parameters associated with transverse normal and shear deformations, c_N and c_S, can be derived in the form

$$c_N = \frac{E_m}{h\bar{V}} \tag{12.27}$$

and

$$c_S = \frac{2G_m}{\pi h} \frac{1}{\left(\delta^2 + h^2 \ln\left(\frac{h^2}{\delta^2 + h^2}\right)\right)}, \tag{12.28}$$

where E_m and G_m are the elastic and shear moduli of the matrix material, respectively, and $\bar{V}$ is the volume of a material point. Detailed derivations of these expressions are given in Appendix C.

Note that the shear bonds have a different characteristic than bonds in fiber and arbitrary directions, and interlayer bonds, because the shear bond constant relates the body force density, **f**, to the change in angle of the bond from its original orientation (shear angle), φ. Therefore, the force density and micropotential expressions for a shear bond can be written as

$$\mathbf{f} = c_S \, \varphi \, (\Delta x)^2 \frac{\mathbf{y}' - \mathbf{y}}{|\mathbf{y}' - \mathbf{y}|} \tag{12.29}$$

and

$$w = \frac{1}{2} c_S \, \varphi^2, \tag{12.30}$$

where (Δx) is the spacing between material points on the plane of the lamina.

Failures of the interlayer and shear bonds correspond to mode-I and mode-II, respectively. Interlayer damage represents the breakage of (interlayer) bonds between a layer and its adjacent layers above and below. Hence, it provides the extent of delamination between the adjacent layers. Therefore, the interlayer bonds are assumed to fail only in tension. The critical stretch value for the interlayer bonds, s_N, can be obtained analytically by equating the energy consumed by an advancing mode-I crack to the work required to break all interlayer bonds as

$$s_{Nc} = \sqrt{\frac{2G_{IC}}{hE_m}}, \tag{12.31}$$

where G_{IC} is the mode-I critical energy release rate of the matrix material.

The shear bonds can fail if the shear angle of the bonds exceeds the critical shear angle value, φ_{Sc}. It can also be obtained analytically by equating the energy consumed by an advancing Mode-II crack to the work required to break all shear bonds as

$$\varphi_{Sc} = \sqrt{\frac{G_{IIC}}{hG_m}}, \tag{12.32}$$

where G_{IIC} is the mode-II critical energy release rate of the matrix material.

Derivations of the relationships between the critical stretch value for the inter-layer bonds, s_{Nc}, and the mode-I critical energy release rate, and between the critical shear angle value, φ_{Sc}, and the mode-II critical energy release rate, respectively, are given in Appendix D.

12.4 Numerical results

A lamina under uniaxial tension and uniform temperature change. A unidirectional thin lamina with a fiber orientation of $\theta = 0°$ is considered, as shown in Fig. 12.7. The length and width of the lamina are specified as $L = 152.4\,\text{mm}$ and $W = 76.2\,\text{mm}$, respectively. It has a thickness of $h = 0.1651\,\text{mm}$. Its elastic moduli in the fiber and transverse directions are $E_{11} = 159.96\,\text{GPa}$ and $E_{22} = 8.96\,\text{GPa}$, respectively. The thermal expansion coefficients in the fiber and transverse directions are $\alpha_{11} = -1.52\,\text{ppm}\,/\,°\text{C}$ and $\alpha_{22} = 34.3\,\text{ppm}\,/\,°\text{C}$, respectively. The PD model is generated using a single layer of material points with a grid size of $\Delta x = 6.35 \times 10^{-4}\,\text{m}$. The horizon radius is specified as $\delta = 3.015\Delta x$. The peridynamic parameters are computed using Eqs. (12.20) and (12.21) as $c_F = 5.72 \times 10^{23}\,\text{N/m}^6$ and $c_A = 1.86 \times 10^{22}\,\text{N/m}^6$. The quasi-static solution is obtained using the adaptive dynamic relaxation technique with a time increment of 1 and a stable mass density value of $7.005 \times 10^{18}\,\text{kg/m}^3$ [12]. Failure is not allowed in order to verify the solution against analytical predictions based on classical continuum mechanics.

Figure 12.7: Loading and geometry of the unidirectional lamina under uniaxial tension and uniform temperature change.

First, a uniaxial tension loading of $P = 159.96\,\text{MPa}$ is applied as a body load of $b_x = 5.95 \times 10^{10}\,\text{N/m}^3$ along the edges of the lamina through a volumetric region with a depth of $b = 2.54 \times 10^{-3}\,\text{m}$. The variation of the horizontal and vertical displacement components along the central axes in the x- and y-directions, respectively, is computed at the end of 8000 time steps and compared against analytical results, as shown in Figs. 12.8 and 12.9. Analytical results are computed by using the relations

$$u_x = \frac{P}{E_1} x \tag{12.33}$$

and

$$u_y = -\nu_{12} \frac{P}{E_1} y. \tag{12.34}$$

For both displacement components, there is a remarkable agreement between analytical and PD results.

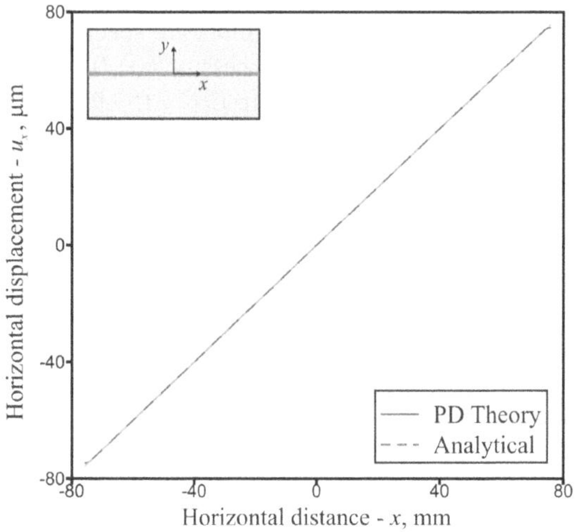

Figure 12.8: Horizontal displacement along the central axis at the end of 8000 time steps.

Next, the lamina is only subjected to a uniform temperature change of $\Delta T = 50°C$. For this loading condition, the analytical horizontal and vertical displacements along the central axes are computed by

$$u_x = \alpha_{11} \Delta T x \tag{12.35}$$

and

$$u_y = \alpha_{22} \Delta T y. \tag{12.36}$$

Comparisons of horizontal and vertical displacements obtained analytically and from PD analysis, shown in Figs. 12.10 and 12.11, indicate a remarkably close agreement.

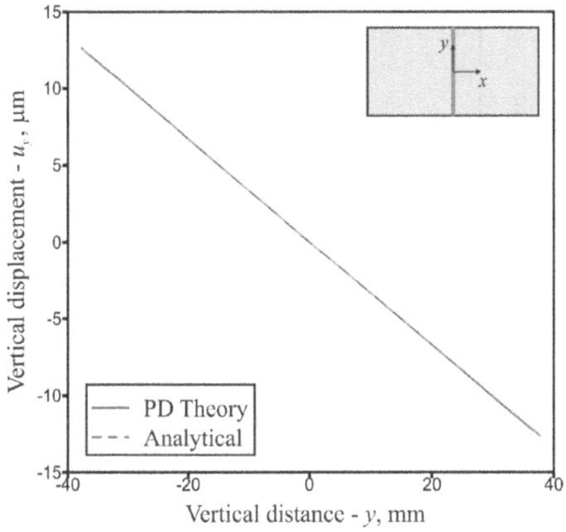

Figure 12.9: Vertical displacement along the central axis at the end of 8000 time steps.

Laminates under uniaxial tension. The validation is continued by considering two different three-ply laminates with stacking sequences of $[0°/90°/0°]$ and $[0°/45°/0°]$ subjected to uniform tension loading, as shown in Fig. 12.12. The geometrical and material properties are the same as those of the lamina. The inter-layer and shear bond constants are computed by using Eqs. (12.27) and (12.28) as $c_N = 3.45 \times 10^{23}\,\mathrm{N/m^6}$ and $c_S = 1.55 \times 10^{18}\,\mathrm{N/m^5}$, respectively. The uniaxial tension loading is applied as a body load of $b_x = 5.95 \times 10^{10}\,\mathrm{N/m^3}$ through a volumetric region with a depth of $b = 2.54 \times 10^{-3}\,\mathrm{m}$. It corresponds to a stress resultant value of $N = 79228.2\,\mathrm{N\,/\,m}$ along the edges of the laminate. During the process, failure is not allowed in order to compare the PD solution to the analytical solution based on the classical laminate theory.

The variation of the horizontal and vertical displacement components along the central axes in the x- and y-directions, respectively, for the 90° ply of the $[0°/90°/0°]$ lay-up are computed at the end of 8000 time steps and compared against analytical results, as shown in Figs. 12.13 and 12.14. Comparisons of the displacement components in the 45° ply of the $[0°/45°/0°]$ lay-up are shown in Figs. 12.15 and 12.16. For both laminates, the agreement between the analytical and PD displacements is remarkably close.

A lamina with a pre-existing central crack under tension. In order to demonstrate the failure prediction capability of the PD approach, the same lamina with a pre-existing central crack is considered for three different fiber orientations, $\theta = 0°$, $90°$, and $45°$. As shown in Fig. 12.17, the crack is aligned with the y-axis and has a length of $2a = 0.01778\,\mathrm{m}$. The lamina is subjected to a velocity boundary condition of $v_0 = 2.02 \times 10^{-7}\,\mathrm{m/s}$ along the edges of the lamina through a volumetric region with a depth of $b = 2.54 \times 10^{-3}\,\mathrm{m}$. The failure is only allowed in tension for bonds in fiber and arbitrary directions. The critical stretch for the bond in arbitrary direction is specified as $s_{A_t} = 0.0135$. For the bond in fiber direction, it is assumed that

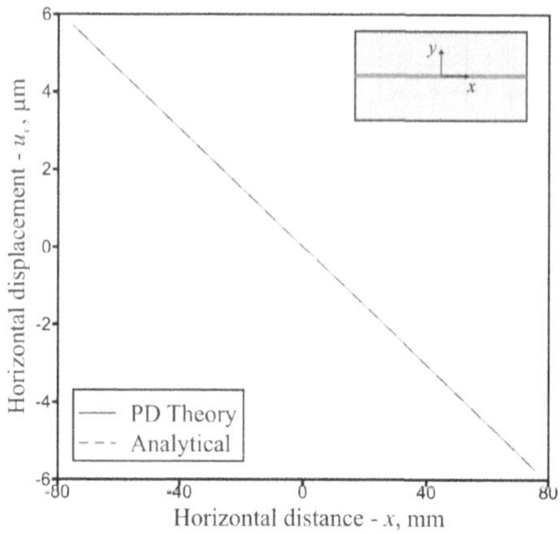

Figure 12.10: Variation of horizontal displacement along the central axis at the end of 8000 time steps when no failure is allowed.

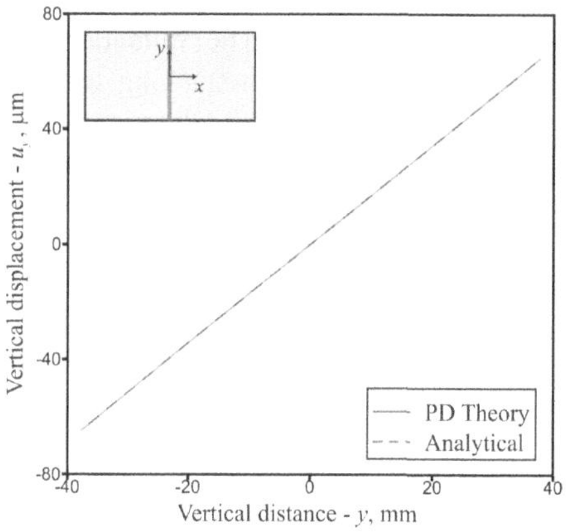

Figure 12.11: Variation of vertical displacement along the central axis at the end of 8000 time steps when no failure is allowed.

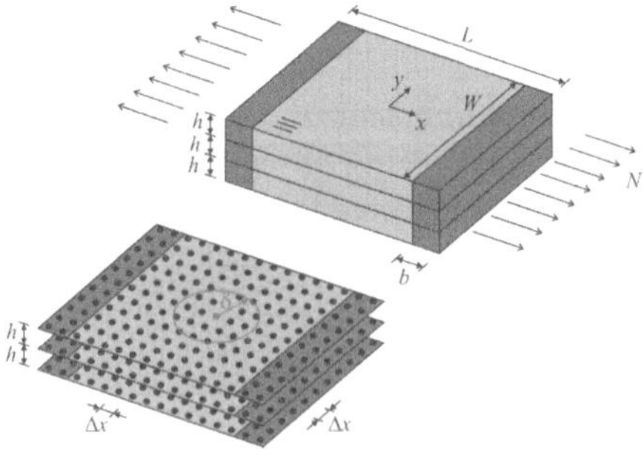

Figure 12.12: Loading and geometry of a composite laminate under uniaxial tension.

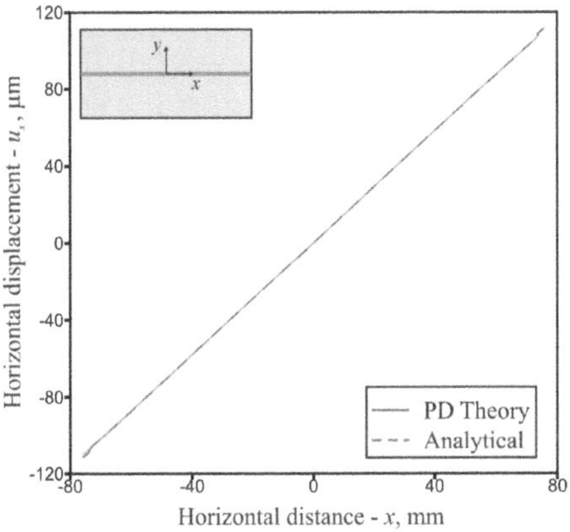

Figure 12.13: Horizontal displacement along the central axis in the 90° ply of the [0°/90°/0°] lay-up at the end of 8000 time steps.

its critical stretch value is twice the critical stretch for the bond in arbitrary direction, i.e., $s_{Ft} = 0.027$.

As shown in Fig. 12.18, the crack propagates in the fiber direction, referred to as the splitting mode in all cases. Similar experimental observations confirm that the current PD model accurately captures the failure modes.

Laminates with a pre-existing central crack under tension. The failure prediction capability of the PD theory is further demonstrated by introducing a central crack in the two laminate lay-ups of [0°/90°/0°] and [0°/45°/0°]. As shown in Fig. 12.19, the crack is aligned with the y-axis and has a length of $2a = 0.01778\,\text{m}$. The laminates are subjected to a velocity boundary condition of $v_0 = 2.02 \times 10^{-7}\,\text{m / s}$. The critical stretch parameters specified for the bonds in fiber and arbitrary directions are $s_{Ft} = 0.027$ and $s_{At} = 0.0135$, respectively. The critical stretch and angle parameters for the interlayer and shear bonds are specified as $s_{Nc} = \varphi_{Sc} = 0.0135$. The procedure for computing these critical stretch and angle values is explained in Appendix D.

For the [0°/90°/0°] laminate, an "H" type splitting failure mode is observed for all plies, as shown in Fig. 12.20. In this case, 0° plies are dominant in the loading direction; therefore, 0° plies determine the failure behavior of the laminate. However, in the [0°/45°/0°] laminate, a "Z" type failure mode is obtained in all plies due to the presence of a 45° ply, as shown in Fig. 12.21. Also observed is the delamination failure mode between the plies due to the breakage of shear bonds around crack tip regions, as shown in Figs. 12.22 and 12.23. These damage patterns are consistent with those observed in a study by Bogert et al. [4].

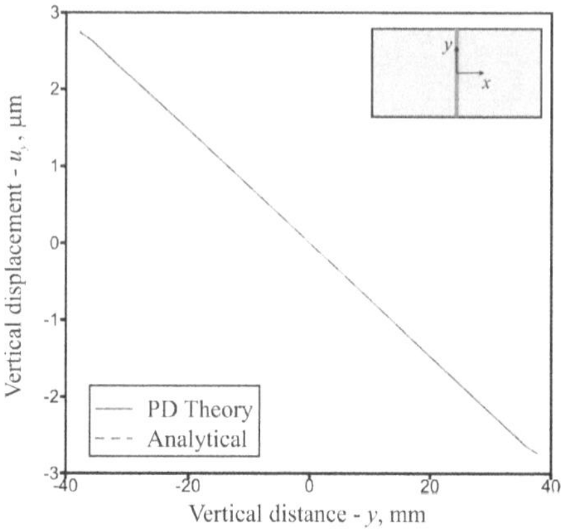

Figure 12.14: Vertical displacement along the central axis in the 90° ply of the [0°/90°/0°] lay-up at the end of 8000 time steps.

12.5 Conclusions

Based on the numerical results, the PD approach successfully predicts the damage growth patterns in fiber-reinforced laminates with a pre-existing crack while considering the distinct properties of the fiber and matrix, as well as the interlayer material between the plies. The predictions capture the correct failure mechanisms of the splitting failure mode and delamination without resorting to any special treatments, and agree with the experimental observations published in the literature. The simulations also capture failure modes among each ply, which are usually distinct; they heavily depend on fiber direction, which is realistically exhibited in the current results. It can be concluded that the peridynamic theory is a powerful method that can be employed for failure analysis of composite materials.

12.6 Appendix A: PD material constants of a lamina

As shown in Fig. 12.24, a lamina is discretized with a single layer of material points in the thickness direction. The PD horizon, H, has the shape of a disk with radius δ and thickness h. The displacements of material points i and p are represented by $\mathbf{u}^{(i)}$ and $\mathbf{u}^{(p)}$, respectively. The initial relative position vector between these material points is denoted by ¸ $= \mathbf{x}^{(p)} - \mathbf{x}^{(i)}$.

Similar to the determination of a PD material constant for an isotropic material [7], equating the strain energy density of a material point in a lamina computed by using the PD theory and classical continuum mechanics yields the relationships be-

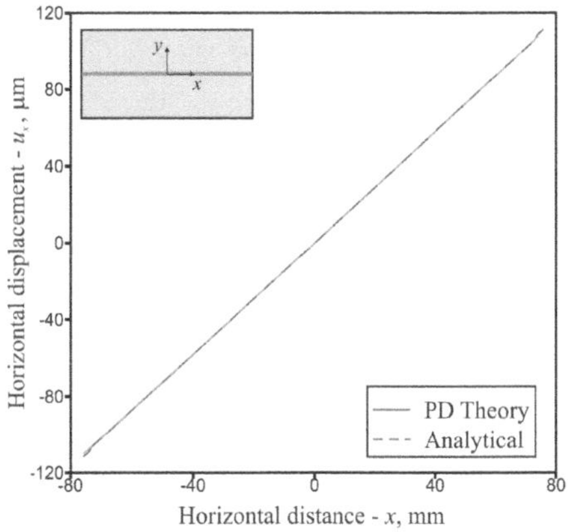

Figure 12.15: Horizontal displacement along the central axis in the 45° ply of the [0°/45°/0°] lay-up at the end of 8000 time steps.

tween PD material constants c_F and c_A in terms of the engineering constants E_1, E_2, G_{12}, and ν_{12}.

The strain energy densities are calculated by considering four different simple loading conditions:

1. Simple shear, $\gamma_{12} = \zeta, \varepsilon_{11} = \varepsilon_{22} = 0$
2. Uniaxial stretch in the fiber direction, $\varepsilon_{11} = \zeta, \varepsilon_{22} = \gamma_{12} = 0$
3. Uniaxial stretch in the transverse direction, $\varepsilon_{22} = \zeta, \varepsilon_{11} = \gamma_{12} = 0$
4. Biaxial stretch, $\varepsilon_{11} = \varepsilon_{22} = \zeta, \gamma_{12} = 0$

12.6.1 Simple shear

As shown in Fig. 12.25, the length of the relative position of material points $\mathbf{y}'$and $\mathbf{y}$in the deformed state becomes

$$|\mathbf{y}' - \mathbf{y}| = [1 + (\sin(\phi)\cos(\phi))\,\zeta]\,|\mathbf{x}' - \mathbf{x}|. \tag{12.37}$$

The mechanical stretch, $\bar{s}$, in the absence of temperature change can be expressed in terms of the relative displacement of the material points i and p, arising from the mechanical loading, as

$$\bar{s} = (\sin(\phi)\cos(\phi))\,\zeta. \tag{12.38}$$

By using Eq. (12.18), the strain energy density based on the PD theory at a material point in a composite lamina can be evaluated as

$$W^{PD} = c_F(0) + \frac{1}{2}\left(\int\limits_H \frac{c_A(\sin(\phi)\cos(\phi)\,\zeta)^2\,\xi}{2}\,dH\right) = \frac{\pi c_A h \delta^3}{48}\zeta^2. \tag{12.39}$$

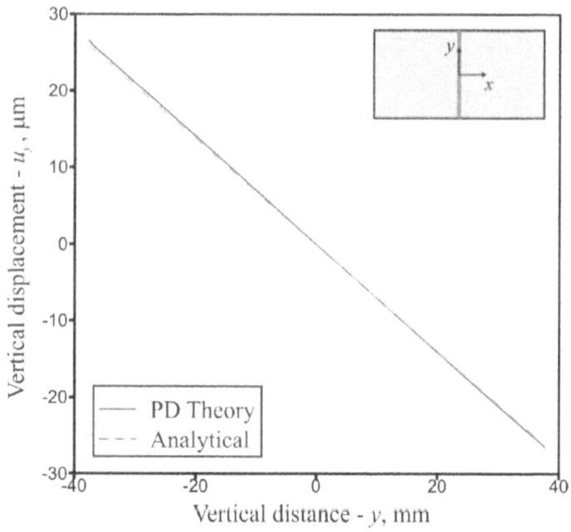

Figure 12.16: Vertical displacement along the central axis in the 45° ply of the [0°/45°/0°] lay-up at the end of 8000 time steps.

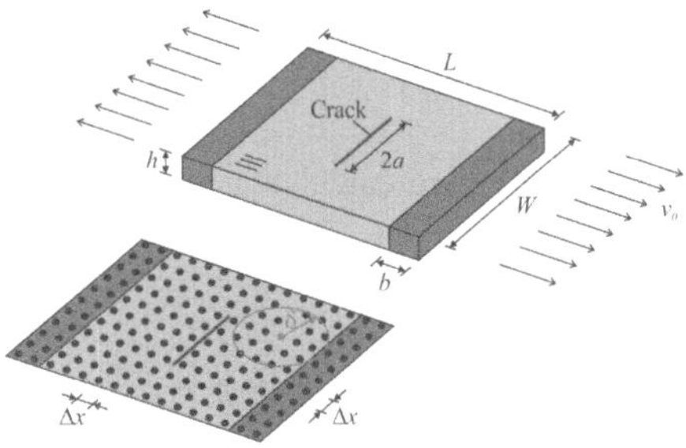

Figure 12.17: Loading and geometry of the unidirectional lamina with a crack under tension loading.

By using Eq. (12.6) in conjunction with Eq. (12.10), the strain energy density of a material point based on classical continuum mechanics can be written as

$$W^{CM} = \frac{1}{2} Q_{66} \zeta^2. \tag{12.40}$$

By equating strain energies obtained from PD theory and classical continuum mechanics, Eq. (12.39) and Eq. (12.40), respectively, lead to

$$c_A = \frac{24 Q_{66}}{\pi h \delta^3}. \tag{12.41}$$

12.6.2 Uniaxial stretch in the fiber direction

As shown in Fig. 12.26, the length of the relative position of material points $\mathbf{y}'$and $\mathbf{y}$ in the deformed state for this loading condition becomes

$$\left| \mathbf{y}' - \mathbf{y} \right| = \left[1 + \left(\cos^2 (\phi) \right) \zeta \right] \left| \mathbf{x}' - \mathbf{x} \right|. \tag{12.42}$$

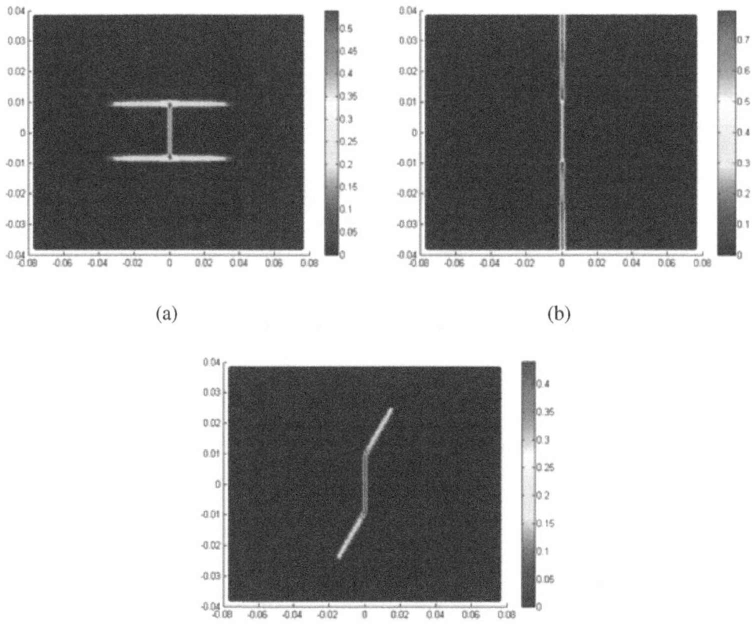

Figure 12.18: Damage plots for laminae having a central crack with fiber orientations of (a) $\theta = 0°$, (b) $\theta = 90°$, and (c) $\theta = 45°$.

The mechanical stretch, $\bar{s}$, in the absence of temperature change can be expressed in terms of the relative displacement of the material points i and p, arising from the mechanical loading, as

$$\bar{s} = \left(\cos^2(\phi)\right)\zeta. \tag{12.43}$$

By using Eq. (12.18), the strain energy density based on the PD theory at a material point in a composite lamina can be evaluated as

$$\begin{aligned} W^{PD} &= \frac{1}{2}\sum_{q=1}^{Q} \frac{c_F \zeta^2 \xi_{qi}}{2} V_q + \frac{1}{2}\left(\int_H \frac{c_A\left(\cos^2(\phi)\zeta\right)^2 \xi}{2} dH\right) \\ &= \frac{c_F \zeta^2}{4}\left(\sum_{q=1}^{Q} \xi_{qi} V_q\right) + \frac{\pi c_A h \delta^3}{16}\zeta^2. \end{aligned} \tag{12.44}$$

By using Eq. (12.6), the strain energy density of a material point based on classical continuum mechanics can be written as

$$W^{CM} = \frac{1}{2} Q_{11} \zeta^2. \tag{12.45}$$

By equating strain energies obtained from PD theory and classical continuum

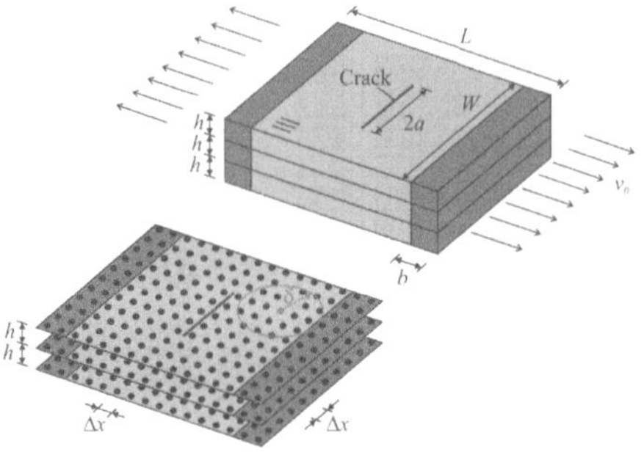

Figure 12.19: Loading and geometry of a composite laminate with a crack under tension loading.

mechanics, Eq. (12.44) and Eq. (12.45), respectively, lead to

$$c_F = \frac{2(Q_{11} - Q_{22})}{\left(\sum_{q=1}^{Q} \xi_{qi} V_q\right)}. \tag{12.46}$$

12.6.3 Uniaxial stretch in the transverse direction

As shown in Fig. 12.27, the length of the relative position of material points $\mathbf{y}'$and $\mathbf{y}$ in the deformed state for this loading condition becomes

$$|\mathbf{y}' - \mathbf{y}| = \left[1 + \left(\sin^2(\phi)\right)\zeta\right] |\mathbf{x}' - \mathbf{x}|. \tag{12.47}$$

The mechanical stretch, $\bar{s}$, in the absence of temperature change can be expressed in terms of the relative displacement of the material points i and p, arising from the mechanical loading, as

$$\bar{s} = \left(\sin^2(\phi)\right)\zeta. \tag{12.48}$$

By using Eq. (12.18), the strain energy density based on the PD theory at a material point in a composite lamina can be evaluated as

$$W^{PD} = c_F(0) + \frac{1}{2}\left(\int_H \frac{c_A\left(\sin^2(\phi)\,\zeta\right)^2 \xi}{2} dH\right) = \frac{\pi c_A h \delta^3}{16}\zeta^2. \tag{12.49}$$

By using Eq. (12.6), the strain energy density of a material point based on classical continuum mechanics can be written as

$$W^{CM} = \frac{1}{2} Q_{22} \zeta^2. \tag{12.50}$$

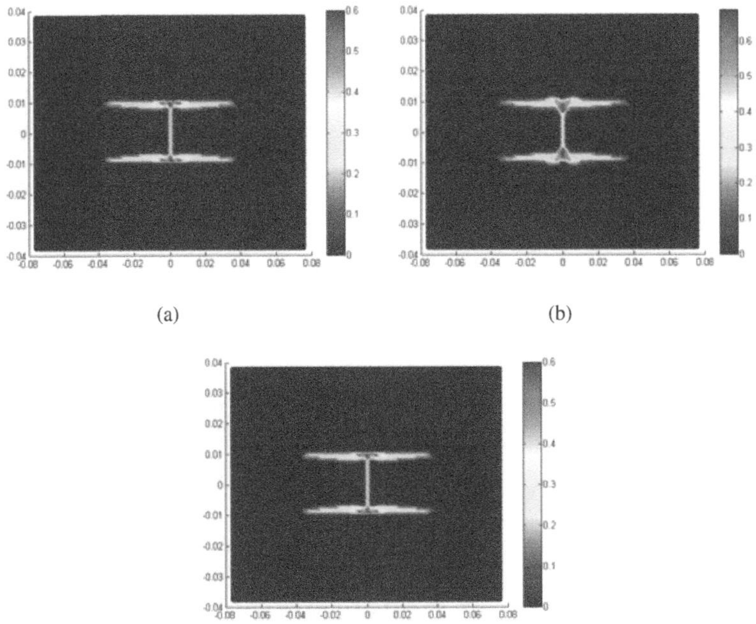

(a) (b)

(c)

Figure 12.20: Damage plots for a laminate of $[0°/90°/0°]$ with a pre-existing crack: (a) bottom ply, 0°; (b) center ply, 90°; and (c) top ply, 0°.

By equating strain energies obtained from PD theory and classical continuum mechanics, Eq. (12.49) and Eq. (12.50), respectively, lead to

$$c_A = \frac{8Q_{22}}{\pi h \delta^3}. \tag{12.51}$$

Comparing Eqs. (12.41) and (12.51) yields

$$Q_{22} = 3Q_{66}. \tag{12.52}$$

12.6.4 Biaxial stretch

As shown in Fig. 12.28, the length of the relative position of material points $\mathbf{y}'$and $\mathbf{y}$ in the deformed state for this loading condition becomes

$$\left|\mathbf{y}' - \mathbf{y}\right| = [1 + \zeta] \left|\mathbf{x}' - \mathbf{x}\right|. \tag{12.53}$$

The mechanical stretch, $\bar{s}$, in the absence of temperature change can be expressed in terms of the relative displacement of the material points i and p, arising from the mechanical loading, as

$$\bar{s} = \zeta. \tag{12.54}$$

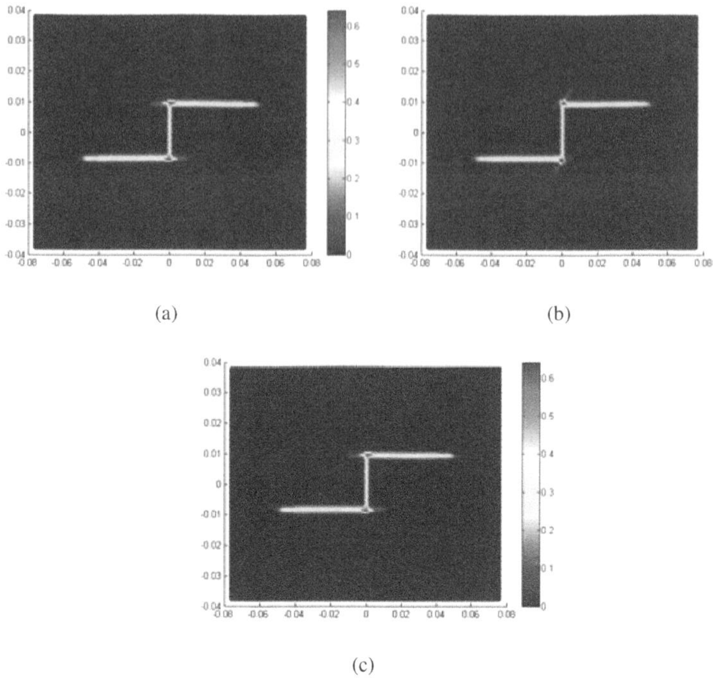

Figure 12.21: Damage plots for a laminate of $[0°/45°/0°]$ with a pre-existing crack: (a) bottom ply, 0°; (b) center ply, 45°; and (c) top ply, 0°.

By using Eq. (12.18), the strain energy density based on the PD theory at a material point in a composite lamina can be evaluated as

$$\begin{aligned} W^{PD} &= \frac{1}{2}\sum_{q=1}^{Q}\frac{c_F\zeta^2\xi_{qi}}{2}V_q + \frac{1}{2}\left(\int_H \frac{c_A(\zeta)^2\xi}{2}\,dH\right) \\ &= \frac{c_F\zeta^2}{4}\left(\sum_{q=1}^{Q}\xi_{qi}V_q\right) + \frac{\pi c_A h\delta^3}{6}\zeta^2. \end{aligned} \tag{12.55}$$

By using Eq. (12.6), the strain energy density of a material point based on classical continuum mechanics can be written as

$$W^{CM} = \frac{1}{2}(Q_{11} + 2Q_{12} + Q_{22})\zeta^2. \tag{12.56}$$

By equating strain energies obtained from PD theory and classical continuum mechanics, Eq. (12.55) and Eq. (12.56), respectively, lead to

$$Q_{22} = 3Q_{12}. \tag{12.57}$$

The expressions for the bond constants c_F and c_A given by Eqs. (12.41) and

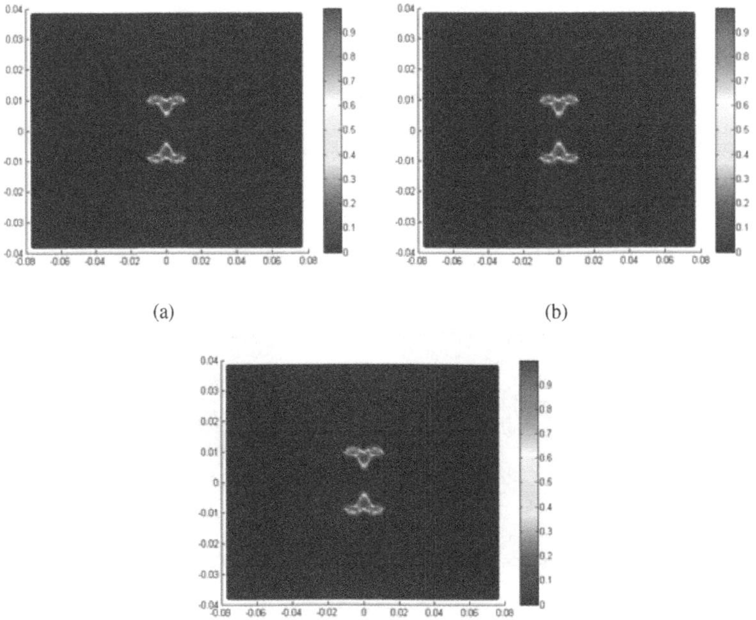

(a) (b)

(c)

Figure 12.22: Shear bond damage plots for a laminate of $[0°/90°/0°]$ with a pre-existing crack: (a) bottom ply, 0°; (b) center ply, 90°; and (c) top ply, 0°.

(12.46), as well as the relations given in Eqs. (12.52) and (12.57), can be rewritten in terms of the engineering constants as

$$c_F = \frac{2E_1\left(E_1 - E_2\right)}{\left(E_1 - \frac{1}{9}E_2\right)\left(\sum_{q=1}^{Q} \xi_{qi} V_q\right)}, \tag{12.58}$$

$$c_A = \frac{8E_1E_2}{\left(E_1 - \frac{1}{9}E_2\right)\pi h \delta^3}, \tag{12.59}$$

$$G_{12} = \frac{\nu_{12}E_2}{1 - \nu_{21}\nu_{12}}, \tag{12.60}$$

and

$$\nu_{12} = \frac{1}{3}. \tag{12.61}$$

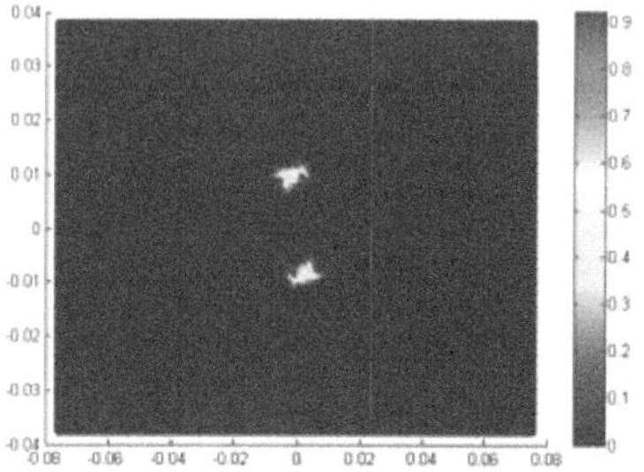

(a)

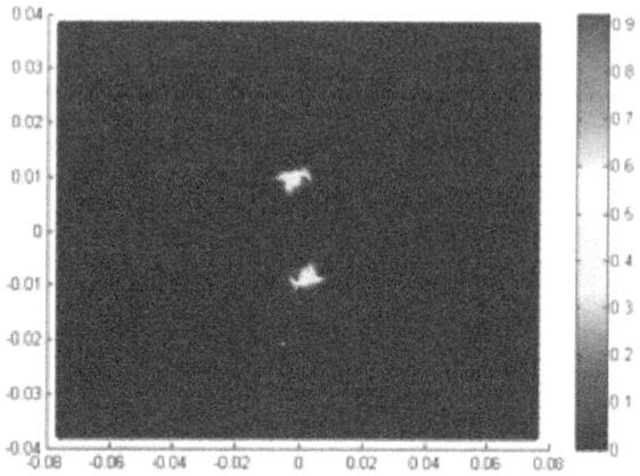

(b)

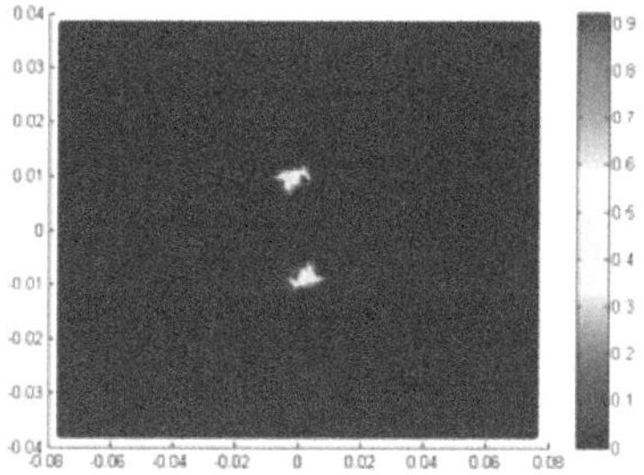

(c)

Figure 12.23: Shear bond damage plots for a laminate of $[0^\circ/45^\circ/0^\circ]$ with a pre-existing crack: (a) bottom ply, 0°; (b) center ply, 45°; and (c) top ply, 0°.

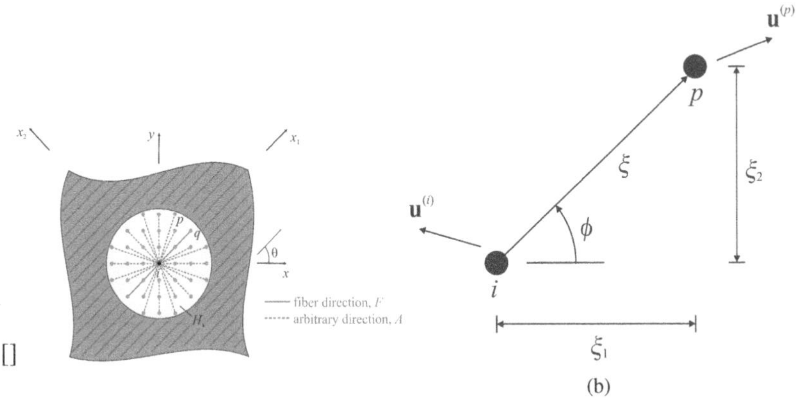

[]

Figure 12.24: (a) PD horizon for a lamina with a fiber orientation of θ and PD bonds between material point i and other material points within its horizon and (b) a PD bond between material points i and p with an orientation of ϕ.

12.7 Appendix B: Surface correction factors for a composite lamina

The surface correction factors for bonds in fiber and arbitrary directions are determined by computing the strain energy density at two distinct material points under uniaxial strain conditions in the x_1 and x_2 directions, i.e., $\varepsilon_{11} \neq 0, \varepsilon_{22} = \gamma_{12} = 0$ and $\varepsilon_{22} \neq 0, \varepsilon_{11} = \gamma_{12} = 0$. The first material point located near an external surface has a truncated horizon, as shown in Fig. 12.29. The second material point is located far away from an external boundary and is completely embedded in a single lamina, as shown in Fig. 12.30.

The strain energy density of a material point at $\mathbf{x}$ is decomposed as

$$W(\mathbf{x}) = W_{(F)}(\mathbf{x}) + W_{(A)}(\mathbf{x}), \tag{12.62}$$

where $W_{(F)}(\mathbf{x})$ and $W_{(A)}(\mathbf{x})$ represent the contribution of bonds in fiber and arbitrary directions, respectively.

Uniaxial strain loading is applied in the x_1- and x_2-directions, respectively, and the resulting displacement field can be expressed at material point $\mathbf{x}$ as

$$\mathbf{u}^T(\mathbf{x}) = \left\{ \begin{array}{cc} \frac{\partial u_1^*}{\partial x_1} x_1 & 0 \end{array} \right\} \tag{12.63}$$

and

$$\mathbf{u}^T(\mathbf{x}) = \left\{ \begin{array}{cc} 0 & \frac{\partial u_2^*}{\partial x_2} x_2 \end{array} \right\}, \tag{12.64}$$

in which $\partial u_\beta^* / \partial x_\beta$ with $\beta = 1, 2$ is the applied constant displacement gradient. The

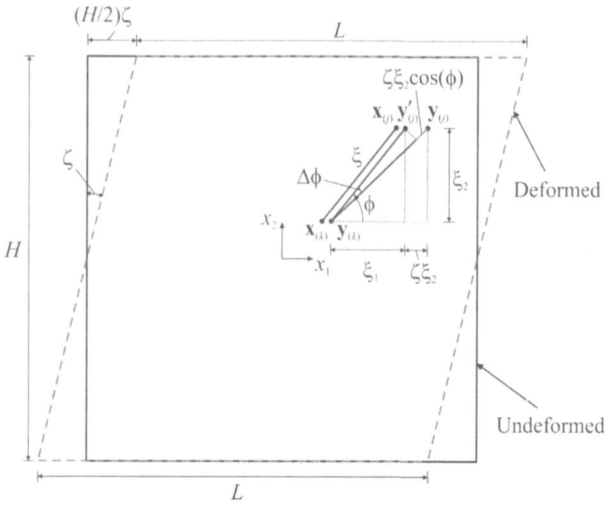

Figure 12.25: Relative displacement between material points i and p.

strain energy density, $W_\beta(\mathbf{x})$, due to the applied displacement gradient is expressed as

$$W_\beta^{PD}(\mathbf{x}) = \int\limits_H w(\mathbf{u}' - \mathbf{u}, \mathbf{x}' - \mathbf{x})\, dH, \tag{12.65}$$

where H represents the horizon of the material point at $\mathbf{x}$ and w represents the strain energy density of the PD bond between material points at $\mathbf{x}$ and $\mathbf{x}'$. The subscripts 1 and 2 denote uniaxial strain loading condition in the x_1- and x_2-directions, respectively.

In accordance with Eq. (12.62), the strain energy density at material point $\mathbf{x}$ due to the applied uniaxial strain loading in the x_1- and x_2-directions can be decomposed as

$$W_1 = W_{(F)1} + W_{(A)1} \tag{12.66}$$

and

$$W_2 = W_{(F)2} + W_{(A)2}. \tag{12.67}$$

With this decomposition, the strain energy density vectors $\mathbf{W}_{(F)}(\mathbf{x})$ and $\mathbf{W}_{(A)}(\mathbf{x})$ can be formed as

$$\mathbf{W}_{(F)}^T(\mathbf{x}) = \{\ W_{(F)1} \quad W_{(F)2}\ \} \tag{12.68}$$

and

$$\mathbf{W}_{(A)}^T(\mathbf{x}) = \{\ W_{(A)1} \quad W_{(A)2}\ \}, \tag{12.69}$$

where $\mathbf{W}_{(F)}(\mathbf{x})$and $\mathbf{W}_{(A)}(\mathbf{x})$ represent the contribution of bonds in fiber and arbitrary directions to the strain energy density of the material point at $\mathbf{x}$, respectively.

For both bonds in fiber and arbitrary directions, the correction factors corresponding to the two loading directions can be defined as the ratio of the strain energy density of the material point embedded far away from an external surface in a lamina,

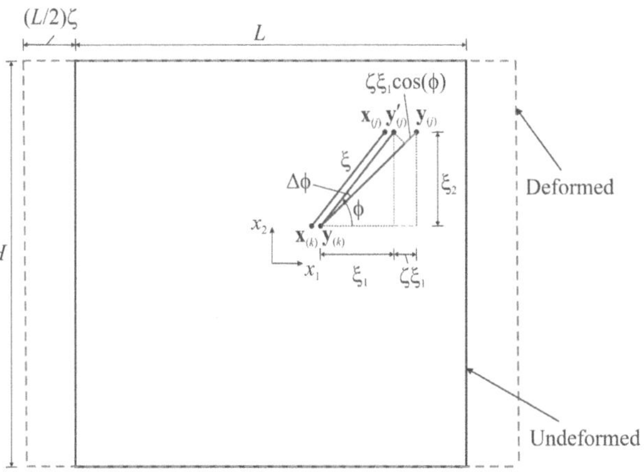

Figure 12.26: Relative displacement between material points i and p.

$W_{(\beta)\alpha}^{CM}$, to that of a material point near an external surface with a truncated horizon, $W_{(\gamma)\beta}^{PD}$, with $\beta = 1,2$ and $\gamma = F, A$. For a material point whose horizon is completely embedded in a single lamina, the strain energy densities for the uniaxial strain loading condition in the x_1- and x_2-directions can be computed by using the classical continuum mechanics as

$$W_1^{CM} = \frac{1}{2} Q_{11}\, \zeta^2 \tag{12.70}$$

and

$$W_2^{CM} = \frac{1}{2} Q_{22}\, \zeta^2, \tag{12.71}$$

where Q_{11} and Q_{22} are the coefficients of the reduced stiffness matrix $\mathbf{Q}$ [11]. The strain energy densities given by Eqs. (12.70) and (12.71) can be decomposed into two parts, which are associated with the deformations of bonds in fiber and arbitrary directions, i.e., $W_{(\beta)\alpha}^{CM}$, with $\beta = 1,2$ and $\gamma = F, A$, as

$$W_1^{CM} = W_{(F)1}^{CM} + W_{(A)1}^{CM} \tag{12.72}$$

and

$$W_2^{CM} = W_{(F)2}^{CM} + W_{(A)2}^{CM}. \tag{12.73}$$

However, the explicit form of this decomposition is not obvious because each lamina is treated as homogenous and orthotropic within the realm of classical continuum mechanics. The decomposition can be obtained by utilizing Eq. (12.14) in conjunction with the bond constant expressions given in Eqs. (12.20) and (12.21). This assumption leads to the following decomposition of strain energies given by Eqs. (12.70) and (12.71) in the form

$$W_{(F)1}^{CM} = \frac{1}{2} (Q_{11} - Q_{22})\, \zeta^2, \tag{12.74}$$

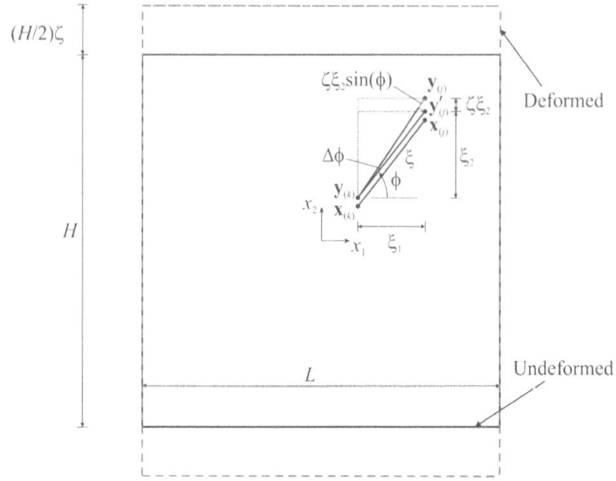

Figure 12.27: Relative displacement between material points i and p.

$$W_{(A)1}^{CM} = \frac{1}{2}Q_{22}\zeta^2, \tag{12.75}$$

$$W_{(F)2}^{CM} = 0, \tag{12.76}$$

and

$$W_{(A)2}^{CM} = \frac{1}{2}Q_{22}\zeta^2. \tag{12.77}$$

With these values, a vector of correction factors for bonds in fiber and arbitrary directions at material point $\mathbf{x}$ can be formed as

$\mathbf{g}_{(\gamma)}(\mathbf{x}) = \{g_{(\gamma)1}, g_{(\gamma)2}\}^T = \left\{W_{(\gamma)1}^{CM}\big/W_{(\gamma)1}^{PD},\ W_{(\gamma)2}^{CM}\big/W_{(\gamma)2}^{PD}\right\}^T$, with $\gamma = F,A$.

These correction factors are only based on loading in the x_1- and x_2-directions. However, they can be used as the principal values of an ellipse in order to approximate the surface correction factor in an arbitrary direction of unit vector, $\mathbf{n}$ (Fig. 12.31).

In the case of a surface correction factor for a PD bond between material points $\mathbf{x}_{(i)}$ and $\mathbf{x}_{(j)}$ under general loading conditions, shown in Fig. 12.32(a), the correction factors in the direction of the relative position vector, $\mathbf{n} = \xi/|\xi| = \{n_1, n_2\}^T$, in the undeformed configuration between these two material points, can be obtained in a similar manner.

A vector of correction factors at material points $\mathbf{x}_{(i)}$ and $\mathbf{x}_{(j)}$ can be formed as

$$\mathbf{g}_{(\gamma)(i)}(\mathbf{x}_{(i)}) = \{g_{(\gamma)1(i)}, g_{(\gamma)2(i)}\}^T = \left\{W_{(\gamma)1}^{CM}\big/W_{(\gamma)1(i)}^{PD},\ W_{(\gamma)2}^{CM}\big/W_{(\gamma)2(i)}^{PD}\right\}^T \tag{12.78}$$

and

$$\mathbf{g}_{(\gamma)(j)}(\mathbf{x}_{(j)}) = \{g_{(\gamma)1(j)}, g_{(\gamma)2(j)}\}^T = \left\{W_{(\gamma)1}^{CM}\big/W_{(\gamma)1(j)}^{PD},\ W_{(\gamma)2}^{CM}\big/W_{(\gamma)2(j)}^{PD}\right\}^T. \tag{12.79}$$

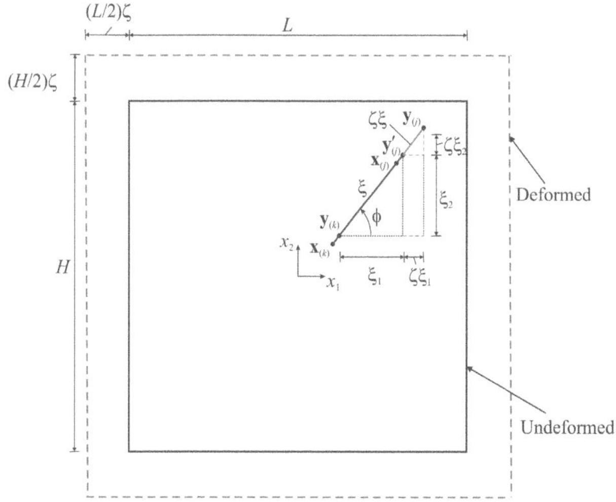

Figure 12.28: Relative displacement between material points i and p.

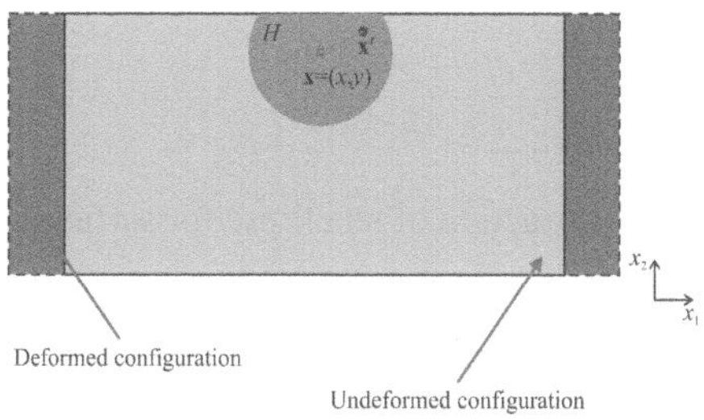

Figure 12.29: Material point **x** with a truncated horizon in a lamina.

These correction factors are, in general, different at material points $\mathbf{x}_{(i)}$ and $\mathbf{x}_{(j)}$. Therefore, the correction factor for a PD bond between material points $\mathbf{x}_{(i)}$ and $\mathbf{x}_{(j)}$ can be obtained by their mean values as

$$\bar{\mathbf{g}}_{(\gamma)(i)(j)} = \left\{\bar{g}_{(\gamma)(i)(j)1}, \bar{g}_{(\gamma)(i)(j)2}\right\}^T = \left(\mathbf{g}_{(\gamma)(i)} + \mathbf{g}_{(\gamma)(j)}\right)\Big/2, \tag{12.80}$$

which can be used as the principal values of an ellipse, as shown in Fig. 12.32. The intersection of the ellipse and a relative position vector of material points $\mathbf{x}_{(i)}$ and $\mathbf{x}_{(j)}$, $\mathbf{n}$, provides the correction factors as

$$G_{(\gamma)(i)(j)} = \left(\left[n_1/\bar{g}_{(\gamma)(i)(j)1}\right]^2 + \left[n_2/\bar{g}_{(\gamma)(i)(j)2}\right]^2\right)^{-1/2}. \tag{12.81}$$

After considering the surface effects, the discrete form of the equations of motion given in Eq. (12.1) can be expressed as

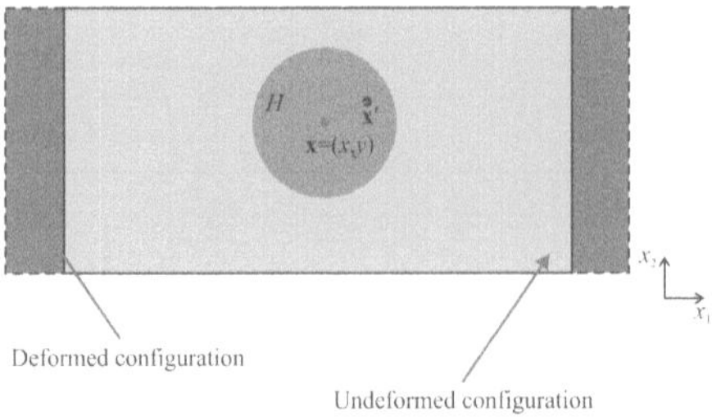

Figure 12.30: Material point **x** far away from external surfaces of a lamina.

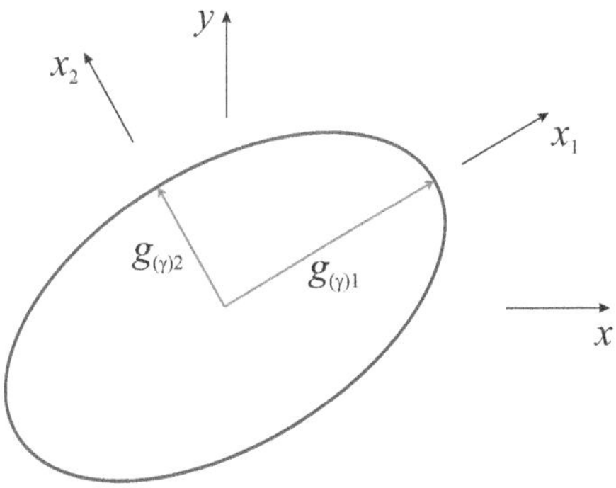

Figure 12.31: Construction of an ellipse for surface correction factors.

$$\rho\left(\mathbf{x}_{(i)}\right)\ddot{\mathbf{u}}\left(\mathbf{x}_{(i)},t\right)= \tag{12.82}$$
$$\sum_{j=1}^{M}\left(\begin{array}{l} a_{(i)(j)}G_{(F)(i)(j)}\mathbf{f}\left(\mathbf{u}\left(\mathbf{x}_{(j)},t\right)-\mathbf{u}\left(\mathbf{x}_{(i)},t\right),\mathbf{x}_{(j)}-\mathbf{x}_{(i)}\right)+ \\ G_{(A)(i)(j)}\mathbf{f}\left(\mathbf{u}\left(\mathbf{x}_{(j)},t\right)-\mathbf{u}\left(\mathbf{x}_{(i)},t\right),\mathbf{x}_{(j)}-\mathbf{x}_{(i)}\right)\end{array}\right)V_{(j)}+\mathbf{b}\left(\mathbf{x}_{(i)},t\right),$$

where the coefficient $a_{(i)(j)}$ takes a value of either 1 or 0 if the interaction between material points $\mathbf{x}_{(i)}$ and $\mathbf{x}_{(j)}$ is in the fiber direction or not, respectively.

12.8 Appendix C: PD interlayer and shear bond constants of a laminate

The interlayer bond constant, c_N, and the shear bond constant, c_S, shown in Fig. 12.33, can be expressed in terms of engineering constants based on the transverse normal and shear deformation response of the isotropic matrix material by equating the total strain energy density of interlayer and shear bonds calculated from peridynamic theory and classical continuum mechanics. The strain energy density of the

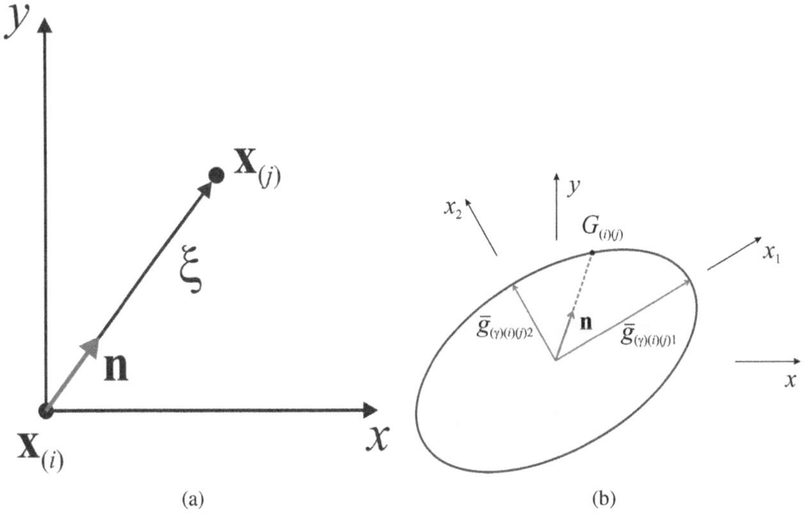

Figure 12.32: (a) PD bond between material points at $\mathbf{x}_{(i)}$ and $\mathbf{x}_{(j)}$and (b) the ellipse for the surface correction factor.

interlayer bonds associated with the material point *a* can be computed by summing up the strain energy density of the two interlayer bonds between material points *d* and *e* and the material point *a* (Fig. 12.33). The total strain energy density due to the interlayer and shear bonds can be computed as

$$W^{PD} = \frac{1}{2} \sum_{j=d,e} \frac{c_N s_{ja}^2 \xi_{ja}}{2} V_j + \frac{1}{2} \int_H \frac{c_S \varphi^2}{2} dH. \tag{12.83}$$

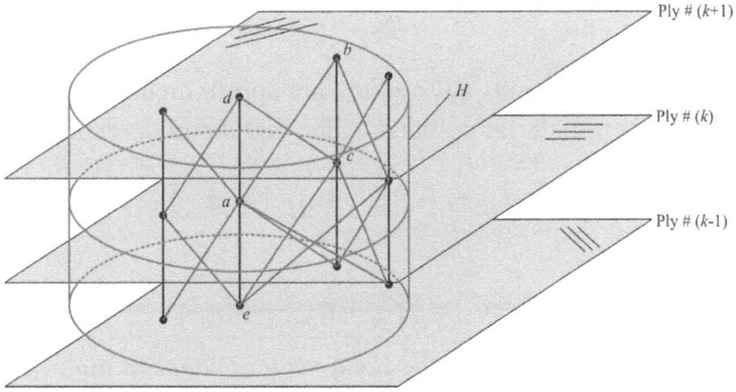

Figure 12.33: Interlayer and shear bonds between neighboring plies (only some of the interactions are depicted explicitly for clarity).

The expression for the shear angle in Eq. (12.83) is obtained by determining the average shear angle inside the quadrilateral formed by material points *a*, *b*, *c*, and *d*, as shown in Fig. 12.34. Averaging is achieved by computing the shear angles along the lines between material points *a* and *d*, and *b* and *c*, which are defined as ϕ_{da} and ϕ_{bc}, respectively.

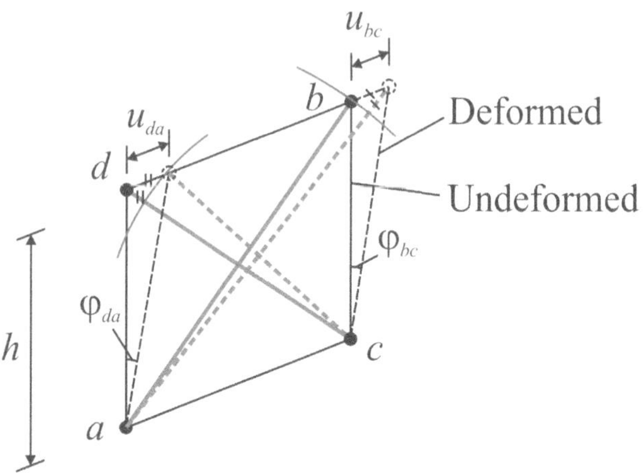

Figure 12.34: Shear bonds between material points *b* and *a* and between material points *d* and *c* in both undeformed and deformed configurations.

These shear angles are obtained from the ratio of the displacements u_{da} and u_{bc} of material points *d* and *b* with respect to *a* and *c*, respectively, to the ply thickness, *h*, as

$$\varphi_{da} = \frac{u_{da}}{h} \tag{12.84}$$

and

$$\varphi_{bc} = \frac{u_{bc}}{h}. \tag{12.85}$$

The relative displacements u_{da} and u_{bc} are approximated as the change in length of the bonds between material points *d* and *c*, and *b* and *a*, respectively,

$$u_{da} = -\left(|\xi_{dc} + \eta_{dc}| - |\xi_{dc}|\right) \tag{12.86}$$

and

$$u_{bc} = |\xi_{ba} + \eta_{ba}| - |\xi_{ba}|, \tag{12.87}$$

where ξ_{ba} and ξ_{dc} correspond to the bond vectors between material points *b* and *a*, and between material points *d* and *c*, respectively. Similarly, the vectors η_{ba} and η_{dc} are the relative displacement vectors between material points *b* and *a*, and between material points *d* and *c*, respectively.

Note that the minus sign in Eq. (12.86) arises due to the contraction of the shear bond between material points d and c, whereas the bond between material points b and a extends. The average value of the shear strains ϕ_{da} and ϕ_{bc} results in

$$\varphi = \frac{\varphi_{da} + \varphi_{bc}}{2} = \frac{(|\xi_{ba} + \eta_{ba}| - |\xi_{ba}|) - (|\xi_{dc} + \eta_{dc}| - |\xi_{dc}|)}{2h}. \tag{12.88}$$

In order to obtain the interlayer bond constant, the laminate is subjected to an isotropic expansion loading of $s = \zeta$, as shown in Fig. 12.35(a).

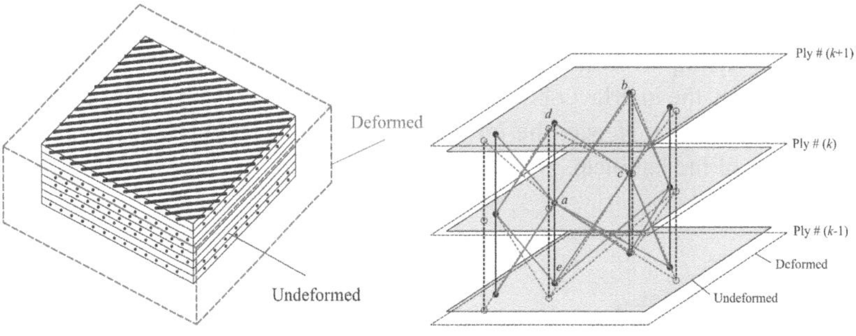

Figure 12.35: (a) A composite laminate subjected to isotropic expansion loading, and (b) deformation of interlayer and shear bonds between neighboring plies (only some of the interactions are depicted explicitly for clarity).

For a material point, a, located in the kth ply of the laminate (Fig. 12.35(b)), the contributions of the interlayer and shear bonds to the strain energy density of the material point due to isotropic expansion loading can be calculated by using Eq. (12.83),

$$W^{PD} = \frac{1}{2} \sum_{j=d,e} \frac{c_N s_{ja}^2 \xi_{ja}}{2} V_j. \tag{12.89}$$

Note that the shear strain, φ, defined in Eq. (12.88) has a zero value for this loading condition because the relative displacements u_{da} and u_{bc} given in Eqs. (12.86) and (12.87) are equal in magnitude with opposite signs. Therefore, shear bonds do not have any contribution to the strain energy density for this loading condition. Both bond lengths ξ_{da} and ξ_{ea} are equivalent to the ply thickness, h. Therefore, for this loading condition, Eq. (12.89) can be evaluated as

$$W^{PD} = \frac{c_N \zeta^2 h \bar{V}}{2}, \tag{12.90}$$

where $\bar{V}$ is equal to the volume of material points d and e, i.e., $\bar{V} = V_d = V_e$.

The corresponding strain energy density of the material point for the same loading condition can be calculated by using classical continuum mechanics as

$$W^{CM} = \frac{1}{2} E_m \zeta^2, \tag{12.91}$$

with E_m representing the elastic modulus of matrix material. Equating strain energy densities from Eqs. (12.90) and (12.91) yields the expression of the interlayer bond constant, c_N, as

$$c_N = \frac{E_m}{h\bar{V}}. \tag{12.92}$$

The shear bond constant, c_S, can be evaluated similarly. In this case, the laminate is subjected to a simple shear loading of $\gamma = \zeta$, as shown in Fig. 12.36(a). For this loading condition, the interlayer bonds do not extend (Fig. 12.36(b)). Therefore, their stretch values are zero. Hence, the interlayer bonds do not contribute to the strain energy density of the laminate.

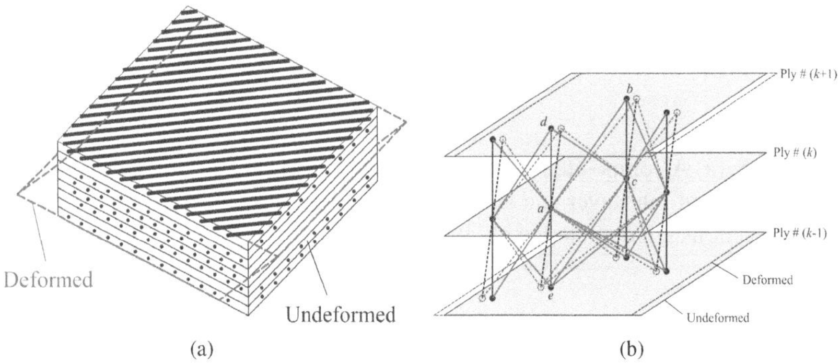

Figure 12.36: (a) A composite laminate subjected to simple shear loading, and (b) deformation of interlayer and shear bonds between neighboring plies (only some of the interactions are depicted explicitly for clarity).

As shown in Fig. 12.37, the original and deformed lengths of the shear bond between material points b and a can be expressed as

$$|\xi_{ba}| = \sqrt{\ell^2 + h^2} \tag{12.93}$$

and

$$|\xi_{ba} + \eta_{ba}| = \sqrt{\bar{\ell}^2 + h^2}, \tag{12.94}$$

where h is the ply thickness.

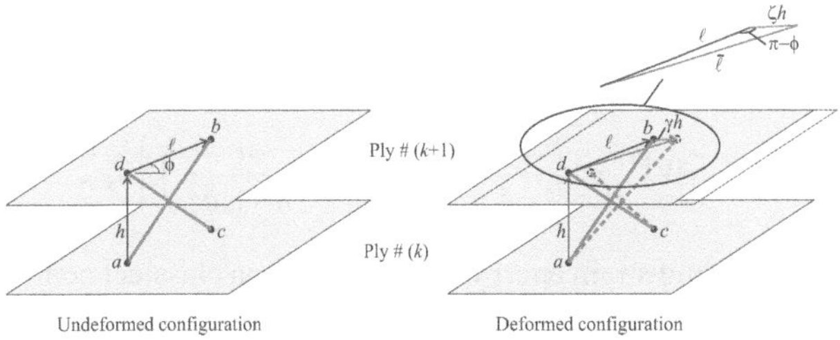

Figure 12.37: Shear bonds between material points b and a and between material points d and c in both undeformed and deformed configurations.

For the triangle depicted in Fig. 12.37, by utilizing the law of cosines, the length of the radial component of the deformed bond vector, $\bar{\ell}$, can be written in terms of the length of the radial component of the original bond vector, ℓ, and magnitude of displacement vector, ζh, as

$$\bar{\ell}^2 = \ell^2 + (\zeta h)^2 - 2\ell\zeta h\cos(\pi - \theta). \tag{12.95}$$

After substituting Eq. (12.95) into Eq. (12.94), the deformed bond length can be rewritten as

$$|\xi_{ba} + \eta_{ba}| = \sqrt{\ell^2 + h^2 + 2\ell\zeta h\cos(\theta)}. \tag{12.96}$$

In deriving Eq. (12.96), the $(\zeta h)^2$ term is neglected with respect to h^2 because ζ is much less than unity. The square root term on the right- hand side of Eq. (12.96) can be further simplified by using the square root approximation

$$\sqrt{N^2 + d} = N + \frac{d}{2N}, \tag{12.97}$$

where $d << N$. Therefore, the deformed bond length expression given in Eq. (12.96) can be rewritten as

$$|\xi_{ba} + \eta_{ba}| = \sqrt{\ell^2 + h^2} + \frac{\ell\zeta h\cos(\theta)}{\sqrt{\ell^2 + h^2}}. \tag{12.98}$$

The original and deformed bond lengths between material points d and c can be computed similarly as

$$|\xi_{dc}| = \sqrt{\ell^2 + h^2} \tag{12.99}$$

and

$$|\xi_{dc} + \eta_{dc}| = \sqrt{\ell^2 + h^2} - \frac{\ell\zeta h\cos(\theta)}{\sqrt{\ell^2 + h^2}}. \tag{12.100}$$

Therefore, the shear angle for this loading condition can be computed by using Eq. (12.88) as

$$\phi = \frac{\ell\zeta\cos(\theta)}{\sqrt{\ell^2 + h^2}}. \tag{12.101}$$

After substituting the shear angle expression given in Eq. (12.101) into the strain energy density expression given in Eq. (12.83) and performing the integration,

$$W^{PD} = \left\{ \frac{\pi c_S h}{4} \left(\delta^2 + h^2 \ln \frac{h^2}{h^2 + \delta^2} \right) \right\} \zeta^2. \tag{12.102}$$

The corresponding strain energy density based on classical continuum mechanics can be computed as

$$W^{CM} = \frac{1}{2} G_m \zeta^2. \tag{12.103}$$

Equating the strain energy densities calculated from PD theory and classical continuum mechanics, i.e., Eqs. (12.102) and (12.103), leads to the explicit form of the shear bond constant in terms of the shear modulus of the matrix material, G_m,

$$c_S = \frac{2 G_m}{\pi h} \frac{1}{\left(\delta^2 + h^2 \ln \left(\frac{h^2}{\delta^2 + h^2} \right) \right)}. \tag{12.104}$$

12.9 Appendix D: Critical Stretch Values for Bond Constants

The critical stretch for the interlayer bond, s_{Nc}, can be computed by equating the energy required to break an interlayer bond between material point $\mathbf{x}$ located at the kth ply and the material point $\mathbf{x}'$ located at the $(k+1)$th ply (see Fig. 12.38) to the mode-I critical energy release rate of the material, G_{IC}, as

$$h \left(\frac{c_N s_{Nc}^2 h}{2} \right) \bar{V} = G_{IC}, \tag{12.105}$$

where h and $\bar{V}$represent the thickness of the ply and the volume of the material point $\mathbf{x}'$, respectively.

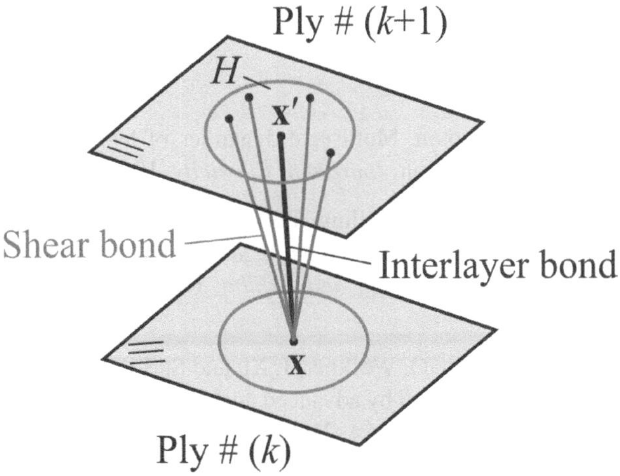

Figure 12.38: Interlayer and shear bonds between material point **x** and other material points located at the $(k+1)$th ply.

By using the relation given by Eq. (12.105) in conjunction with the bond constant expression given by Eq. (12.27), the critical stretch expression for the interlayer bond can be expressed as

$$s_{Nc} = \sqrt{\frac{2G_{IC}}{hE_m}}. \tag{12.106}$$

As opposed to interlayer bonds, multiple shear bonds exist between the material point **x** and other material points in the $(k+1)$th ply, as shown in Fig. 12.38. The failure of these shear bonds corresponds to a mode-II type of failure. Therefore, the energy required to break all of these shear bonds can be equated to the mode-II critical energy release rate of the material, G_{IIC}, as

$$h\int_H \frac{c_S\,\varphi_{Sc}^2}{2}\,dH = G_{IIC}, \tag{12.107}$$

where φ_{Sc} is the critical shear angle. This equation can be rewritten after splitting the domain of integration as

$$h\left(h\frac{c_S\,\varphi_{Sc}^2}{2}\int_0^\delta \ell\,d\ell\left(\int_0^{2\pi} d\varphi\right)\right) = G_{IIC}. \tag{12.108}$$

Performing the integrations results in the critical shear angle expression as

$$\varphi_{Sc} = \sqrt{\frac{G_{IIC}}{hG_m}}. \tag{12.109}$$

References

[1] B. Alali and R. Lipton. Multiscale dynamics of heterogeneous media in the peridynamic formulation. *Journal of Elasticity*, 106(1): 71–103, 2012.

[2] E. Askari, J. Xu and S. Silling. Peridynamic analysis of damage and failure in composites. In *Proceedings of the 44th AIAA/ASME/ASCE/AHS/ASC Aerospace Sciences Meeting and Exhibit*. Reno, NV, USA, AIAA 2006-88, 2006.

[3] A. Askari, K. Nelson, O. Weckner, J. Xu and S. Silling. Hail impact characteristics of a hybrid material by advanced analysis techniques and testing. *Journal of Aerospace Engineering*, 24: 210–217, 2011.

[4] P.B. Bogert, A. Satyanarayana and P.B. Chuncu. Comparison of damage path predictions for composite laminates by explicit and standard finite element analysis tools. In *Proceedings of the* 47th *AIAA/ASME/ASCE/AHS/ASC Structures, Structural Dynamics, and Materials Conference*. Newport, RI, USA, AIAA 2006-1750, 2006.

[5] K.W. Colavito, B. Kilic, E. Celik, E. Madenci, E. Askari and S. Silling. Effect of void content on stiffness and strength of composites by a peridynamic analysis and static indentation test. In *Proceedings of the* 48th *AIAA/ASME/ASCE/AHS/ASC Structures, Structural Dynamics, and Materials Conference*. Honolulu, HI, USA, AIAA 1007-2251, 2007.

[6] K.W. Colavito, B. Kilic, E. Celik, E. Madenci, E. Askari and S. Silling. Effect of nano particles on stiffness and impact strength of composites. In *Proceedings of the* 48th *AIAA/ASME/ASCE/AHS/ASC Structures, Structural Dynamics, and Materials Conference*, Honolulu, HI, USA, AIAA 2007-2021, 2007.

[7] W. Gerstle, N. Sau and S. Silling. Peridynamic modeling of plain and reinforced concrete structures. In *Proceedings of the* 8th *International Conference on Structural Mechanics in Reactor Technology (SMiRT 18)*. Beijing, China, SMiRT18-B01-02, 2005.

[8] B.G. Green, M.R. Wisnom and S.R. Hallet. An experimental investigation into the tensile strength scaling of notched composites. *Composites Part A: Applied Science and Manufacturing*, 38(3): 867–878, 2007.

[9] S.R. Hallett and M.R. Wisnom. Experimental investigation of progressive damage and the effect of layup in notched tensile tests. *Journal of Composite Materials*, 40(2): 119–141, 2006.

[10] W. Hu, Y.D. Ha and F. Bobaru. Peridynamic model for dynamic fracture in unidirectional fiber-reinforced composites. *Computer Methods in Applied Mechanics and Engineering*, 217–220: 247–261, 2012.

[11] A.K. Kaw. *Mechanics of Composite Materials*. 2nd ed., CRC Press, Boca Raton, FL, USA, 2006.

[12] B. Kilic. Peridynamic theory for progressive failure prediction in homogeneous and heterogeneous materials. Ph.D Dissertation, Department of Aerospace and Mechanical Engineering, University of Arizona, Tucson, AZ, USA, 2008.

[13] B. Kilic, A. Agwai and E. Madenci. Peridynamic theory for progressive damage prediction in centre-cracked composite laminates. *Composite Structures*, 90(2): 141–151, 2009.

[14] E. Madenci and E. Oterkus. *Peridynamic Theory and Its Applications*. Springer, New York, USA, 2014.

[15] E. Oterkus, A. Barut and E. Madenci. Damage growth prediction from loaded composite fastener holes by using peridynamic theory. In *Proceedings of the 51st AIAA/ASME/ASCE/AHS/ASC Structures, Structural Dynamics, and Materials Conference*. Orlando, FL, USA, AIAA 2010-3026, 2010.

[16] E. Oterkus and E. Madenci. Peridynamic analysis of fiber-reinforced composite materials. *Journal of Mechanics of Materials and Structures*, 7(1): 45–84, 2012.

[17] E. Oterkus, E. Madenci, O. Weckner, S. Silling, P. Bogert and A. Tessler. Combined finite element and peridynamic analyses for predicting failure in a stiffened composite curved panel with a central slot. *Composite Structures*, 94(3): 839–850, 2012.

[18] S.A. Silling. Reformulation of elasticity theory for discontinuities and long-range forces. *Journal of the Mechanics and Physics of Solids*, 48(1): 175–209, 2000.

[19] E.M. Wu. Fracture mechanics of anisotropic plates. In S. W. Tsai, J. C. Halphin, N. J. Pagano, editors. *Composite materials workshop*, pages 20-43. Technomic Publishing Co., Inc., Lanchaster, PA, USA, 1968.

[20] J. Xu, A. Askari, O. Weckner, H. Razi and S. Silling. Damage and failure analysis of composite laminates under biaxial loads. In *Proceedings of the 48th AIAA/ASME/ASCE/AHS/ASC Structures, Structural Dynamics, and Materials Conference*. Honolulu, HI, USA, AIAA 2007-2315, 2007.

Chapter 13

Peridynamic Modeling of Impact and Fragmentation

Florin Bobaru

University of Nebraska-Lincoln

Zhanping Xu

University of Nebraska-Lincoln

Yenan Wang

University of Nebraska-Lincoln

CONTENTS

13.1 Introduction 380
13.2 Convergence studies and damage models that influence the damage behavior 382
13.2.1 Damage-dependent critical bond strain 382
13.2.2 Critical bond strain dependence on compressive strains along other directions 383
13.2.3 Surface effect in impact problems 383
13.2.4 Convergence study for impact on a glass plate 384
13.3 Impact on a multilayered glass system 388
13.3.1 Model description 389
13.3.2 A comparison between FEM and peridynamics for the elastic response of a multilayered system to impact 391
13.4 Computational results for damage progression in the seven-layer glass system 393
13.4.1 Damage evolution for the cross section 394
13.4.2 Damage evolution in the first layer 396
13.4.3 Damage evolution in the second layer 398
13.4.4 Damage evolution in the fourth layer 399
13.4.5 Damage evolution in the seventh layer 400
13.5 Conclusions 402
Acknowledgments 403

References 403

13.1 Introduction

In this chapter we consider impact problems in which the material response (for the range of impact speeds analyzed, and assuming isothermal conditions) of the target can be considered brittle (i.e. plastic dissipation is negligible). Many material systems, like inorganic glasses, ceramic materials, concrete, and certain metals or metallic alloys, do behave in a brittle way when loaded dynamically. Common dynamic loadings are projectile impact or plate impact.

The question of how damage progresses in time and how materials fail when impacted by a projectile is an important one. Understanding this progression can help with improving design of material systems and structures like safety windows, protective shields and armor, and even screens of mobile devices like cell phones. These types of problems have been difficult to model in the past, and to our knowledge, no model can accurately describe the entire evolution of damage, from the initiation stages to the post-fracture and fragmentation phase. More often than not, to even obtain some partial solutions to the problem of predicting brittle damage from impact, one has to "insert" parts of solution into the model, by ways of, for example, criteria for failure, etc., that might work for crack initiation but might not work well when the process leads to fragmentation, etc. The aim of the peridynamic model is to rely on a simple, basic damage mechanism and be able to encompass a variety of failure modes from fine cracks, to fragmentation, and anything in-between.

This chapter covers a discussion on several choices that can be made when modeling brittle damage with peridynamics and presents results on δ-convergence and m-convergence (see [6]) for damage and fragmentation patterns seen when a small projectile hits a glass plate. The remainder of the chapter is based on results reported in [3], which provide details of peridynamic simulations of impact on a multilayered glass system (see also [13]). For a full explanation of how and why cracks initiate and propagate in the complex patterns observed in impact on brittle targets, the reader should consult [22].

Glass laminates have become important in recent years and several experimental studies have been dedicated to the behavior of these structures under dynamic loading (see [1, 15, 10]). Experiments for high-velocity impact on multilayered glass structures demonstrate very complex damage patterns with significant differences between the types of fracture and characteristics of damage in each layer. A seven-layer glass laminate system has been used in the experiments in [1] with the impact projectile hitting at speeds above 1 km/s. While some common features of damage are observed between the glass layers, especially for layers 3 to 6, distinct damage morphologies are seen in the first layer, the second, and the last layer (see [1]).

In itself a complex material, glass exhibits a variety of damage patterns under impact, including: crack branching (bifurcation of a crack), crack-path instability and crack curving, successive branching events, circumferential or ripple cracking,

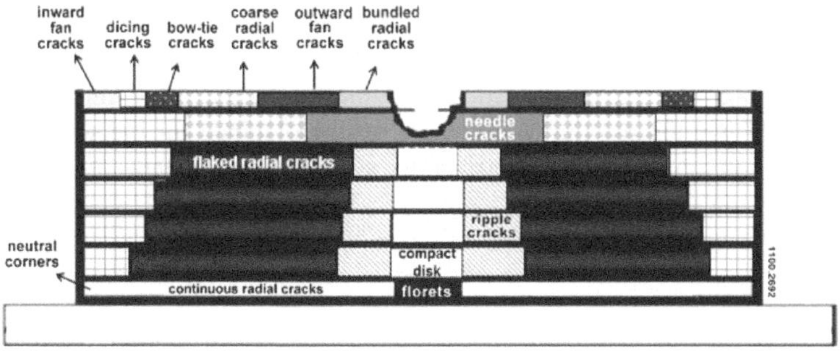

Figure 13.1: A schematic of damage observed after impact on a seven-layer glass system (from [1]).

micro-cracking, etc. The work published in [1] is a recent attempt of describing in detail the impact damage from a high-velocity projectile onto a glass laminate with the goal of serving as a guide and a calibration for numerical models for penetration-resistance of glass. In this chapter we use a peridynamic model for a simplified version of the physical system used in [1], while maintaining the true physical scales (dimensions) from the experiment. While we employ the simplest peridynamic constitutive model for a generic brittle material and the boundary conditions imposed in the model are slightly different from the conditions in the experiments, the results show that, at least to first order, the peridynamic model captures most of the damage morphologies observed in the experiments. Future enhancements for more accurate representation of the boundary conditions as well modeling the effect of presence of binding polyurethane thin layers between the glass sheets (not included in the simplified model discussed here but present in the experimental setting) could lead to quantitative predictive simulations of penetration resistance of multi-layer glass laminates using peridynamics.

The peridynamic model used in what follows is the same as the one described in Chapter 10, with the following exceptions:

- here we use the 3D model;
- surface corrections (for the micromodulus and critical bond strain values) are applied for nodes near the boundaries of the intact specimens;
- a damage-dependent critical bond strain (like the one used in [12]) is used to adjust this value for nodes that have incurred a certain level of damage;
- a correction of the critical bond strain for bonds at nodes which, for example, may be in tension in one direction and highly compressed in the other direction (as suggested in [17]).

In the next section we discuss the various modeling options and perform convergence studies in terms of damage patterns observed when a projectile hits a single glass plate. In section 13.3 we describe a multi-layer glass system whose impact damage and fragmentation behavior has been described in detail in [1]. In the same section we show that the peridynamic model used matches well FEM-based results for some deformation characteristics in the elastic regime of the impact problem. In section 13.4 we discuss the peridynamic results obtained for the seven-layer glass system and compare the damage patterns with the experimental observations from [1].

13.2 Convergence studies and damage models that influence the damage behavior

Bond-based peridynamics has been widely used to simulate dynamic fracture in brittle materials. However, special care has to be taken in order to correctly capture the damage and fragmentation in a brittle target from impact. The original constant critical bond strain (or relative elongation) model [17] treats the bond strength to be the same everywhere inside the material. Because of the surface effect in peridynamics (see, e.g., discussions in [2], [12]), using this model leads to the formation of a larger number of fragments than observed from impact problems on brittle targets. See, for example, the results in Fig. 13.4 below and compare with experimental results from [13]). Here, we demonstrate that the correct damage and fragmentation patterns can be achieved by a few simple corrections applied: one to correct for the surface effect, one to adjust the critical bond strain for bond attached to nodes that have accumulated a certain level of damage, and one to consider an increase in the critical bond strain for bonds attached to nodes that also have compressed bonds. The last two types of corrections are motivated here by what could be considered reasonable assumptions, rather than by precise derivations. This is not to say that such physics-based derivations for these corrections cannot be introduced.

13.2.1 Damage-dependent critical bond strain

In prototype microelastic brittle (PMB) material (e.g. glass), the critical strain for bond failure in 3D is:

$$s_0 = \sqrt{\frac{5G_0}{9\kappa\delta}} \tag{13.1}$$

where G_0 is the fracture energy required for the complete separation of the body into two halves, κ stands for the bulk modulus, and δ defines the horizon size.

If the constant critical stretch s_0 is applied over the entire domain, effectively weaker regions are created near the location of any damage that gets generated, because the integration volume used to compute s_0 by matching the material fracture energy is smaller for regions where damage is present, compared to a location in the bulk of the undamaged body. Especially in impact problems in which many cracks can pack close to each other, such weaker regions lead to over-fragmentation. In bond-based peridynamics, these weaker bonds can be "strengthened" by allowing s_0 to increase for bonds connected to nodes that have suffered certain levels of damage. We adopt here the simple formula from the EMU user manual (also used in [12]) which modifies s_0 into $\widehat{s_0}$ when $D = \max(D_1, D_2)$, the maximum damage index between the two nodes connected by the current bond, is larger than some pre-defined value d_1:

$$\widehat{s_0} = s_0 \times \min\left(d_3, 1 + d_2 \frac{D - d_1}{1 - D}\right), \quad \text{where} \quad D = \max(D_1, D_2). \tag{13.2}$$

The (d_1, d_2, d_3)-tuple is set to $(0.35, 1, 2)$ whenever damage-dependence of critical bond strain is turned on in this study. With the selection $(0, 0, 1)$, the above formula reduces to the damage-independent critical bond strain model.

13.2.2 Critical bond strain dependence on compressive strains along other directions

In the model discussed here, bonds can only fail in tension. One can consider bond failure due to compressive strains as well, but this is not pursued here. It is not difficult to see that, in many types of deformations, at a node there will be bonds in tension and bonds in compression. In such cases, it is natural to assume that the critical bond strain for the bonds in tension depends on how much compression there is in the bonds connected to the same node, and increase that value to account for the amount of compression. With such a dependency, better agreement with experimental results can be seen for fracture in glass, for example (see below for some evidence of this). The dependency proposed in [17] and used in EMU modifies s_0 to depend on $s_{\min}$, the minimum non-positive stretch among all bonds (i) connected to the two nodes linked by the current bond:

$$\widetilde{s_0(t)} = s_0 - \alpha s_{\min}(t), \quad \text{where} \quad s_{\min}(t) = \min_i\{0, s_i\}, \tag{13.3}$$

where α is a constant typically chosen as 1/4, s_i is the bond strain for bonds with the index i going over all of the bonds connected to one of the two nodes linked by the current bond. Thus, the critical stretch s_0 on an extended bond (recall that only extended bonds are allowed to break) increases if any nearby compressed bonds exist.

13.2.3 Surface effect in impact problems

The peridynamic surface effect has been discussed in many works before (see, e.g., [2], [14]) and in many applications it can be ignored as long as the horizon size is sufficiently small. However, in impact problems, correcting it is essential because many cracks are produced by surface waves interactions with other waves (see, e.g., [13]). Initiation of cracks due to impact is also affected strongly by the surface effect.

This is why, as we shall show below, it is important to use surface correction methods for modeling of brittle failure induced by impact. The correction has to be performed on both the bond micromodulus and the critical bond strain. A review of various methods that can be used for surface correction is presented in [14]. Out of those reviewed there, the force normalization method is used here. For example, the critical bond strain computed for the bulk gets multiplied by a correction or normalization coefficient f_{norm} as follows:

$$s_0 = s_0 \times f_{\text{norm}}, \quad \text{where} \quad f_{\text{norm}} = \frac{1}{2}(f_{\text{norm},1} + f_{\text{norm},2}), \tag{13.4}$$

where f_{norm} for a bond is taken as the average of the two normalization factors of the two nodes forming the bond. For more details, please consult [14]. Here, we compute the force normalization factor for a node in the same way as is done in EMU, as the ratio of strain energy density of a point in the bulk w_{bulk} to the strain energy density $w_{\text{grid},k}$ at the current grid node k under some test loads:

$$f_{\text{norm},k} = \frac{w_{\text{bulk}}}{w_{\text{grid},k}}. \tag{13.5}$$

13.2.4 Convergence study for impact on a glass plate

In a peridynamic simulation of brittle fracture, all three correction factors discussed above are applied, but in order to understand the actual effect they introduce compared to the case when none is used, we perform a series of tests. We start with convergence studies damage and fragmentation produced by impact on a single thin glass plate. The problem setup is as follows: we consider the impact of a rigid spherical projectile (440C stainless steel) onto the center of a thin soda-lime glass plate. The projectile has mass 0.0865 g, diameter 0.278 cm, and its impact velocity is 100 m/s normal to the plate surface. The target soda-lime glass plate has dimension of 3 cm × 3 cm × 0.3 cm, density 2440 kg/m^3, Young's modulus E = 72 GPa, Poisson ratio 0.22 (modeled in our bond-based peridynamic simulations as 0.25), and fracture energy 8.0 J/m^2. Free boundary conditions are used in order to avoid any unnecessary boundary effects on the computed fracture and fragmentation patterns. All simulations in this section were performed on an Intel Xeon E5-2670 2.60GHz CPU core + a Tesla K20m GPU using an in-house developed GPU code.

We first perform a convergence study in terms of the size of the nonlocal region (the horizon) (δ-convergence) and in terms of the grid density relative to the horizon size (m-convergence), when all corrections are applied. These computational tests are performed in order to provide the appropriate size of the horizon, knowing that the length-scale introduced by the horizon leads to wave dispersion and a large horizon size induces unrealistic dispersion of waves which can then influence the initiation and propagation of cracks in the brittle target. Moreover, these studies allow us to determine what is an appropriate grid density (relative to the horizon size) to capture fracture and fragmentation patterns with sufficient accuracy.

For a horizon size of 1 mm (recall that the sample thickness is 3 mm), Fig. 13.2

shows the damage map for the top surface of the glass plate obtained with finer and finer discretizations: with $m = 3$ (82810 nodes), $m = 4$ (190333 nodes), and $m = 5$ (364816 nodes). The same colorbar is used for all damage map plots (see the colorbar in Fig. 13.4). Between the results in Fig. 13.2, similar characteristics of the damage created by impact are observed: radial cracks, circumferential cracks, cracks parallel to the sample boundaries, and significant fragmentation near the impact location. These types of cracks are observed in experiments (see, e.g., [13]). The differences between these results are rather subtle, one of them being that the crack paths are more clearly defined and damage is more localized when m is larger compared to when m is smaller. From this point of view, larger m is slightly better; however, the computational effort increases substantially with m, as the scaling is $O(m^3)$. Because of these reasons, we adopt $m = 4$ for the remaining computations in this chapter, as it appears to strike a good balance between accuracy and efficiency.

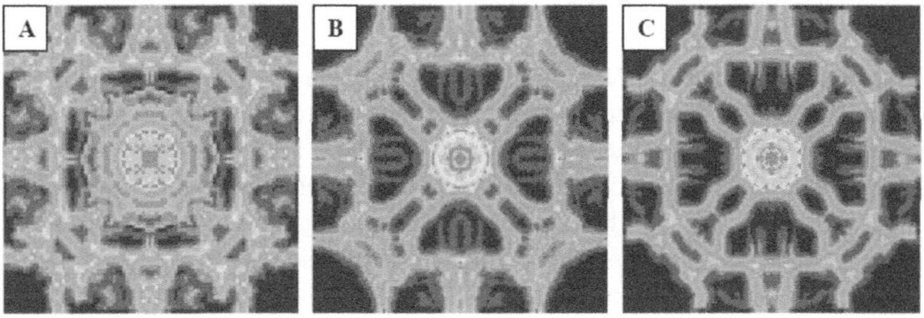

Figure 13.2: Damage maps on the impact surface at 15 μs from the time of impact. Horizon size is fixed, δ=1 mm (m-convergence study) and m=3 in A, m=4 in B, and m=5 in C.

With a fixed horizon factor $m = 4$, we perform a δ-convergence study by observing the damage patterns produced by the use of different horizon sizes. Since m is fixed, this means that as δ decreases, the number of nodes used in the discretization increases. Figure 13.3 illustrates the top surface damage maps at 15 μs for $m = 4$ and with $\delta = 1.5$ mm (59049 nodes), $\delta = 1.0$ mm (190333 nodes), and $\delta = 0.5$ mm (1452025 nodes). Similar damage patterns are observed in all three cases, but the “resolution” of cracks increases with the decrease of horizon size. It is interesting to notice the extra details seen around the impact region, or in the region of the cracks parallel to the boundaries of the plate. In this case as well, decreasing the horizon size costs more, the increase in the computational time being proportional to $O(1/\delta^3)$. To strike a balance between accuracy and efficiency, we select $\delta = 1.0$ mm for the remaining computations in this chapter.

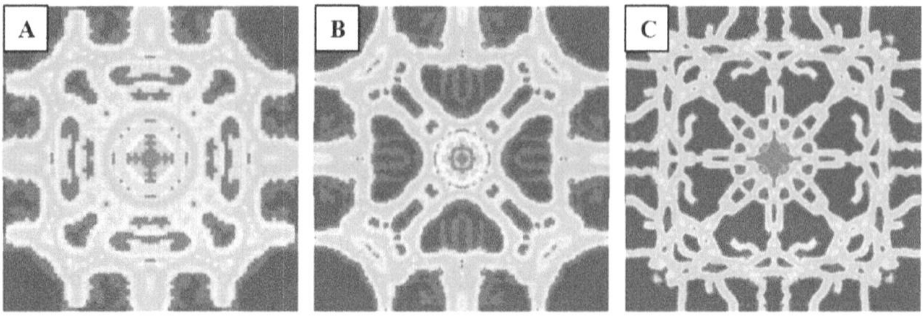

Figure 13.3: Damage maps on the impact surface at 15 μs from the time of impact. Horizon factor $m = \frac{\delta}{\Delta x}$ is kept fixed (equal to 4), and the horizon varies: in (A) δ = 1.5 mm; in (B) δ = 1.0 mm; in (C) δ = 0.5 mm.

We next study the effect on damage patterns of turning off one or all of the corrections types mentioned in the previous sections. The plots in Fig. 13.4 show damage maps for the top surface of the thin glass plate obtained using different critical bond strain correction factor combinations. If surface correction (SC), compressive-strain dependence (CSD), and damage-dependence (DD) are all disabled, the amount of fragmentation is high, and near the impact location the material is comminuted over a wide area (see Fig. 13.4E). This type of behavior, at these impact speeds and sample thickness, does not happen in comparable experiments. As seen in Fig. 11 in Reference [13], similar type of impact produces much less fragmentation in the impact region, and even if the conditions are slightly different (there is a polycarbonate backing plate, impact location is significantly off-center, tape is applied along the sides of the glass-PC two plate system, and the dimensions are larger in the in-plane directions), one can still observe some cracks parallel to the plate's boundaries, caused by waves, reflected from the sides, interacting with incident Rayleigh waves generated by the impact.

When CSD and DD are applied, but SC is left out, fragmentation is severe and a multitude of circumferential cracks appear (see Fig. 13.4)A. Again, this is not observed in glass plates of this thickness (3 mm) and impacted at these velocities (100 m/s). Leaving out only the CSD (see Fig. 13.4B), fragmentation is slightly reduced compared to the previous case, but it is still far from realistic. Only when DD alone is left out (see Fig. 13.4C) do the results start to get closer to the type of damage that is observed in experiments. With all three corrections applied (see Fig. 13.4D), the improvements compared with those in Fig. 13.4C are along the crack lines and in the impact region. Localized cracks are "cleaner" in D, while in C they exhibit some small fragments (red * nodes are nodes that lost all of their bonds). Near the impact region, while in C everything gets smashed into little pieces, in D there is some clearer structure and one can distinguish some band cracks, talked about in [8].

*FOr color figures please see the electronic version of the handbook.

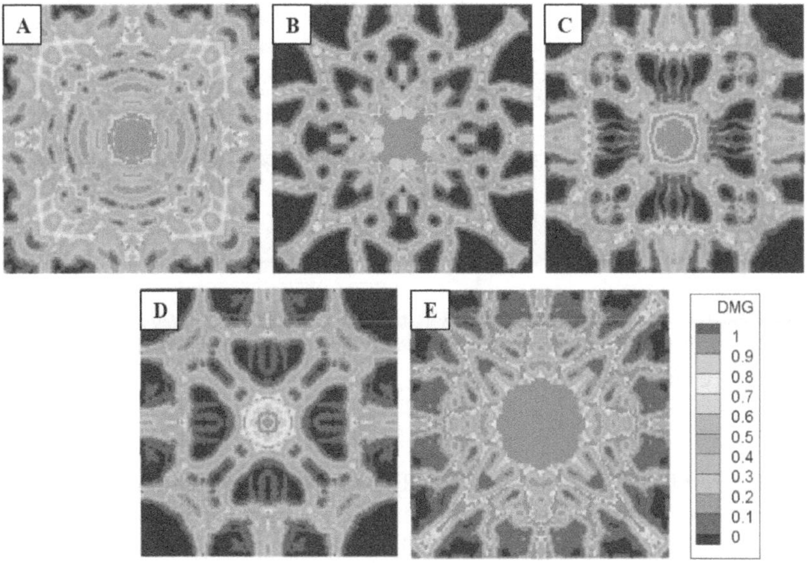

Figure 13.4: Damage map on the strike face at 15 μs for different combinations of critical bond-strain corrections applied: in (A) CSD and DD, but no SC; in (B) SC and DD, but no CSD; in (C) SC and CSD, but no DD; (D) all corrections applied; in (E) all corrections disabled.

We can conclude that damage-dependence of critical bond strain model helps in reducing extra/non-physical damage on surfaces of localized cracks, while the surface correction prevents easy initiation and excessive growth of surface circumferential cracks. The compression-dependent critical bond strain allows for the correct evolution of cracks generated by propagation and interaction between waves (see experiments in [8]) and the absence of this correction leads to unrealistically high fragmentation.

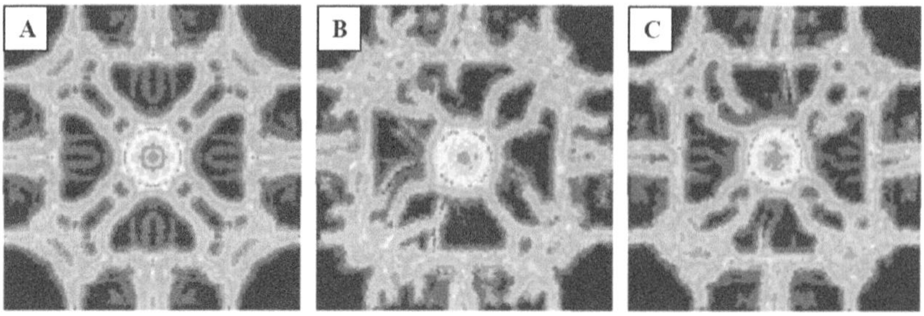

Figure 13.5: Strike face damage map for different perturbations of the grid or the impact location: (A) no perturbations used; (B) random perturbations of nodal coordinates; (C) off-center impact.

The results obtained so far were all using perfect grid symmetry and with the projectile hitting the center of the plate. To study what happens when small perturbations are introduced either in the grid, or the impact location, we perform the following simulations (see Fig. 13.5): in one case (see Fig. 13.5B), all nodes are given a random perturbation of their coordinates of up to 0.025 mm (10% of the grid spacing for $\delta = 1$ mm, with $m = 4$); in another case (see Fig. 13.5B), we set the projectile to hit slightly off-center, at (0.0125, 0.0375) mm, away from the plate center at (0, 0). While, qualitatively, such small perturbations leave the results unchanged, they may eliminate some cracks, or connect cracks that were not connected in a perfect-symmetry case, or create a crack where one was not present in the ideal case. As glass material and experimental conditions are never ideal or perfectly symmetric, the results from simulations performed on grids with perfect symmetry, for example, have to be understood as qualitative tests when compared with actual experimental data. The overall structure of damage and the types of cracks seen are what one should be looking to confirm in computational modeling of brittle fracture from impact, not the match of location of every single crack surface seen in an experiment.

13.3 Impact on a multilayered glass system

In this section we investigate the damage behavior of a complex structure: seven layers of glass backed by a polycarbonate plate. All of the corrections mentioned in the previous section are used in the peridynamic simulations shown below. While the results in the previous section have been obtained with our in-house GPU-enhanced code, due to the memory limitations of the GPU system, the treatment of this larger system would not be feasible. The parallel code EMU (from Sandia National Labs) was used to obtain all of the results in the remaining part of this chapter.

13.3.1 Model description

A simplified three-dimensional model of the multi-layered glass system is created using the geometry and dimensions used in [1] and shown in Fig. 13.6. While in our model, the geometry of the seven-layer glass system with a polycarbonate backing is the same as in the experiments, the boundary conditions are slightly different from experiments. For example, in the experiments, the impact face of the top layer is taped so that the fragments are not lost upon impact. The membrane effect induced by the tape certainly influences the mechanical response, but whether this influence is of first or second order remains to be seen. We do not impose conditions that mimic the tape presence on the front face. In addition, in the experiments the sides of the multi-layered system are also taped. We, instead, enforce zero in-plane displacements for nodes on this boundary. These conditions are more "rigid" than the experimental conditions and we expect less dissipation of waves reflecting from the boundaries (which would cause more damage and fragmentation) than in the real test. Moreover, in our simulations the polycarbonate layer has zero-vertical displacements imposed over where the polycarbonate extends beyond the glass (in the form of a square frame; see Fig. 13.6), while in the experiments, the polycarbonate is supported by wood blocks (more dissipative than the rigid supports) over the same region. We can conclude that the experimental boundary conditions are more "absorbing" (or dissipative) than our zero-displacements conditions. Note also that the boundary conditions used in [4] enforced zero-vertical displacements everywhere on the bottom of the polycarbonate layer.

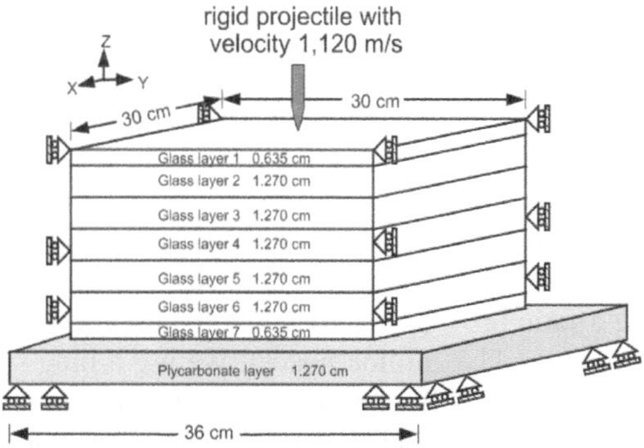

Figure 13.6: Sample geometry of the multi-layer glass target (with a polycarbonate backing plate) and boundary conditions used in the peridynamic model (see [1] and [3]).

In the experimental sample, the glass layers are bonded together with thin polyurethane layers (0.6 mm thick) which are not damaged, except for that between the first and second layers that gets punctured by the projectile. Our current computational model does not include these thin layers, but modeling their presence with special types of peridynamic bonds is possible and will be attempted in the future. We, instead, use simple non-penetrating contact conditions at the interfaces between the glass layers with no bonding. Because of these slight differences in the boundary conditions and the structure of the glass laminate (the absence of tape and binders between the layers) we expect damage patterns to differ to a certain extent from those reported in the experiments.

The simplest peridynamic linear-elastic with damage model is correlated to material properties for plate glass. The material properties for the glass material are: $\rho =$ 2440 kg/m3, Youngs modulus E = 72 GPa, and energy release rate G_0 = 15.47 J/m^2. Note that the results in Reference [4] used slightly different boundary conditions and a significantly higher fracture energy, one that corresponds to the fracture energy for soda-lime glass at crack branching (reported in [9]), instead of at crack initiation. A discussion comparing the present results with those in [4] is given in Section 13.3.2 below. The bond-based peridynamic model used in this work leads to a fixed Poisson ratio of 0.25 (in 3D), which is close to that of the soda-lime glass of 0.22. For the polycarbonate (PC) layer we use a density of 1200 kg/m^3 and Young's modulus of 2 GPa. Since we use the bond-based version of peridynamics, the Poisson ratio of the PC backing material, which is about 0.37, is replaced with the fixed value 0.25. This will slightly affect the stress wave speed in the layer, but, as we shall see, no damage initiates in this layer and thus we believe the consequence of this approximation is secondary and the qualitative picture of damage progression in the multi-layered glass structure is not significantly affected.

A standard 12.7-mm, 13.4-g projectile with an 18-mm long cylindrical body is used in the experiments with an initial velocity of 1 120 m/s ([1]). In our simplified numerical model, we use a *rigid* projectile of the same dimensions (ogive tip). Note that in the experiments, the projectile deforms from its 18 mm initial length to 8 mm, measured after being ejected from the target. The problem analysis with a deformable projectile is left for the future. The code EMU from Sandia, with some modifications described below, is used in all the peridynamic computational results shown in the remainder of this chapter. For the sample with the geometry described above, we use a peridynamic horizon size of about 2.73 mm, and a discretization size Δx of about 0.91 mm, giving a ratio $m = \frac{\delta}{\Delta x} = 3.01$ and a total number of degrees of freedom of around 35 million ($\approx$11.7 million nodes). We use a time step of about 0.07 μs which ensures numerical stability for the forward Euler time-integration used in this work. For the numerical spatial integration of the peridynamic equations of motion, we use the mid-point integration with a modified approximation of the nodal volume as proposed in [2]. For the damage model we use all three "corrections" discussed in Section 13.2.

The nonlocal region size, the horizon, is probably still a little large for the small thickness of some of the glass layers, and a δ-convergence study might be useful; however, we were limited by the computational resources to try with a smaller horizon size (and, implicitly, a finer grid). From the point of view of being able to capture the correct crack propagation speed, the results in [7] show that for glass, a horizon in the range of 1 mm is sufficiently small to lead to crack propagation speeds similar to those measured in experiments. Note that the authors of [1] only discuss the damage morphology in the glass layers after the impact; the propagation speed of the failure fronts or of individual cracks is not monitored. The reason for the dependence of the propagation speed on the horizon size is discussed in [5], where it is argued that when dynamic crack growth in brittle materials is induced by stress waves interacting with the fracture, then, due to trailing waves behind the stress wave front (which depend on the size of the nonlocal region and break bonds), the crack propagation speed is higher the larger the horizon is (see [11] and [12]). When, however, the crack propagation is not influenced by stress waves (as in the case when loads are applied on the crack faces and the propagation is monitored until the stress waves return from the boundaries of the structure to meet the crack tip), then the crack propagation speed is independent of the horizon size ([5]). We also note that the damage map is less sensitive to the horizon size (see [4]) than the crack propagation speed. Here, therefore, we report on damage characteristics obtained with the peridynamic model and compare the results with the discussion on the experiments presented in [1].

It is important to note that we do not use any explicit material strain-rate dependency in this model. Any strain-rate dependence in the results will be due to the intrinsic dynamics of wave propagation, damage, and fracture in the brittle layered system. In a sense, the results shown below will serve as a test for the idea that, at high strain-rates, glass is effectively behaving as a quasi-brittle material and that no explicit material strain-rate dependence is needed in modeling its mechanical behavior under such loading conditions.

13.3.2 A comparison between FEM and peridynamics for the elastic response of a multilayered system to impact

Before analyzing the damage problem, we answer whether the nonlocal solution for the problem of elastic impact, in which no damage is allowed to initiate in the computational model, with the horizon size specified above, leads to an elastic bending deformation field similar to one obtained from a dynamic (explicit) Finite Element Analysis (FEM) using Abaqus. For this purpose, we monitor the bending deformation in the problem of elastic impact of a spherical rigid projectile onto a multi-layered plate system similar to the one described above. In both the FEM and the peridynamic models, we exclude the possibility of damage. Using the horizon size and grid spacing mentioned in the previous section, which will be used for the seven-layer structure as well, we consider a "reduced" system comprised of only two layers of glass (30 cm by 30 cm by 1.27 cm) and the polycarbonate backing layer (36 cm by 36 cm by 1.27 cm). The same boundary conditions as in the seven-layer system are used (see Fig. 13.6), and linear elastic and linear micro-elastic constitutive models are used for both the FEM and the peridynamic solutions, respectively. For the FEM solution, we perform a convergence study with quadratic elements in which the mesh size (in the xy-plane) varies from 1.8 cm (coarse) to 0.9 cm (medium-coarse) to 0.45 cm (medium-fine) to 0.15 cm (fine). The first two meshes use only one element through the thickness of each glass layer and PC backing, while the medium-fine and fine meshes have three and eight elements through the thickness of each layer, respectively.

A rigid sphere of 2.5 cm radius, and the mass-density of steel, is impacting the center of the system at 100 m/s. Fig. 13.7 shows the vertical displacement of the centroid of the first glass layer, in time, obtained by the several FEM solutions from Abaqus (with finer and finer meshes) and from the peridynamic model that uses the same horizon and grid spacing as those used for the seven-layer structure analyzed below in Section 13.4. We observe that the peridynamic solution with this horizon is nearly identical with the converged FEM solution in the initial stages of the deformation caused by the impact. Some differences of about 10% appear at the later times, possibly due to the differences between the contact algorithms used in EMU and Abaqus. It is important to note, however, that at the deformation levels obtained after a few microseconds or tens of microseconds from the impact, the actual glass layer would start to fracture. Therefore, our attention in this verification test of the elastic behavior only needs to focus on the first few tens of microseconds from the time of impact. During this interval, the match between the converged FEM solution and the peridynamic result is very good. We conclude that the peridynamic horizon and the discretization mentioned in the previous section, and to be used in modeling the impact on the seven-layer in section 13.4, are sufficiently small to capture the correct elastic bending deformation of a multi-layer structure, in the absence of damage. The comparison for the projectile speed between the FEM results with the finest mesh we tried, and the peridynamic solution, is also shown in Fig. 13.7.

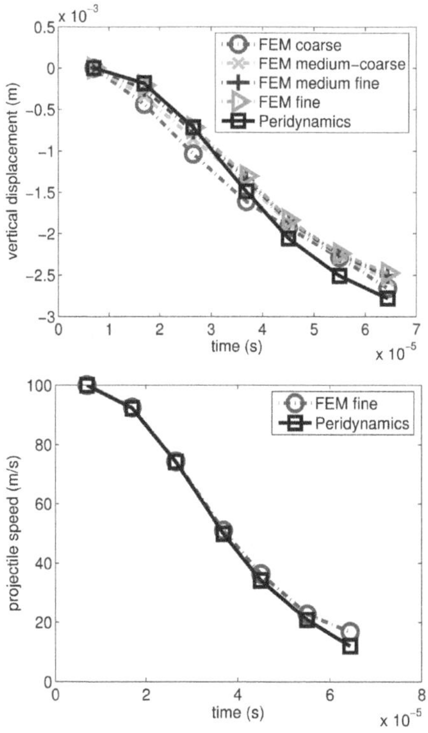

Figure 13.7: Elastic impact of a rigid sphere onto a two-layer glass with a polycarbonate backing. Top: vertical displacement of the glass layer centroid obtained with the FEM or peridynamics. Bottom: evolution of the projectile velocity over 60 μs time span (from [3]).

13.4 Computational results for damage progression in the seven-layer glass system

The projectile impact generates strong elastic stress waves that propagate in the seven-layer glass laminate. These waves interact with, and reflect from, the boundaries, the interfaces between the layers, and the newly created crack surfaces. We now present and examine the peridynamic damage evolution results for this problem in terms of the damage index. The peridynamic damage index is computed at every discretization node and at every time step, as the number of broken peridynamic bonds divided by the total number of original bonds for that particular node. This definition means that the damage index is between zero and one. The legend shown in Fig. 13.8 is used in all damage plots shown below. A damage index of about 0.5 at points aligned along a certain direction means that a crack separating two surfaces has been created along that direction. If, instead, the damage index map shows no localization in a certain area, that means that a diffuse-type damage has occurred there.

We compare our results for the simplified model discussed in section 13.3.1 of the sample used in the experiments of [1] with those discussed in [1], where a detailed post mortem analysis of the damage that has occurred after impact is provided. In contrast with the experiments, the simulation results allow us to observe not only the final damaged stage, but also how damage and fracture evolve in time. This is an important advantage of simulation results since accurate modeling of dynamic behavior under such extreme loading conditions can lead to better designs for multi-layered brittle systems. A further analysis of why various types of cracks and damage initiate and propagate the way they do in a glass plate is presented in [22].

13.4.1 Damage evolution for the cross section

A cross-sectional cut plane (xz or yz plane in Fig. 13.6) through the middle of the sample (the impact point) is used to plot the damage map. Figure 13.8 shows damage progression, in time, through the seven layers of glass backed by the polycarbonate plate. Displacements are amplified by a factor of 2 in this figure. An interesting feature is revealed: the through-thickness damage appears to progress, for layers from 2 to 7, from the bottom up, indicating an important role played by the interfaces (and wave reflections from them) and by the bending deformation, which helps in propagating cracks from the bottom surface to the top face of each layer. We observe the formation of a Hertz-type conical crack similar to what is observed in the experiments (see Fig. 3 in [1]). *The projectile is ejected from the sample at around 23 μs, but most of the damage takes place after this event.* In experiments, the projectile stops in the second layer after penetrating about 19 mm and is ejected, while in our peridynamic simulations the projectile stops in the second layer after penetrating a depth of about 17 mm, and is ejected.

Notice that no damage initiates in the polycarbonate layer, similar to what is observed in the experiments. The bending deformation is clearly visible and it is interesting to compare these results with those in [4] where, besides the tougher glass used, the bottom of the polycarbonate (PC) layer is constrained from moving in the z-direction. Due to limitations of the computational resources we only monitor the damage progression for the first about 200 μs from the moment of impact. The elastic "snap-back" of the polycarbonate layer that follows after that will likely lead to further evolution of damage but this will be confined to enlarging the already-formed cracks since most of the energy has already been consumed in creating the damage observed until the end-time of our simulation. Note that the projectile is not ejected by the rebound of the PC backing plate, which happens much later, but by the rebound of the first few glass layers in the first few tens of microseconds.

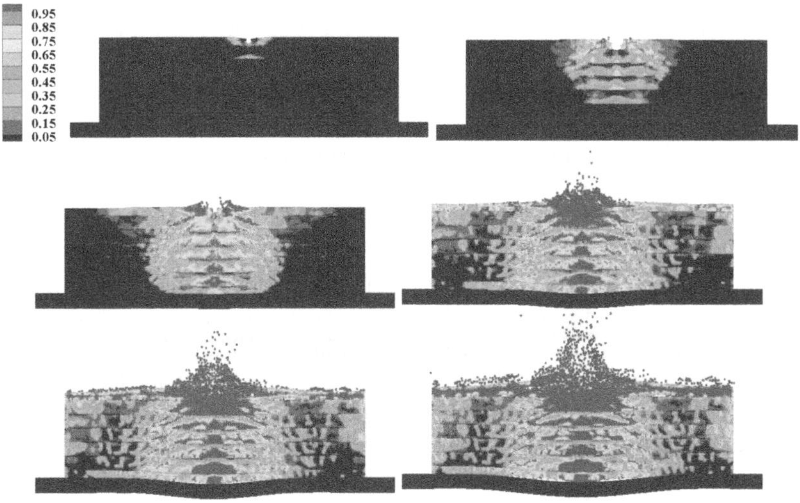

Figure 13.8: Time evolution of damage for the through-thickness cross section through the impact point for the system in Fig. 13.6. From left to right, and top to bottom, snapshots taken at 12, 27, 43, 79, 110, and 161 μs from the time of contact between the projectile and the glass laminate (from [3]).

In our peridynamic model we do not apply any constraints to mimic the presence of tape in experiments on the impact face, nor do we "tape" the sides of the glass layers together as in experiments: instead, we use zero in-plane displacements on the sides, a slightly more rigid boundary condition. Therefore, fragments from the first layer fly away after impact. Note, however, that the first layer does not bounce away from the rest of the structure as in the case in which the polycarbonate layer is constrained over its entire bottom face, as is done in [4]. Recall that we do not have any "glue" between the layers, different from the experiments.

Spall-type fractures in the cross section of the first layer are observed along its length. These are cracks parallel to the impact surface of the layer. The absence of the taping on the impact face in our model likely contributes to the formation of these cracks. In the experiments, cracks parallel to the impact surface are seen in the deeper layers, including the last layer. We attribute these differences between the computations and the experiments to the absence of bonding between the layers in our simplified peridynamic model compared to the polyurethane bonding used in the experiments. The experiments indicate that surface chips are ejected from some of the cracks in the front plate, which the authors of [1] attribute to the late-time bending from the rebound. They also attribute transverse cracks to the rebound bending. The cross-sectional views in Fig. 13.8 show that some of these fragments from the top layer, and some of the transverse cracks (see more about transverse cracks below), form in the initial bending. The rebound bending certainly adds to these types of failure. We also note that since we do not bond the layers together, as is done in the experiments, the interdependence of the rebound between the layers is weaker in our simulations than it is in the experiments. In fact, the rebound of the PC backing layer does not even take place in the first 200 s after the impact. The cause for the slight departures from perfect symmetry seen in the computational damage map in Fig. 13.8 and all subsequent figures has been discussed in [12].

13.4.2 Damage evolution in the first layer

It is interesting to observe the evolution of damage in each individual layer and discuss it in terms of the corresponding experimental results reported in [1]. For the first layer, the evolution of damage is shown in Fig. 13.9, and, for convenience, we reproduce below the figure from [1] showing the post mortem state of damage in the first glass layer (see Fig. 13.10). Only the last column of plots from Fig. 13.9 should be compared with the experimental result in Fig. 13.10, and we should keep in mind that in experiments, tape was applied on top of the impact surface, while our computational model does not include such constraints.

Figure 13.9: Time evolution of damage for the first glass layer in the laminate: top row shows the impact face, bottom row shows the back of this first layer. From left to right, snapshots taken at 9, 27, 43, 60, and 127 μs from the time of contact between the projectile and the glass laminate (from [3]).

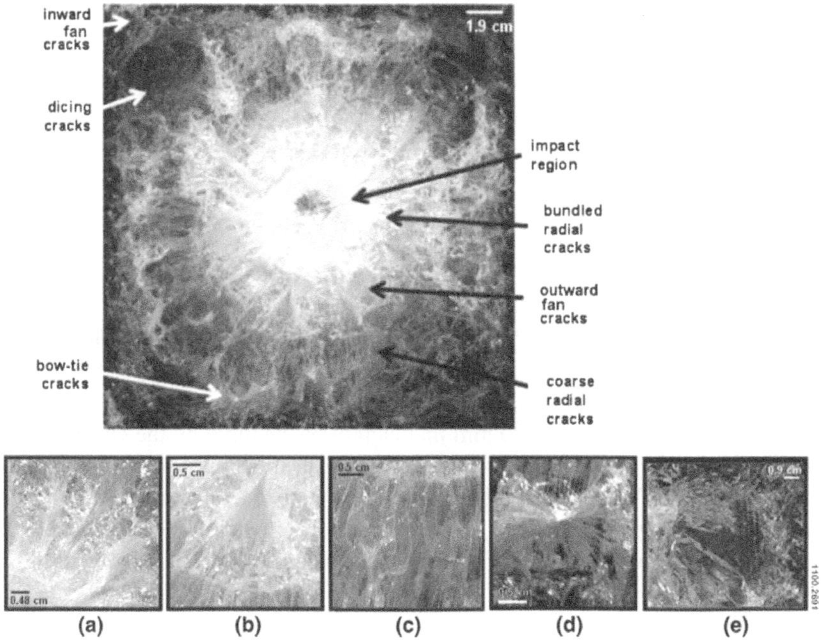

Figure 13.10: Damage in the impact layer first layer). The small figures show: (a) bundled radial cracks, (b) fan crack, (c) coarse radial cracks, (d) bow-tie crack, and (e) dicing cracks (from [1]).

We observe that damage evolves differently between the top face and the bottom face of the first glass layer, and, as we shall see, this is happening in every glass layer. In this first layer, radial cracks start on the bottom face; they continue on both faces as bundled cracks that start to fan out as they propagate towards the boundaries of the layer. This is very similar to the schematic description from the experiments, shown in Figure 7 in [1].

Notice also that the "floret"-type crack on the bottom face is the same as that observed in [1] on the bottom of the last layer. As we shall see, the failure mode at the bottom of each layer starts as an asterisk. We conclude that this type of crack is a result of the stress waves that propagate through the thickness and interact with the surface of the layer. Of course, the floret crack in the first layer is subsequently "erased" by the progression of the perforation through the first and part of the second layer, while in the other layers it is engulfed by further damage. Note that the cracks that have a similar location on the top and bottom faces of a glass layer are straight through-thickness cracks. If, on the other hand, the trace of damage on the top and bottom surface of a layer is similar but not perfectly aligned, then the conclusion is that these cracks are tilted in the thickness direction.

Some notable differences are also seen with the results in [4] that were obtained with a much higher fracture energy for the glass (at branching) and zero vertical displacement on the entire bottom face of the PC backing layer. In the tougher glass case, there are significantly fewer dicing cracks than we observe now, and much less damage in general. The different boundary conditions below the polycarbonate layer used in [4] led to a rebound of the first layer which is visibly pushed away from the rest of the structure.

13.4.3 Damage evolution in the second layer

The second layer is experiencing very different types of damage compared with the first layer. The bundled radial cracks that fan out are not present in this layer; instead, a wider central region experiences a more diffuse-type damage (see Fig. 13.11). This type of damage may be indicative of the needle-type cracks observed in the experiments in this region only (see [1]). In the experiment, these cracks create fragments with dimensions of $1 \times 1 \times 20$ mm that cannot be captured by the current computational resolution of our model. In this second layer the projectile stops and rebounds from the target. We mention that damage continues to evolve even after the 127 μs mark shown last in Fig. 13.11, but its structure remains basically unchanged. At around 200 μs all damage activity in the peridynamic model quiets down.

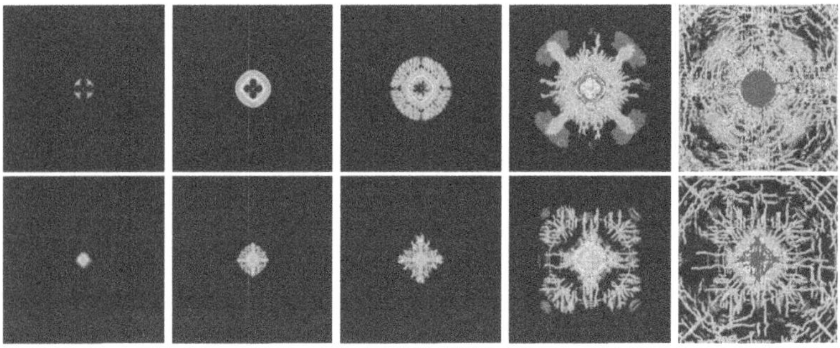

Figure 13.11: Time evolution of damage for the second glass layer in the laminate: top row shows the impact face, bottom row shows the back of this second layer. From left to right, snapshots taken at 12, 18, 24, 43, and 127 μs from the time of contact between the projectile and the glass laminate (from [3]).

Away from the central diffuse-damage zone a region with coarser radial cracks forms, and even closer to the edges, dicing cracks become prevalent. This damage morphology closely resembles that described in the experiments in [1]. Some of the cracks grow from the boundaries of the structure, and especially easy to identify are the cracks in the corners, at 45° angles. These cracks are consistent with wave reflections from the corners and they are also seen in Figure 10 from [1] which is a picture of the third layer. The corners of the second layer are not shown in [1].

13.4.4 Damage evolution in the fourth layer

From the inner layers, third to sixth, we select the fourth as a representative layer. Indeed, the computational results show that much similarity between the damage patterns in these layers exists, and this aspect is also noted in the experiments from [1]. Circumferential cracks (called *"ripple cracks"* in [1]) are highly prominent in these inner layers and they surround a central region of initially undamaged glass on the top face, as reported in the experiments. The dicing cracks reported in the experiments (see Figure 10 in [1]) are also obtained by the peridynamic simulations near the boundaries of the layer. Our computations indicate that these cracks have the same signature on the top and bottom faces of the layer; therefore, they are vertical through-thickness cracks.

The central region, with a diameter similar to the crater in the second layer, on the top surface of this layer and all the layers from third to seventh, shows significantly less damage than the corresponding region on the bottom face of the layer (see top row of plots in Figs. 13.12 and 13.13). These features correspond to the compact disks found in the experiments of Bless and Chen [1]. The lower parts of these regions are heavily damaged. The cause for these features is, likely, wave propagation and reflection across the interfaces.

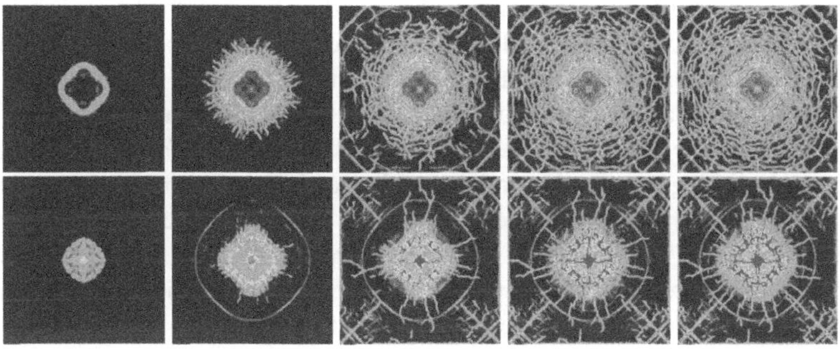

Figure 13.12: Time evolution of damage for the fourth glass layer in the laminate: top row shows the impact face, bottom row shows the back of this fourth layer. From left to right, snapshots taken at 27, 43, 79, 127, and 161 μs from the time of contact between the projectile and the glass laminate (from [3]).

Something that is not captured by our peridynamic simulation is the massive quadrant cracks (radial cracks along the horizontal and vertical directions) shown schematically in Figure 11 in [1]. One explanation for this is the absence of bonding between the layers in our simplified computational model. The bonding used in the experiments is likely to lead to a more coordinated bending motion between the layers, thus creating conditions for quadrant cracks to be produced after the rebound of the entire system. As observed from Fig. 13.8, by the end of our simulation, the backing PC layer has not yet rebounded, while the top layers have already done so.

Had we continued the simulations until the PC layer rebounds, some further quadrant cracking would have been created. However, in the absence of bonding between layers in our simulations, the more likely scenario is that layers would be pushed apart from each other. In the future we plan to include special peridynamic bonds between the layers to mimic the presence of the thin polyurethane layers used in the experiments to bond the glass plates together. We also note that quadrant cracks do appear in our simulations in the seventh layer (see below) but they are not massive crevices, and stop short of reaching the boundaries, at least by the end of our computations.

13.4.5 Damage evolution in the seventh layer

In the last layer of the multi-layered glass systems, which has the same thickness as the first one and half the thickness of the other layers, our computations reveal the generation of a number of very straight radial cracks, along the quadrant cracks directions seen in the experiments (see Fig. 13.13).

Bless and Chen noted (see [1]): *"unlike the strike plate, radial cracks on the last plate were very straight, were continuous, and extended all the way to the edges."* Unlike the experiments, however, our results show such radial cracks running to the edges only for the bisector directions (those ending at the corners) and, to a lesser extent, the cracks along the $x-$ and $y-$ directions. This difference, compared with experiments, is likely due to the absence of bonding between the layers in our simplified computational model, and also due to boundary conditions (no in-plane displacement allowed) that differ from the taping of the lateral sides of the system used in the experiments, and which reflect elastic stress waves differently compared to what happened in the experiment. The order in which these cracks form in the simulations is also interesting to note: the straight cracks running to the corners appear last, and, in fact, they grow from the corners towards the center! * Especially on the top face, we see growth of parallel cracks with regular spacing of about 1 cm in the late stages of the damage process. Similar regular spacing between continuous radial cracks is reported in [1] in bands that bisect the edges into quadrants.

In the center of the back face of this last layer, an asterisk-like feature is seen to develop. Such features are reported in [1], where it is noted that these do not appear to be through-thickness cracks. The same is seen from our results in Fig. 13.13. Dicing cracks develop in this layer as well and they tend to be more prevalent near the edges of the layer.

Compared to the results obtained with a much tougher glass from [4], here the results indicate an amount and spread of damage similar with what is observed from Fig. 13.14, reproduced from [1].

* A similar type of edge-to-center cracks has been observed and explained in detail in [22].

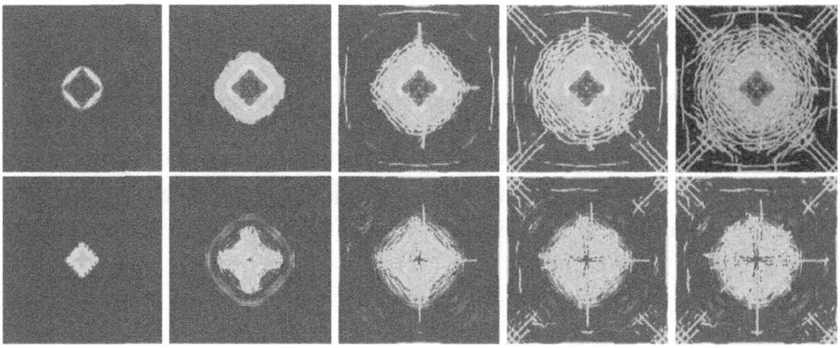

Figure 13.13: Time evolution of damage for the last layer of glass (7th) in the multilayered system: top row shows the impact face, bottom row shows the back of the this seventh layer. From left to right, snapshots taken at 36, 50, 79, 110, and 161 μs from the time of contact between the projectile and the glass laminate (from [3]).

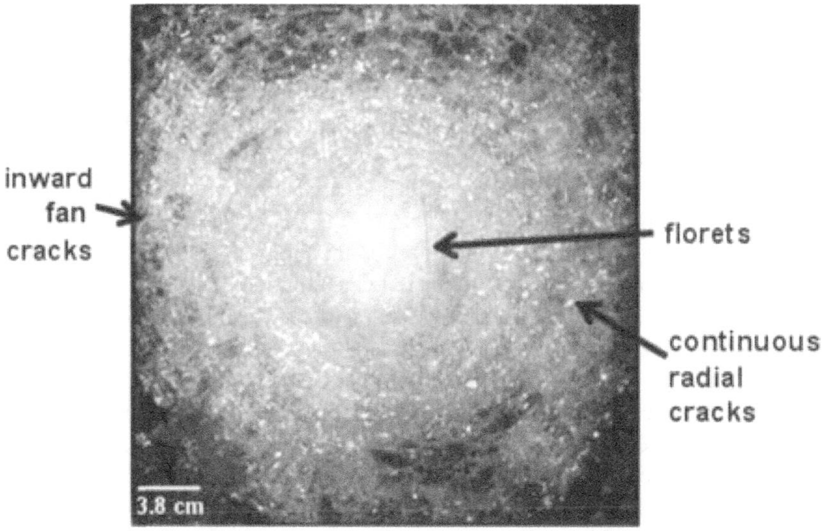

Figure 13.14: Fracture regions in the last (7th) glass layer (from [1]).

13.5 Conclusions

High-velocity impact on multi-layer glass systems produces a variety of damage types with distinct patterns in every layer of the glass laminate. We use a simplified peridynamic model of a multi-layered glass structure for which experiments have been reported in the literature.

We first test the peridynamic model in terms of types of cracks produced in a single glass plate from impact. We discuss various corrections to the peridynamic damage model: the surface correction, the dependence on compressive strains correction, and the damage-dependence of the critical bond strain. We explain why, in problems of impact on brittle targets, such corrections are necessary. We then proceed and investigate a multi-layered system in terms of elastic deformations. We test the peridynamic model against a dynamic finite element solution for elastic impact on a multi-layered system. We then use brittle damage in the peridynamic model and compare the results with damage morphologies observed in experiments conducted on a seven-layer glass system with a polycarbonate layer backing. Layer-by-layer, most of the fracture patterns observed in experiments are well reproduced by the simplified peridynamic model. The peridynamic simulation shows that the dominant mechanisms that induce and influence the damage evolution in the impact problem on a multi-layered brittle structure are: (a) the dynamics of stress waves and their interactions with the interfaces between layers, the boundaries, and the newly created crack surfaces, and (b) the structural response (flexural waves) of the glass layers and polycarbonate backing plate. The brittle damage progression is sensitive to the boundary conditions.

Compared to experimental conditions on the multi-layered glass block, our simplified peridynamic model does not include the taping of the strike face of the strike face, and it replaces the taping of the sides of the layers with zero in-plane displacements. Moreover, in our computational model, the layers are not bonded by the polyurethane layers used in experiments and, because of that, the rebound behavior of the glass layers and the polycarbonate backing plate is different than in the experiments. These differences, however, do not preclude the simplified peridynamic model to capture some essential damage and fragmentation characteristics that take place in high-velocity impact on a multi-layered glass system. Some finer features of damage observed in experiments, such as bow-tie cracks and needle cracks, are not captured, due to the insufficient resolution of the numerical model. It is important to note that the results obtained here do not use any explicit material strain-rate dependencies. All the strain-rate dependence in the results is due only to the implicit dynamics of wave propagation, damage, and fracture in the brittle multilayer system. The closeness between the peridynamic results presented here and the experimental results can be also interpreted as a validation of the point that, at the strain rates induced by high-velocity impact, glass behaves very closely as a quasi-brittle material. Therefore, no explicit material strain-rate dependence is needed in modeling the mechanical behavior under such loading conditions.

The peridynamic results presented here demonstrate that it is possible to obtain predictive simulations of dynamic fracture and damage in multi-layered brittle materials from high-velocity impact. Future plans include modeling the bonding between the glass plates, which has an important influence on the bending and rebound of the structure and on stress wave dissipation, which influence the damage evolution in the multi-layered system, and using a deformable projectile instead of a rigid one.

Acknowledgments

This research has been supported by ARO/ARL (grant number W911NF1010431), program manager Dr. Larry Russell (ARO) and Dr. Chian-Fong Yen (ARL), and by the AFOSR's MURI Center for Material Failure Prediction Through Peridynamics, program managers Dr. David Stargel, Dr. Ali Sayir, and Dr. Fariba Fahroo.

References

[1] S. Bless and T. Chen. Impact damage in layered glass. *International Journal of Fracture*, 162(1): 151–158, 2010.

[2] F. Bobaru and Y.D. Ha. Adaptive refinement and multiscale modeling in 2D peridynamics. *International Journal for Multiscale Computational Engineering*, 9(6): 635–659, 2011.

[3] F. Bobaru, Y.D. Ha, and W. Hu. Damage progression from impact in layered glass modeled with peridynamics. *Open Engineering*, 2(4): 551–561, 2012.

[4] F. Bobaru, Y.D. Ha, and W. Hu. Peridynamic modeling of dynamic brittle fracture in a multi-layered glass system. In: Proceedings of SAMPE (May 21–24, 2012, Baltimore USA), 2012.

[5] F. Bobaru and W. Hu. The meaning, selection, and use of the peridynamic horizon and its relation to crack branching in brittle materials. *International Journal of Fracture*, 176: 215–222, 2012.

[6] F. Bobaru, M. Yang, L.F. Alves, S.A. Silling, E. Askari, and J. Xu. Convergence, adaptive refinement, and scaling in 1D peridynamics. *International Journal for Numerical Methods in Engineering*, 77(6): 852–877, 2009.

[7] F. Bobaru and G. Zhang. Why do cracks branch? A peridynamic investigation of dynamic brittle fracture. *International Journal of Fracture*, 196(1): 59–98, 2015.

[8] F.P. Bowden and J.E. Field. The brittle fracture of solids by liquid impact, by solid impact, and by shock. *Proceedings of the Royal Society of London A: Mathematical, Physical and Engineering Sciences*, 282(1390): 331–352, 1964.

[9] F.P. Bowden, J.H. Brunton, J.E. Field, and A.D. Heyes. Controlled fracture of brittle solids and interruption of electrical current. *Nature*, 216: 38–42, 1967.

[10] M.A. Grinfeld, J.W. McCauley, S.I. Schoenfeld, and T.W. Wright. Failure patterns in brittle ceramics and glasses. In: Proceedings of the 23rd international symposium on ballistics, Tarragona, April 2007.

[11] Y.D. Ha and F. Bobaru. Studies of dynamic crack propagation and crack branching with peridynamics. *International Journal of Fracture*, 162(1):229–244, 2010.

[12] Y.D. Ha and F. Bobaru. Characteristics of dynamic brittle fracture captured with peridynamics. *Engineering Fracture Mechanics*, 78: 1156–1168, 2011.

[13] W. Hu, Y. Wang, J. Yu, C.F. Yen and F. Bobaru. Impact damage on a thin glass with a thin polycarbonate backing. *International Journal of Impact Engineering*, 62: 152–165, 2013.

[14] Q.V. Le and F. Bobaru. Surface corrections in peridynamic models in elasticity and fracture. *International Journal of Solids and Structures*, In review, 2016.

[15] P. J. Patel, G. A. Gilde, P. G. Dehmer, and J. W. McCauley. Transparent ceramics for armor and EM window applications. In: Proceedings of SPIE, 4102, 2000.

[16] S.A. Silling. Reformulation of elasticity theory for discontinuities and long-range forces. *Journal of the Mechanics and Physics of Solids*, 48(1): 175–209, 2000.

[17] S.A. Silling and E. Askari. A meshfree method based on the peridynamic model of solid mechanics. *Computers and Structures*, 83(17-18): 1526–1535, 2005.

[18] S.A. Silling and F. Bobaru. Peridynamic modeling of membranes and fibers. *International Journal of Non-Linear Mechanics*, 40: 395–409, 2005.

[19] S.A. Silling, M. Epton, O. Weckner, J. Xu, and E. Askari. Peridynamic states and constitutive modeling. *Journal of Elasticity*, 88(2): 151–184, 2007.

[20] S.A. Silling and R.B. Lehoucq. Peridynamic theory of solid mechanics. *Advances in Applied Mechanics*, 44: 73–168, 2010.

[21] S.A. Silling, M. Zimmermann, and R. Abeyaratne. Deformation of a peridynamic bar. *Journal of Elasticity*, 73(1):173–190, 2003.

[22] Y. Wang, F. Bobaru, J. Wright, J. Yu, and C.F. Yen. Fracture mechanisms in a thin glass plate under impact. (in review), 2016.

Part V

Multiphysics and Multiscale Modeling

Chapter 14

Coupling Local and Nonlocal Models

Yan Azdoud

COHMAS Laboratory, King Abdullah University of Science and Technology

Fei Han

COHMAS Laboratory, King Abdullah University of Science and Technology

David J. Littlewood

Sandia National Laboratories

Gilles Lubineau

COHMAS Laboratory, King Abdullah University of Science and Technology

Pablo Seleson

Oak Ridge National Laboratory

CONTENTS

14.1 Introduction 408
14.2 Energy-based blending schemes 410
14.2.1 The Arlequin method 410
14.2.1.1 Description of the coupling model 410
14.2.1.2 A numerical example 413
14.2.2 The morphing method 415
14.2.2.1 Overview 415
14.2.2.2 Description of the morphing method 417
14.2.2.3 One-dimensional analysis of ghost forces 419
14.2.2.4 Numerical examples 419
14.3 Force-based blending schemes * 423
14.3.1 Convergence of peridynamic models to classical models 424

*First published in P. Seleson, Y. D. Ha, and S. Beneddine, Concurrent coupling of bond-based peridynamics and the Navier equation of classical elasticity by blending, *International Journal for Multiscale Computational Engineering*, 13(2): 91113 (2015). Copyright 2015 by Begell House, Inc.

14.3.2 Derivation of force-based blending schemes 425
14.3.3 A numerical example 427
14.4 Summary 431
Acknowlegments 431
References 431

14.1 Introduction

Techniques for coupling local and nonlocal models can greatly expand the application space of peridynamics (PD). A foremost concern is computational expense, which in practice limits the use of PD in large-scale, geometrically complex simulations. As demonstrated in Chapter 7, certain discretizations of PD models possess a similar computational structure to molecular dynamics. For discretizations of this type, the computational cost and memory requirements are strongly tied to the number of interactions in a model, which depends on the ratio between the horizon and the distance between neighboring nodes. When this ratio becomes large, simulations can be intractable. The computational burden of PD is also evident in the structure of tangent stiffness matrices for implicit time integration. The number of nonzero elements in a PD stiffness matrix grows rapidly as the above-mentioned ratio increases, resulting in a bandwidth that is typically much larger than that of the stiffness matrix associated with a corresponding local model in classical continuum mechanics (CCM). The same observation applies to finite element discretizations of PD models, when the horizon is large relative to the mesh size [33]. A second motivation for local-nonlocal coupling is the desire to combine the strengths of PD with those of CCM, for example as implemented in a traditional finite-element analysis code. The most obvious strength of PD models is their ability to capture pervasive material failure. Other traits of PD models may or may not be desirable in practice; for example, PD models introduce wave dispersion not always represented in CCM models [41, 38]. In addition, many classical finite element codes contain an array of features not widely available in PD codes, for example structural elements such as beams and joints. A third motivation for concurrent coupling of local and nonlocal models is the challenge posed by the imposition of nonlocal boundary conditions. Nonlocal models require the extension of classical boundary conditions, so that instead of these conditions being imposed on boundary surfaces, they are imposed in volumetric regions [36, 20, 1]. Many times, such an extension is not well-defined. Concurrent coupling methods, however, offer the possibility to solve nonlocal problems equipped with classical local boundary conditions.

In [46], it was shown that elastic PD models converge to corresponding CCM models in the limit of vanishing nonlocality. Moreover, it can be shown that linear PD models exhibit the same material response as corresponding CCM models in problems characterized by smooth deformations (cf. Section 14.3.1). As a consequence, it may be possible to replace PD governing equations with corresponding CCM equations in domains where the models behave similarly. In that case, PD models are applied in regions where discontinuities may be present, while CCM models are applied far from such regions, where displacement fields are smooth. Simulations based on a coupled PD/CCM model can result in efficient and accurate solutions of PD problems. The main challenge lies in the design of algorithms that properly couple the two models across interfaces.

There is a vast literature in the field of concurrent multiscale material modeling concerning methods for coupling nonlocal and local models. Specifically, approaches to concurrently couple nonlocal atomistic (discrete) models with local continuum models have been proposed. Such approaches are commonly referred to as atomistic-to-continuum (AtC) coupling methods. Common examples of AtC coupling methods include the Arlequin method [6, 31, 9, 10], the quasicontinuum method [27, 39, 48], blending methods [4, 5, 14, 21], and bridging domain methods [8, 50, 35], among others. A known problem, commonly present in these types of methods, is the appearance of spurious effects across transition regions connecting the nonlocal and local models. These effects may include the following: presence of "ghost forces" (i.e., forces acting on particles in equilibrium positions), failure to pass "patch tests" defined by uniform strain solutions, lack of energy preservation, breakdown of Newton's third law, and appearance of artificial wave reflections. Such effects are not present in the reference models and are a result of the enforced coupling. In the case of AtC coupling, the atomistic and continuum models possess very disparate mathematical natures: the atomistic model is discrete, whereas the other model is continuum; the atomistic model is nonlocal, whereas the other model is local. Furthermore, the length and time scales at which atomistic and continuum models are typically applied differ by orders of magnitude. The situation is different in the coupling of PD and CCM models. First, both PD and CCM are continuum theories. Second, nonlocal PD models can be generally connected to local CCM counterparts, through a limiting process. Third, many of the problems that PD models are applied to are characterized by the same length and time scales as in CCM. These properties facilitate the derivation and analysis of coupling schemes for PD and CCM models. Research on the coupling of PD and CCM is underway; however, we present below some coupling schemes which have been published in recent years. For simplicity, we focus in this chapter on linear isotropic microelastic bond-based PD models, and we assume a pairwise equilibrated reference configuration [41]. Accordingly, we employ corresponding CCM models provided by classical linear elasticity.

Although there are many different possible approaches to couple local and nonlocal models, we focus in this chapter on a certain class of methods which can be referred to, in a broad sense, as *blending methods*. The domain of interest is decomposed into subdomains described by different models and a blending region (also called overlapping or morphing or transition region) where the models are blended; a blending function (also called weight or morphing function) is introduced to characterize each domain as well as to weight the contribution of each model across the blending region. Such approaches typically appear in two flavors. In *energy-based blending schemes*, we employ blended energy partitions, possibly with certain constraints; we present these methods in Section14.2. In contrast, in *force-based blending schemes* the coupling is performed directly at the force level, using blended governing equations; these methods are presented in Section14.3. We note that blending methods have recently drawn interest also in the quasicontinuum community due to their potential to improve the accuracy of calculations [49, 30, 28].

14.2 Energy-based blending schemes

14.2.1 The Arlequin method

The Arlequin approach is a very versatile coupling strategy that has been used with success in various model-coupling configurations [13, 7, 12]. It relies on a simple but robust two-part idea: (1) two descriptions of a material system coexist in a finite overlapping region and (2) the energy of the material system within the overlapping region is built by weighting the energy of each description following a partition of unity approach. For applications in solid mechanics, two representations of the displacement field are assumed to coexist in the overlapping region, one for each of the coupled models. The relationship between the displacement fields is dictated by the specific Arlequin strategy employed. Typically, a weak kinematic compatibility is enforced via Lagrange multipliers. Literature in this area includes the application of the Arlequin method for coupling disparate CCM models [13, 11] and for coupling CCM and molecular dynamics models [8, 6].

14.2.1.1 Description of the coupling model

Let $\mathcal{B}$ be a continuum body subjected to a body force density $\mathbf{b}$ (*cf.* Figure 14.1), in which two overlapping subdomains $\mathcal{B}_1$ and $\mathcal{B}_2$ coexist such that $\mathcal{B}_1 \cup \mathcal{B}_2 = \mathcal{B}$ and $\mathcal{B}_1 \cap \mathcal{B}_2 = \mathcal{B}_c$, where $\mathcal{B}_c$ is the overlapping region. The measure of $\mathcal{B}_c$, $\mathbf{meas}(\mathcal{B}_c)$, is such that $\mathbf{meas}(\mathcal{B}_c) > 0$. For the solid mechanics applications considered here, a CCM model involving contact forces is prescribed in the subdomain $\mathcal{B}_1$, and a bond-based PD model involving nonlocal interactions is used in the subdomain $\mathcal{B}_2$. The displacement $\overline{\mathbf{u}}$ is imposed on the Dirichlet boundary $\partial\mathcal{B}_u \subset \partial\mathcal{B}$ and the traction $\overline{\mathbf{T}}$ is imposed on the Neumann boundary $\partial\mathcal{B}_T \subset \partial\mathcal{B}$. We denote by $\mathbf{n}$ the outward unit normal vector to $\partial\mathcal{B}_T$. We assume that $\mathcal{B}_2$ is embedded in $\mathcal{B}_1$ and $(\partial\mathcal{B}_u \cup \partial\mathcal{B}_T) \subset \partial\mathcal{B}_1$, which means that the boundary conditions are only prescribed in the CCM model. This is not a major restriction and largely simplifies the definition of the boundary conditions.

We introduce the space $\mathcal{V}_1 := \left\{\mathbf{v} \in H^1(\mathcal{B}_1),\ \mathbf{v} = \overline{\mathbf{u}} \text{ on } \partial\mathcal{B}_u\right\}$ for admissible solutions and the space $\mathcal{V}_1^0 := \left\{\mathbf{v} \in H^1(\mathcal{B}_1),\ \mathbf{v} = \mathbf{0} \text{ on } \partial\mathcal{B}_u\right\}$ for test functions. The weak formulation corresponding to the CCM model in $\mathcal{B}_1$ reads as follows:

$$\text{Find } \mathbf{u} \in \mathcal{V}_1 \text{ such that } a_c(\mathbf{u},\mathbf{v}) = l_c(\mathbf{v}) \quad \forall \mathbf{v} \in \mathcal{V}_1^0, \tag{14.1}$$

with

$$a_c(\mathbf{u},\mathbf{v}) := \int_{\mathcal{B}_1} \sigma(\mathbf{u}) : \varepsilon(\mathbf{v})\, dV, \tag{14.2}$$

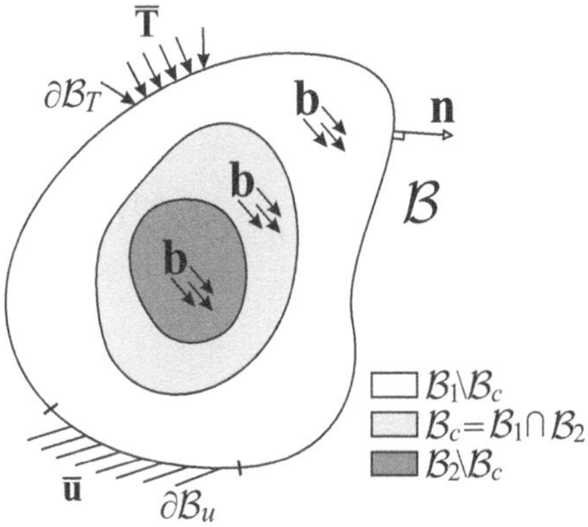

Figure 14.1: Continuum body $\mathcal{B}$ composed of two overlapping subdomains $\mathcal{B}_1$ and $\mathcal{B}_2$, such that $\mathcal{B}_1 \cup \mathcal{B}_2 = \mathcal{B}$ and $\mathcal{B}_1 \cap \mathcal{B}_2 = \mathcal{B}_c$.

and

$$l_c(\mathbf{v}) := \int_{\mathcal{B}_1} \mathbf{b}\cdot\mathbf{v}dV + \int_{\partial\mathcal{B}_T} \overline{\mathbf{T}}\cdot\mathbf{v}dS. \tag{14.3}$$

In (14.2), ε and σ denote the infinitesimal strain tensor and the Cauchy stress tensor, respectively.

The weak formulation corresponding to the PD model in $\mathcal{B}_2$ reads as follows [18, 24]:

$$\text{Find } \mathbf{u} \in \mathcal{V}_2 \text{ such that } a_p(\mathbf{u},\mathbf{v}) = l_p(\mathbf{v}) \quad \forall \mathbf{v} \in \mathcal{V}_2^0, \tag{14.4}$$

with

$$a_p(\mathbf{u},\mathbf{v}) := \frac{1}{2}\int_{\mathcal{B}_2}\int_{\mathcal{B}_2} (\mathbf{v}(\mathbf{x}') - \mathbf{v}(\mathbf{x}))^{\mathrm{T}}\mathbf{C}(\|\xi\|)(\mathbf{u}(\mathbf{x}') - \mathbf{u}(\mathbf{x}))dV_{\mathbf{x}'}dV_{\mathbf{x}} \tag{14.5}$$

and

$$l_p(\mathbf{v}) := \int_{\mathcal{B}_2} \mathbf{b}\cdot\mathbf{v}dV. \tag{14.6}$$

In (14.5), the micromodulus $\mathbf{C}(\|\xi\|) = \lambda(\|\xi\|)\xi\otimes\xi$ is a second-order symmetric tensor, where $\xi := \mathbf{x}' - \mathbf{x}$. We denote in (14.4) by $\mathcal{V}_2$ the admissible displacement field space that satisfies the Dirichlet boundary conditions, and by $\mathcal{V}_2^0$ the subset of displacements that vanish on the Dirichlet boundary. We should note that $\mathcal{V}_2$ and $\mathcal{V}_2^0$ do not have the same regularity requirements as $\mathcal{V}_1$ and $\mathcal{V}_1^0$. Indeed, they can possibly include discontinuous displacement fields. This is the primary motivation for using a PD model.

The key idea of the coupling strategy is to implement a partition of energy between the two models, within the overlapping region, using complementary weight functions. This approach falls into the category of a partition of unity in terms of energy (this is classical when using an Arlequin approach [11, 16]). The weighted internal and external virtual work are defined, respectively, by

$$a_1(\mathbf{u}_1,\mathbf{v}_1) := \int_{\mathcal{B}_1} \alpha_1 \sigma(\mathbf{u}_1) : \varepsilon(\mathbf{v}_1)\, dV, \quad (\mathbf{u}_1,\mathbf{v}_1) \in \mathcal{V}_1 \times \mathcal{V}_1^0, \tag{14.7}$$

$$a_2(\mathbf{u}_2,\mathbf{v}_2) := \frac{1}{2}\int_{\mathcal{B}_2}\int_{\mathcal{B}_2} \alpha_2(\mathbf{x},\mathbf{x}')\left(\mathbf{v}_2(\mathbf{x}') - \mathbf{v}_2(\mathbf{x})\right)^{\mathrm{T}} \mathbf{C}(\|\xi\|)(\mathbf{u}_2(\mathbf{x}') - \mathbf{u}_2(\mathbf{x}))\, dV_{\mathbf{x}'} dV_{\mathbf{x}},$$
$$(\mathbf{u}_2,\mathbf{v}_2) \in \hat{\mathcal{V}}_2 \times \hat{\mathcal{V}}_2, \tag{14.8}$$

and

$$l_1(\mathbf{v}_1) := \int_{\mathcal{B}_1} \beta_1 \mathbf{b}\cdot \mathbf{v}_1 dV + \int_{\partial\mathcal{B}_T} \beta_1 \overline{\mathbf{T}} \cdot \mathbf{v}_1 dS, \quad \mathbf{v}_1 \in \mathcal{V}_1^0, \tag{14.9}$$

$$l_2(\mathbf{v}_2) := \int_{\mathcal{B}_2} \beta_2 \mathbf{b} \cdot \mathbf{v}_2 dV, \quad \mathbf{v}_2 \in \hat{\mathcal{V}}_2, \tag{14.10}$$

where $\mathcal{V}_1, \mathcal{V}_1^0$ have been defined above and $\hat{\mathcal{V}}_2$ is the admissible displacement field spaces on $\mathcal{B}_2$. It is worth to note that no boundary conditions are prescribed to $\mathcal{B}_2$ in the above coupling formulation. The weight functions (α_1, β_1) and (α_2, β_2) are defined over $\mathcal{B}_1$ and $\mathcal{B}_2$, respectively, and satisfy

$$\alpha_1(\mathbf{x}) = \beta_1(\mathbf{x}) = \begin{cases} 1 & \mathbf{x} \in \mathcal{B}_1 \backslash \mathcal{B}_c, \\ 0 & \mathbf{x} \in \mathcal{B}_2 \backslash \mathcal{B}_c. \end{cases} \tag{14.11}$$

In $\mathcal{B}_c$, we can choose $\alpha_1(\mathbf{x})$ and $\beta_1(\mathbf{x})$ as constant, linear, or higher-order polynomial functions. Following a partition of unity,

$$\begin{cases} \alpha_2(\mathbf{x},\mathbf{x}') = 1 - \alpha_1\left(\dfrac{\mathbf{x}+\mathbf{x}'}{2}\right) \quad \forall \mathbf{x},\mathbf{x}' \in \mathcal{B}, & \text{(14.12a)} \\ \beta_2(\mathbf{x}) = 1 - \beta_1(\mathbf{x}) \quad \forall \mathbf{x} \in \mathcal{B}. & \text{(14.12b)} \end{cases}$$

A weighting approach for discrete particles [6, 16] gives some insight into (14.12a), and permits α_2 to be symmetric with respect to interchanging $\mathbf{x}$ and $\mathbf{x}'$.

Different choices of weight functions produce different Arlequin solutions (*cf.* (14.17)). These solutions are predominantly affected by Lagrange multipliers, which are introduced in the overlapping region to couple the PD and CCM models. The Arlequin problem can be written as follows:

$$\text{Find } (\mathbf{u}_1,\mathbf{u}_2,\varphi) \in \mathcal{V}_1 \times \hat{\mathcal{V}}_2 \times \mathcal{M}, \quad \forall (\mathbf{v}_1,\mathbf{v}_2,\psi) \in \mathcal{V}_1^0 \times \hat{\mathcal{V}}_2 \times \mathcal{M},$$

such that

$$a_1(\mathbf{u}_1,\mathbf{v}_1) + M(\varphi,\mathbf{v}_1) = l_1(\mathbf{v}_1), \tag{14.13}$$

$$a_2(\mathbf{u}_2,\mathbf{v}_2) - M(\varphi,\mathbf{v}_2) = l_2(\mathbf{v}_2), \tag{14.14}$$

$$M(\psi,\mathbf{u}_1 - \mathbf{u}_2) = 0, \tag{14.15}$$

where a_1, a_2, l_1, and l_2 are defined by (14.7) – (14.10), respectively, with the weight functions satisfying (14.11), (14.12a), and (14.12b), and $\mathcal{M}$ denotes the space of Lagrange multipliers in $\mathcal{B}_c$. The coupling operator $M(\cdot,\cdot)$ is given by

$$M(\psi,\mathbf{v}) = \int_{\mathcal{B}_c} \kappa_0\psi\cdot\mathbf{v} + \kappa_1\varepsilon(\psi):\varepsilon(\mathbf{v})\,dV, \quad \forall(\psi,\mathbf{v})\in\mathcal{M}\times\mathcal{M}. \tag{14.16}$$

The parameters (κ_0,κ_1) are non-negative and determine the type of coupling operator. For example, $(\kappa_0,\kappa_1)=(1,1)$ defines the H^1 coupling operator, whereas $(\kappa_0,\kappa_1)=(1,0)$ yields L^2 coupling [22, 23]. These operators are compared in the numerical examples in §14.2.1.2.

The displacement solutions in the subdomains $\mathcal{B}_1$ and $\mathcal{B}_2$, corresponding to the CCM model and the PD model, respectively, can be obtained by solving (14.13)–(14.15). However, neither of them separately describes a complete solution in the overlapping region $\mathcal{B}_c$. Nevertheless, a full displacement field $\mathbf{u}$ can be constructed by coupling both solutions in $\mathcal{B}_c$ as follows:

$$\mathbf{u}(\mathbf{x}) = \begin{cases} \mathbf{u}_1(\mathbf{x}) & \mathbf{x}\in\mathcal{B}_1\backslash\mathcal{B}_c, \\ \mathbf{u}_2(\mathbf{x}) & \mathbf{x}\in\mathcal{B}_2\backslash\mathcal{B}_c, \\ \alpha_1(\mathbf{x})\mathbf{u}_1(\mathbf{x})+(1-\alpha_1(\mathbf{x}))\mathbf{u}_2(\mathbf{x}) & \mathbf{x}\in\mathcal{B}_c. \end{cases} \tag{14.17}$$

The constructed displacement field $\mathbf{u}$ in (14.17) is convenient for post-processing simulation results.

14.2.1.2 A numerical example

We consider a two-dimensional cracked plate under normal and shear traction with a plane strain assumption. A PD model given by (14.4–14.6) is prescribed around the crack in $\mathcal{B}_2$ with a fine mesh, while a CCM model given by the Navier-Cauchy equation of classical elasticity (cf. (14.49)) is used outside the region of the crack in $\mathcal{B}_1$ with a coarse mesh. Lagrange multipliers are applied in $\mathcal{B}_c$ using a fine mesh with the same resolution as the one in $\mathcal{B}_2$ (cf. Figure 14.2). Dirichlet boundary conditions are imposed on the upper and lower edges of the structure. The dimensions of the domain, the boundary conditions, and the discretization meshes are indicated in Figure 14.2.

In the PD model, we choose $\lambda(\|\xi\|) = c^0\exp(-\|\xi\|/\ell)/\|\xi\|^2$, where c^0 is a constant parameter and ℓ is an internal length scale. Following [24], we choose $c^0 = 400\,\mathrm{N\,m}^{-7}$ and $\ell = 0.1\,\mu\mathrm{m}$. The PD horizon (*cf.* (14.19)) is chosen as $\delta = 1.875\mu\mathrm{m}$; this value guarantees that the micromodulus is negligible for neighbors further than δ (i.e. $e^{-\delta/\ell} < 1.0\times10^{-8}$). The associated Young's modulus and Poisson's ratio are $E = 251.3\,\mathrm{GPa}$ and $\nu = 1/3$, respectively. These parameters in the CCM model depend on the nature of the material but should be equivalent to the PD parameters in the homogenization sense (*cf.* (14.24)) [16], so that the CCM model is compatible with the PD model.

The structure deformation and contour plots of the strain component ε_{xx} are displayed in Figure 14.3.The solutions calculated by the Arlequin method with a piecewise constant weight function ($\alpha_1 = 1$ in $\mathcal{B}_1\backslash\mathcal{B}_c$, $\alpha_1 = 0.5$ in $\mathcal{B}_c$, and $\alpha_1 = 0$ in $\mathcal{B}_2\backslash\mathcal{B}_c$) are shown in Figures 14.3(a) and 14.3(b), using an H^1 coupling operator and an L^2

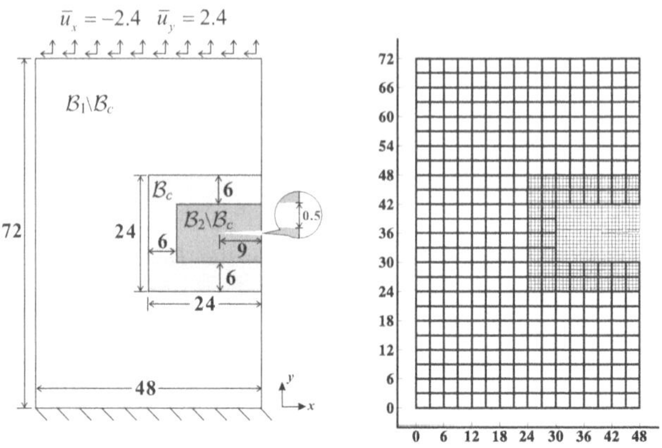

Figure 14.2: Dimensions and boundary conditions (left plot) and meshes (right plot) for a plate with a crack (unit of length: μm).

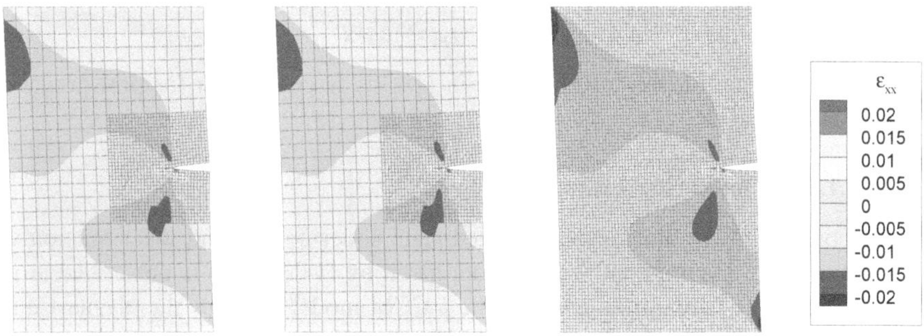

(a) (b) (c)

Figure 14.3: Structure deformation and strain component ε_{xx} of (a) the Arlequin solution using the H^1 coupling operator, (b) the Arlequin solution using the L^2 coupling operator, and (c) the PD solution.

coupling operator in $\mathcal{B}_c$, respectively. Both Arlequin solutions closely agree with that obtained by the PD model in Figure 14.3(c). However, an L^2 coupling method results in an ill-posed problem when the mesh size goes to zero [9, 32]. Indeed, an L^2 coupling method induces singularities in the Lagrange multiplier field [13, 22]. To observe this effect in our example, we plot in Figure 14.4 the Lagrange multipliers in the overlapping region along the crack line (from the point 24 to 30 in the x-direction in Figure 14.2 (right)). While the Lagrange multiplier field is smooth in the H^1 coupling method, the L^2 coupling method results in perturbations in the Lagrange multiplier field, especially at the boundaries of the overlapping region. These observations and comparisons suggest that the H^1 coupling method is a better choice than the L^2 coupling method, because the H^1 coupling operator guarantees a well-posed problem and a regular Lagrange multiplier field [22].

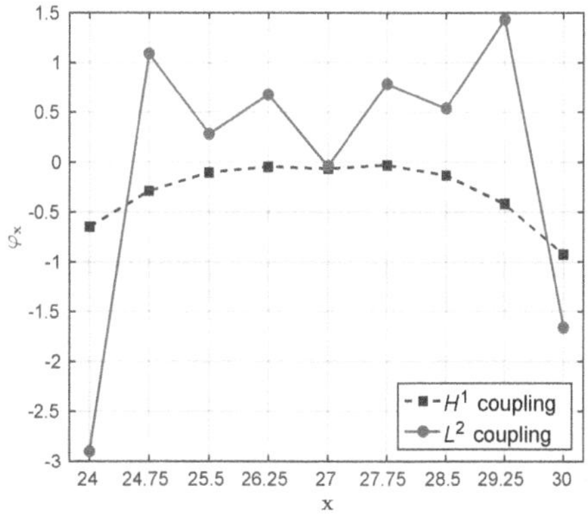

Figure 14.4: The x-component of the Lagrange multiplier field, $\varphi_{\mathbf{x}}$, using the H^1 and L^2 coupling methods with constant weight functions in $\mathcal{B}_c$ (unit of length: μm).

14.2.2 The morphing method

14.2.2.1 Overview

A bond-based PD model. We introduce here the PD model to which the coupling strategy is applied. Let $\mathcal{B}$ be the domain of interest. The bond-based PD equilibrium equation is

$$\int_{\mathcal{H}(\mathbf{x},\delta)\cap\mathcal{B}} \mathbf{f}(\mathbf{x}',\mathbf{x})dV_{\mathbf{x}'} + \mathbf{b}(\mathbf{x}) = \mathbf{0} \quad \forall \mathbf{x} \in \mathcal{B}, \tag{14.18}$$

where $\mathbf{b}$ is a body force density and $\mathbf{f}(\mathbf{x}',\mathbf{x})$ is the pairwise force function. The point $\mathbf{x}$ is assumed to interact with its *neighborhood*,

$$\mathcal{H}(\mathbf{x},\delta) := \left\{\mathbf{x}' \in \mathbb{R}^d, d = 1,2,\text{or } 3 : \|\mathbf{x}'-\mathbf{x}\| \leqslant \delta\right\}, \tag{14.19}$$

which defines the region beyond which interactions are truncated. Consequently, $\mathbf{f}(\mathbf{x}',\mathbf{x}) = \mathbf{0}$ for $\||\mathbf{x}'-\mathbf{x}\| > \delta$. In (14.19), δ is a length scale referred to as the PD *horizon* [41] (cf. Figure 14.5).

Following [45], we can define the pairwise force function as:

$$\mathbf{f}(\mathbf{x}',\mathbf{x}) = \mathbf{t}(\mathbf{x}',\mathbf{x}) - \mathbf{t}(\mathbf{x},\mathbf{x}'), \tag{14.20}$$

where $\mathbf{t}(\mathbf{x}',\mathbf{x})$ is the partial interaction due to the action of the point $\mathbf{x}'$ on the point $\mathbf{x}$,

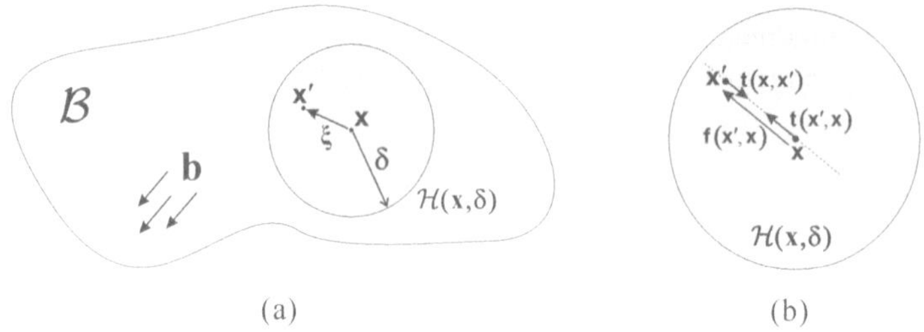

Figure 14.5: (a) Continuum body $\mathcal{B}$ and neighborhood of $\mathbf{x}$, $\mathcal{H}(\mathbf{x},\delta)$, which depends on the horizon δ. The vector $\xi := \mathbf{x}' - \mathbf{x}$. (b) Decomposition of the pairwise force function, $\mathbf{f}(\mathbf{x}',\mathbf{x})$, into partial interactions.

and $\mathbf{t}(\mathbf{x},\mathbf{x}')$ is the partial interaction due to the action of the point $\mathbf{x}$ on the point $\mathbf{x}'$. For small deformations, a linearized constitutive model can be written as (*cf.* [41, 42, 45])

$$\mathbf{t}(\mathbf{x}',\mathbf{x}) = \frac{\mathbf{C}(\mathbf{x},\|\xi\|)}{2}(\mathbf{u}(\mathbf{x}') - \mathbf{u}(\mathbf{x})), \tag{14.21}$$

where $\mathbf{u}$ is the displacement field, $\xi := \mathbf{x}' - \mathbf{x}$ is the relative position vector in the reference configuration, and $\mathbf{C}(\mathbf{x},\|\xi\|)$ is the micromodulus function (cf. (14.33)). It is assumed that $\mathbf{C}(\mathbf{x},\|\xi\|) = \mathbf{0}$ for $\|\xi\| > \delta$. Together (14.20) and (14.21) lead to

$$\mathbf{f}(\mathbf{x}',\mathbf{x}) = \frac{\mathbf{C}(\mathbf{x},\|\xi\|) + \mathbf{C}(\mathbf{x}',\|\xi\|)}{2}\{\mathbf{u}(\mathbf{x}') - \mathbf{u}(\mathbf{x})\}. \tag{14.22}$$

Under the assumption of a homogeneous material (i.e., $\mathbf{C}(\mathbf{x},\|\xi\|) = \mathbf{C}(\mathbf{x}',\|\xi\|) = \mathbf{C}(\|\xi\|)$), the macroelastic energy density at a point $\mathbf{x}$ is given by (*cf.* [18])

$$W^p(\mathbf{x}) = \frac{1}{4}\int_{\mathcal{H}(\mathbf{x},\delta)\cap\mathcal{B}} (\mathbf{u}(\mathbf{x}') - \mathbf{u}(\mathbf{x}))^{\mathrm{T}}\mathbf{C}(\|\xi\|)\,(\mathbf{u}(\mathbf{x}') - \mathbf{u}(\mathbf{x}))\,dV_{\mathbf{x}'}, \quad \forall \mathbf{x} \in \mathcal{B}. \tag{14.23}$$

An energy equivalence. When a homogeneous deformation is prescribed within the neighborhood of a point, in a homogeneous material, it is possible to derive an equivalent local stiffness tensor from the nonlocal model. The strain energy density of an equivalent local model, at a point $\mathbf{x}$ in the bulk of the body (cf. (14.43)), can be written as

$$W^c(\mathbf{x}) = \frac{1}{2}\varepsilon(\mathbf{x}) : \mathbf{K}^0 : \varepsilon(\mathbf{x}) \simeq W^p(\mathbf{x}) \quad \text{with} \quad \mathbf{K}^0 = \frac{1}{2}\int_{\mathcal{H}(\mathbf{0},\delta)} \lambda(\|\xi\|)\xi\otimes\xi\otimes\xi\otimes\xi\, dV_\xi, \tag{14.24}$$

where $(\otimes)$ is the dyadic product and $\mathbf{C}(\|\xi\|) = \lambda(\|\xi\|)\xi\otimes\xi$ with $\lambda(\|\xi\|)$ a scalar-valued function satisfying $\lambda(\|\xi\|) = 0$ for $\|\xi\| > \delta$. In (14.24), the PD macroelastic energy density, $W^p(\mathbf{x})$, is similar to the strain energy density of a CCM model, $W^c(\mathbf{x})$, with a linear elastic material defined by the stiffness tensor $\mathbf{K}^0$ at the point $\mathbf{x}$

in the bulk of the body. From this equivalence, we can obtain (cf. (14.47) and (14.48)) that the elastic moduli (in 3D) of the equivalent local model are

$$l = \mu = \frac{2\pi}{15}\int_0^{\delta} \lambda(r) r^6 dr, \tag{14.25}$$

with l the first Lamé parameter and μ the shear modulus. The energy equivalence (14.24) is at the root of the morphing method presented in Section 14.2.2.2, as it allows creation of an energy constraint over the morphing region.

14.2.2.2 Description of the morphing method

We assume that the material response in the domain of interest $\mathcal{B}$ is represented by a PD model with a micromodulus function defined by the scalar-valued function $\lambda^0(\|\xi\|)$. We can define an equivalent CCM model characterized by the stiffness tensor $\mathbf{K}^0$ in (14.24). Rather than using a PD model in the whole $\mathcal{B}$, we propose a hybrid local/nonlocal approach in which the PD model is used only where needed. The pure CCM model is prescribed in the subdomain $\mathcal{B}_1 \backslash \mathcal{B}_c$ (cf. Figure 14.1), where the deformation is smooth. The pure PD model is used in the subdomain $\mathcal{B}_2 \backslash \mathcal{B}_c$ (cf. Figure 14.1). A morphing between the two models, i.e., the hybrid model, is constructed in a morphing region $\mathcal{B}_c$ located between the two subdomains. Without a major restriction, we use the same boundary conditions as in the Arlequin method (cf. Figure 14.1).

A morphing between local and nonlocal models. We build a unified system of governing equations in $\mathcal{B}$ using a hybrid local/nonlocal model.

☐ **Kinematic admissibility and compatibility**

$$\varepsilon(\mathbf{x}) = \frac{1}{2}\left(\nabla\mathbf{u}(\mathbf{x}) + (\nabla\mathbf{u})^{\mathrm{T}}(\mathbf{x})\right) \qquad \forall\mathbf{x} \in \mathcal{B}_1, \tag{14.26}$$

$$\eta(\mathbf{x}',\mathbf{x}) = \mathbf{u}(\mathbf{x}') - \mathbf{u}(\mathbf{x}) \qquad \forall\mathbf{x},\mathbf{x}' \in \mathcal{B}_2, \tag{14.27}$$

$$\mathbf{u}(\mathbf{x}) = \overline{\mathbf{u}}(\mathbf{x}) \qquad \forall\mathbf{x} \in \partial\mathcal{B}_u. \tag{14.28}$$

☐ **Static admissibility**

$$div\,\sigma(\mathbf{x}) + \int_{\mathcal{H}(\mathbf{x},\delta)\cap\mathcal{B}} \{\mathbf{t}(\mathbf{x}',\mathbf{x}) - \mathbf{t}(\mathbf{x},\mathbf{x}')\} dV_{\mathbf{x}'} = -\mathbf{b}(\mathbf{x}) \quad \forall\mathbf{x} \in \mathcal{B}, \tag{14.29}$$

$$\sigma(\mathbf{x})\mathbf{n}(\mathbf{x}) = \overline{\mathbf{T}}(\mathbf{x}) \qquad \forall\mathbf{x} \in \partial\mathcal{B}_T. \tag{14.30}$$

☐ **Constitutive equations**

$$\sigma(\mathbf{x}) = \mathbf{K}(\mathbf{x}) : \varepsilon(\mathbf{x}) \qquad \forall\mathbf{x} \in \mathcal{B}_1, \tag{14.31}$$

$$\mathbf{t}(\mathbf{x}',\mathbf{x}) = \frac{\mathbf{C}(\mathbf{x},\|\xi\|)}{2}\eta(\mathbf{x}',\mathbf{x}) \qquad \forall\mathbf{x},\mathbf{x}' \in \mathcal{B}_2. \tag{14.32}$$

In (14.28), the displacement $\bar{\mathbf{u}}$ is imposed on the Dirichlet boundary $\partial\mathcal{B}_u$ (cf. Figure 14.1). In (14.30), the traction $\overline{\mathbf{T}}$ is imposed on the Neumann boundary $\partial\mathcal{B}_T$ (cf. Figure 14.1) with σ the classical stress tensor and $\mathbf{n}$ an outward unit normal. In (14.31), $\mathbf{K}(\mathbf{x})$ is the stiffness tensor of CCM at the point $\mathbf{x}$. In (14.32), $\mathbf{C}(\mathbf{x}, \|\xi\|)$ is the PD micromodulus function at the point $\mathbf{x}$. We define $\mathbf{C}(\mathbf{x}, \|\xi\|)$ by introducing a morphing function, α, such that

$$\mathbf{C}(\mathbf{x}, \|\xi\|) = \alpha(\mathbf{x})\lambda^0(\|\xi\|)\xi \otimes \xi. \tag{14.33}$$

Both the morphing function, α, and the stiffness tensor, $\mathbf{K}$, drive the morphing between the PD model and the CCM model: a fully PD model at a point $\mathbf{x}$ is obtained when $\mathbf{K}(\mathbf{x}) = \mathbf{0}$ and $\alpha(\mathbf{x}') = 1$, $\forall \mathbf{x}' \in \mathcal{H}(\mathbf{x}, \delta)$; a fully CCM model is obtained when $\mathbf{K}(\mathbf{x}) = \mathbf{K}^0$ and $\alpha(\mathbf{x}') = 0$, $\forall \mathbf{x}' \in \mathcal{H}(\mathbf{x}, \delta)$; and a relation between $\alpha(\mathbf{x})$ and $\mathbf{K}(\mathbf{x})$ is then prescribed to guarantee that both models are compatible and complementary to each other throughout the morphing region.

Defining the morphing: α and K. The morphing function α is defined *a priori* and is used to determine a PD/CCM partition. Then, based on the function α, we find the proper $\mathbf{K}$ to minimize coupling artifacts.

Constraints based on the local strain energy density. At any point $\mathbf{x} \in \mathcal{B}_c$, the elastic energy density of the hybrid local/nonlocal model can be written as

$$W(\mathbf{x}) = \frac{1}{2}\sigma(\mathbf{x}) : \varepsilon(\mathbf{x}) + \frac{1}{4}\int_{\mathcal{H}(\mathbf{x},\delta)\cap\mathcal{B}} \left(\mathbf{t}(\mathbf{x}', \mathbf{x}) - \mathbf{t}(\mathbf{x}, \mathbf{x}')\right) \cdot \eta(\mathbf{x}', \mathbf{x}) dV_{\mathbf{x}'}. \tag{14.34}$$

Using (14.31), (14.32), and (14.33) in (14.34) leads to

$$W(\mathbf{x}) = \frac{1}{2}\varepsilon(\mathbf{x}) : \mathbf{K}(\mathbf{x}) : \varepsilon(\mathbf{x}) + \int_{\mathcal{H}(\mathbf{x},\delta)\cap\mathcal{B}} \lambda^0(\|\xi\|)\frac{\alpha(\mathbf{x}) + \alpha(\mathbf{x}')}{2}\frac{(\xi \cdot \eta(\mathbf{x}', \mathbf{x}))^2}{4} dV_{\mathbf{x}'}. \tag{14.35}$$

We can then constrain the elastic energy density with a general function:

$$W(\mathbf{x}) = W^m(\mathbf{x}, \mathbf{u}|_{\mathcal{B}}). \tag{14.36}$$

Equation (14.36) ensures that the elastic energy density at a point $\mathbf{x}$ in the morphing region depends on a global function, W^m, of the displacement field over the whole body. Various choices for W^m exist, where the simpler choices are

- If we define

$$W^m(\mathbf{x}, \mathbf{u}|_{\mathcal{B}}) = \frac{1}{2}\varepsilon(\mathbf{x}) : \mathbf{K}^0 : \varepsilon(\mathbf{x}), \tag{14.37}$$

  then the elastic density at the point $\mathbf{x}$ of the hybrid local/nonlocal model is chosen as the strain energy density of the equivalent CCM model.

- If we define

$$W^m(\mathbf{x}, \mathbf{u}|_{\mathcal{B}}) = \int_{\mathcal{H}(\mathbf{x},\delta)\cap\mathcal{B}} \lambda^0(\|\xi\|)\frac{(\xi \cdot \eta(\mathbf{x}', \mathbf{x}))^2}{4} dV_{\mathbf{x}'}, \tag{14.38}$$

then the elastic energy density at the point $\mathbf{x}$ of the hybrid local/nonlocal model is chosen as the macroelastic energy density of the equivalent PD model.

The constraints (14.37) and (14.38) are, in general, not identical.

The constraint for a smooth coupling. We assume that the strain field varies slowly over the morphing region. This assumption is reasonable since the morphing should take place far from regions with strong gradients (the regions of interest). Thus, let us assume a homogeneous deformation over the neighborhood of the point $\mathbf{x} \in \mathcal{B}_c$, such that

$$\mathbf{F}(\mathbf{x}') \simeq \mathbf{F}(\mathbf{x}) = \overline{\mathbf{F}} \quad \forall \mathbf{x}' \in \mathcal{H}(\mathbf{x}, \delta). \tag{14.39}$$

Using (14.36) and (14.37), we obtain a general relation between the local stiffness tensor and the morphing function [29, 2]:

$$\mathbf{K}(\mathbf{x}) = (1-\alpha(\mathbf{x}))\mathbf{K}^0 + \frac{1}{2}\int_{\mathcal{H}(\mathbf{x},\delta)\cap\mathcal{B}} \lambda^0(\|\xi\|)\frac{\alpha(\mathbf{x})-\alpha(\mathbf{x}')}{2}\xi\otimes\xi\otimes\xi\otimes\xi\, dV_{\mathbf{x}'}. \tag{14.40}$$

14.2.2.3 One-dimensional analysis of ghost forces

While the constraint (14.40) guarantees energy conservation for a smooth strain field in the morphing region, it is not sufficient to eliminate all spurious effects. For example, we observe the presence of "ghost forces," which are common artifacts in AtC coupling methods [16, 19]. The magnitude of these ghost forces can be evaluated by solving an inverse problem in which a homogeneous deformation is prescribed. In Lubineau *et al.* [29], the ghost force f_g that has the dimension of stress has been identified for various choices of the morphing function, α, in a one-dimensional test, providing a micromodulus function $\lambda^0(\|\xi\|) = c^0/\|\xi\|^2$ or $\|\xi\| \leqslant \delta$ with c^0 constant. It is possible to relate the ghost force to the horizon, δ, and the length of the morphing region, L_c. For a piecewise linear choice of α, i.e., when α evolves linearly in the morphing region, it was shown that $\frac{f_g}{\sigma} = o(\frac{\delta}{L_c})$, where σ is a homogeneous stress in the body. Additionally, it was found that the magnitude of the ghost force reduces when the smoothness of α increases. Indeed, when α has a cubic evolution in the morphing region, $\frac{f_g}{\sigma} = o(\frac{\delta^2}{L_c^2})$. It was noted that the induced spurious effect was negligible in most cases, and it was restricted to the morphing region. Additional details are given in [29].

14.2.2.4 Numerical examples

Benchmark simulations. The examples in this section demonstrate the quality of the morphing method in three dimensions for homogeneous deformations. We study a cube subjected to traction and shear boundary conditions. Figure 14.6 provides the dimensions of the cube and its boundary conditions. In Figure 14.6(a), the spherical subdomain $\mathcal{B}_2\backslash\mathcal{B}_c$, colored dark gray, is described by a PD model, i.e.,

$\mathcal{B}_2\backslash\mathcal{B}_c = \{\mathbf{x} \mid \alpha(\mathbf{x}') = 1, \forall\mathbf{x}' \in \mathcal{H}(\mathbf{x},\delta)\}$; the white subdomain, $\mathcal{B}_1\backslash\mathcal{B}_c$, is described by a CCM model, i.e., $\mathcal{B}_1\backslash\mathcal{B}_c = \{\mathbf{x} \mid \alpha(\mathbf{x}') = 0, \forall\mathbf{x}' \in \mathcal{H}(\mathbf{x},\delta)\}$; the remaining light-gray subdomain, $\mathcal{B}_c$, is the morphing region where α is chosen as a piecewise linear function. In Figure 14.6(b), Dirichlet boundary conditions are imposed on the top surface of the cube with $\overline{u}_z = 0.25\mu$m, while keeping the bottom surface fixed, resulting in a state of constant traction in the z-direction. In Figure 14.6(c), Dirichlet boundary conditions given by

$$\begin{cases} \overline{u}_x = 0.2y + 0.2z, \\ \overline{u}_y = 0.2x + 0.2z, \\ \overline{u}_z = 0.2x + 0.2y, \end{cases} \tag{14.41}$$

are imposed on all the faces of the cube, resulting in a state of shear. For these benchmark simulations, the Young's modulus and Poisson's ratio are $E = 100\,$GPa and $\nu = 1/4$, respectively. For post-processing purposes, we extend the definition of the strain tensor (cf. (14.26)) to the whole body after the displacement solution is obtained, so that the strain field in the deformed configuration can be shown and compared conveniently over the whole region.

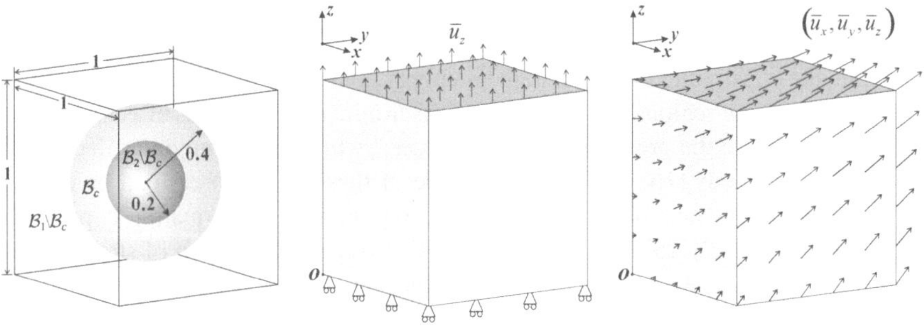

(a) (b) (c)

Figure 14.6: Dimensions and boundary conditions of the cube for the benchmark simulations (unit of length: μm).

The relative error in the strain for the principal component ε_{ZZ} for the traction test is given in Figure 14.7. We evaluate the relative error using the formula $\frac{\varepsilon_{ZZ}-\varepsilon_{ZZ}^A}{\varepsilon_{ZZ}^A}$, where ε_{ZZ}^A is the average value of the principal strain component ε_{ZZ}. Figure 14.7(a) displays the isosurfaces of the relative error in the strain at $+0.05\%$ and -0.05%. This reveals that the strain field is uniform across most of the deformed cube, except inside the morphing region. More details are given by the slice in Figure 14.7(b), where the relative error in the strain appears well contained inside the morphing region and is bounded by values below 0.25%.

In Figure 14.8, we present the results of the shear test for the principal strain com-

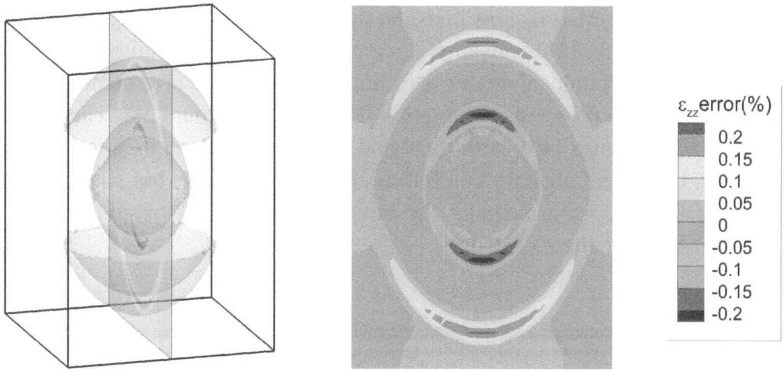

(a) (b)

Figure 14.7: Isosurfaces (a) and a contour slice (b) of the relative error in the strain component ε_{zz} for a three-dimensional cube under traction.

ponent ε_{yz}. The error in this case is comparable to the one described in Figure 14.7. Isosurfaces in Figure 14.8(a) are set at relative errors of 0.02% and −0.06%, respectively. In Figure 14.8(b), we display a slice which is a diagonal plane in the deformed bulk. Most of the relative errors are less than 0.14% and greater than −0.18%. Overall, the relative error in the strain produced by the ghost forces is restricted to the morphing region and remains small for a piecewise linear choice of α.

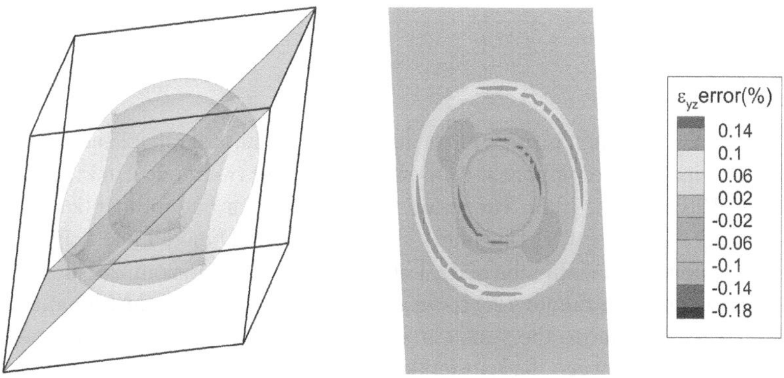

(a) (b)

Figure 14.8: Isosurfaces (a) and a contour slice (b) of the relative error of the strain component ε_{yz} for a three-dimensional cube under shear.

A cracked block. Let us now consider an application of the morphing method for a cube with a crack. In this example, we compare the results given by a PD model and a hybrid local/nonlocal model. The aim of this example is to demonstrate that the coupling approach preserves the quality of the displacement field. The cracked cube is presented in Figure 14.9, with its boundary conditions and the local and nonlocal partitions. Dirichlet boundary conditions are imposed on the top surface of the cube with $\bar{u}_z = 0.1\mu$m, while keeping the bottom surface fixed, resulting in a state of constant traction in the z-direction.

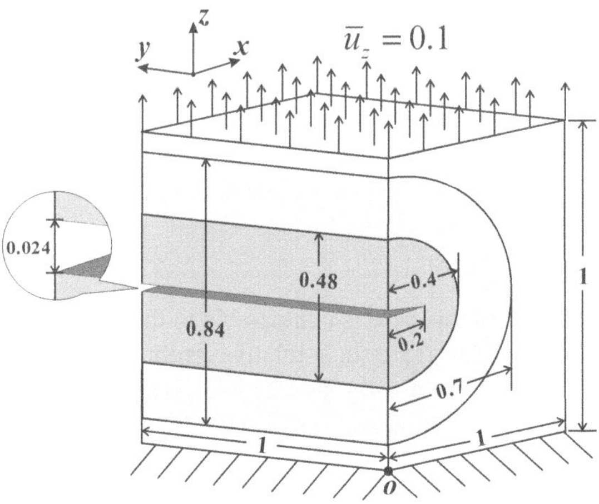

Figure 14.9: Dimensions and boundary conditions of a cracked cube (unit of length: μm).

A static simulation is run with the hybrid local/nonlocal model and the PD model. The principal strain component ε_{xx} is shown for these two models in Figures 14.10(a) and 14.10(b), respectively. The strain fields are similar but not identical on the surface. This is because for points near surfaces in the PD model, the neighborhood is not fully contained within the body. Let us display a slice to compare the strain fields within the cube. In Figure 14.11, we display the principal strain component ε_{xx} for slices perpendicular to the y-axis at $y = 0.5\mu$m (cf. Figure 14.9). A clear similarity is achieved between the two models, which suggests that the morphing method performs well under these conditions.

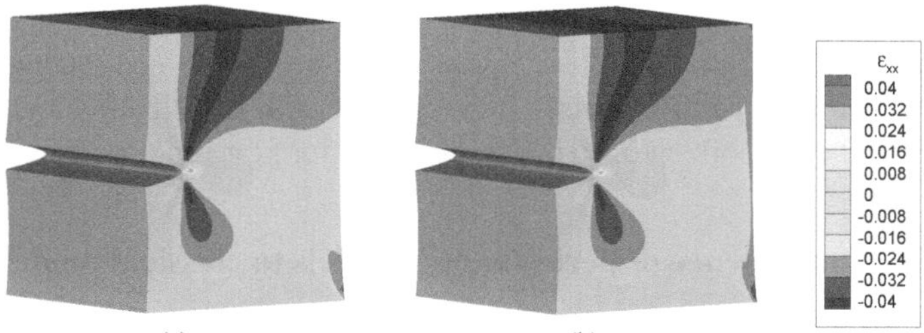

(a) (b)

Figure 14.10: The strain component ε_{xx} calculated by (a) the hybrid local/nonlocal model and (b) the PD model.

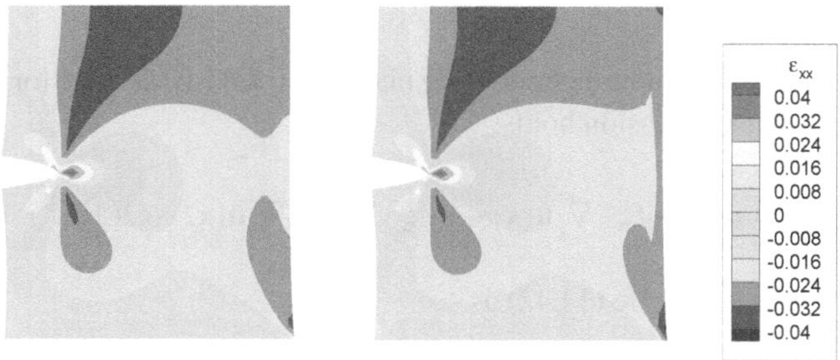

(a) (b)

Figure 14.11: Contour slices of the strain component ε_{xx} calculated by (a) the hybrid local/nonlocal model and (b) the PD model.

14.3 Force-based blending schemes *

In this section, we present coupling schemes based on blended governing equations. These coupling schemes are referred to as *force-based blending schemes*. As opposed to common force-based blending schemes appearing in the AtC coupling literature, where a coupling between nonlocal and local models is enforced across a bridge region, we attempt instead to derive a coupling scheme between PD and CCM using inherent connections between them. This section is based on the work presented in [34, 37].

Let $\mathcal{B}$ be a bounded body. The PD equilibrium equation is

$$\int_{\mathcal{B}} \lambda(\|\xi\|)(\xi \otimes \xi)\left(\mathbf{u}(\mathbf{x}') - \mathbf{u}(\mathbf{x})\right) dV_{\mathbf{x}'} + \mathbf{b}(\mathbf{x}) = \mathbf{0} \quad \forall \mathbf{x} \in \mathcal{B}, \tag{14.42}$$

*First published in P. Seleson, Y. D. Ha, and S. Beneddine, Concurrent coupling of bond-based peridynamics and the Navier equation of classical elasticity by blending, *International Journal for Multiscale Computational Engineering*, 13(2): 91113 (2015). Copyright 2015 by Begell House, Inc.

with $\mathbf{u}$ the displacement field, $\xi := \mathbf{x}' - \mathbf{x}$ the relative position vector in the reference configuration, $\mathbf{b}$ a body force density, and λ a scalar-valued function [41]. It is assumed that each point $\mathbf{x}$ directly interacts only with its neighborhood, defined in (14.19) by the PD horizon δ, so that $\lambda(\|\xi\|) = 0$ for $\|\xi\| > \delta$.

14.3.1 Convergence of peridynamic models to classical models

We now demonstrate the convergence of (14.42) to its CCM counterpart: the Navier–Cauchy equation of classical elasticity. This property is paramount to the derivation of the proposed force-based blending scheme in Section 14.3.2. The connection between PD and CCM models is established for points in the bulk of a material, i.e., far from boundaries. For that purpose, we define the internal subregion of $\mathcal{B}$ as

$$\mathcal{B}^{0,\delta} := \{\mathbf{x} \in \mathcal{B} : \mathcal{H}(\mathbf{x}, \delta) \subset \mathcal{B}\}. \tag{14.43}$$

Let $\mathbf{x} \in \mathcal{B}^{0,\delta}$ and assume a smooth displacement field $\mathbf{u}$, so that for $\|\xi\| \leqslant \delta$ the following Taylor expansion holds:

$$\mathbf{u}(\mathbf{x}') = \mathbf{u}(\mathbf{x}) + (\xi \cdot \nabla)\mathbf{u}(\mathbf{x}) + \frac{1}{2}(\xi \cdot \nabla)(\xi \cdot \nabla)\mathbf{u}(\mathbf{x}) + \mathcal{O}(\|\xi\|^3). \tag{14.44}$$

In this case, we can write (14.42) as

$$\left[\int_{\mathcal{H}(\mathbf{0},\delta)} \lambda(\|\xi\|)\xi_i\xi_j\xi_k dV_\xi\right] \frac{\partial u_j}{\partial x_k}(\mathbf{x})\hat{\mathbf{e}}_i + \frac{1}{2}\left[\int_{\mathcal{H}(\mathbf{0},\delta)} \lambda(\|\xi\|)\xi_i\xi_j\xi_k\xi_\ell dV_\xi\right] \frac{\partial^2 u_j}{\partial x_k \partial x_\ell}(\mathbf{x})\hat{\mathbf{e}}_i$$
$$+ \mathbf{b}(\mathbf{x}) + \mathcal{O}(\delta^2) = \mathbf{0}, \tag{14.45}$$

where we used a change of integration variable, the Einstein summation convention for repeated indices, and $\hat{\mathbf{e}}_i$ as a unit vector in the ith direction. Due to the symmetry of the integration range, the first term on the left-hand side of (14.45) vanishes, because its integrand is antisymmetric. In a further Taylor expansion, a term containing third partial derivatives would vanish as well. Consequently, the next term in the expansion, which contains fourth partial derivatives, is the leading term contributing to the truncation error. This term is the origin of the $\mathcal{O}(\delta^2)$ error in (14.45). We define a fourth-order tensor with components

$$C_{ijk\ell} := \frac{1}{2}\int_{\mathcal{H}(\mathbf{0},\delta)} \lambda(\|\xi\|)\xi_i\xi_j\xi_k\xi_\ell dV_\xi. \tag{14.46}$$

A more explicit expression for (14.46) can be derived as

$$C_{ijk\ell} = \left(\delta_{ij}\delta_{k\ell} + \delta_{ik}\delta_{j\ell} + \delta_{i\ell}\delta_{jk}\right)\frac{\Lambda}{3} \tag{14.47}$$

with

$$\Lambda := \frac{3}{2}\frac{|\mathcal{H}(\mathbf{0},1)|}{d+2}\int_0^\delta \lambda(r)r^{d+3}dr \tag{14.48}$$

and where δ_{ij} is the Kronecker delta and $|\mathcal{H}(\mathbf{0},1)|$ is the size of a neighborhood of unit horizon, i.e., its volume, area, or length for $d = 3, 2$, or 1, respectively, with d the dimension. In the limit as $\delta \to 0$, assuming $\Lambda < \infty$, Eq. (14.45) becomes

$$\frac{2\Lambda}{3}\left(\frac{\partial^2 u_j}{\partial x_i \partial x_j}(\mathbf{x}) + \frac{1}{2}\frac{\partial^2 u_i}{\partial x_k \partial x_k}(\mathbf{x})\right)\hat{\mathbf{e}}_i + \mathbf{b}(\mathbf{x}) = \mathbf{0}. \tag{14.49}$$

Equation (14.49) is the Navier–Cauchy equation of classical elasticity, for a fixed Poisson's ratio of $\nu = 1/4$ in 3D or $\nu = 1/3$ in 2D, with Young's modulus $E = 5\Lambda/6$ in 3D, $E = 8\Lambda/9$ in 2D, or $E = \Lambda$ in 1D.

Remark 14.1 Equation (14.42) corresponds to a bond-based PD model; as such, it is restricted to a Poisson's ratio of $\nu = 1/4$ in 3D or $\nu = 1/3$ in 2D. As a consequence, it converges to a CCM model with a single independent elastic modulus. An isotropic linear elastic state-based PD model would converge instead to a classical elastic model with two independent elastic moduli [41, 44]. ∎

Remark 14.2 The Taylor expansion (14.44) and the resulting governing equation (14.45) suggest that, for solutions with enough regularity, a PD model is equivalent to a higher-order gradient model, with leading terms corresponding to a CCM model. However, smooth solutions in PD may contain, in principle, information corresponding to a model with an infinite number of higher-order gradients. ∎

14.3.2 Derivation of force-based blending schemes

The derivations in Section14.3.1 demonstrated that the material response in the PD model (14.42) can be represented by the Navier–Cauchy equation of classical elasticity (14.49), in regions where displacement fields are smooth and the contribution of higher-order gradients of displacements (of order higher than three) is negligible. We thus exploit this connection to develop concurrent coupling schemes for PD and CCM, based on blended governing equations.

Following traditional blending methods, we divide the domain into three non-overlapping regions: $\mathcal{B}_{\text{PD}}$, $\mathcal{B}_{\text{CCM}}$, and $\mathcal{B}_{\text{b}}$, so that $\overline{\mathcal{B}} = \overline{\mathcal{B}_{\text{PD}}} \cup \overline{\mathcal{B}_{\text{CCM}}} \cup \overline{\mathcal{B}_{\text{b}}}$. A scalar-valued *blending function* is then defined as

$$\beta(\mathbf{x}) = \begin{cases} 1 & \mathbf{x} \in \mathcal{B}_{\text{PD}}, \\ 0 & \mathbf{x} \in \mathcal{B}_{\text{CCM}}, \\ \hat{\beta}(\mathbf{x}) & \mathbf{x} \in \mathcal{B}_{\text{b}}, \end{cases} \tag{14.50}$$

where $\hat{\beta} : \mathbb{R}^d \to (0,1)$ is commonly taken as a continuous polynomial (d is the dimension). We define the mean value of the blending function at $\mathbf{x} \in \mathcal{B}$ as

$$\langle \beta \rangle(\mathbf{x}) := \frac{\displaystyle\int_{\mathcal{H}(\mathbf{x},\delta)\cap\mathcal{B}} \beta(\mathbf{x}')dV_{\mathbf{x}'}}{\displaystyle\int_{\mathcal{H}(\mathbf{x},\delta)\cap\mathcal{B}} dV_{\mathbf{x}'}}. \tag{14.51}$$

The function $\langle\beta\rangle$ takes a value of 1 in $\mathcal{B}_{\mathrm{PD}}^{0,\delta}$ and a value of 0 in $\mathcal{B}_{\mathrm{CCM}}^{0,\delta}$. Let the *transition region* be

$$\mathcal{B}_{\mathrm{t}} := \{\mathbf{x} \in \mathcal{B} : 0 < \langle\beta\rangle(\mathbf{x}) < 1\}. \tag{14.52}$$

We assume the material response in $\mathcal{B}_{\mathrm{PD}} \setminus \mathcal{B}_{\mathrm{t}}$ to be described by the PD model (14.42). In $\mathcal{B}_{\mathrm{CCM}} \setminus \mathcal{B}_{\mathrm{t}}$, we assume displacement fields are smooth, so that we can use instead the Navier–Cauchy equation of classical elasticity (14.49). These two regions do not interact directly; they communicate through the transition region $\mathcal{B}_{\mathrm{t}}$. Such region is larger than $\mathcal{B}_{\mathrm{b}}$, which is usually referred to as the bridge region.

To derive a force-based blending scheme, we take the PD governing equation (14.42) and begin by splitting the contribution of the internal force density into two terms, through the introduction of symmetric weights based on the blending function. The resulting equation is

$$\begin{aligned}&\int_{\mathcal{B}} \frac{\beta(\mathbf{x})+\beta(\mathbf{x}')}{2} \lambda(\|\xi\|)(\xi \otimes \xi)\left(\mathbf{u}(\mathbf{x}')-\mathbf{u}(\mathbf{x})\right) dV_{\mathbf{x}'} \\ &+\int_{\mathcal{B}}\left(1-\frac{\beta(\mathbf{x})+\beta(\mathbf{x}')}{2}\right) \lambda(\|\xi\|)(\xi \otimes \xi)\left(\mathbf{u}(\mathbf{x}')-\mathbf{u}(\mathbf{x})\right) dV_{\mathbf{x}'}+\mathbf{b}(\mathbf{x})=\mathbf{0}.\end{aligned} \tag{14.53}$$

By introducing weights which are symmetric with respect to interchanging $\mathbf{x}$ and $\mathbf{x}'$, we preserve the antisymmetry of the integrand in each of the first two terms on the left-hand side of (14.53). We now observe that the second term on the left-hand side of (14.53) vanishes for $\mathbf{x} \in \mathcal{B}_{\mathrm{PD}}^{0,\delta}$, because $\beta(\mathbf{x}) = \beta(\mathbf{x}') = 1, \forall \mathbf{x}' \in \mathcal{H}(\mathbf{x}, \delta)$. Therefore, let us assume that displacement fields in $\mathcal{B} \setminus \mathcal{B}_{\mathrm{PD}}^{0,\delta}$ are smooth enough such that the Taylor expansion (14.44) holds for all $\mathbf{x} \in \mathcal{B} \setminus \mathcal{B}_{\mathrm{PD}}^{0,\delta}$ and $\|\xi\| \leqslant \delta$. In this case, we can write (14.53) as

$$\begin{aligned}&\int_{\mathcal{B}} \frac{\beta(\mathbf{x})+\beta(\mathbf{x}')}{2} \lambda(\|\xi\|)(\xi \otimes \xi)\left(\mathbf{u}(\mathbf{x}')-\mathbf{u}(\mathbf{x})\right) dV_{\mathbf{x}'} \\ &+\int_{\mathcal{B}}\left(1-\frac{\beta(\mathbf{x})+\beta(\mathbf{x}')}{2}\right) \lambda(\|\xi\|)(\xi \otimes \xi)\left[(\xi \cdot \nabla)\mathbf{u}(\mathbf{x})+\frac{1}{2}(\xi \cdot \nabla)(\xi \cdot \nabla)\mathbf{u}(\mathbf{x})\right] dV_{\mathbf{x}'} \\ &+\mathbf{b}(\mathbf{x})+E(\mathbf{x})=\mathbf{0}.\end{aligned} \tag{14.54}$$

The term $E(\mathbf{x})$ in (14.54) contains higher-order gradients of displacement fields originated in the Taylor expansion (14.44). We assume this term to be negligible; a bound for its magnitude will be provided in Remark 14.3 below. Reordering terms, we obtain the blending scheme

$$\begin{aligned}&\int_{\mathcal{B}} \frac{\beta(\mathbf{x})+\beta(\mathbf{x}')}{2} \lambda(\|\xi\|)(\xi \otimes \xi)\left(\mathbf{u}(\mathbf{x}')-\mathbf{u}(\mathbf{x})\right) dV_{\mathbf{x}'} \\ &+\left[\int_{\mathcal{B}}\left(1-\frac{\beta(\mathbf{x})+\beta(\mathbf{x}')}{2}\right) \lambda(\|\xi\|)\, \xi_i \xi_j \xi_k dV_{\mathbf{x}'}\right] \frac{\partial u_j}{\partial x_k}(\mathbf{x})\, \hat{e}_i \\ &+\left[\int_{\mathcal{B}}\left(1-\frac{\beta(\mathbf{x})+\beta(\mathbf{x}')}{2}\right) \lambda(\|\xi\|)\, \xi_i \xi_j \xi_k \xi_\ell \, dV_{\mathbf{x}'}\right] \frac{1}{2} \frac{\partial^2 u_j}{\partial x_k \partial x_\ell}(\mathbf{x})\, \hat{e}_i \\ &+\mathbf{b}(\mathbf{x})=\mathbf{0}.\end{aligned} \tag{14.55}$$

We observe that for $\mathbf{x} \in \mathcal{B}^{0,\delta}_{\text{PD}}$, (14.55) reduces to the PD model (14.42). Furthermore, for $\mathbf{x} \in \mathcal{B}^{0,\delta}_{\text{CCM}}$, (14.55) reduces to the Navier-Cauchy equation of classical elasticity (14.49). Note that, in this case, the integration is over the whole neighborhood of $\mathbf{x}$, which is symmetric, so that the second term on the left-hand side of (14.55) vanishes (its integrand is antisymmetric).

We now demonstrate that the blending scheme (14.55) is patch-test consistent. A model is said to be *patch-test consistent* if homogeneous deformations represent static solutions for that model, in the absence of external loads. Let $\mathbf{u} = G\mathbf{x}$, where G is a constant displacement gradient. In this case,

$$\mathbf{u}(\mathbf{x}') - \mathbf{u}(\mathbf{x}) = G\xi, \quad \frac{\partial u_j}{\partial x_k}(\mathbf{x}) = G_{jk}, \quad \frac{\partial^2 u_j}{\partial x_\ell \partial x_k}(\mathbf{x}) = 0.$$

A direct substitution shows that the blending scheme (14.55) is patch-test consistent.

Remark 14.3 The truncation error $E(\mathbf{x})$ in (14.54) at a point $\mathbf{x} \in \mathcal{B}^{0,\delta}$ can be bounded, given a PD model with $\lambda(\|\xi\|) \geqslant 0 \ \forall \xi \in \mathbb{R}^d$ [37], by considering higher-order gradients in the Taylor expansion (14.44), as

$$\begin{aligned}\|E(\mathbf{x})\| \leqslant & \frac{d^6(d+2)}{12}\frac{\Lambda}{3}\left\{\left[\overline{\beta}(\mathbf{x}) - \underline{\beta}(\mathbf{x})\right]\delta \sup_{j,k,\ell,m}\left|\frac{\partial^3 u_j}{\partial x_m \partial x_\ell \partial x_k}(\mathbf{x})\right|\right. \\ & \left. + \left[1 - \frac{\beta(\mathbf{x}) + \underline{\beta}(\mathbf{x})}{2}\right] d\,\delta^2 \sup_{\substack{\mathbf{x}' \in \mathcal{H}(\mathbf{x},\delta) \\ j,k,\ell,m,n}}\left|\frac{\partial^4 u_j}{\partial x_n \partial x_m \partial x_\ell \partial x_k}(\mathbf{x}')\right|\right\}\end{aligned} \tag{14.56}$$

with d the dimension, Λ given in (14.48), and

$$\overline{\beta}(\mathbf{x}) := \max_{\mathbf{x}' \in \mathcal{H}(\mathbf{x},\delta)} \beta(\mathbf{x}'), \quad \underline{\beta}(\mathbf{x}) := \min_{\mathbf{x}' \in \mathcal{H}(\mathbf{x},\delta)} \beta(\mathbf{x}').$$

The error estimate (14.56) provides two pieces of information. First, it bounds the error incurred across transition regions, due to the truncation of the Taylor expansion in (14.44). Second, it quantifies the suitability of the classical local model to describe the material response in the classical region. The first term inside the curly brackets on the right-hand side of (14.56) only contributes for points in transition regions, for which $\overline{\beta}(\mathbf{x}) - \underline{\beta}(\mathbf{x}) \neq 0$. The second term, on the other hand, may also give a nonzero contribution outside transition regions, in the classical region. However, wherever displacement fields can be approximated by a quadratic profile within the neighborhood of a point, both of these terms are small. ∎

14.3.3 A numerical example

We perform static simulations comparing the blending scheme (14.55) and the PD model (14.42). Analytical results showed that the blending scheme is patch-test consistent. Therefore, we do not present here numerical studies of those types of problems; instead, we focus on an application example. We compare the material response of a square plate with a horizontal rectangular slit at its center under both normal and shear loading using the PD and the blended models. Let the square plate be defined by $\mathcal{B}_{\text{plate}} = (-\frac{1}{2}, \frac{1}{2}) \times (-\frac{1}{2}, \frac{1}{2})$ and the rectangular slit by $\mathcal{B}_{\text{slit}} = (-\frac{1}{10}, \frac{1}{10}) \times (-\frac{1}{120}, \frac{1}{120})$. The domain of interest is thus $\mathcal{B} = \mathcal{B}_{\text{plate}} \setminus \mathcal{B}_{\text{slit}}$. In this case, the domain has two boundaries: the outer boundary $\partial\mathcal{B}_{\text{plate}}$ and the inner boundary $\partial\mathcal{B}_{\text{slit}}$. An illustration is given in Figure 14.12(a).

The boundary conditions in the blended problem combine local boundary conditions for the outer boundary consistent with the CCM model, and nonlocal boundary conditions for the inner boundary consistent with the PD model. For the outer boundary, we impose normal and shear traction through Dirichlet boundary conditions on the top edge, while keeping the bottom edge fixed. The inner boundary is taken as a free surface (cf. Figure 14.12(c)). To avoid surface effects in the PD problem, in comparison to the blended problem, we use the numerical solution of the blended problem to impose nonlocal Dirichlet-like boundary conditions in the boundary layer $\mathcal{B} \setminus \mathcal{B}^{0,\delta}_{\text{plate}}$, adjacent to the outer boundary (cf. Figure 14.12(b)). The inner boundary in the PD problem is treated as a free surface.

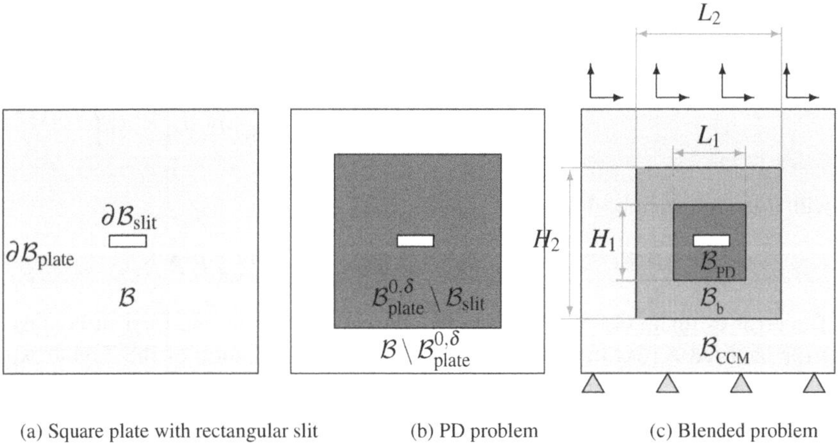

Figure 14.12: (a) Geometry of a square plate with a horizontal rectangular slit in its center. (b) Decomposition of the domain in the PD problem into the domain of interest $\mathcal{B}^{0,\delta}_{\text{plate}} \setminus \mathcal{B}_{\text{slit}}$ and the boundary layer $\mathcal{B} \setminus \mathcal{B}^{0,\delta}_{\text{plate}}$; nonlocal Dirichlet-like boundary conditions, consistent with the blended problem, are imposed in the boundary layer. (c) Domain decomposition and boundary conditions for the blended problem.

For the PD problem, we assume a two-dimensional version of (14.42) with $\lambda(\|\xi\|) = c/\|\xi\|^3$ or $\|\xi\| \leqslant \delta$ and 0 otherwise, and c a constant; using (14.48), we obtain $c = 9E/\pi\delta^3$. We take $b(x) = 0$ and let $g(\mathbf{x})$ be an approximation to the solution of the blended problem (14.58), obtained through a numerical simulation. The PD problem is given by

$$\begin{cases} \int_{\mathcal{B}} \frac{c}{\|\xi\|^3}(\xi \otimes \xi)(\mathbf{u}(\mathbf{x}') - \mathbf{u}(\mathbf{x}))dV_{\mathbf{x}'} = 0 & \forall \mathbf{x} \in \mathcal{B}^{0,\delta}_{\text{plate}} \setminus \mathcal{B}_{\text{slit}}, \\ \mathbf{u}(\mathbf{x}) = g(\mathbf{x}) & \forall \mathbf{x} \in \mathcal{B} \setminus \mathcal{B}^{0,\delta}_{\text{plate}}. \end{cases} \tag{14.57}$$

For the blended problem, the domain $\mathcal{B}$ is decomposed into the subdomains

$$\begin{aligned} \mathcal{B}_{\text{PD}} &= (-\tfrac{1}{2}L_1, \tfrac{1}{2}L_1) \times (-\tfrac{1}{2}H_1, \tfrac{1}{2}H_1) \setminus \mathcal{B}_{\text{slit}}, \\ \mathcal{B}_{\text{b}} &= [-\tfrac{1}{2}L_2, \tfrac{1}{2}L_2] \times [-\tfrac{1}{2}H_2, \tfrac{1}{2}H_2] \setminus (\mathcal{B}_{\text{PD}} \cup \mathcal{B}_{\text{slit}}), \\ \mathcal{B}_{\text{CCM}} &= \mathcal{B} \setminus (\mathcal{B}_{\text{b}} \cup \mathcal{B}_{\text{PD}}), \end{aligned}$$

as illustrated in Figure 14.12(c); the transition region $\mathcal{B}_{\text{t}}$ is given by (14.52). The blended problem is given by

$$\begin{cases} \int_{\mathcal{B}} \frac{c}{\|\xi\|^3}(\xi \otimes \xi)(\mathbf{u}(\mathbf{x}') - \mathbf{u}(\mathbf{x}))dV_{\mathbf{x}'} = 0, & \forall \mathbf{x} \in \mathcal{B}_{\text{PD}} \setminus \mathcal{B}_{\text{t}}, \\ \int_{\mathcal{H}(\mathbf{x},\delta)} \left(\frac{\beta(\mathbf{x}) + \beta(\mathbf{x}')}{2}\right) \frac{c}{\|\xi\|^3}(\xi \otimes \xi)(\mathbf{u}(\mathbf{x}') - \mathbf{u}(\mathbf{x}))dV_{\mathbf{x}'} & \\ -\frac{1}{2}\left[\int_{\mathcal{H}(\mathbf{0},\delta)} \beta(\mathbf{x}+\xi)\frac{c}{\|\xi\|^3}\xi_i\xi_j\xi_k dV_\xi\right]\frac{\partial u_j}{\partial x_k}(\mathbf{x})\,\hat{e}_i & \\ +\left[\int_{\mathcal{H}(\mathbf{0},\delta)} \left(1 - \frac{\beta(\mathbf{x}) + \beta(\mathbf{x}+\xi)}{2}\right)\frac{c}{2\|\xi\|^3}\xi_i\xi_j\xi_k\xi_\ell\, dV_\xi\right]\frac{\partial^2 u_j}{\partial x_k \partial x_\ell}(\mathbf{x})\,\hat{e}_i = 0, & \forall \mathbf{x} \in \mathcal{B}_{\text{t}}, \\ \frac{3E}{4}\left[\nabla(\nabla \cdot \mathbf{u})(\mathbf{x}) + \frac{1}{2}\nabla^2 \mathbf{u}(\mathbf{x})\right] = 0, & \forall \mathbf{x} \in \mathcal{B}_{\text{CCM}} \setminus \mathcal{B}_{\text{t}}, \\ \mathbf{u}(\mathbf{x}) = 0 & \forall \mathbf{x} \in \mathcal{B} : y = -0.5, \\ \mathbf{u}(\mathbf{x}) = (0.05, 0.05) & \forall \mathbf{x} \in \mathcal{B} : y = 0.5, \\ t(\mathbf{x}) = 0 & \forall \mathbf{x} \in \mathcal{B} : x = \pm 0.5, \end{cases} \tag{14.58}$$

with $\mathbf{x} = (x, y)$, the traction vector $t(\mathbf{x}) = \sigma(\mathbf{x})n(\mathbf{x})$, and where the stress tensor is given by

$$\sigma = \frac{3E}{4}\left[\frac{1}{2}\left(\nabla\mathbf{u} + (\nabla\mathbf{u})^{\text{T}}\right) + \frac{1}{2}\text{tr}(\nabla\mathbf{u})I\right], \tag{14.59}$$

n is an outward unit normal, and I is a unit tensor.

We use a meshfree method [43] to discretize the PD model, a finite-difference method to discretize the CCM model, and a combined meshfree/finite-difference approach to discretize the blending scheme. The blending function is chosen as piecewise constant with $\hat{\beta} = 0.5$ in $\mathcal{B}_{\text{b}}$ (cf. (14.50)). The simulation results, using the parameters given in Table 14.1, are presented in Figure 14.13, the values of the strain component ε_{xx} in the deformed configuration. A comparison between the solutions of the PD and blended problems demonstrates that the blending scheme recovers the PD solution at a reduced computational cost. In this case, the blended simulation incurs in a speedup of approximately $11\times$ over the PD simulation. In addition, we observe that the PD model and the CCM model (as observed in the classical region of the blended simulation) have the same material response outside the region of the slit, as expected.

It should be mentioned that the efficiency of the blending scheme can be significantly improved as follows. First, it is possible to optimize the domain decomposition geometry, as well as the functional form of the blending function, in a way that reduces the computational cost, without significantly affecting accuracy. Second, the number of degrees of freedom in the classical local region can be reduced, by coarsening the underlying mesh, while preserving the resolution in the nonlocal and transition regions. This is expected to highly decrease the computational time, without significantly affecting the accuracy of solutions, as demonstrated in [34].

Table 14.1: Parameters for the rectangular slit simulations.

E	δ	Δx	L_1	L_2	H_1	H_2
1	0.05	1/120	$1/5+4\delta$	$1/5+6\delta$	$1/60+4\delta$	$1/60+6\delta$

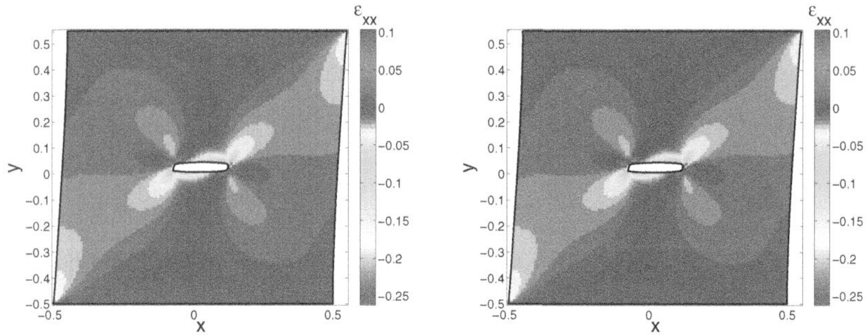

(a) PD model (b) Blended model

Figure 14.13: Profiles for the strain component ε_{xx}, in the deformed configuration, for the PD model (a) and the blended model (b).

14.4 Summary

We presented in this chapter two main approaches to couple peridynamics (PD) and classical continuum mechanics (CCM): energy-based blending schemes and force-based blending schemes. We focused here on the coupling of linear elastic models. We demonstrated that, even for such simple constitutive models, special techniques are required in the coupling process to preserve characteristic properties (such as the energy) of the reference models, and to prevent the appearance of spurious effects across blending regions. Connections between PD and CCM models have been used to obtain proper blending schemes. Numerical results showed that coupled PD/CCM models may provide a means to achieve efficient and accurate solutions of PD problems.

The research outcomes are relatively new and the field of PD/CCM coupling is expected to be an active area of research, the success of which will be key to the application of PD to large-scale, complex engineering simulations. Up to now, the morphing method has been developed to couple state-based PD with CCM [26]. Applications of the proposed coupling methods to time-dependent problems require additional studies focusing on potential wave reflections across blending regions. The adaptive coupling technique is an alternative to restrict the PD model adaptively to key areas, so as to reduce computational cost [3]. Extensions of the current coupling methods are required to treat problems beyond linear elasticity, such as those involving plasticity, visco-elasticity/plasticity, or damage. An important and very recent step forward is a new way to look at PD as a regularization tool in a more general damage to PD framework [25].

Acknowlegments

The work of DJL was supported by Sandia. Sandia National Laboratories is a multi-program laboratory managed and operated by Sandia Corporation, a wholly owned subsidiary of Lockheed Martin Corporation, for the U.S. Department of Energy's National Nuclear Security Administration under contract DE-AC04-94AL85000. SAND 2014-4214P.

References

[1] B. Aksoylu and M. L. Parks. Variational theory and domain decomposition for nonlocal problems. *Applied Mathematics and Computation*, 217: 6498–6515,

2011.

[2] Y. Azdoud, F. Han , and G. Lubineau. A morphing framework to couple non-local and local anisotropic continua. *International Journal of Solids and Structures*, 50(9): 1332–1341, 2013.

[3] Y. Azdoud, F. Han, and G. Lubineau. The morphing method as a flexible tool for adaptive local/non-local simulation of static fracture. *Computational Mechanics*, 54(3): 711–722, 2014.

[4] S. Badia, P. Bochev, R. Lehoucq, M. Parks, J. Fish, M. A. Nuggehally, and M. Gunzburger. A force-based blending model for atomistic-to-continuum coupling. *International Journal for Multiscale Computational Engineering*, 5: 387–406, 2007.

[5] S. Badia, M. Parks, P. Bochev, M. Gunzburger, and R. Lehoucq. On atomistic-to-continuum coupling by blending. *Multiscale Modeling and Simulation*, 7(1): 381–406, 2008.

[6] P. T. Bauman, H. Ben Dhia, N. Elkhodja, J.T. Oden, and S. Prudhomme. On the application of the Arlequin method to the coupling of particle and continuum models. *Computational Mechanics*, 42: 511–530, 2008.

[7] P.T. Bauman, J.T. Oden, and S. Prudhomme. Adaptive multiscale modeling of polymeric materials with Arlequin coupling and Goals algorithms. *Computer Methods in Applied Mechanics and Engineering*, 198(5–8): 799–818, 2009.

[8] T. Belytschko and S. P. Xiao. Coupling methods for continuum model with molecular model. *International Journal for Multiscale Computational Engineering*, 1(1): 115–126, 2003.

[9] H. Ben Dhia. Problèmes mécaniques multi-échelles: La méthode Arlequin. *Comptes Rendus de l'Académie des Sciences - Series IIB - Mechanics-Physics-Astronomy*, 326(12): 899–904, 1998.

[10] H. Ben Dhia. Numerical modeling of multiscale problems: The Arlequin method. In *Proceedings of the First European Conference on Computational Mechanics*, 1999.

[11] H. Ben Dhia. Further insights by theoretical investigations of the multiscale Arlequin method. *International Journal for Multiscale Computational Engineering*, 6(3): 215–232, 2008.

[12] H. Ben Dhia and O. Jamond. On the use of XFEM within the Arlequin framework for the simulation of crack propagation. *Computer Methods in Applied Mechanics and Engineering*, 199(21–22): 1403–1414, 2010.

[13] H. Ben Dhia and G. Rateau. The Arlequin method as a flexible engineering design tool. *International Journal for Numerical Methods in Engineering*, 62(11):1442–1462, 2005.

[14] P. Bochev, R. Lehoucq, M. Parks, S. Badia, and M. Gunzburger. Blending methods for coupling atomistic and continuum models. In *Multiscale Methods: Bridging the Scales in Science and Engineering*, pages 165–189. Oxford University Press, Oxford, 2009.

[15] G. Bontempi and Y. Le Borgne. An adaptive modular approach to the mining of sensor network data. In *Proceedings of the Workshop on Data Mining in Sensor Networks, SIAM SDM*, pages 3–9. SIAM Press, 2005.

[16] L. Chamoin, S. Prudhomme, H. Ben Dhia, and J.T. Oden. Ghost forces and spurious effects in atomic-to-continuum coupling methods by the Arlequin approach. *International Journal for Numerical Methods in Engineering*, 83(8–9):1081–1113, 2010.

[17] W. A. Curtin and R. E. Miller. Atomistic/continuum coupling in materials modeling. *Modelling and Simulation in Materials Science and Engineering*, 11: R33–R68, 2003.

[18] M. Di Paola, G. Failla, and M. Zingales. The mechanically-based approach to 3D non-local linear elasticity theory: Long-range central interactions. *International Journal of Solids and Structures*, 47(18–19):2347–2358, 2010.

[19] M. Dobson and M. Luskin. An analysis of the effect of ghost force oscillation on the quasicontinuum error. *ESAIM: Mathematical Modelling and Numerical Analysis*, 43: 591–604, 2009.

[20] Q. Du, M. Gunzburger, R. B. Lehoucq, and K. Zhou. A nonlocal vector calculus, nonlocal volume-constrained problems, and nonlocal balance laws. *Mathematical Models and Methods in Applied Sciences*, 23(3): 493–540, 2013.

[21] J. Fish, M. A. Nuggehally, M. S. Shephard, C. R. Picu, S. Badia, M. L. Parks, and M. Gunzburger. Concurrent AtC coupling based on a blend of the continuum stress and the atomistic force. *Computer Methods in Applied Mechanics and Engineering*, 196(45–48): 4548–4560, 2007.

[22] P.-A. Guidault and T. Belytschko. On the L^2 and the H^1 couplings for an overlapping domain decomposition method using Lagrange multipliers. *International Journal for Numerical Methods in Engineering*, 70(3): 322–350, 2007.

[23] P.-A. Guidault and T. Belytschko. Bridging domain methods for coupled atomistic-continuum models with L^2 or H^1 couplings. *International Journal for Numerical Methods in Engineering*, 77(11): 1566–1592, 2009.

[24] F. Han and G. Lubineau. Coupling of nonlocal and local continuum models by the Arlequin approach. *International Journal for Numerical Methods in Engineering*, 89(6):671–685, 2012.

[25] F. Han, G. Lubineau, and Y. Azdoud. Adaptive coupling between damage mechanics and peridynamics: a route for objective simulation of material degra-

dation up to complete failure. *Journal of the Mechanics and Physics of Solids*, In press, 2016.

[26] F. Han, G. Lubineau, Y. Azdoud, and A. Askari. A morphing approach to couple state-based peridynamics with classical continuum mechanics. *Computer Methods in Applied Mechanics and Engineering*, 301: 336–358, 2016.

[27] J. Knap and M. Ortiz. An analysis of the quasicontinuum method. *Journal of the Mechanics and Physics of Solids*, 49(9): 1899–1923, 2001.

[28] X. H. Li, M. Luskin, C. Ortner, and A. Shapeev. Theory-based benchmarking of the blended force-based quasicontinuum method. *Computer Methods in Applied Mechanics and Engineering*, 268(1): 763–781, 2014.

[29] G. Lubineau, Y. Azdoud, F. Han, C. Rey, and A. Askari. A morphing strategy to couple non-local to local continuum. *Journal of the Mechanics and Physics of Solids*, 60(6): 1088–1102, 2012.

[30] M. Luskin, C. Ortner, and B. Van Koten. Formulation and optimization of the energy-based blended quasicontinuum method. *Computer Methods in Applied Mechanics and Engineering*, 253: 160–168, 2013.

[31] S. Prudhomme, H. Ben Dhia, P. T. Bauman, N. Elkhodja, and J.T. Oden. Computational analysis of modeling error for the coupling of particle and continuum models by the Arlequin method. *Computer Methods in Applied Mechanics and Engineering*, 197(41–42): 3399–3409, 2008.

[32] G. Rateau. *Méthode Arlequin pour les problèmes mécaniques multi-échelles*. PhD thesis, 2003.

[33] P. D. Seleson. *Peridynamic Multiscale Models for the Mechanics of Materials: Constitutive Relations, Upscaling from Atomistic Systems, and Interface Problems*. PhD thesis, Florida State University, 2010. Electronic Theses, Treatises and Dissertations. Paper 273.

[34] P. Seleson, S. Beneddine, and S. Prudhomme. A force-based coupling scheme for peridynamics and classical elasticity. *Computational Materials Science*, 66:34–49, 2013.

[35] P. Seleson and M. Gunzburger. Bridging methods for atomistic-to-continuum coupling and their implementation. *Communications in Computational Physics*, 7(4): 831–876, 2010.

[36] P. Seleson, M. Gunzburger, and M. L. Parks. Interface problems in nonlocal diffusion and sharp transitions between local and nonlocal domains. *Computer Methods in Applied Mechanics and Enginering*, 266: 185–204, 2013.

[37] P. Seleson, Y.-D. Ha, and S. Beneddine. Concurrent coupling of bond-based peridynamics and the Navier equation of classical elasticity by blending. *International Journal for Multiscale Computational Engineering*, 13(2): 91–113, 2015..

[38] P. Seleson, M. L. Parks, M. Gunzburger, and R. B. Lehoucq. Peridynamics as an upscaling of molecular dynamics. *Multiscale Modeling and Simulation*, 8(1): 204–227, 2009.

[39] V.B. Shenoy, R. Miller, E.B. Tadmor, R. Phillips, and M. Ortiz. Quasicontinuum models of interfacial structure and deformation. *Physical Review Letters*, 80(4): 742–745, 1998.

[40] V.B. Shenoy, R. Miller, E.B. Tadmor, D. Rodney, R. Phillips, and M. Ortiz. An adaptive finite element approach to atomic-scale mechanics – The quasicontinuum method. *Journal of the Mechanics and Physics of Solids*, 47: 611–642, 1999.

[41] S.A. Silling. Reformulation of elasticity theory for discontinuities and long-range forces. *Journal of the Mechanics and Physics of Solids*, 48: 175–209, 2000.

[42] S.A. Silling. Linearized theory of peridynamic states. *Journal of Elasticity*, 99: 85–111, 2010.

[43] S.A. Silling and E. Askari. A meshfree method based on the peridynamic model of solid mechanics. *Computers & Structures*, 83:1526–1535, 2005.

[44] S.A. Silling, M. Epton, O. Weckner, J. Xu, and E. Askari. Peridynamic states and constitutive modeling. *Journal of Elasticity*, 88:151–184, 2007.

[45] S.A. Silling and R.B. Lehoucq. Peridynamic theory of solid mechanics. *Advances in Applied Mechanics*, 44:73–168, 2010.

[46] S. A. Silling and R. B. Lehoucq. Convergence of peridynamics to classical elasticity theory. *Journal of Elasticity*, 93: 1337, 2008.

[47] E.B. Tadmor. *The quasicontinuum method.* PhD thesis, 1996.

[48] E.B. Tadmor, M. Ortiz, and R. Phillips. Quasicontinuum analysis of defects in solids. *Philosophical Magazine A*, 73(6): 1529–1563, 1996.

[49] B. Van Koten and M. Luskin. Analysis of energy-based blended quasi-continuum approximations. *SIAM Journal on Numerical Analysis*, 49(5): 2182–2209, 2011.

[50] S. P. Xiao and T. Belytschko. A bridging domain method for coupling continua with molecular dynamics. *Computer Methods in Applied Mechanics and Engineering*, 193(17–20): 1645–1669, 2004.

Chapter 15

A Peridynamic Model for Corrosion Damage

Ziguang Chen

University of Nebraska-Lincoln

Florin Bobaru

University of Nebraska-Lincoln

CONTENTS

Abstract 438
15.1 Introduction 438
15.2 Electrochemical kinetics 443
15.3 Problem formulation of 1D pitting corrosion 446
15.4 The peridynamic formulation for 1D pitting corrosion 449
15.5 Results and discussion of 1D pitting corrosion 453
15.5.1 Pit corrosion depth proportional to $\sqrt{t}$ 453
15.5.2 Activation-controlled, diffusion-controlled, and IR-controlled corrosion 456
15.6 Corrosion damage and the Concentration-Dependent Damage (CDD) model 459
15.6.1 Damage evolution 462
15.6.2 Saturated concentration 464
15.7 Formulation and results of 2D and 3D pitting corrosion 465
15.7.1 PD formulation of 2D and 3D pitting corrosion 466
15.7.2 The Concentration-Dependent Damage (CDD) model for pitting corrosion: example in 2D 469
15.7.3 A coupled corrosion/damage model for pitting corrosion: 2D example 472
15.7.4 Diffusivity affects the corrosion rate 474
15.7.5 Pitting corrosion with the CDD+DDC model in 3D 475
15.8 Pitting corrosion in heterogeneous materials: examples in 2D 476
15.8.1 Pitting corrosion in layer structures 476
15.8.2 Pitting corrosion in a material with inclusions: a 2D example 479
15.9 Conclusions 480

Acknowledgments ... 481
15.10 Appendix ... 481
15.10.1 Convergence study for 1D diffusion-controlled corrosion ... 481
15.10.2 Convergence study for 2D activation-controlled corrosion with Concentration-Dependent Damage model ... 482
References ... 484

Abstract

In this chapter we introduce a peridynamic model for the evolution of damage from pitting corrosion capable of capturing subsurface damage. The model is based on the recent work that has appeared in [12]. Further extension and validation of the model has appeared in [14]. The anodic reaction in corrosion processes (in which electroplating is negligible) is modeled as an effective peridynamic diffusion process in the electrolyte/solid system coupled with a phase-change mechanism that allows for autonomous evolution of the moving interface. In order to simulate creation of subsurface damage, we introduce a corrosion damage model based on a stochastic relationship that connects the concentration in the metal to the damage of peridynamic mechanical-bonds that are superposed onto diffusion-bonds. We study convergence of this formulation for the diffusion-dominated stage. The model leads to formation of a subsurface damage layer, seen in experiments. We validate results against experiments on pit growth rate and polarization data for pitting corrosion. We extend the 1D model to the 2D and 3D, and introduce a new damage-dependent corrosion model to account for broken mechanical bonds that enhance the corrosion rate. This coupled model can predict the pit shape and damage profile in materials with microstructural heterogeneities, such as defects, interfaces, inclusions, and grain boundaries.

15.1 Introduction

Corrosion is an extremely widespread danger to the durability and safety of structures, and there is considerable interest in this phenomenon, reflected in a large number of studies e.g. [21, 19, 20, 28, 39]. Pitting corrosion is a localized form of corrosion that rapidly penetrates the substrate thickness and produces deep, narrow cavities or holes in the material (see Figure 15.1 *). Pitting typically develops at weak spots in the surface film. A small area becomes anodic while the larger surrounding area remains cathodic, and corrosion is driven by the potential difference between the anodic area inside the pit and the surrounding cathodic area.

Several factors can lead to pitting corrosion, including localized chemical or me-

*For the color figure please see the electronic version of the handbook.

chanical damage to the protective oxide film; water chemistry factors that can cause breakdown of a passive film; low dissolved oxygen concentrations, which tend to render a protective oxide film less stable; and high concentrations of chlorides, as in seawater. Pitting can also occur in the presence of non-uniformities in the metal structure of the component, such as nonmetallic inclusions, or when a protective coating is poorly applied or has localized damage.

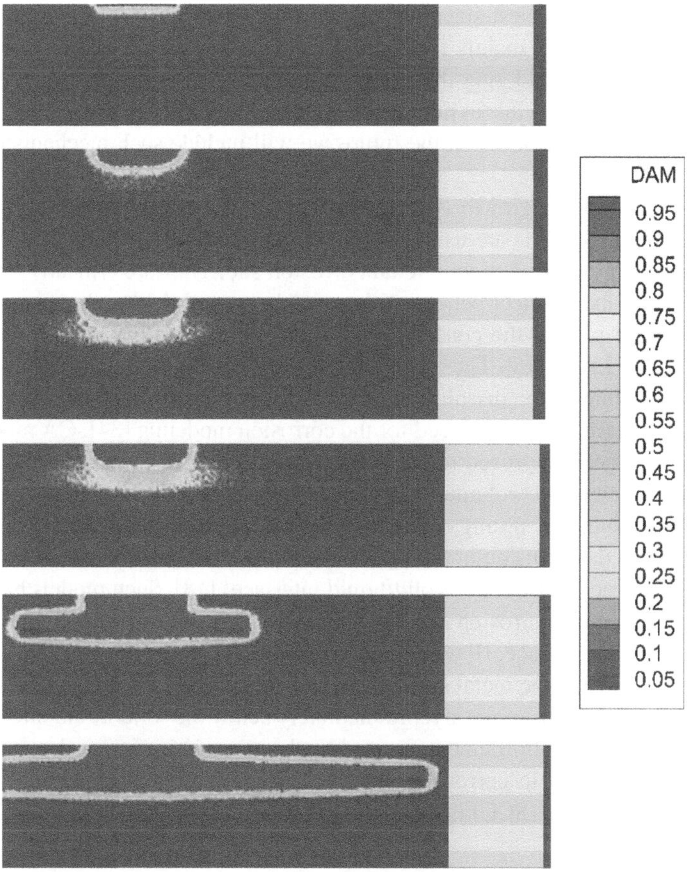

Figure 15.1: Snapshots from a peridynamic simulation of corrosion damage. The initial size of the defect in the top layer has a length of $60\ \mu$m. The legend represents the mechanical degradation or damage caused by the corrosion process. Red corresponds to completely corroded material, while blue shows pristine metal without any corrosion damage. The colors between blue and red describe degrees of partially damaged material.

Considered to be more hazardous than uniform corrosion damage, pitting is more difficult to detect, predict, and design against. A small, narrow pit with minimal overall metal loss can lead to the failure of an entire engineering system [21]. Pitting corrosion is a common denominator in almost all types of localized corrosion attack, and may assume different shapes, such as open, uncovered pits or pits covered with a semi-permeable membrane of corrosion products. Pits can be either hemispherical or cup-shaped.

Fundamental studies on pitting corrosion have focused on one of the following stages of the pitting corrosion process [17, 18, 23, 31, 46]: characteristics of the passive film, the earliest stages of passive film breakdown, and the pit growth. In the present work, we mainly focus on modeling the anodic reaction involved in corrosion, without considering the passive film breakdown (which we assume already broken, see [21] for a discussion on the breakdown of passive film) or explicitly considering the repassivation. In the future we will include such mechanisms involved in pitting corrosion.

The study of pit growth is the main focus in the community of corrosion modeling. Existing models are usually of deterministic nature, aiming at describing the propagation process of an individual corrosion site. They are typically based on diffusion laws and on electrochemical kinetics laws, and deliver the geometrical evolution of the pit and the concentration of dissolved ions throughout the corrosion site [17, 18, 23, 31, 46]. The movement of the electrode boundary was not included in the modeling of the dissolution process. More recently, Cellular Automata (CA) techniques have been introduced for the corrosion modeling [34]. CA models of corrosion have been improved to capture the roughening of the reaction front [49]. In addition to CA, finite volume model has also been used to capture the diffusion and activation-controlled pitting corrosion [42, 43]. Most of these deterministic models are based on the diffusion in the solution, and the reaction is considered to only happen on the reaction front (the solid/liquid interface) [18]. Such models have successfully predicted the corrosion rate and some have been able to capture the roughening of the corrosion surface. However, since these models consider that the corrosion reaction only affects the metal surface, they cannot capture changes in the mechanical properties in the subsurface layer immediately below the solid/liquid interface. These changes, such as embrittlement induced by the corrosion or stress-dependence of the diffusion processes in corrosion, are determining factors in explaining how Stress Corrosion Cracking (SCC) is triggered and how it progresses in time [52].

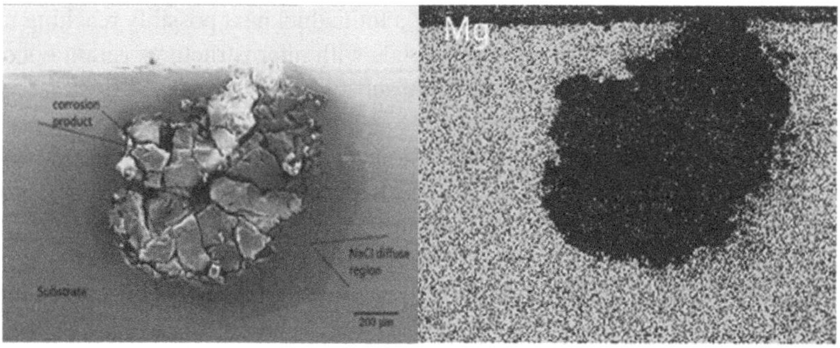

Figure 15.2: Left: SEM of the cross section of a corrosion pit on a Mg-alloy sample exposed to the immersion corrosion environment after 60 h; the wetted layer is hundreds of microns thick. Right: X-ray mapping of the distribution of Mg after corrosion shows diffuse damage near the pit surface (reprinted from [48]).

As evidenced by recent experiments, corrosion affects a subsurface region below the liquid/solid interface [48, 15, 51]. For example, while there are no obvious surface changes due to corrosion, a steel structure can become brittle and fail catastrophically, without going through the safer, necking deformation phase [15]. Such a phenomenon is due to embrittlement/toughness reduction induced by corrosion in a certain layer near the material surface. The reasons behind this effect are not fully understood at the moment. However, based on the SEM results and the energy dispersive X-ray spectroscopy (EDS) mapping (see Figure 15.2) from [48], it is reasonable to assume that a layer of a certain thickness near the surface participates in the corrosion reaction. Experiments ([48]) show that the solution wets" a subsurface metal layer of the corrosion pit. It is then reasonable to assume that the corrosion reaction does take place in this wetted layer. Metal ions may diffuse out from this region into the solution. This can be concluded from the X-ray maps of the cross section of a corrosion pit [48], which indicates a gradient in the metal concentration, meaning that metal diffuses out from this layer near the pit surface (see Figure 15.2, right panel). The reaction rate in the corrosion layer is controlled by the ionic flux. Hence, the corrosion reaction (anodic reaction) can be expressed as an effective diffusion process, in which the metal ions diffusing out into the solution and the diffusion flux (equivalent to the current density) depend on the overpotential. We assume that this dependency is similar with the one between the current density and the overpotential. We call this diffusion process, to avoid confusion, an *effective diffusion*. This idea constitutes the basis for the new model of corrosion damage we introduce in this chapter.

One possible mechanism for out-diffusion of metal ions is that the solution diffused in the metal provides pathways for the metal ions to out-diffuse. The SEM results (see Figure 15.2, left panel) in [48] show a region of solution diffusion inside the metal and, over the same region, a reduction in metal content (see Figure 15.2, right panel). This reveals that corrosion does not happen on a mathematically sharp metal-solution interface, but in a layer of a finite thickness possibly reaching microns to perhaps hundreds of microns in materials with microstructures (grain boundaries, etc.). Another possible mechanism is the substitutional vacancy diffusion (see [40]) in which the vacancy diffuse into metal from the interface, while the metal atoms are moving out. Although the vacancy diffusion coefficients in metals are very small (for example, $3.0 \times 10^{-9} \mu m^2/s$ in α-Fe at 500 °C [9]) compared to the ionic diffusion in solution (around $10^4 \mu m^2/s$ [18, 29]), they can be larger by many orders of magnitude when the substitutional vacancy path is along defects such as dislocations and grain boundaries (see [40]). For example, the grain boundary substitutional diffusion coefficient of Ga in AL-bicrystal is 10^{16} higher than that for the bulk of the crystal [51]. The combination of these two mechanisms is expected to result in corrosion reaction taking place hundreds of microns under what could be considered as the "pit surface." To our knowledge, no models of corrosion take this subsurface process into account. In modeling corrosion damage, the details of how damage evolves below the pit surface is critical for predicting where cracks will start from, especially in polycrystalline materials [35]. Moreover, when a metal with defects (vacancies, dislocations, grain boundaries, inclusions, cracks) is subjected to mechanical loading, the stress field can enhance diffusivity by orders of magnitude (see e.g. [32]), especially in regions around geometrical defects such as around the tip of a crack, where high stress concentrations occur. Consequently, investigations of the subsurface damage produced by corrosion are necessary.

With the goal of capturing subsurface corrosion damage, in this chapter we model the anodic reaction in corrosion by a nonlocal diffusion process [3, 4, 36] in the metal-solution bi-material and couple it with a material damage model by superposing mechanical bonds to the diffusion bonds between material points. We introduce a novel relationship between damage (understood here as material disintegration due to corrosion processes) and concentration in the solid that allows the formation of corrosion damage over a layer *inside* the solid. Different from the solution-diffusion models, in our model for the metal-solution bi-material we consider an effective diffusion of metal into the solution. The metal concentration in both solid and solution is the unknown. Although we can apply a classical model for this moving-boundary problem, the main reason for using a nonlocal peridynamic (PD) model is its ability to simulate the evolution of corrosion damage and material disintegration in a mathematically consistent manner [47], via a new coupling between concentration (carried by the diffusion bonds) and damage (carried by the superposed mechanical bonds). This model will allow us to observe material damage caused by corrosion through a certain thickness inside the metal, and thus be able to model the subsurface damage observed in experiments discussed above. A second advantage is that in the nonlocal model, interface conditions (Stefan conditions) in bi-material diffusion are not necessary, the motion of the solid-liquid interface being implicitly determined by the effective diffusion properties of the bi-material, while in the classical models a flux continuity condition is required at the interface. In the PD model, the effective diffusion properties contain the overall response of the specific, but complex, chemical interactions that take place at the interface and through the subsurface.

The chapter is organized as follows: in Section 15.2, the electrochemical kinetics is briefly reviewed and the procedure for evaluating the diffusion constant in the solid (based on the overpotential) is given. The 1D ions diffusion in the solution-solid system is formulated in Section 15.3, followed by the PD formulation of diffusion for corrosion in Section 15.4. In Section 15.5, we compare experimental results on 1D pit corrosion behavior with computational results obtained with our PD model. We show how model calibration in the activation-controlled stage leads to predictive results for the polarization map over a wide range of overpotentials, covering activation-controlled, IR-, and diffusion-controlled corrosion. To introduce material damage caused by corrosion, a new concentration-drop/corrosion-damage model is presented in Section 15.6. In this model, mechanical bonds, which carry material damage information (and, in a future model, mechanical deformation), are superposed onto the diffusion bonds. We study the effects of diffusivity/saturated concentration on corrosion damage and thickness of the subsurface damage. In Section 15.7, we extend our 1D model to 2D and 3D, and apply the model to activation-controlled corrosion. Different from the 1D case, we employ a two-way-coupled concentration-drop/corrosion-damage model. In Section 15.8, we investigate pitting corrosion in heterogeneous materials, and show how microstructural heterogeneity affects pit shape and damage evolution.

15.2 Electrochemical kinetics

In this chapter we only consider the anodic reaction since this controls the evolution of damage in the metal on pitting corrosion. Consider an anodic reaction: $M(s) \rightarrow M(aq)^{n+} + ne^{-}$, shown in Figure 15.3. The reaction rate is expressed in terms of the electric current density as follows [2]:

$$i_a = i_0 \exp(\frac{\alpha n F \eta}{RT}) \tag{15.1}$$

where F is the Faraday constant; i_0 is the exchange current and is a constant; α, the transfer coefficient, can range from zero to unity; n is the charge number of the dissolving reactant; R is the gas constant; T is the absolute temperature; $\eta = E - E_e$ is the overpotential; E is the applied potential, and E_e is the equilibrium potential. This equation is essentially transferred from the Butler–Volmer equation by assuming the overpotential is positive. The symmetric term with a negative exponential in the Butler–Volmer equation is not considered in the present study but will be in the future. The present model is strictly a corrosion model; the possibility of a zero corrosion rate at a particular over-potential or reversal to electroplating is precluded. The Butler–Volmer equation is usually expressed as the following Tafel law:

$$\eta = b_a \log(\frac{i_a}{i_0}) \tag{15.2}$$

where $b_a = 2.303\frac{RT}{\alpha n F \eta}$ is the Tafel slope. This parameter can be experimentally measured, and a more detailed discussion is provided in Section 15.5.2 for model calibration.

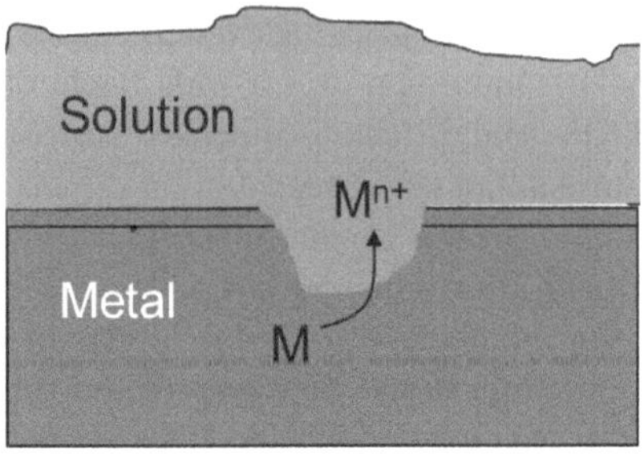

Figure 15.3: During anodic reaction in a corrosion process, metal atoms lose electrons and diffuse into the solution. The blue color * depicts a passive layer on the metal surface (redraw from [30]).

As mentioned in the introduction (see [48, 15, 51]) the corrosion reaction not only happens at the interface, but also in a subsurface layer of the corrosion pit. Specifically, [48] shows a gradient in the metal concentration near the interface. This evidence supports the following assumptions in our modeling of the corrosion reaction: we employ an "effective diffusion" model in the metal-liquid bi-material to capture the kinetics of the metal ions out-diffusion into and through the solution; because of the very low diffusivity in the solid, this reaction, practically, only takes place in the thin metal layer near the interface. In the following, we reduce the field equations governing corrosion to a classical diffusion equation, in which flux represents transport of metal ions with the driving force being the concentration gradient, and a phase-change of solid diffusivity into liquid diffusivity that happens when concentration at a point reaches the saturated concentration in the liquid. Electrochemical effects are approximated by making the diffusion coefficient in the metal a function of the over-potential. The mismatch in diffusion coefficient, together with appropriate boundary conditions at the far-field, and the phase-change mechanism mentioned, will produce solutions that resemble the classical Stefan moving boundary problem. The peridynamic model for this approach does not require interface conditions and the motion of the interface is autonomous, being part of the solution, not part of the problem. The peridynamic model of corrosion is extended to capture subsurface damage, observed in experiments, in Sections 15.6 and 15.7.3.

According to Faraday's second law, the number of electrons moving to the cathode is equal to the number of metal ions diffusing out multiplied by the charge number of the dissolving reactant. In other words, the metal ionic flux is proportional to the current density. Therefore, the ionic flux out from the solid shares a similar Arrhenius equation as the current density (Eq. 15.1). Note that Nernst—Planck equation is usually employed to investigate the diffusion near a charged surface [44]. The Nernst–Planck equation is a conservation of mass equation utilized to describe the motion of chemical species. The total flux per area J_i (in 1D) through the interface includes contributions from the diffusion flux (induced by the concentration gradient), the electromigration in the electric field, and from the convective flux:

$$J_i = -K\nabla C - \frac{nF}{RT}KC\nabla\Phi + \mu C. \tag{15.3}$$

Here K is the diffusivity in the solution; C is the metal concentration in the solution; Φ is the electric potential; u is the convective velocity. The potential gradient $\nabla\Phi$, which depends on the overpotential, only exists in a thin solution layer near the interface. This thin layer is the so-called electrical double layer (EDL), the thickness (H_{edl}) of which, less than 10 nm in our case, depends on the electrolyte concentration [45]. In EDL, the potential gradient, roughly inverse-proportional to H_{edl}, is the driving force for the ionic motion. In pitting corrosion, the metal/electrolyte interface is not a sharp mathematical surface. For instance, a diffusion region of NaCl solution inside the Mg alloy is observed in [48] (see Figure 15.2). This diffusion region can be treated as a nanoporous or subnano porous structure, where the pores filled with solution provide "pathways" for the metal ionic motion. Thus, the ionic motion inside the solid near the interface is affected by the potential gradient.

In this work, because of the discussion on the existence of diffusion in the solid layer near the metal surface, we assume that Eq. (15.3) can also be applied on the solid side (wetted) of the interface as follows: in the solid, convection is negligible, so $u = 0$, and we choose to combine the effects of the EDL into an "effective diffusivity" of metal ions in the solid, that depends on the overpotential and the diffusivity of metal ions through the "pathways" On the liquid side, we can also simplify the Nernst–Planck equation by combining the EDL effect and the convective flux into the diffusion flux term because the diffusivity of the solution is orders of magnitude larger than that in the solid and it will dominate the metal ion concentration evolution independent of adding or not these terms. Therefore, our simplified model of corrosion uses Fick's Law for the metal ionic concentration in the bi-material solid-liquid to describe the corrosion kinetics:

$$J_i = \begin{cases} -K(\eta)\nabla C & \text{Solid side} \\ -K_L\nabla C & \text{Solution side} \end{cases} \tag{15.4}$$

where $K(\eta)$ is the effective diffusivity in the wetted solid, while K_L is the diffusivity inside the solution. We explain below how the effective diffusivity is evaluated. In Eq. (15.4), the flux linearly depends on the diffusivity (solid side). Therefore, to ensure the linear dependence between the ionic flux and the current density (implied by Faraday's second law), a linear dependence between the effective diffusivity and the current density should be imposed. Substituting this linear dependence into Eq. (15.1), we obtained a similar Arrhenius equation for the effective diffusivity in the metal solid (K_1) corresponding to the overpotential η:

$$K_1 = K(\eta) = K_0 \exp\left(\frac{\alpha nF\eta}{RT}\right) \tag{15.5}$$

where K_0 is the effective diffusivity in the metal at zero overpotential and this can be obtained from experiments by calibration (see Section 15.5.2). In a heterogeneous material, K_0 is a location-dependent variable and we study this case in Section 15.8.

Similar to Tafel's law of overpotential/current density, we can also write Eq. 15.5 as:

$$\eta = b_a \log\left(\frac{K_1}{K_0}\right). \tag{15.6}$$

In summary, our hypothesis is that we represent the corrosion process via an effective out-difffusion of the metal ions from a layer near the pit surface. This effective diffusion is calibrated (see Section 15.5.2) to experimental results from an anodic pitting corrosion. Note that this model is a general corrosion model, and does not restrict to the type of pitting corrosion.

15.3 Problem formulation of 1D pitting corrosion

Since the reaction is considered as an effective diffusion process here, we can model pitting corrosion as diffusion in a bi-material: metal ions out-diffusing from the metal (effective diffusion), and metal ions diffusing in the pitting electrolyte. Let the diffusivities corresponding to the metal be K_1 and that of the fluid be K_2. As shown in Figure 15.4, $h(t)$ denotes the location of the solid/liquid interface, or the depth of pitting corrosion. Essentially, the evolution of this interface is controlled by the competition between the fluxes through the interface (J_1) and at the boundary of the liquid solution (J_2). Flux J_1 represents the rate of dissolution, while the flux J_2 denotes the diffusive transport in the liquid. In Figure 15.4, we schematically describe three possible approaches to modeling corrosion with a diffusion-based model: the most common model found in the corrosion literature is that which considers diffusion to take place only in the solution, with the interface being a sharp liquid-solid interface. Such a model cannot capture subsurface damage, for example. The alternative we introduce is diffusion in the bi-material liquid-solid system. We prefer the PD model to a classical model for this because of the natural way in which material damage evolution is handled in PD. The damage model is given in Section 15.6, where mechanical bonds are superposed to the diffusion bonds to track material damage caused by corrosion in a layer of a certain thickness. Damage and fracture are treated in a unified way by peridynamics, and, in the future, the mechanical bonds will track mechanical deformation as well and SCC will be modeled in a mathematically consistent way.

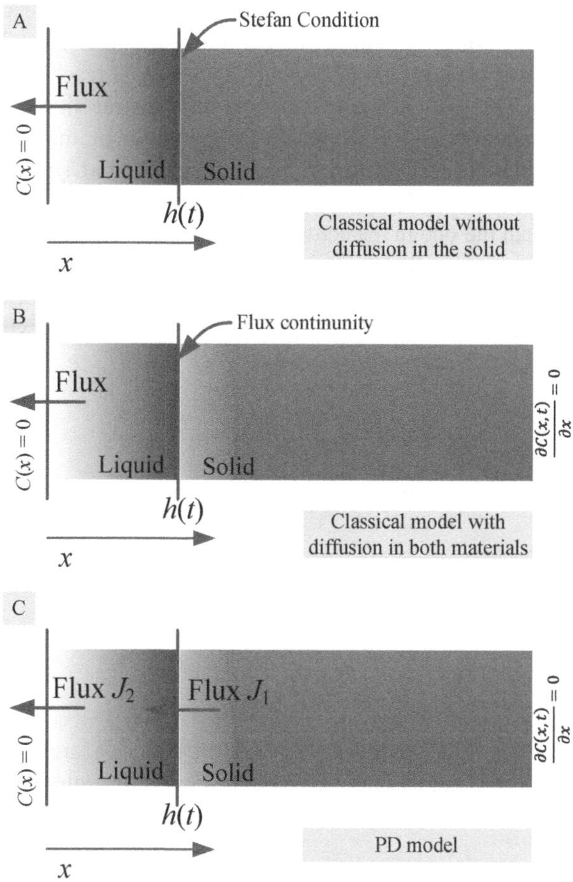

Figure 15.4: Corrosion modeled by diffusion with a moving interface. The evolution of the interface ($h(t)$) is controlled by the fluxes through both liquid/solid interface and the pitting hole surface. In the classical model (A), the diffusion in the solid is not considered and a Stefan condition is imposed at the interface. When the diffusion in the solid is considered, a flux continuity condition at the interface is required in a classical model (B). In our PD model (C), the motion of the interface is part of the solution.

If the initial pitting depth is zero and we assume the concentration outside the pitting hole is also zero, at the initial stage the short pitting hole has the ability to transport the ions dissolved out from the pitting hole, meanwhile keeping the ions concentration smaller than the saturated concentration. This is the activation-controlled corrosion. In this stage, the interface movement is controlled by the diffusion in the solid and the moving speed is a constant exponentially dependent on the overpotential. With the pitting depth increasing, the diffusive transport becomes slower (J_2 decreases), and the concentration in the liquid near the interface ($C(h^-)$) increases. There exists a critical pitting time, beyond which $C(h^-)$ reaches the saturated concentration (C_{sat}). This saturated concentration stops the reaction in the solid. Corrosion resumes again after $C(h^-)$ decreases. At that moment, the interface movement is controlled by the diffusive transport in the liquid. When $C(h^-)$ stays at the value C_{sat}, this stage is called diffusion-controlled corrosion. In this stage, the corrosion rate is smaller than the one in the activation-controlled corrosion stage.

In the diffusion-controlled corrosion, the interface concentration (at both liquid side and solid side) is fixed at the saturated concentration value. Consider a 1D bar with a zero initial corrosion depth ($h(0) = 0$), we have the following classical governing equations for bi-material diffusion:

$$\begin{aligned} \frac{\partial C(x,t)}{\partial t} &= K(x)\frac{\partial^2 C(x,t)}{\partial x^2} \\ K(x) &= \begin{cases} K_1, & \text{if } C(x) > C_{\text{sat}} \\ K_2, & \text{if } C(x) \leq C_{\text{sat}} \end{cases} \end{aligned} \tag{15.7}$$

where L is the total length of the bar, and $C(x,t)$ is the concentration at position x and time t. The boundary conditions are:

$$\begin{aligned} C(0,t) &= 0 \\ \frac{\partial C(x,t)}{\partial x}\Big|_{x=L} &= 0. \end{aligned} \tag{15.8}$$

To reduce the effect of the far end of the metal, we choose a large L in our calculation so that this end is sufficiently far away from the corrosion region. At the interface we need to have flux continuity and concentration conditions:

$$\begin{aligned} \frac{\partial C(x,t)}{\partial x}\Big|_{x=h^-} &= \frac{\partial C(x,t)}{\partial x}\Big|_{x=h^+} \\ 0 < C(x,t) \leq C_{\text{sat}} & \quad (0 < x < h(t)) \\ C_{\text{sat}} < C(x,t) \leq C_{\text{solid}} & \quad (h(t) < x < L). \end{aligned} \tag{15.9}$$

C_{solid} is the mole concentration of undamaged metal. h^- and h^+ denote the liquid side and solid side of the interface. In the diffusion-controlled corrosion, the concentration at the interface is continuous:

$$C(h^-,t) = C(h^+,t) = C_{\text{sat}}. \tag{15.10}$$

Initially, the whole bar is uncorroded; hence the initial condition is:

$$C(x,0) = C_{\text{solid}} \tag{15.11}$$

where $0 < x < L$, and corrosion starts as activation-controlled.

The first equation in Eq. 15.7 is normally used to model diffusion in the solution [42, 43], while the interface motion is controlled by a Stefan condition (see Figure 15.4A). In our approach, since we are after the subsurface changes in the metal, we need to model the out-diffusion of the metal component (the second equation in Eq. 15.7). In the classical model of solving this solution/solid diffusion, a flux continuity condition (the first equation in Eq. 15.9) is required at the interface (see Figure 15.4B). Note that similar to the modeling in [42, 43], we assume that the passive film is broken, and part of the metal surface is directly in contact with the solution. In the 1D case, this model leads to what can be considered uniform corrosion. The 2D and 3D extension of this model is presented in Section 15.7.

The PD model is helpful in capturing autonomous damage evolution [6, 25, 26] and this is the main reason we choose it for this problem. In addition, the interface conditions discussed above are not necessary in the PD model, the motion of the solid-liquid interface being implicitly determined by the effective diffusion properties of the bi-material. Therefore, the peridynamic method is applied to solve the coupled diffusion/corrosion damage problem in a bi-material in Sections 15.4, 15.5, and 15.6.

15.4 The peridynamic formulation for 1D pitting corrosion

Similar to our peridynamic formulation for heat transfer problem [3, 4], the domain is considered as a collection of nodes (material points) connected with peridynamic *diffusion bonds* (see Figure 15.5). At each node, a concentration of metal ions is present. Through the diffusion bonds, the ions diffuse from the high concentration nodes to the nodes with low concentration.

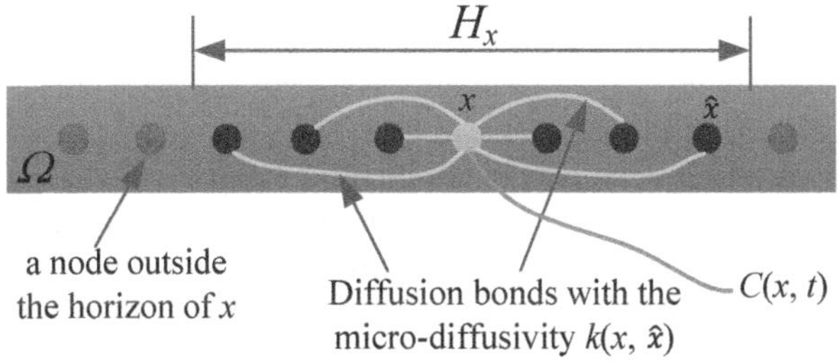

Figure 15.5: Peridynamic description of the diffusion model. The region denoted by H_x is the horizon (nonlocal region around the point) of node x, which has concentration $C(x,t)$. Bonds connecting points have micro-diffusivity $k(\hat{x},x)$.

To explain the idea in detail, we present a 1D case here. Consider a body Ω and a region H_x around node x (Figure 15.5). We call this region the horizon of node x, and diffusion bonds connect x with all $\hat{x}$ in H_x. The point x is characterized by its metal concentration, $C(x,t)$. As shown in Figure 15.5, the bond between x and $\hat{x}$ is described by micro-diffusivity $k(x,\hat{x})$. Note that for modeling a moving interface, as is the case in corrosion problems, the bi-material diffusion model has to be enhanced by a phase-change mechanism that transforms a solid node into a liquid node due to the corrosion progression; the micro-diffusivity associated with this node will change from the one in solid to that of the liquid. From our previous work on heat transfer [3, 4], we have the following evolution of metal concentration:

$$\frac{\partial C(x,t)}{\partial t} = \int_H J(\hat{x},x,t)\mathrm{d}V_{\hat{x}} \tag{15.12}$$

where the kernel, the micro-flux (J) versus the micro-diffusivity, is expressed as [26]:

$$J(\hat{x},x,t) = \begin{cases} k(x,\hat{x})\frac{C(\hat{x},t)-C(x,t)}{|\hat{x}-x|} & |\hat{x}-x| \leq \delta \\ 0 & |\hat{x}-x| > \delta \end{cases} \tag{15.13}$$

where δ is the horizon size.

It is noted here that there are other options for the kernel in this equation, for example, one without denominator $|\hat{x}-x|$, another with the denominator $(\hat{x}-x)^2$ [3, 4]. For the 1D corrosion, we will use the denominator also used in [36]. However, we have recently shown that convergence to the classical model, in the limit of the horizon size going to zero, is faster with the $(\hat{x}-x)^2$ denominator (see [13]), and we will use that kernel form in the extensions to 2D and 3D from Sections 15.7 and 15.8.

Note that in the corrosion problem, the metal concentration inside the metal cannot be smaller than the saturated concentration in the solution. Once the concentration drops below the saturated value, the metal material point dissolves and becomes a liquid point (the diffusion constant switches from K_1 to K_2, as in a phase change). This is how the interface moves in the PD model of corrosion.

To implement the peridynamic method for this diffusion process in 1D, we discretize the domain using a uniform grid. Non-uniform grids are possible in peridynamics [5, 24], but this is not pursued here. We use a mid-point integration scheme to approximate the integral over H_x in Eq. 15.12. Each node has a "volume" in 1D given by the length of the segment associated with it, in our case Δx. We call the ratio between the horizon size and the grid spacing *the horizon factor* $m = \delta/\Delta x$. The nodes at the ends have the coordinates $\Delta x/2$ and $L - \Delta x/2$. Other options can be selected: for example, we can place nodes at the actual ends of the bar in which case these end nodes will have half-volumes. The results we report below use the first method, with all nodes having the same volume (length).

The spatial discretization of the PD equation is:

$$J(x_p,x_i,t) = k(x_p,x_i)\frac{C(x_p,t)-C(x_i,t)}{|x_p-x_i|} \tag{15.14}$$

and

$$\frac{\partial C(x_i,t)}{\partial t} = \sum_p k(x_p,x_i)\frac{C(x_p,t)-C(x_i,t)}{|x_p-x_i|}V_p \tag{15.15}$$

where the summation is after all nodes x_p inside the horizon of node x_i, and V_p is the portion of the volume of the node x_p actually covered by the horizon of node x_i.

The micro-diffusivity can be determined by equating the peridynamic flux to the classical flux through a surface (a point in 1D case) arising from a simple linear concentration profile [3]. The micro-diffusivity function over the horizon region can be selected in various forms. If it is constant over the horizon, then we obtain:

$$k(x_p, x_i) = \frac{2K}{\delta^2}. \tag{15.16}$$

For a linear ("triangular profile" over the horizon (see [3])), we get:

$$k(x_p, x_i) = \frac{6K}{\delta^2}\left(1 - \frac{|x_p - x_i|}{\delta}\right). \tag{15.17}$$

For the results in this chapter we select the micro-diffusivity to be constant over the horizon H_x region (see Figure 15.5). Thus, the micro-diffusivities k_1 (solid) and k_2 (liquid) are calculated based on Eq. 15.16, using the corresponding diffusivities K_1 and K_2. If a bond has one end in the liquid and the other end in the solid, then the micro-diffusivity of the bond k is related to the distance between the solid nodes to the interface (l_1) and the distance between the liquid node to the interface (l_2) as follows [37]:

$$k = \frac{l_1 + l_2}{\frac{l_1}{k_1} + \frac{l_2}{k_2}} = \frac{l}{\frac{l_1}{k_1} + \frac{l_2}{k_2}} \tag{15.18}$$

where l is length of the bond.

The PD model for diffusion in a bi-material (solution/solid) *with a moving interface* is:

$$\begin{cases} \frac{\partial C(x,t)}{\partial t} = \int_H k(x,\hat{x}) \frac{C(\hat{x},t) - C(x,t)}{|\hat{x}-x|} \mathrm{d}V_{\hat{x}}, (0 < x < L, |\hat{x} - x| \leq \delta) \\ C(x,0) = C_{\text{solid}} \\ \frac{\partial C(x,t)}{\partial x}|_{x=L} = 0 \\ k(x,\hat{x}) = \begin{cases} k_1, & \text{if } C(x) \text{ and } C(\hat{x}) > C_{sat} \\ k_2, & \text{if } C(x) \text{ and } C(\hat{x}) > C_{sat} \\ \frac{l}{\frac{l_1}{k_1} + \frac{l_2}{k_2}}, & \text{if } C(x) \geq C_{sat} \geq C(\hat{x}) \text{ or } C(x) \leq C_{sat} \leq C(\hat{x}). \end{cases} \end{cases} \tag{15.19}$$

Different from the classical model [11] we do not need to include flux continuity at the interface [16]. The motion of the interface (given by the location $h(t)$ of the interface at time t) is obtained as part of the solution of the problem by the transformation of diffusivity at a point from that of the solid into that of the liquid when its concentration drops below the saturated concentration value. In this 1D example, the interface is the location of the material point with a concentration equal to C_{sat}. The mechanism we introduced to allow for autonomous evolution of the interface acts like a material phase-change induced by the drop in concentration.

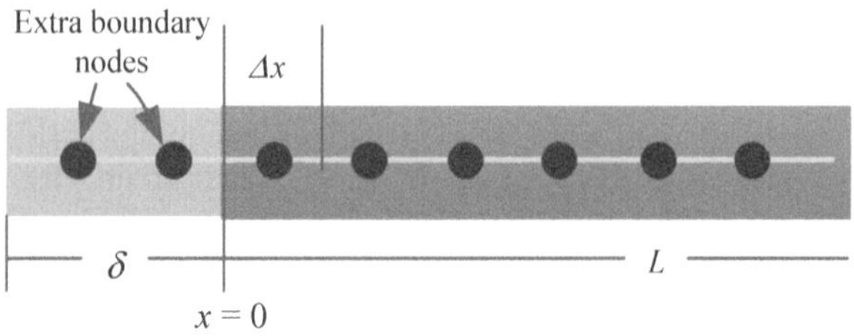

Figure 15.6: Peridynamic nodes distribution. An extra boundary with length δ is added onto the left end of the 1D bar for taking care of the surface effect (see [36]).

The evolution of metal concentration can be described by the following steps:

1. ***Mesh and initial condition:*** *Input nodes* $(x, \delta, C(x))$. $C = C_{solid}$. *Besides all of the nodes in the bar, we add extra boundary nodes at the end:* $x = 0$ *(see Figure 15.6) to fix the surface effect. These nodes are used to impose the zero concentration Dirichlet boundary condition [33]. Other ways to impose the boundary conditions in PD are shown in [3, 4, 16].*

2. ***Micro-diffusivity:*** *Compute the micro-diffusivities based on Eqs. 15.16 and 15.18.*

3. ***Concentration calculation:*** *Compute concentration distribution* $C(x)$, *at the first time step.*

4. ***Interface advancement:*** *The liquid/solid "interface" moves autonomously when the left-most solid node (the current interface location) is at a concentration smaller than* C_{sat} *and is switched to a liquid node (its diffusivity constant changes to that of the liquid).*

5. *Repeat steps 2–4 for all nodes and all time steps.*

The interface starts to move once the concentration at the solid boundary drops to the saturated concentration. Theoretically, the movements should start at $t = 0$ s. In our calculation, the starting time depends on the discretization, since the boundary (node) is of finite volume (length) and a certain time is needed for the average concentration in this volume to drop from C_{solid} to C_{sat}. Until the interface concentration at the liquid side reaches C_{sat}, we have activation-controlled corrosion, and after that we get diffusion-controlled corrosion. Therefore, if we choose a small K_1 (small overpotential), we get activation-controlled corrosion over a significant time interval, while when we choose a large K_1, we almost immediately transition to diffusion-controlled corrosion.

15.5 Results and discussion of 1D pitting corrosion

In this section, a real case from experiments will be modelled by our PD approach. The linear relationship between the corrosion depth and the square root of time [21] will be examined; we will study the interplay between activation & diffusion-based corrosion.

With the above models established, we consider the following example (see [23]) for a 1D bar with total length of 100 μm. The material is 304SS. Material properties are taken from [42]: $C_{\text{solid}} = 143000$ mol/m^3, $C_{\text{sat}} = 5100$ mol/m^3. The average charge number of stainless steel 304SS is obtained from the charge numbers of Fe, Ni, Cr, and their mole fractions within the alloy, $n = 2.19$ (see [43]). We consider the immersion of stainless steel 304 into a 1 M NaCl solution at an absolute temperature $T = 298.15$ K. The ions diffusivity in the liquid $K_2 = 8.5 \times 10^{-6}$ cm^2/s, while the effective diffusivity in the solid (K_1) is controlled by the overpotential (see Eq. 15.5). Different values ranging from 0.001 to 8 μm^2/s are chosen for K_1 in the following calculation.

15.5.1 Pit corrosion depth proportional to $\sqrt{t}$

When $K_1 = 2.0$ μm^2/s, Figure 15.7 shows the results for a case when the horizon size is δ=0.05 μm, and with a horizon factor $m = \delta/\Delta x = 4$: subfigure (A) shows concentration and interface evolution within 20 s. Figure 15.7B gives a zoom-in near the saturated concentration level. The concentration gap between the solid boundary node and liquid boundary node decreases to near zero after a while. As the gap is reduced, the pitting corrosion switches from activation-controlled (up to 2 s) to diffusion-controlled (after 5 s). Between these two stages, the corrosion is IR (current-resistance)-controlled.

The m-convergence results (which considers the horizon fixed and varies the grid spacing; see, e.g. [3, 7]) for the location of the interface are shown in Figure 15.8. The m-convergence results for other horizon sizes and the answer to the question on "How do results change as the horizon size goes to zero while m is fixed?" are given in Appendix 15.10.1. The slopes to the curves plotted in Figure 15.8A define the interface moving speed. This speed converges in the diffusion-controlled corrosion stage. At the initial stage, the interface moving speed is dependent on the discretization. This is due to the concentration jump at the interface at the initial stage (Figure 15.7). Recall that the micro-diffusivities in the PD model are calibrated to actual diffusivities based on constant flux state (linear concentration behavior). Therefore, highly nonlinear or discontinuous concentration profiles over the size of the horizon are affected by the discretization size, since a finer discretization leads to steeper approximation of the flux near the steep gradient or the jump discontinuity in the concentration. This is why the solution depends on the discretization when the concentration has a jump discontinuity. When the corrosion goes into the diffusion controlled phase, there is no longer a jump in concentration. Our results for the diffusion controlled corrosion, including the interface advancing and the damage profiles, are independent of discretization size.

In order to obtain numerical convergence for the activation-controlled pitting corrosion, we have to reformulate the model to avoid the concentration jump at the interface. One solution is that instead of modeling the liquid/solid system, a sole analysis on the solid part with a "fixed" flux may be able to capture the interface advancing and the damage evolution in the solid for the activation-controlled corrosion. This will be studied in our future work.

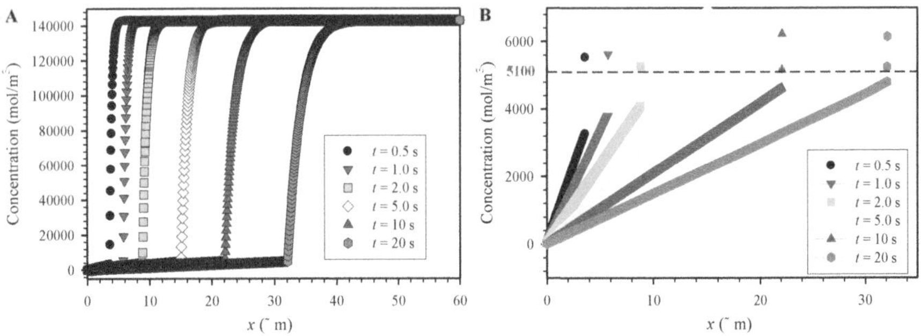

Figure 15.7: (A) Concentration and interface evolution within 20 s. (B) Zoom-in near the saturated concentration location. The location of the jump discontinuity is the location of the interface (the nodes with concentration lower than 5100 mol/m^3 are liquid nodes, while the others are solid nodes). Concentration gap between solid boundary node and liquid boundary node decreases to near zero in time ($K_1 = 2.0\ \mu m^2/s$, $\delta = 0.05\ \mu m$, $m = 4$).

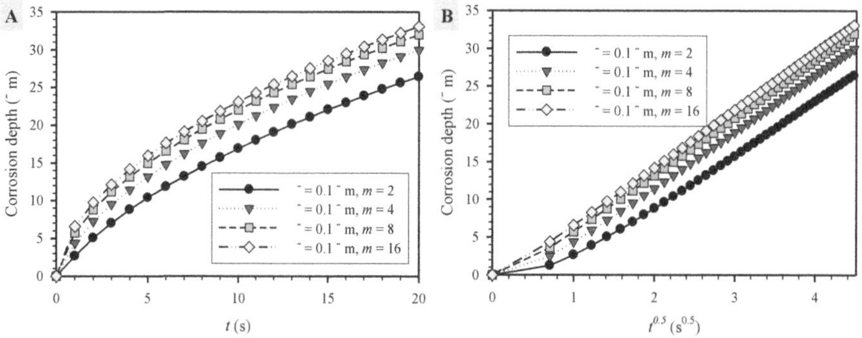

Figure 15.8: (A) The m-convergence results for the location of the interface at different times. (B) Corrosion depth as a function of $\sqrt{t}$. Except for the early time (when corrosion is activation controlled), the dependence is linear (in $\sqrt{t}$), matching perfectly the experimental results (see [21])

A plot of corrosion depth as a function of $\sqrt{t}$ shown in Figure 15.8B indicates that $h(t) \propto \sqrt{t}$. We can write this as:

$$h(t) = s\sqrt{t} \tag{15.20}$$

where s is a constant. This linear dependence perfectly matches with the experimental results on the diffusion-controlled corrosion [21]. Note that the square-root time-dependence of the corrosion depth is associated with the diffusion-controlled corrosion stage. For the activation-controlled or IR-controlled corrosion, there is no such dependence. For a different metal and/or a different experiment setup, the time needed for the corrosion to transition into the diffusion-controlled stage may vary. For example, if a large convection exists inside the solution, the square-root time-dependence will never be observed. Another example is a metal with high corrosion resistance (low solid diffusivity in our model).

For a quantitative analysis, for the material sample we use (SS304 metal bar); the value for the s parameter, extracted from the results shown in Figure 15.30 in Appendix 15.10.1, is $s = 7.7\ \mu\text{m/s}^{0.5}$ (when $\delta = 0.05\ \mu$m and with $m = 8$). This value is close to the experimental results in the diffusion-controlled corrosion on the same sample given in [19, 20]: $s = 8.1\ \mu\text{m/s}^{0.5}$ at15 °C, and $s = 9.4\ \mu\text{m/s}^{0.5}$ at 28 °C. The model in this section does not include some of the additional physical mechanisms present in the corrosion process, like the temperature dependence, or the dependence on damage. The coupling corrosion-damage is discussed in Sections 15.6 and 15.7. Interface speed smaller than the measured speed is also reported in [43] using a different model.

Since our model shows the correct pitting-depth/time evolution, it also correctly predicts the inverse dependence between current density and the pitting depth observed in experiments (see, e.g. [23]). Indeed, since the interface moving speed is the time derivative of the corrosion depth:

$$v(t) = \frac{s}{2\sqrt{t}} \tag{15.21}$$

and this speed is proportional to the current density, because it obeys Faraday's second law (the electric current at an electrode is proportional to the mass being dissolved per time instant),

$$i_a = nFC_{\text{solid}}v(t). \tag{15.22}$$

We have, by substituting Eqs. (15.20) and (15.21) into Eq. (15.22),

$$i_a = \frac{nFC_{\text{solid}}s^2}{2}\frac{1}{h(t)}. \tag{15.23}$$

This inverse dependence between current density with the pitting depth is confirmed by the experimental results in [23].

15.5.2 Activation-controlled, diffusion-controlled, and IR-controlled corrosion

Different corrosion types are corresponding to different values of applied voltages. In this subsection, we will use our PD model to map the relationship between the overpotential and the current density.

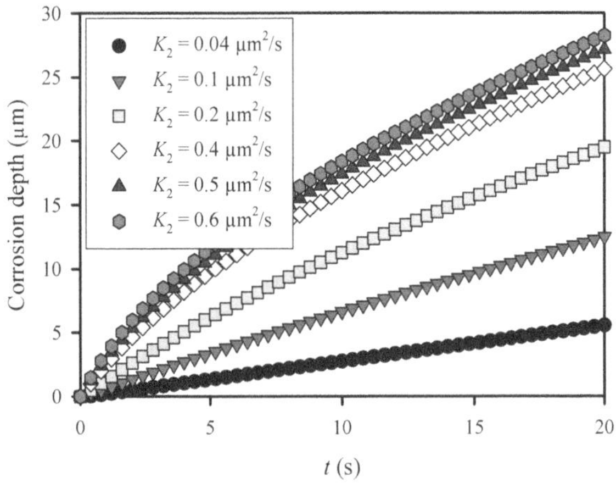

Figure 15.9: Corrosion depth versus time at different effective diffusivities in the solid. When the diffusivity is small (small applied voltage), the corrosion process over the time interval shown is activation controlled, and the interface moving speed is linearly dependent on the diffusivity. When the diffusivity is large (larger than 0.4 $\mu m^2/s$), the corrosion process over this time interval is dominated by the diffusion-controlled corrosion with a constant ratio s (see Eq. 15.20)

Figure 15.9 shows the dependence between the interface evolution and the effective diffusivity K_1 in the solid (which is connected to the overpotential through Eq. 15.6). When a small overpotential/effective diffusivity is applied/used, the corrosion process shown in the given time interval is activation controlled, and the interface moving speed ($v(t)$) is linearly dependent on the diffusivity K_1. In this corrosion type, the current density is governed by Tafel's law (Eq. 15.2). When the overpotential is large (corresponding to a diffusivity larger than 0.4 μm^2/s), the corrosion process over the 20 s time interval is dominated by diffusion-controlled type, with a constant ratio s. Between these two corrosion types, there exists a transition corrosion type without a constant $v(t)$ or a constant s. More details about this transition region are shown in Figure 15.10.

Before the discussion of the relationship between current density and overpotential in Figure 15.10, we first describe the way we calibrate the relation between the effective solid diffusivity and the overpontential. For that we need the values of the Tafel slope b_a and the zero-overpotential diffusivity K_0 in Eq. (15.6). After calibration, our model will be able to compute the current density based on an input overpotential value, instead of an effective solid diffusivity. We calibrate the model using the experimental results shown in [23]. Since our model only considers the anodic reaction, we employ the data points corresponding to the pitting without repassivation, although repassivation could be realized by decreasing the applied potential, as shown in [23].

Tafel slope b_a is obtained by fitting the experimental results (see Figure 15.10, where $1/b_a$ is the slope of the fitting line). The way we calculate the zero-overpotential diffusivity K_0 is described as follows:

1. *Input a small effective diffusivity K_1 (to ensure activation-controlled corrosion regime) into our PD model (for a given horizon size and given discretization size), and solve the model to obtain the interface advancing speed (v).*

2. *Apply Eq.(15.22) to obtain the current density (i_a).*

3. *From the experimental VI measurements, use this current density to extract the corresponding overpotential (η).*

4. *Find the calibrated zero-overpotential diffusivity K_0, by substituting the Tafel slope b_a, the input effective diffusivity and the obtained overpotential into Eq. (15.6).*

With the K_0 value calibrated from one data point from the measured overpotential/current density VI data, we can now compute (see Eq. 15.6) the effective diffusivity from a given overpotential. We then solve the PD corrosion model (with the same horizon size and discretization used to obtain K_0) and find the interface velocity. This value, used in Eq. 15.22, gives the PD-based current density corresponding to the given overpotential. This new data point is then plotted onto the VI graph (see triangles in Figure 15.10).

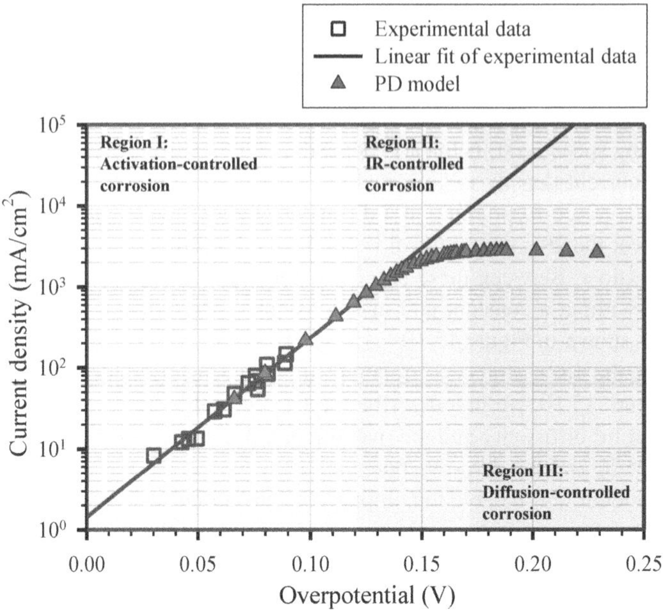

Figure 15.10: Experimental polarization data from [23] (squares) and prediction for the polarization computed with our PD model (triangles). The experimental data Tafel slope and exchange current are determined by the Tafel's law according to the experimental results in the activation-controlled corrosion [23]: $b_a = 45.1$ mV, $i_0 = 1.422$ mA/cm^2. Using just one experimental data-point in the activation-controlled region for calibrating the model, we predict the entire profile of the polarization curve.

For example, when $K_1 = 0.04$ μm^2/s is used in the PD computations, the model gives a constant interface moving speed $v = 0.2771$ μm/s (over a time period in which the liquid does not reach saturated concentration). Substituting this value into Eq. (15.22), the current density is found as: $i_a = 837.3$ mA/cm^2 (see Section 15.2 for F value and Section 15.5 for n, C_{solid} values). From the fitting of the experimental results (Figure 15.10), we obtain: b_a=45.1 mV, and the overpotential corresponding to i_a is $\eta = 124.9$ mV. Substituting $K_1 = 0.04$ μm^2/s, $b_a = 45.1$ mV and $\eta = 124.9$ mV into Eq. (15.6), we can get the zero-overpotential diffusivity $K_0 = 6.8 \times 10^{-5}$ μm^2/s. With this calibrated value, given any other overpotential, we find the corresponding K_1 from Eq. (15.6) with which we compute the new solution, find the new interface speed of this numerical solution and estimate, from Eq. (15.22), the current density. We observe (see triangles in Figure 15.10) that using this *single data-point calibration* (over the activation-controlled stage), we obtain the full range of behavior in corrosion: activation-controlled, IR-controlled, and diffusion-controlled regions. To the best of our knowledge, there no existing model can capture the overpotential/current density (VI) relationship in the entire region including activation-controlled, IR-controlled, and diffusion-controlled regions.

Note that, besides the activation-controlled corrosion, our model reveals diffusion-controlled corrosion when the overpotential is large. Between these two corrosion regions, a transition region is observed. This region is called IR(current-resistance)-controlled corrosion [23]. It should be noted that the model data are corresponding to the total time $t = 20$ s. Shorter time intervals would give a longer linear region (activation-controlled corrosion).

From the computational results, we can see that for low potentials the logarithm of current density is almost linearly dependent on them. At high potentials the current density reaches a limiting value. A transition region links these two. This matches very well with the experimental observations in [22].

In our PD model, the length-scale introduced by the horizon size influences the interfacial speed in activation-controlled corrosion, while it does not influence the interfacial speed in diffusion-controlled corrosion. We are currently working on calibrating the horizon size for activation-controlled corrosion to parameters obtained experimentally, such as thickness of the damage layer, and /or the measured current. This will be reported in a future publication.

The material breakdown caused by the corrosion process leads to damage formation. In the peridynamic theory of mechanics, damage is a result of mechanical bonds stretched beyond a critical value calibrated to the material's fracture energy. Here we extend that type of damage to that that can account for material disintegration induced by corrosion processes. For this, in the next section we will superpose onto the diffusion bonds mechanical bonds that carry damage information, caused by either mechanical or chemical/corrosion processes. For now, only the corrosion effects are considered in our model. In a future model we will report on an extension of the present model to SCC, thus allowing both corrosion progression and mechanical loading to damage, through which mechanical bonds are broken. This generalized damage notion couples corrosion and mechanical behavior of materials since corrosion influences mechanical properties of materials, and breakdown of material is associated with dangling chemical bonds or higher surface energy, which influences corrosion rates.

In this work, we only discuss the corrosion problem, so the mechanical bonds only serve the purpose of monitoring corrosion-induced damage. In the next section, a new concentration drop/damage model is introduced for this purpose.

15.6 Corrosion damage and the Concentration-Dependent Damage (CDD) model

As metal ions out-diffuse from the metal, material damage ensues. Experiments show damage takes place in a significant region under the pit bottoms [48]. Modeling of this subsurface damage is important in understanding stress-corrosion cracking [8]. In this section, we introduce a relationship between concentration and damage of mechanical bonds that we superpose to the diffusion bonds from the previous section. In the first step, we only consider the dependence of damage on concentration. We call this the Concentration-Dependent Damage (or CDD) model. In reality, damage also, in return, affects the effective diffusion properties in the corrosion model we introduced above. A comprehensive model combining the CDD model and the effect of damage on the diffusion is introduced in Section 15.7 for the 2D and 3D modeling. In this section we use the CDD model to investigate the dependencies of the damage profile on the applied potential and on the saturated concentration.

Both corrosion and cracking are related to bond rupture. Corrosion is due to the chemical reactions, while cracking is induced by the mechanical loading. The rate of bond-rupture events in a chemical reaction is mainly dependent on the activation energy. These atomic bond ruptures are modeled, as an effective process, via breaking peridynamic mechanical bonds connected with this point (see Figure 15.11). Therefore, in addition to the metal concentration information transferred between nodes via the diffusivity bonds, we now consider that a point x is also characterized by its damage index $d(x,t)$ of mechanical bonds superposed onto the diffusivity bonds.

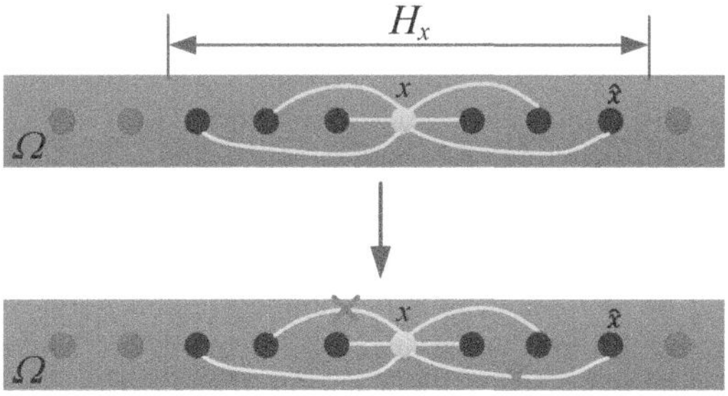

Figure 15.11: Concentration-dependent damage (CDD) model. The damage index for the mechanical bonds is linked to the ratio between the total concentration drop and the difference of C_{solid} and C_{sat}. When the concentration drops to the saturated concentration, the damage index is 1, i.e. mechanical bonds (lower picture) are totally damaged. The diffusivity bonds, upper picture, are not affected by damage, but they can suffer changes if their ends goes through a phase change, from solid to liquid, when concentration equals C_{sat} (the PD model for corrosion). When the concentration is C_{solid}, no bonds are damaged, i.e., the damage index is 0.

The damage index at a node is computed as the ratio between the numbers of broken (or failed) mechanical bonds (N_f) and the total mechanical bonds (N) originally associated with this node ($d(x,t) = \frac{N_f}{N}$). In this work, mechanical bonds only carry information on damage caused by the corrosion process. In our corrosion diffusion model, the diffusion takes out metal ions from their original position. Associated with this metal out-diffusion, part of the mechanical bonds connected with this material point will break. When the concentration reaches the saturated concentration in the liquid, C_{sat}, that is an indication of total material dissolution, which tells us that we need to break all of the mechanical bonds at that point, which means the damage index is 1. Recall that our bi-material diffusion PD model already assumes that solid nodes change into liquid nodes when their metal-ion concentration equals C_{sat} (see Section 15.4). When the concentration is that of the intact metal, the damage index should be zero. We propose here a particular model, a linear dependence of the corrosion damage index with the metal concentration, as follows:

$$d(x,t) = \frac{C_{\text{solid}} - C(x,t)}{C_{\text{solid}} - C_{\text{sat}}}. \tag{15.24}$$

To generate this type of damage we need to specify how and when bonds break, as a result of the concentration drop. The procedure for mechanical bonds breaking due to corrosion is as follows: we impose that any intact mechanical bond connected with a node x has a chance $0 \leq P \leq 1$ to break at the current time step, with P being related (see below) to the concentration drop at node x from the previous time step. In this work we assume that damage due to corrosion is "isotropic"; therefore, the breaking chance (P) for all of the intact bonds at this node and current time step is the same. Gradation of damage is induced then by the gradation in concentration. Special *orientation-dependent damage evolution* can be also formulated, for example for microstructures in which material resistance to damage/cracking from corrosion is anisotropic, but this is not considered here. The damage index obtained from the algorithm should obey the relation in Eq. (15.24). The detailed steps of this algorithm, at a given time step and given node, are:

1. *for each intact bond, generate a random number from a uniform distribution in (0,1);*

2. *if this number is smaller than P, then we break this mechanical bond. Note that each mechanical bond connects two nodes. Therefore, in addition to updating the damage information of node x, we also update the damage information of the node at the other end of this broken mechanical bond;*

3. *go to next bond in the family of bonds of node x.*

From the above algorithm, we get the relationship between the damage index at the current time step (d_i) and the damage index at the previous time step (d_{i-1}):

$$d_i = d_{i-1} + (1 - d_{i-1})P \tag{15.25}$$

where $(1 - d_{i-1})$ is the number of intact bonds at time step $i-1$, and P is the probability that any such bond will break at time step i.

We now find the relationship between P and the concentration drop at the node. There are two ways of solving P: mathematically from Eqs. (15.24) and (15.25), or mechanism-based way. Both approaches give the same relation between P and the concentration value.

The damage index at any time step should satisfy the damage/concentration relation in Eq. (15.24), i.e., $d_i = \frac{C_{\text{solid}} - C_i}{C_{\text{solid}} - C_{\text{sat}}}$ and $d_{i-1} = \frac{C_{\text{solid}} - C_{i-1}}{C_{\text{solid}} - C_{\text{sat}}}$ where C_i and C_{i-1} are the nodes concentrations at the ith and $(i-1)^{\text{th}}$ time steps, respectively. Substituting these damage/concentration relations back into Eq. (15.25) and solving for P, we have:

$$P = \frac{1}{1 - d_{i-1}} \left(\frac{\Delta C_i}{C_{\text{solid}} - C_{\text{sat}}} \right) \tag{15.26}$$

where $\Delta C_i = C_{i-1} - C_i$ is the concentration drop from the previous time step. Note that the bond breaking chance is *not* the ratio between the concentration drop and the difference between C_{solid} and C_{sat}.

We can also understand Eq. (15.26) from a mechanism-based viewpoint. The number of the broken mechanical bonds increases at the ith time step by a certain amount, $N_{fi} - N_{f(i-1)}$, which should depend linearly on the concentration drop from the previous time step, according to Eq. (15.24):

$$N_{fi} - N_{f(i-1)} = N \frac{\Delta C_i}{C_{\text{solid}} - C_{\text{sat}}} \tag{15.27}$$

The number of the intact bonds from the previous time step is:

$$N_{(i-1)} = N(1 - d_{i-1}) \tag{15.28}$$

In order to break $N_{fi} - N_{f(i-1)}$ bonds at time step i, the breaking chance P has to equal the ratio between $N_{fi} - N_{f(i-1)}$ and N_{i-1}, that is,

$$P = \frac{N_{fi} - N_{f(i-1)}}{N_{i-1}}. \tag{15.29}$$

Substituting Eqs. (15.27) and (15.28) into (15.29), we obtain the relation given in Eq. (15.26). We will apply Eq. (15.26) as the bond breaking chance P in our computations.

15.6.1 Damage evolution

The major benefit of our PD model for studying corrosion is the ability to capture subsurface damage evolution. In this subsection, damage evolution for different diffusivities is investigated.

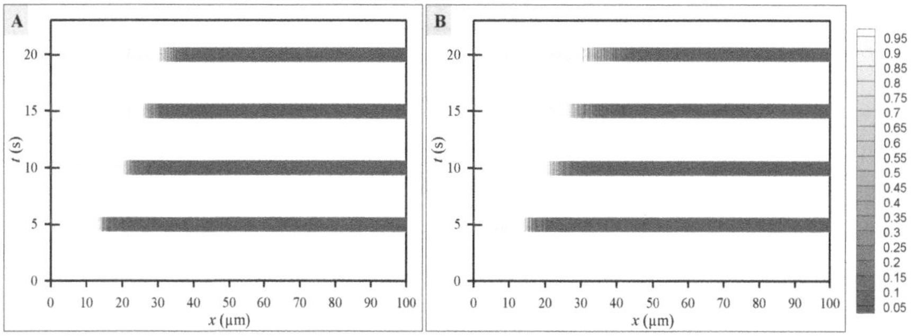

Figure 15.12: The evolution of damage profiles at different diffusivity: (A) $K_1 = 2.0\ \mu m^2/s$; (B) $K_1 = 4.0\ \mu m^2/s$. In this map, dark regions correspond to 0% damage in metal, while lighter regions to 100% damaged metal. The size of the damage layer does not change much during the diffusion-controlled pitting corrosion. Results computed with: $\delta = 0.05\ \mu m$, $m = 4$.

The evolution of damage profiles is shown in Figure 15.12. Two different diffusivities ((A) $K_1 = 2.0\ \mu m^2/s$; (B) $K_1 = 4.0\ \mu m^2/s$) are employed here. It is noted here that along with the corrosion evolution, a graded damage zone appears inside the metal. This finite-thickness damage zone is the most important advantage of our model compared with other models in which a sharp interface is used for representing pitting corrosion. The thickness of the damage layer does not change much in diffusion-controlled pitting corrosion regime.

To capture the dependence between the damage profile and the overpotential, damage maps at a certain time for different diffusivities are shown in Figure 15.13. From Eq. (15.5), we know that the diffusivity is exponentially dependent on the overpotential. Hence, a larger diffusivity corresponds to a larger overpotential. In Figure 15.13, it is revealed that the thickness of the damage layer is linearly dependent on the diffusivity. This figure predicts that although the current density tends to be a constant when the overpotential reaches a certain value, the damage profile is highly dependent on the overpotential, i.e., the mechanical properties of metal corroded under different overpotential would be different in a certain layer near the pit surface.

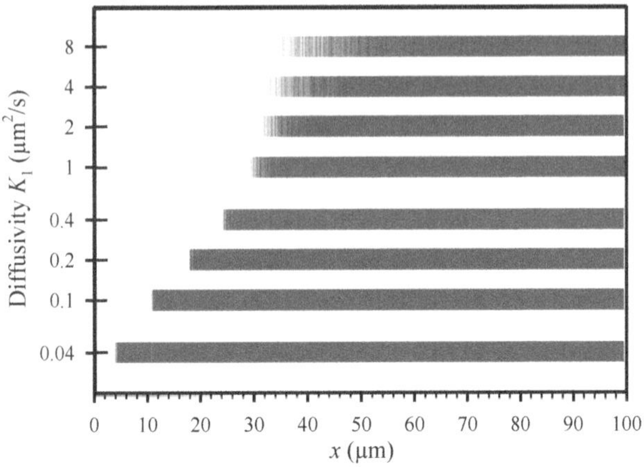

Figure 15.13: Damage maps at 20 s showing that the size of the damage layer is linearly dependent on the diffusivity. Although the interface moving speed is weakly related to the diffusivity in diffusion-controlled corrosion stage, the profile of the damaged layer is highly dependent on the diffusivity value. Since the diffusivity is an exponential function of overpotential, increasing the overpotential changes the damage profile.

Note that in 1D cases, we did not consider the effect of damage on the diffusion. A comprehensive model coupling our CDD model and the effect of damage on the diffusion makes more sense in 2D or 3D cases, and it is introduced in Section 15.7.

15.6.2 Saturated concentration

Saturated concentration is another parameter controlling the corrosion rate and the profile of the damage layer. In this subsection, the effects of the saturated concentration on the pitting speed and structure of the damage layer are investigated by applying different saturated concentration values in our PD model. In the experiments, this saturated concentration value is controllable via regulating the pH value, the concentration of anions, or charged nanoparticles [38].

Figure 15.14 shows the dependence of s on the saturated concentration. As shown in Eq. (15.21), larger s corresponds to a faster pitting speed. Therefore, as revealed in Figure 15.14, increasing the saturated concentration can promote the pitting speed. The season for this dependence is that when the saturated concentration is larger, the slope of the linearly distributed concentration in the liquid is larger (see Figure 15.7B, the slope is linearly dependent on the saturated concentration) and the flux through the pit surface is consequently higher. In addition, the thickness of the damage layer is also dependent on the saturated concentration. The damage profiles corresponding to four different saturated concentrations are shown in the inset of Figure 15.14. All of these profiles are taken at time $t = 20$ s. It is revealed that the smaller the saturated concentration is, the thicker the damage layer is.

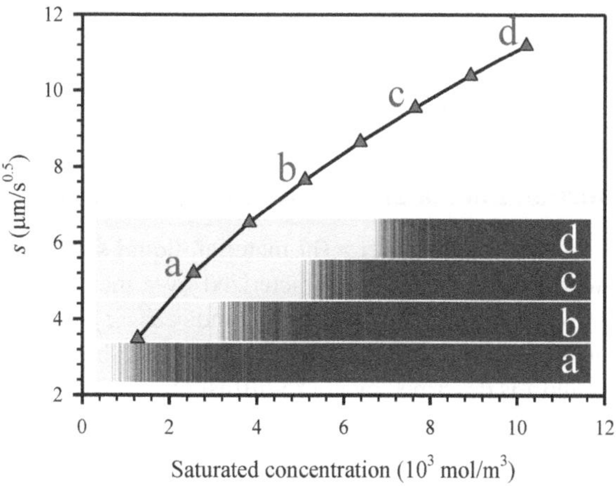

Figure 15.14: The dependence of s and damage profile on the saturated concentration. Larger saturated concentration leads to: larger speed of corrosion progression but smaller size (thickness) of the damaged region.

15.7 Formulation and results of 2D and 3D pitting corrosion

In this section, we extend the 1D modeling of pitting corrosion to 2D and 3D cases. In the simple 1D modeling, we have illustrated the effective out-diffusion process in the metal solid, captured the damage layer underneath the pit surface, and validated our modeling by comparing with experimental results. The subsurface damage is important in stress-corrosion cracking (SCC). Although the 1D modeling can capture the surface damage, to investigate SCC, higher dimensional models of corrosion are needed. Therefore, in this section, we study the PD modeling of 2D and 3D pitting corrosion, including the evolution of the subsurface damage layer from corrosion-induced damage.

We first formulate the 2D and 3D models. Then, for the 2D activation-controlled pitting corrosion regime, we investigate the effect of the nonlocal parameters (horizon factor and horizon size in the PD modeling) and the material parameter (effective diffusivity in the PD modeling) on the corrosion damage. The results will show that, for the activation-controlled regime, both the horizon size and the diffusivity can alter the corrosion rate, while the thickness of the damage layer is controlled by the horizon size only. These dependencies provide us ways to calibrate our model. For example, we can obtain the damage thickness from experimental characterization of the pit cross section. The measured thickness can lead to a certain horizon size in our PD model. We then could use the measured corrosion rate combined with this fixed horizon size to figure out the effective diffusivity, as we did for the 1D model in Section 15.5.2. At the end of this section, a 3D case study is also shown.

15.7.1 PD formulation of 2D and 3D pitting corrosion

Similar as the 1D case, we consider a (bi-material, liquid-solid) body Ω, and a material point x with horizon H_x: x is characterized by a metal concentration, $C(x,t)$ (mediated by the diffusion bonds) and damage index $d(x,t)$ (mediated by the superposed mechanical bonds introduced in Section 15.6). As shown in Figure 15.15, the bond between x and $\hat{x}$ is described by micro-diffusivity $k(x,\hat{x})$. We have the following evolution of metal concentration (mol/m^3):

$$\frac{\partial C(x,t)}{\partial t} = \int_{H_x} J(\hat{x},x,t)\mathrm{d}V_{\hat{x}} \tag{15.30}$$

where the calculation kernel, the flux (J) versus the micro-diffusivity, is expressed as:

$$J(\hat{x},x,t) = \begin{cases} k(x,\hat{x})\frac{C(\hat{x},t)-C(x,t)}{\|\hat{x}-x\|^2}, & \|\hat{x}-x\| \leq \delta \\ 0, & \|\hat{x}-x\| > \delta. \end{cases} \tag{15.31}$$

Note that we use the square of the bond length $\|\hat{x}-x\|$ as the denominator, instead of the value of the bond length as used in the 1D model. As we mentioned in Section 15.4, both denominators are valid, but in some cases the one with the quadratic form in the denominator has certain advantages (see[13]). Equations (15.30) and (15.31) are independent of the dimension (1D, 2D, or 3D), but obviously, the units for the micro-diffusivity change depending on the dimension in which we work.

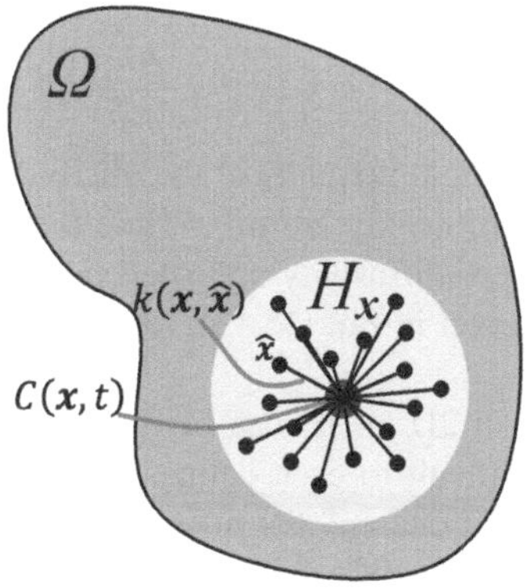

Figure 15.15: Peridynamic description of the diffusion model in 2D. The region denoted by H_x is the horizon of node x. The concentration at x is $C(x,t)$. Bonds connecting points have micro-diffusivity $k(x,\hat{x})$.

To implement the peridynamic method in 2D or 3D, we discretize the domain using a uniform grid. We still use the mid-point integration scheme to approximate the integral over H_x as we did in the 1D case. Each node has a "volume," in 2D (3D) given by the area (volume) associated with it.

The spatial discretization of the PD equation in 2D or 3D is:

$$J(x_p,x_i,t) = k(x_p,x_i)\frac{C(x_p,t)-C(x_i,t)}{(x_p-x_i)^2} \tag{15.32}$$

and

$$\frac{\partial C(x_i,t)}{\partial t} = \sum_p k(x_p,x_i)\frac{C(x_p,t)-C(x_i,t)}{(x_p-x_i)^2}V_p \tag{15.33}$$

where the summation is after all nodes x_p inside the horizon of node x_i, and V_p is the portion of the volume of the node x_p actually covered by the horizon of node x_i. For the nodes crossing the boundary of the horizon, we use the same algorithm to calculate the volume as we used in [4, 26].

The micro-diffusivity can be determined by equating the peridynamic flux to the classical flux through a surface (a line in 2D case) arising from a simple linear concentration profile [3]. The micro-diffusivity function over the horizon region can be selected in various forms. We select the simplest form, a constant over the horizon. With the form of the kernel given in Eq. (15.31), in 1D we then obtain:

$$k(x_p,x_i) = \frac{K}{\delta}. \tag{15.34}$$

While in 2D, we get:

$$k(x_p,x_i) = \frac{4K}{\pi\delta^2} \tag{15.35}$$

and for the 3D case:

$$k(x_p, x_i) = \frac{9K}{2\pi\delta^3}. \tag{15.36}$$

Depending on the dimension, the units of the micro-diffusivity differ. For instance, the units of 1D micro-diffusivity are m/s, while the ones for 2D and 3D are 1/s, 1/m·s. Note that the micro-diffusivity units depend on the form of the kernel used, for instance with the kernel in Eq. (15.13), for the 1D model the units are 1/s (see Eq. 15.16).

Since the interface in 2D or 3D can be rough and in the presence of damage the interface may be difficult to even define, if the ends of a bond are located in two different materials (at one end the node has the metal diffusivity, while at the other end the node has liquid diffusivity), then we define its micro-diffusivity to be the harmonic average of the micro-diffusivities corresponding to the solid and liquid diffusivities (see [37]):

$$k = \frac{2k_1 k_2}{k_1 + k_2}. \tag{15.37}$$

To use the above model in corrosion, in which the solid-liquid interface is a moving interface, we consider the bi-material diffusion model together with the phase-change discussed in Section 15.4. Once the concentration at a solid node reaches C_{sat}, its diffusivity changes to that of the liquid. So:

$$k(\hat{x}, x) = \begin{cases} k_1, & \text{if } C(x) \text{ and } C(\hat{x}) > C_{sat} \\ k_2, & \text{if } C(x) \text{ and } C(\hat{x}) > C_{sat} \\ \frac{l}{\frac{l_1}{k_1} + \frac{l_2}{k_2}}, & \text{if } C(x) \geq C_{sat} \geq C(\hat{x}) \text{ or } C(x) \leq C_{sat} \leq C(\hat{x}). \end{cases} \tag{15.38}$$

We use the same damage model introduced in Section 15.6, the CDD model, to track the evolution of the corrosion damage (see Figure 15.16). We set the damage index ($d(x,t)$) according to Eq. (15.24). When the concentration drops to the saturated concentration, the damage index is 1, i.e. mechanical bonds are totally damaged (the diffusion bonds are unaffected). When the concentration is C_{solid}, no bonds are damaged, i.e. the damage index is 0.

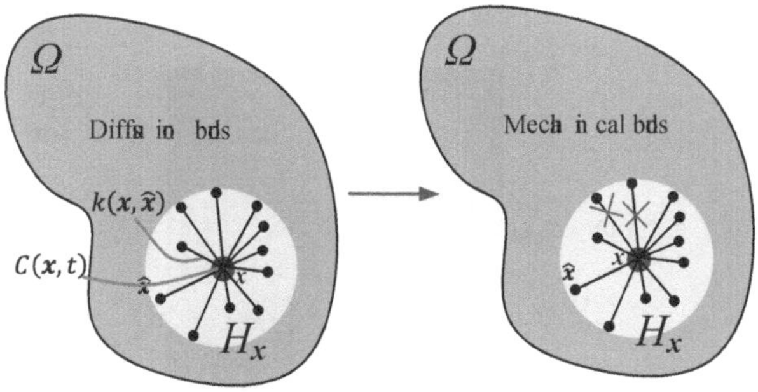

Figure 15.16: Concentration-dependent damage (CDD) model in 2D. Diffusion bonds (are affected by phase change, but cannot break), and mechanical bonds that break due to the concentration drop.

Following the calculation steps given in Section 15.6, we can capture the concentration and damage evolution in 2D and 3D pitting corrosion. Note that in 1D model, we added extra boundary nodes as the ends of the bi-material domain. In the 2D and 3D models, we do not add extra boundary nodes and so the nodes on the boundaries have smaller areas (volumes) than nodes in the bulk.

15.7.2 The Concentration-Dependent Damage (CDD) model for pitting corrosion: example in 2D

In this subsection, we represent a 2D pitting corrosion case: a 2D plate of area 100 μm $\times$ 100 μm, with a half-circular initial pit of diameter 20 μm on top-center. The surface of this metal plate is covered by passive film except the pit area. Thus the mass exchange (metal ions diffusing out from the pit and the solution flowing into the pit) can only happen through the pit. The boundary conditions and initial conditions are shown in Figure 15.17. The material parameters are the same as shown in the 1D case in Section 15.5. We assume the fresh solution flushes the pit surface all the time, mimicking corrosion conditions for the bottom of a boat in water. Thus, the concentration inside the liquid can always be treated as zero at all time steps. Different from the diffusion-controlled corrosion, in which corrosion is controlled by the ions' diffusion inside the solution, the pit growth in this case is governed by the out-diffusion of metal ions from the solid only. This is activation-controlled corrosion.

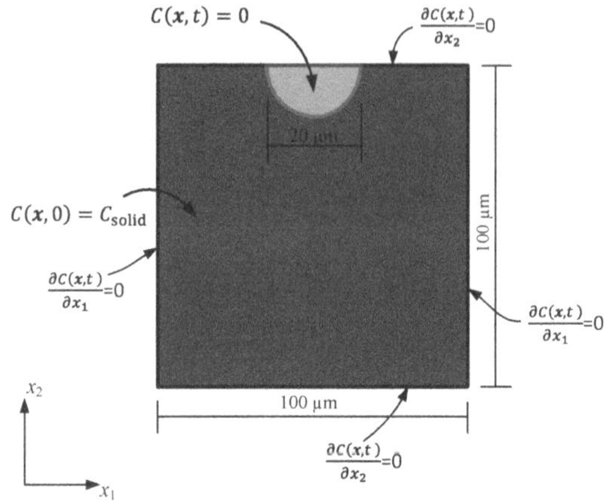

Figure 15.17: A 2D example of bi-material diffusion: boundary conditions and initial conditions. There is no flux around the boundary of the metal, except over the solid/liquid interface. In the PD model, no conditions have to be imposed at the solid/liquid interface.

In the activation-controlled corrosion, there is always a concentration jump (equal to the saturated concentration, see Figure 15.18). In classical diffusion theory, this jump leads to infinite flux. In reality, the flux at that point is a constant that depends on the overpotential. In our PD model, this jump is, effectively, linearized over the two nodes on both sides of the interface. Thus, in the model, the pit growth rate depends on both the overpotential (effective diffusivity) and the grid spacing (see the discussion in Section 15.5.1, Figure 15.7).

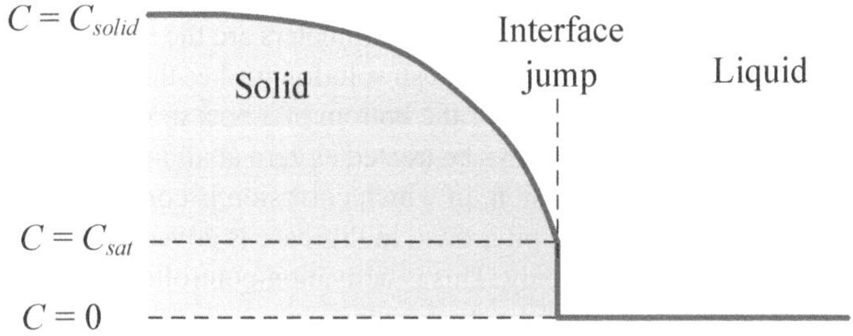

Figure 15.18: The concentration distribution around solid/liquid interface within activation-controlled corrosion. The concentration inside the liquid is always zero, while the one inside the solid is always larger than the saturated concentration inside the liquid. Thus, there is a concentration jump at the interface.

Coupling the bi-material diffusion model with the CDD model (see Section 15.6), we obtain the damage profiles shown in Figure 15.19 (at $t = 20$ s), corresponding to effective diffusivities of 2.0 $\mu m^2/s$. Only the solid nodes are shown here and in the remaining damage maps. Note that even though there is not flux through the top boundary, except for the surface of the liquid in the original pit, damage spreads uniformly in all directions.

Convergence studies for the damage profile in terms of grid refinement for a fixed horizon size (m-convergence) and in terms of horizon size going to zero, for a fixed horizon factor m, are shown in Section 15.10.2. Although the damage profile suffers little change as the respective parameters vary for the m-convergence and δ-convergence studies, as seen in Figures 15.31 and 15.32, the solid/liquid interface for all these cases does not advance within the 20 s of the physical time simulated. There are solid ranges near the solid/liquid interface with totally damaged nodes (nodes lose all of the mechanical bonds). In reality, this part of metal is loosely attached on the metal solid, and can be easily rushed away by the liquid. In addition, the mechanical damage can decrease the corrosion resistance (enhance the pitting corrosion). Hence, to capture this coupled damage/corrosion dependence, we propose to switch all of the solid nodes with damage index 1 to liquid nodes, even if they have a concentration larger than C_{sat}. This operation represents a phase change-like transformation, and is only one possible approach for coupling damage and corrosion. We will refer to this model as the Damage-Dependent Corrosion (DDC) model. The reason we add the DDC model to the CDD model in 2D and 3D, but not in the 1D case, is that, in two or three dimensions, there are a multitude of possible pathways through which the solution can bypass a metal material point. In the next sections, we describe the implementation of this combined model (CDD+DDC) and show results of corrosion damage in 2D and 3D.

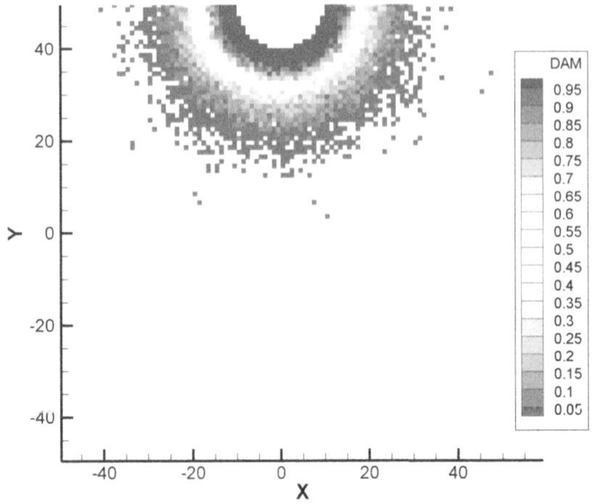

Figure 15.19: The damage profiles around a half-circular pit after corrosion in NaCl solution for 20 s. The horizon factor is 4.02 and the horizon size is 4.02 μm. The effective diffusivities are 2.0 μm^2/s, which is corresponding to an overpotential around 0.2 V (units on both axes are μm).

15.7.3 A coupled corrosion/damage model for pitting corrosion: 2D example

With mechanical damage developing inside the metal solid, like porosity that gets filled by liquid, the effective diffusivity should increase since metal points have more contact with the solution. Therefore, in addition to the original phase-change mechanism introduced in the PD bi-material diffusion model to allow for autonomous interface motion in corrosion, we now introduce a mechanism that produces a phase change from solid to liquid even if the concentration is larger than the saturated one, as long as damage index is one. This can happen in our model because of the stochastic nature of the model we introduced for breaking of mechanical bonds (the CDD model in Section 15.6). Indeed, we will have points at which the concentration is close to the saturated concentration, and that could lead to a damage index value of 1. Rather than wait for the concentration to drop to the saturated value, we perform the phase change (thus increasing the diffusivity/corrosion rate at that point). To capture the dependence of corrosion on mechanical damage, we use the combination of these two models: CDD+DDC. In this coupled CDD+DDC model, the evolution of metal concentration and damage can be described as follows:

1. *Generate mesh, initial, and boundary conditions: Input nodal information $(x, \delta, C(x))$, and initial and boundary conditions.*

2. *Compute the micro-diffusivities based on Eqs. (15.35) and (15.37).*

3. *Compute concentration distribution $C(x)$, at the first time step.*

4. *Generate damage in the mechanical bonds by using the CDD model.*

5. *If a solid node drops its concentration to a value smaller than C_{sat} (the CDD model) or if its damage index reaches 1, independent of its concentration (the DDC model), then switch it to a liquid node by replacing its micro-diffusivity with the micro-diffusivity corresponding to the solution.*

6. *Repeat steps 2–5 for all nodes and all time-steps.*

This algorithm results in an autonomous evolution of the liquid/solid interface, as well as the development of subsurface damage. In Figure 15.20 and Figure 15.21 we show the results corresponding to this coupled CDD+DDC model. With the same diffusivity (2.0 $\mu m^2/s$) used in Figure 15.19 where the CDD model was used, the pit now, with the CDD+DDC model, advances much faster over the same time period. In Figure 15.20, different horizon sizes are used and the corrosion depths are about 18 μm, 35 μm and 67 μm, corresponding to horizon size 8.04 μm, 4.02 μm and 2.01 μm, respectively. The thicknesses of the damage layer are about 5 μm, 10 μm, and 20 μm, respectively. We conclude that in the PD modeling of *activation-controlled corrosion*, the corrosion progression is strongly affected by the horizon size. The corrosion depth is inversely proportional to the horizon size, while the damage thickness linearly depends on it. The study on the horizon factor (see Figure 15.21) shows the m-convergence behavior. The corrosion depth and damage thickness tend to be influenced less by the horizon factor (m) variations than by variation in the horizon size (compare Figure 15.21 with Figure 15.20).

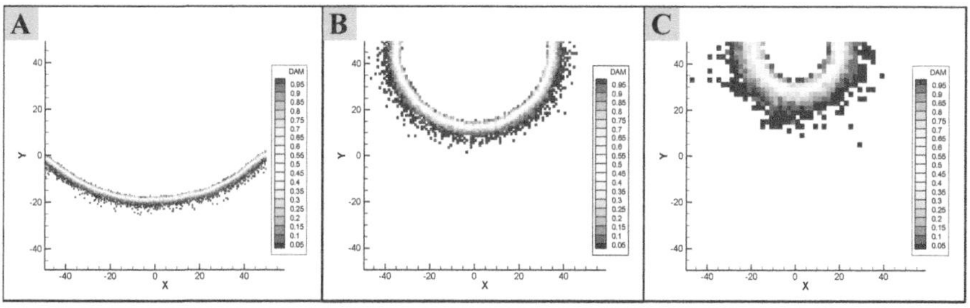

Figure 15.20: The horizon size effect for the CDD+DDC model. The horizon factors in three panels are 4.02. The horizon sizes are different: (A) 2.01 μm; (B) 4.02 μm; (C) 8.04 μm. The effective diffusivities are 2.0 $\mu m^2/s$, which is corresponding to an overpotential around 0.2 V.

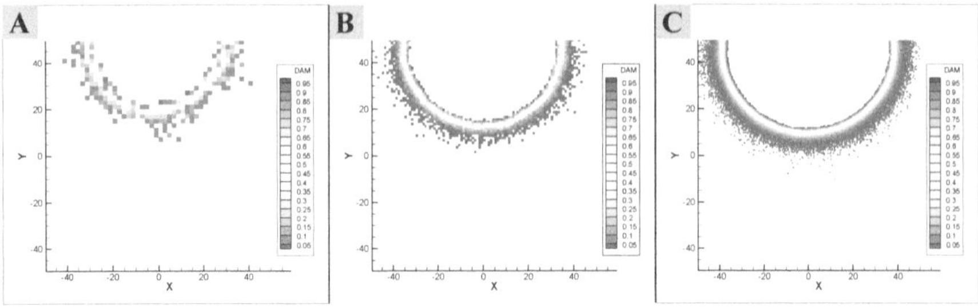

Figure 15.21: m-convergence for the CDD+DDC model. The horizon sizes are kept as 4.02 μm, while the horizon factors are doubled each time: (A) 2.01; (B) 4.02; (C) 8.04.

In the remaining part of this section, we investigate how the effective diffusivity influences the progression of corrosion damage. For these calculations we select a horizon size of 4.02 μm, and a numerical grid corresponding to $m = 4.02$. In Section 15.7.5 we show 3D results, and in Section 15.8 we study some heterogeneous materials.

15.7.4 Diffusivity affects the corrosion rate

Figure 15.22 shows the corrosion damage dependence on the diffusivity, which, in turn, could be calibrated from the polarization curve, as we presented in Section 15.5.2 for the 1D case. The thicknesses of the damage layer corresponding to different diffusivities are the same. However, the corrosion rate (the interface moving speed) is linearly depending on the diffusivity. A more detailed study on the dependence between diffusivity, horizon size, and the corrosion evolution will lead to ways to calibrate the effective diffusivity and the horizon size.

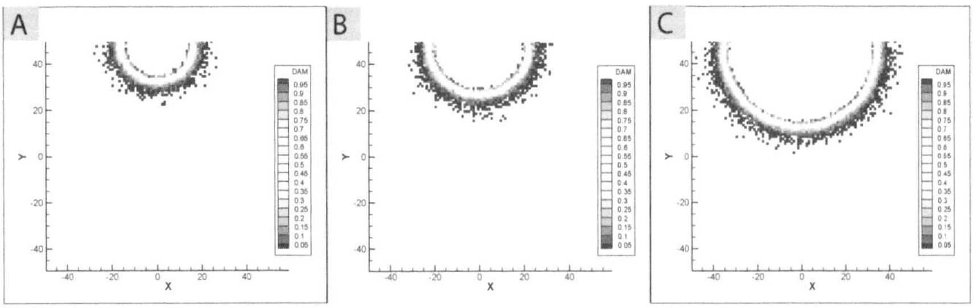

Figure 15.22: The damage profile after 20 s of corrosion process for different diffusivities: (A) 0.5 $\mu m^2/s$; (B) 1.0 $\mu m^2/s$; (C) 2.0 $\mu m^2/s$. The corrosion depths are 16 μm, 21 μm, and 35 μm.

15.7.5 Pitting corrosion with the CDD+DDC model in 3D

Due to the complexity of 3D modeling, the pitting corrosion models are usually restricted to 1D or 2D problems. Pitting corrosion 3D modeling has rarely been reported [50], because of the cost and complexity of the modeling and simulations. Here, we show that our PD model can be easily extended to 3D. The formulation for the 3D problems is given in Eqs. (15.30), (15.31), and (15.38), and the corresponding constant micro-diffusivity is shown in Eq. (15.36).

We apply the model under activation-controlled corrosion conditions, similar to those used in the 2D case presented in Section 15.7.2. The geometry, initial, and boundary conditions are shown in Figure 15.23.

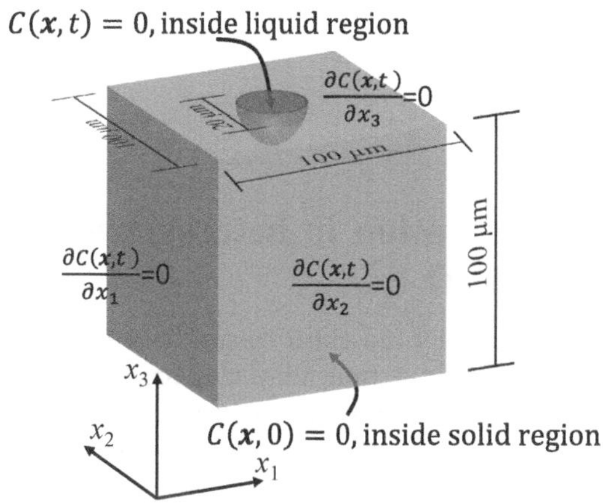

Figure 15.23: Geometry, initial, and boundary conditions for a 3D sample. There is zero-flux on the boundary of the metal cubic sample except over the solid/liquid interface disk region on the top-center.

Figure 15.24 shows the damage evolution in 3D activation-controlled pitting corrosion. In this 3D case, the material parameters are the same as introduced in 1D case in Section 15.5. The metal sample considered is a cube of volume 100 μm $\times$ 100 μm $\times$ 100 μm, with a half-spherical initial pit of diameter 20 μm on top-center (see Figure 15.23). The effective diffusivity is 2.0 μm^2/s. We choose CDD+DDC model as the damage/corrosion dependence. At the beginning of the pitting corrosion, only the nodes near the solution are damaged. During the corrosion, the thickness of the damage layer becomes larger and later reaches a stable value. In the case shown in Figure 15.24, the damage thickness is about 30 μm. As we discussed in previous sections, this thickness can be affected by the horizon size, the saturated concentration in the solution. At $t = 20$ s, the corrosion reaches the walls of the cubic sample geometry. This example shows that our PD model can be used to study 3D pitting corrosion.

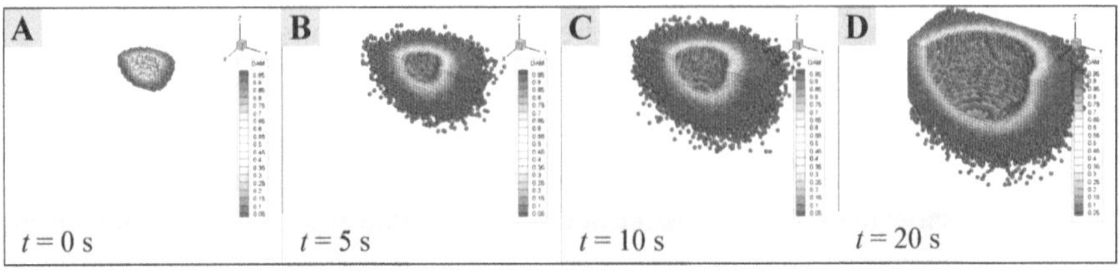

Figure 15.24: The damage evolution of 3D pitting corrosion. Sample is cut through the center to allow better viewing of the progression of corrosion damage. Only the solid nodes are shown. The horizon size and horizon factor are 8.04 μm and 4.02, respectively.

15.8 Pitting corrosion in heterogeneous materials: examples in 2D

Microstructural heterogeneity can either enhance or subdue pitting corrosion [41]. This heterogeneity can originate from alloy additions and impurities which combine to produce both the desired microstructure as well as undesired, large particles, called constituent particles (and residual impurity particles) which have a range of compositions [27]. Heterogeneity can also lead to irregular pit shapes [1, 10], which can influence stress concentration and fracture behavior. Our PD model can be used in modeling of corrosion in heterogeneous materials, by assigning certain diffusivities to the PD nodes, depending on whether these nodes are in one phase or another of the material. Bonds that have one diffusivity at one end, a different one at the other, have their micro-diffusivity computed based on Eq. (15.37).

In this section, we study three examples of pitting corrosion in heterogeneous double-phase materials: (A) a layered structure with the stripe direction parallel to the direction of the pit depth (see Figure 15.25, Case A); (B) a layered structure with the stripe direction perpendicular to the direction of the pit depth (see Figure 15.25, Case B); (C) metal with randomly distributed inclusions (see Figure 15.28).

15.8.1 Pitting corrosion in layer structures

We consider the pitting corrosion in a two-phase metal alloy, phase 1 (light/orange *) and phase 2 (dark/brown) (see Figure 15.25). Corresponding to these two phases, two effective diffusivities are given, $K_{①}$ and $K_{②}$. We define heterogeneity contrast α as the ratio between the diffusivities of phase 1 and phase 2. Figure 15.26 and

*For colors, please see the electronic version of the handbook, or reference [12].

Figure 15.27 show the damage profile after 20 s corrosion for different α values. The material used in Figure 15.26 is the Case A shown in Figure 15.25, and the one in Figure 15.27 is corresponding to the Case B in Figure 15.25.

The heterogeneity constant controls the pit shape. When α equals to 1, the material is homogeneous and the resulting pit is of uniform, half-circular shape, as obtained earlier. When α is larger than 1, the corrosion resistant of phase 1 is less than phase 2, and the pit grows into phase 1 than phase 2, as shown in Figure 15.26 and Figure 15.27; if we consider phase 1 to be the mother phase of this metal, the addition of phase 2 subdues the pitting corrosion, in either horizontal direction (Figure 15.26) or vertical direction (Figure 15.27), depending on architecture of the structure. When α is less than 1, the addition of phase 2 enhances the pitting corrosion.

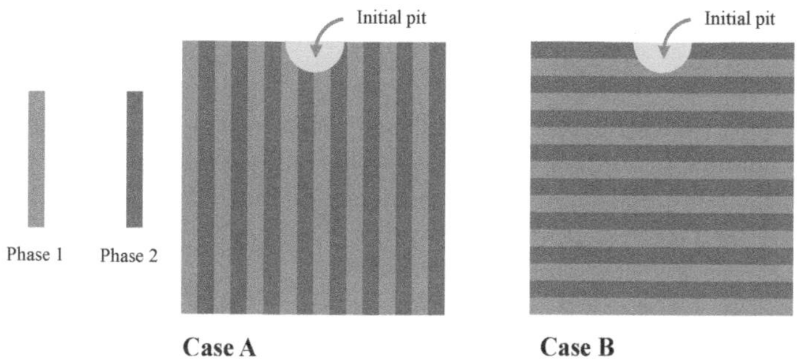

Figure 15.25: Illustrations of two cases of double-phase heterogeneous materials used to study pitting corrosion. The geometry of the metal sample and the initial pit are the same as given in Section 15.7.2. The corrosion resistances are different for phase 1 from phase 2.

The heterogeneity not only determines the pit profile (the shape), but also affects the damage distribution. As shown in Figure 15.26 and Figure 15.27, when α is larger than 1, the subsurface damage is non-uniformly distributed. In Figure 15.27, for example when α is 8, the thickness of the damage layer at the bottom of the pit is larger than other locations. However, when α increases to 64, phase 2 prevents the corrosion damage penetrating downward; thus, the thickness of subsurface damage at the bottom of the pit is smaller than at other locations.

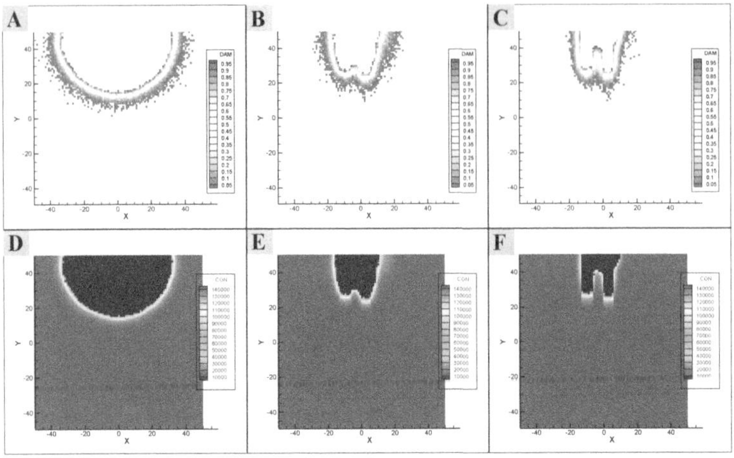

Figure 15.26: The damage profiles (top panels) and the corresponding concentration distributions (bottom panels) after 20 s corrosion for Case A. The material heterogeneity contrast α from left to right: 1 (A and D; this is the homogeneous case), 8 (B and E), and 64 (C and F). $K_{\textcircled{1}}$ fixed at 2.0 μm^2/s. In the damage maps only the solid nodes are shown.

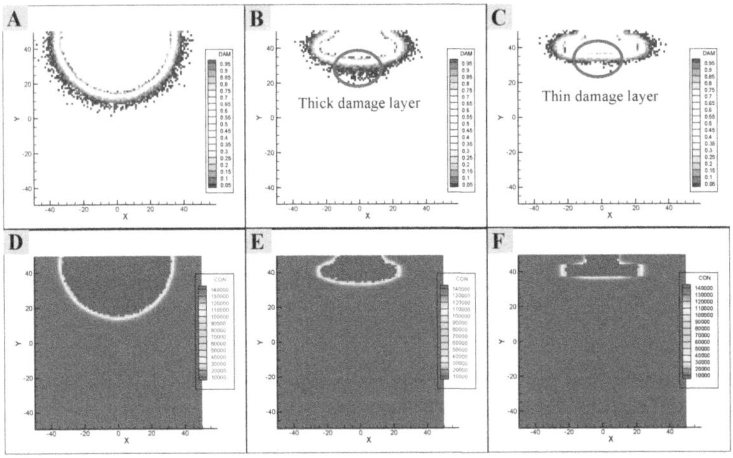

Figure 15.27: The damage profiles (top panels) and the corresponding concentration distributions (bottom panels) after 20 s corrosion for Case B. The material heterogeneity contrast α from left to right: 1 (A and D; this is the homogeneous case), 8 (B and E), and 64 (C and F). $K_{\textcircled{1}}$ is fixed at 2.0 μm^2/s. In the damage maps only the solid nodes are shown.

15.8.2 Pitting corrosion in a material with inclusions: a 2D example

Other types of microstructural heterogeneities that influence corrosion damage are: inclusions, grain boundaries, etc. In this section, we show the corrosion damage evolution in a metal sample with randomly distributed inclusions that have different corrosion resistance than the metal matrix (see Figure 15.28).

Assuming the inclusions have different corrosion resistance translates, in our PD model, into the inclusions having different effective diffusivity from the metal matrix. The locations of the inclusions are random, while the sizes are fixed as the same. The heterogeneity contrast α defined in the previous section monitors the ratio of corrosion resistance between the matrix and the inclusions. Figure 15.29 shows the effect of the inclusions on the damage evolution. In this example, the horizon size is 4.02 μm, smaller than the radius of the inclusions. If the horizon size (δ) is larger than the radius of the inclusions (R_{in}), the effect of the inclusions could be averaged by the matrix material. In the future, we will investigate the dependence on the R_{in}/δ ratio. When the heterogeneity constants are larger than 1, the inclusions prevent the pitting corrosion (the progression of the solid/liquid interface). It is interesting to observe how corrosion damage evolves around the inclusions (see Figure 15.29C).

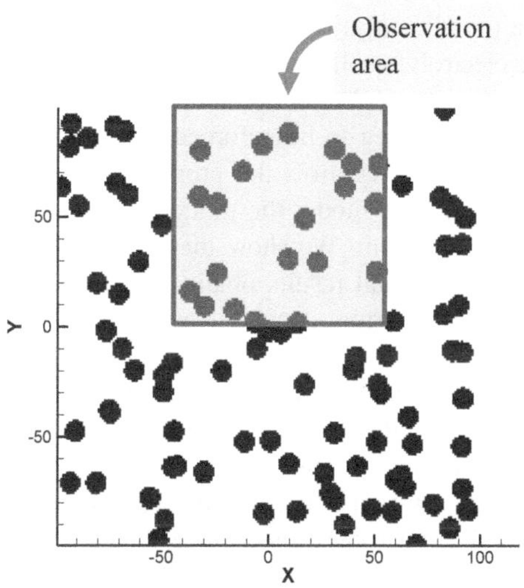

Figure 15.28: A two-phase material with randomly distributed inclusions. Within a 200 μm × 200 μm area, 100 circular inclusions with diameter 10 μm are randomly distributed. The diffusivity of the metal is $K_{①}$, and of the inclusions is $K_{②}$. Only the top-center part (100 μm × 100 μm) is shown in Figure 15.29. $K_{①}$ is fixed at 2.0 μm^2/s

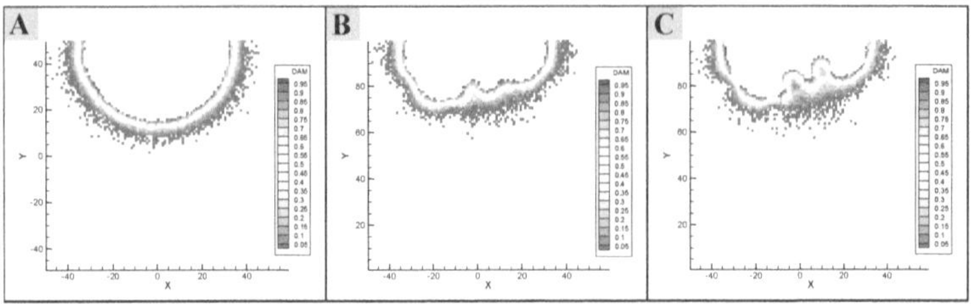

Figure 15.29: Damage profiles after 20 s of corrosion for the case shown in Figure 15.28 (only the observation area is shown and only solid nodes are shown). The heterogeneity contrast α are: (A) 1 (homogeneous case), (B) 8, and (C) 64.

15.9 Conclusions

Corrosion reaction (anodic only) is modeled here by a peridynamic effective diffusion process in the electrolyte-solid system coupled with a phase-change mechanism that allows for autonomous evolution of the moving interface. In this model, no explicit interface conditions need to be enforced. In contrast to the classical models in which a Stefan condition controls the propagation speed of the solution/metal interface, in our peridynamic model the motion of the interface is part of the solution, not part of the problem. We show that our model, once it is calibrated to a data point from experimental results obtained in the activation-controlled stage, can predict the connection current-density/overpotential over an entire range of values, including activation-controlled, IR-controlled, and diffusion-controlled regions. The model predicts that, in the 1D diffusion-controlled corrosion, the current density tends to be a constant when the overpotential goes beyond a certain value.

In order to capture subsurface damage, we introduce new corrosion damage models based on couplings between concentration (carried by the diffusion bonds) and damage (carried by the superposed mechanical bonds). Subsurface damage is important in stress-corrosion cracking. Compared with other models, which assume a sharp metal/solution interface, the present formulation leads to a structure of the corroded region similar to what is observed in experiments. Our corrosion damage model shows that the subsurface damage profile is highly dependent on the overpotential, i.e., the mechanical properties of metal corroded under different overpotentials would be different. This will have consequences for studying stress-corrosion cracking, which we plan to do in the future.

We extend the model to 2D and 3D and apply it for activation-controlled corrosion. Under such conditions, the progression of corrosion damage is affected by both the effective diffusivity (overpotential) and the horizon size (nonlocal interactive region). We have proposed ways to calibrate these parameters from experimental measurements. Results for diffusion-controlled corrosion in 2D and 3D will be reported in the future.

In addition to studying corrosion damage in homogeneous materials, we use the model to analyze pitting corrosion in metals with microstructural heterogeneities. We obtain pit shapes that are strongly influenced by the particular microstructures and comment on the evolution of the damage profiles. Using this model, computing corrosion damage in polycrystalline materials becomes achievable.

Acknowledgments

This work has been supported by NAVAIR, Contract number NASC-M801-UNL (program manager Nam Phan, Branch Head, Patrol/Transport/RotaryWing), by ONR award #11594837 (program manager: William Nickerson), and by AFOSR MURI Center for Materials Failure Prediction through Peridynamics (program managers David Stargel, James Fillerup, Ali Sayir, and Fariba Fahroo). FB also acknowledges support from the STTR N13A-T007 Phase I Contract# N68335-13-C-0344 (Dr. Nagaraja Iyyer, TDA Inc.).

15.10 Appendix

15.10.1 Convergence study for 1D diffusion-controlled corrosion

We study the m-convergence and the δ-convergence [7] for the diffusion-dominated process. We recall that in the m-convergence we consider the horizon δ fixed and take $m \to \infty$. The numerical PD approximation will converge to the exact nonlocal PD solution for the given δ. In the case of δ-convergence, the horizon $\delta \to 0$ while m is fixed or increases with decreasing δ, but at a slower rate than that of δ. For problems with no singularities, the numerical PD approximation converges to the classical local solution.

The m-convergence results in Section 15.5.1 are for a certain horizon δ. The question we answer in this appendix is: how do the results in 1D corrosion depend on the nonlocal region size? For heat transfer, results on δ-convergence were presented in [3, 4].

Since the corrosion advancement in the diffusion-controlled stage can be characterized by a constant s (see Eq. 15.20), we perform convergence study by comparing this value for different horizon size (δ) and horizon factor ($m = \delta/\Delta x$, where Δx is the space between the nodes) for the particular case with $K_1 = 2.0\ \mu\text{m}^2/\text{s}$. The results are shown in Figure 15.30. For each horizon size, the m values are chosen as 2, 4, 8, and 16. A larger m denotes a finer discretization for a given horizon size. m-convergence is also observed for each horizon size. With the horizon size decreasing, s converges to a value close to 8 μm/s$^{0.5}$. This convergence behavior is similar to the m-convergence curves in [3, 16]. From this convergence study for calculations in which diffusion-controlled corrosion is the dominant regime, we conclude that $\delta = 0.05$ μm, and $m = 4$ are good choices to ensure that results do not change when the grid is refined further (m larger than 4) or when the horizon size is decreased further (while keeping m fixed).

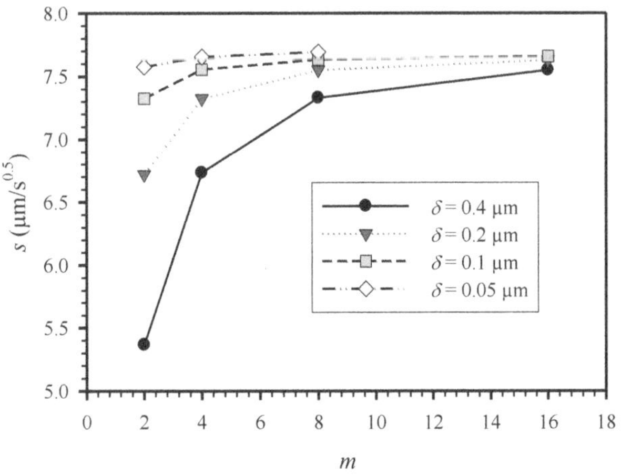

Figure 15.30: Convergence study for the slope of depth vs. $\sqrt{t}$ (s, see Eq. 15.20). With the horizon size decreasing, the ratio between the interface position and the square root of the time (s) converges to a value near 8 μm/s$^{0.5}$. In the calculation of s, we fit the data pointsfrom the diffusion-controlled region with $t \in [15, 20]$ seconds ($\sqrt{t}$ between $\sqrt{15s}$ and $\sqrt{20s}$ in Figure 15.8B).

15.10.2 Convergence study for 2D activation-controlled corrosion with Concentration-Dependent Damage model

The m-convergence and δ-convergence results for 2D activation-controlled corrosion with the CDD model are shown in Figures 15.31 and 15.32. Figure 15.31 shows the damage profiles (at $t = 20$ s) corresponding to different horizon sizes for the horizon factor $m = 4.02$. With smaller horizon size, the thickness of the highly damaged region (shown in color green * on the damage profile map) changes slightly. Figure 15.32 shows the effect of different horizon factors (m) on the damage profile.

*For color please see the electronic version of the handbook or reference [12].

Similar as the horizon size effect, when a larger horizon factor is used (meaning a finer grid spacing), the thickness of the damaged region is slightly larger.

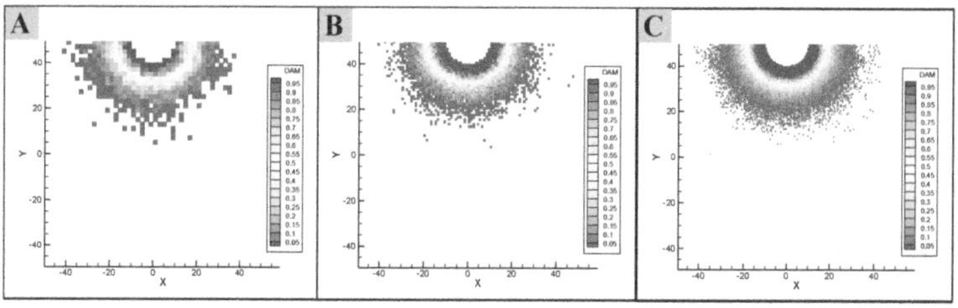

Figure 15.31: The damage profiles from the half-disk original pit after corrosion in NaCl solution for 20 seconds. The horizon factor is the same in all three cases, $m = 4.02$. The horizon sizes are: (A) 8.04 μm; (B) 4.02 μm; (C) 2.01 μm. The effective diffusivities are 2.0 $\mu m^2/s$, corresponding to an overpotential around 0.2 V. Units on both axes are μm, and the color represents the damage index (see the legend).

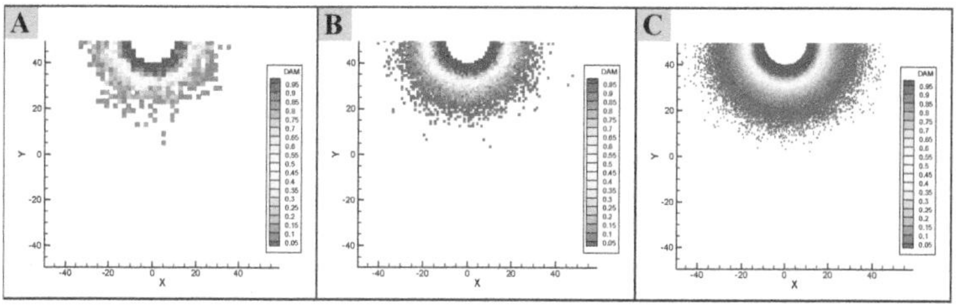

Figure 15.32: The effect of horizon factor m for a constant horizon size (δ = 4.02 μm) on the corrosion damage profile at 20 seconds. The horizon factors are: (A) 2.01; (B) 4.02; (C) 8.04. The effective diffusivity is 2.0 $\mu m^2/s$, corresponding to an overpotential around 0.2 V. Units on both axes are μm.

References

[1] R. Ambat, N. N. Aung, and W. Zhou. Evaluation of microstructural effects on corrosion behaviour of az91d magnesium alloy. *Corrosion Science*, 42(8): 1433–1455, 2000.

[2] A. J. Bard and L. R. Faulkner. *Electrochemical Methods: Fundamentals and Applications*. New York: John Wiley & Sons, 2005.

[3] F. Bobaru and M. Duangpanya. The peridynamic formulation for transient heat conduction. *International Journal of Heat and Mass Transfer*, 53(19-20): 4047–4059, 2010.

[4] F. Bobaru and M. Duangpanya. A peridynamic formulation for transient heat conduction in bodies with evolving discontinuities. *Journal of Computational Physics*, 231(7): 2764–2785, 2012.

[5] F. Bobaru and Y. D. Ha. Adaptive refinement and multiscale modeling in 2D peridynamics. *International Journal for Multiscale Computational Engineering*, 9(6): 635–659, 2011.

[6] F. Bobaru, Y. D. Ha, and W. Hu. Damage progression from impact in layered glass modeled with peridynamics. *Open Engineering*, 2(4): 551–561, 2012.

[7] F. Bobaru, M. J. Yang, L. F. Alves, S. A. Silling, E. Askari, and J. F. Xu. Convergence, adaptive refinement, and scaling in 1d peridynamics. *International Journal for Numerical Methods in Engineering*, 77(6): 852–877, 2009.

[8] J. T. Burns, S. Kim, and R. P. Gangloff. Effect of corrosion severity on fatigue evolution in Al-Zn-Mg-Cu. *Corrosion Science*, 52(2): 498–508, 2010.

[9] W. D. Callister and D. G. Rethwisch. *Materials Science and Engineering: An Introduction*. John Wiley and Sons, 7th ed., 2007.

[10] P. Campestrini, E. P. M. van Westing, H. W. van Rooijen, and J. H. W. de Wit. Relation between microstructural aspects of aa2024 and its corrosion behaviour investigated using afm scanning potential technique. *Corrosion Science*, 42(11): 1853–1861, 2000.

[11] H. S. Carslaw and J. C. Jaeger. *Conduction of Heat in Solids*. London: Oxford University Press, 2nd ed., 1959.

[12] Z. Chen and F. Bobaru. Peridynamics modeling of pitting corrosion damage. *Journal of the Mechanics and Physics of Solids*, 78: 352–381, 2015.

[13] Z. G. Chen and F. Bobaru. Selecting the kernel in a peridynamic formulation: A study for transient heat diffusion. *Computer Physics Communications*, 197: 51–60, 2015.

[14] Z. Chen, G. Zhang, and F. Bobaru. The influence of passive film damage on pitting corrosion. *Journal of The Electrochemical Society*, 163(2): C19–C24, 2016.

[15] A. Contreras, S. Hernández, R. Orozco-Cruz, and R. Galvan-Martínez. Mechanical and environmental effects on stress corrosion cracking of low carbon pipeline steel in a soil solution. *Materials & Design*, 35: 281–289, 2012.

[16] M. Duangpanya, *A Peridynamic Formulation for Transient Heat Conduction in Bodies with Evolving Discontinuities*. PhD thesis, University of Nerbaska - Lincoln, 2011.

[17] G. Engelhardt and D. D. Macdonald. Unification of the deterministic and statistical approaches for predicting localized corrosion damage. I. Theoretical foundation. *Corrosion Science*, 46(11): 2755–2780, 2004.

[18] G. Engelhardt, M. Urquidi-Macdonald, and D. D. Macdonald. A simplified method for estimating corrosion cavity growth rates. *Corrosion Science*, 39(3): 419–441, 1997.

[19] P. Ernst and R. C. Newman. Pit growth studies in stainless steel foils. I. Introduction and pit growth kinetics. *Corrosion Science*, 44(5): 927–941, 2002.

[20] P. Ernst and R. C. Newman. Pit growth studies in stainless steel foils. II. Effect of temperature, chloride concentration and sulphate addition. *Corrosion Science*, 44(5): 943–954, 2002.

[21] G. S. Frankel. Pitting corrosion of metals - a review of the critical factors. *Journal of the Electrochemical Society*, 145(6): 2186–2198, 1998.

[22] G. S. Frankel, J. O. Dukovic, V. Brusic, B. M. Rush, and C. V. Jahnes. Pit growth in nife thin-films. *Journal of the Electrochemical Society*, 139(8): 2196–2201, 1992.

[23] G. T. Gaudet, W. T. Mo, T. A. Hatton, J. W. Tester, J. Tilly, H. S. Isaacs, and R. C. Newman. Mass-transfer and electrochemical kinetic interactions in localized pitting corrosion. *AIChE Journal*, 32(6): 949–958, 1986.

[24] S. F. Henke and S. Shanbhag. Mesh sensitivity in peridynamic simulations. *Computer Physics Communications*, 185(1): 181–193, 2014.

[25] W. K. Hu, Y. D. Ha, and F. Bobaru. Modeling dynamic fracture and damage in a fiber-reinforced composite lamina with peridynamics. *International Journal for Multiscale Computational Engineering*, 9(6): 707–726, 2011.

[26] W. K. Hu, Y. D. Ha, and F. Bobaru. Peridynamic model for dynamic fracture in unidirectional fiber-reinforced composites. *Computer Methods in Applied Mechanics and Engineering*, 217: 247–261, 2012.

[27] A. E. Hughes, N. Birbilis, J. M. C. Mol, S. J. Garcia, X. Zhou, and G. E. Thompson. High strength al-alloys: Microstructure, corrosion and principles

of protection, in *Recent Trends in Processing and Degradation of Aluminium Alloys*, pp. 223–262, 2011.

[28] F. Hunkeler and H. Bohni. Determination of pit growth-rates on aluminum using a metal foil technique. *Corrosion*, 37(11): 645–650, 1981.

[29] A. J. Jacobs and G. P. Wozadlo. Irradiation-assisted stress-corrosion cracking as a factor in nuclear-power plant aging. *Journal of Materials Engineering*, 9(4): 345–351, 1988.

[30] L. J. Korb and D. L. Olson. *Corrosion*, vol. 13 of *ASM Handbook*. ASM International, 1996.

[31] N. J. Laycock, J. S. Noh, S. P. White, and D. P. Krouse. Computer simulation of pitting potential measurements. *Corrosion Science*, 47(12): 3140–3177, 2005.

[32] F. Lechenault, C. L. Rountree, F. Cousin, J. P. Bouchaud, L. Ponson, and E. Bouchaud. Evidence of deep water penetration in silica during stress corrosion fracture. *Physical Review Letters*, 106(16): 165504, 2011.

[33] E. Madenci and E. Oterkus, *Peridynamic Theory and Its Applications*. New York: Springer, 2014.

[34] B. Malki and B. Baroux. Computer simulation of the corrosion pit growth. *Corrosion Science*, 47(1): 171–182, 2005.

[35] Y. Nakai, Y. Shimizu, S. Fujiwara, and T. Ogawa. Atomic-force microscopy on corrosion pits and crack initiation in fatigue metals, in *ICF10, Honolulu, USA, 2001*, 2013.

[36] S. Oterkus, E. Madenci, and A. Agwai. Peridynamic thermal diffusion. *Journal of Computational Physics*, 265: 71–96, 2014.

[37] S. V. Patankar, *Numerical Heat Transfer and Fluid Flow*. Washington: Hemisphere Publishing, 1980.

[38] L. Pauling, *General Chemistry*, 3rd ed. New York: Dover Publications, 1988.

[39] P. C. Pistorius and G. T. Burstein. Metastable pitting corrosion of stainless-steel and the transition to stability. *Philosophical Transactions of the Royal Society of London, Series A: Physical Sciences and Engineering*, 341(1662): 531–559, 1992.

[40] D. A. Porter and K. E. Easterling, *Phase Transformations in Metals and Alloys*, 2nd ed. London: Chapman & Hall, 1992.

[41] S. M. Qidwai, V. G. DeGiorgi, and A. C. Leung. A review of microstructural effects on pitting corrosion in stainless steels, in *ASME 2011 International Design Engineering Technical Conferences and Computers and Information in Engineering Conference*, pp. 161–169, American Society of Mechanical Engineers, 2011.

[42] S. Scheiner and C. Hellmich. Stable pitting corrosion of stainless steel as diffusion-controlled dissolution process with a sharp moving electrode boundary, *Corrosion Science*, 49(2): 319–346, 2007.

[43] S. Scheiner and C. Hellmich. Finite volume model for diffusion- and activation-controlled pitting corrosion of stainless steel. *Computer Methods in Applied Mechanics and Engineering*, 198(37–40): 2898–2910, 2009.

[44] R. B. Schoch, J. Y. Han, and P. Renaud. Transport phenomena in nanofluidics. *Reviews of Modern Physics*, 80(3): 839–883, 2008.

[45] F. Scholz, *Electroanalytical Methods*. New York: Springer, 2010.

[46] S. M. Sharland. A review of the theoretical modeling of crevice and pitting corrosion. *Corrosion Science*, 27(3): 289–323, 1987.

[47] S. A. Silling. Reformulation of elasticity theory for discontinuities and long-range forces. *Journal of the Mechanics and Physics of Solids*, 48(1): 175–209, 2000.

[48] W. W. Song, H. J. Martin, A. Hicks, D. Seely, C. A. Walton, W. B. Lawrimore, P. T. Wang, and M. F. Horstemeyer. Corrosion behaviour of extruded am30 magnesium alloy under salt-spray and immersion environments. *Corrosion Science*, 78: 353–368, 2014.

[49] J. Stafiej, D. di Caprio, and L. Bartosik. Corrosion-passivation processes in a cellular automata based simulation study. *Journal of Supercomputing*, 65(2): 697–709, 2013.

[50] P. Van der Weeen, A. M. Zimer, E. C. Pereira, L. H. Mascaro, O. M. Bruno, and B. De Baets, "Modeling pitting corrosion by means of a 3d discrete stochastic model," *Corrosion Science*, vol. 82, pp. 133–144, 2014.

[51] A. K. Vasudevan. Applied stress affecting the environmentally assisted cracking. *Metallurgical and Materials Transactions A: Physical Metallurgy and Materials Science*, 44A(3): 1254–1267, 2013.

[52] N. Winzer, A. Atrens, G. L. Song, E. Ghali, W. Dietzel, K. U. Kainer, N. Hort, and C. Blawert. A critical review of the stress corrosion cracking (scc) of magnesium alloys. *Advanced Engineering Materials*, 7(8): 659–693, 2005.

Chapter 16

Peridynamics for Coupled Field Equations

Erdogan Madenci

University of Arizona

Selda Oterkus

University of Strathclyde

CONTENTS

Abstract 490
16.1 Introduction 490
16.2 Diffusion equation 492
16.2.1 Thermal diffusion 492
16.2.2 Moisture diffusion 493
16.2.3 Electrical conduction 493
16.3 Coupled field equations 494
16.3.1 Thermomechanics 494
16.3.1.1 Thermal diffusion with a structural coupling term 494
16.3.1.2 Equation of motion with a thermal coupling term 495
16.3.2 Porelasticity 495
16.3.2.1 Mechanical deformation due to fluid pressure . 495
16.3.2.2 Fluid flow in porous medium 496
16.3.3 Electromigration 497
16.3.4 Hygrothermomechanics 499
16.4 Numerical solution to peridynamic field equations 500
16.4.1 Correction of PD material parameters 500
16.4.2 Boundary conditions 502
16.4.2.1 Essential boundary conditions 502
16.4.2.2 Natural boundary conditions 503
16.4.2.3 Example 1 506
16.4.2.4 Example 2 507
16.4.2.5 Example 3 509

16.5 Applications 511
16.5.1 Coupled nonuniform heating and deformation 511
16.5.2 Coupled nonuniform moisture and deformation in a square plate 514
16.5.3 Coupled fluid pore pressure and deformation 519
16.5.4 Coupled electrical, temperature, deformation, and vacancy diffusion 523
16.6 Remarks 527
References 528

Abstract

Although the peridynamic (PD) theory is originally introduced for the solution of deformation field equations, it is applicable to other fields, including thermal diffusion, moisture diffusion, electric potential distribution, vacancy diffusion, and neutronic diffusion in either an uncoupled or a coupled manner. It is a nonlocal theory with an internal length parameter, and thus permits the determination of the field variable in spite of discontinuities. This chapter presents the PD representation of the other field equations, and their coupling with the deformation field. Also, it describes in detail the application of boundary conditions, and establishes the accuracy of their implementation. Solutions to specific coupled field problems concern thermomechanics, hygrothermomechanics, poroelasticity, and electromigration.

16.1 Introduction

Energy transportation is inherently a nonlocal process because it occurs through various particles or excitations, and they arrive at a point in space having brought energy from other points. If the mean free path of particles or excitations is short with respect to the distance scale of variations in the driving force, then a local (classical) theory is sufficient to describe the process. However, if the mean free path of particles or excitations is long, then a nonlocal theory of transport is required.

The nonlocal effects in heat and mass transfer occur in micro- and nanoscales and during ultrafast processes (see [26]). Thus, Fourier's law may not represent the actual behavior because of the boundary and interface scattering and the finite relaxation time of heat carriers. In order to include such nonlocal effects, Chen (see [6]) established a non-Fourier conduction model to derive the ballistic-diffusive heat conduction equation to describe the particle behavior of electrons and photons by using the Boltzmann equation. The ballistic and diffusive components of the heat flux account for the effect of boundaries and scattered and excited carriers, respectively. Therefore, it captures heat transfer in nanoscale microelectromechanical systems (MEMS) and nano- electromechanical systems (NEMS), as well as ultrafast processes. Alternatively, Ardito and Comi ([2]) suggested an integral form of the nonlocal model to include the effect of the microstructure on the dynamic response of MEMS, and captured the thermoelastic dissipation mechanisms.

Furthermore, Fourier's law leads to a parabolic-type equation when describing the local heat conduction process. It yields a physically unacceptable, infinitely large heat wave propagation speed. In reality, the wave speed is finite, although it has a significantly large value ([25]). Similarly, the conventional Fick's law becomes inapplicable if the time scales and length scales of the transport process are not much larger than the scales of the variations of the velocity field. It is not simply proportional to the concentration gradient. Therefore, Koch and Brady (see [11]) calculated the mass flux through a convolution integral over space and all previous times of concentration gradient and spatial and temporal dependent diffusivity. Alternatively, Sen and Ramos ([23]) introduced a mathematical model for nonlocal transport in which the flow model is assumed to have a network of interconnected channels with impermeable walls.

In addition to the presence of nonlocality, as the electronic device sizes decrease, the microstructure of constituent materials makes a significant impact on all of the relevant fields, i.e., deformation, heat transfer, electrical potential distribution, and vacancy diffusion. Thus, the effect of coupling between these fields becomes more pronounced. The nonlocal peridynamic (PD) theory enables the coupling of different fields in the presence of discontinuities.

Gerstle et al. ([8]) applied the PD theory to simulate, for the first time, a coupled phenomenon of electromigration in a copper line. Later, Kilic and Madenci ([10]) included the effect of thermal field on deformation, and performed a thermomechanical analysis. Bobaru and Duangpanya ([4]) developed the PD solution for thermal diffusion. Bobaru and Duangpanya ([5]) extended this approach to thermal diffusion in bodies with growing discontinuities, such as cracks. Based on the previous formulations, Agwai ([1]) developed the PD solution for one-dimensional fully coupled thermomechanics. Recently, Oterkus et al. ([20]) extended it to three-dimensional analysis. Oterkus and Madenci ([17]) captured the effect of cooling and heating ahead of the crack tip due to dynamic loading. It accurately predicts the temperature rise at the crack tips as the crack propagates. Also, Oterkus and Madenci ([18]) extended this approach to consider heat conduction and the fully coupled thermomechanical analysis of fiber-reinforced composites. Furthermore, Oterkus et al. ([19]) extended the one-dimensional analysis from [8] to a three-dimensional form and investigated electromigration in two-dimensional copper lines. Based on the variational principles, Oterkus et al. ([21]) derived the ordinary state-based PD heat conduction equation, and presented a detailed approach for imposing the relevant boundary conditions, as well as the treatment of dissimilar material interface. Similarly, Katiyar et al. ([9]) developed the PD field equation for fluid flow in porous media. Recently, Oterkus et al. ([22]) coupled moisture, temperature, vapor pressure, and deformation field equations to predict damage initiation and its propagation in layered material systems, such as electronic packages.

This chapter focuses on the multiscale and multiphysics modeling of problems with peridynamics. It addresses specific coupled fields, such as thermomechanics, poroelasticity, and electromigration.

16.2 Diffusion equation

The diffusion process occurs in many different physical phenomena, and it can be described by using the classical (local) formulations as

$$m_1 \dot{\psi}(\mathbf{x},t) = m_2 \nabla^2 \psi(\mathbf{x},t) + s(\mathbf{x},t) \tag{16.1}$$

where $\psi(\mathbf{x},t)$ is the field variable (potential), m_1 and m_2 represent the isotropic material properties, and $s(\mathbf{x},t)$ represents the source if it exists. The dot over a variable denotes differentiation with respect to time, and ∇^2 is the Laplacian operator.

Within the peridynamic framework, the interaction between material points is nonlocal. Therefore, a material point is influenced by the other material points within its neighborhood, defined by the horizon. As shown by Oterkus et al. (2014b), the PD form of Eq. (16.1) can be derived as

$$m_1 \dot{\psi}(\mathbf{x},t) = \int_H f(\psi', \psi, \mathbf{x}', \mathbf{x}, t)\, dV_{\mathbf{x}'} + s(\mathbf{x},t) \tag{16.2}$$

where f is the response function that governs the interaction between material points $\mathbf{x}$ and $\mathbf{x}'$. In Eq. (16.2), the parameter H represents the domain of influence region for the material point at $\mathbf{x}$. Its extent is defined by the parameter δ, referred to as the horizon. The response function, f, is zero for material points outside the horizon, i.e., $|\mathbf{x}' - \mathbf{x}| > \delta$. The pairwise response function can be defined as

$$f(\psi', \psi, \mathbf{x}', \mathbf{x}, t) = m \frac{\psi'(\mathbf{x}',t) - \psi(\mathbf{x},t)}{|\mathbf{x}' - \mathbf{x}|} \tag{16.3}$$

where m is the peridynamic material parameter that is dependent on the material properties and horizon. It can be determined by equating the peridynamic "field" potential to the classical "field" potential at a point arising from a simple linear field variation, as explained in [21].

16.2.1 Thermal diffusion

If the potential, ψ, in Eq. (16.1) represents temperature, T, the parameters in Eq. (16.1) are defined as $m_1 = \rho c_v$ and $m_2 = k_T$, with k_T, c_v, and ρ corresponding to the thermal conductivity, specific heat, and density, respectively, and $s = q_T$ is the volumetric heat generation. The thermal response function is denoted by $f_T(T', T, \mathbf{x}', \mathbf{x}, t)$. The response function enables the exchange of heat between material points that are connected through thermal bonds. The PD material parameter, m, corresponds to the micro thermal conductivity, κ_T, as

$$\kappa_T = \frac{2k_T}{A\delta^2} \text{ (1-D)}, \quad \kappa_T = \frac{6k_T}{\pi h\delta^3} \text{ (2-D)} \quad \text{and } \kappa_T = \frac{6k_T}{\pi\delta^4} \text{ (3-D)} \tag{16.4}$$

in which h is the thickness and A is the cross-sectional area. The heat flux, $\mathbf{q}$, which is the rate of flow of heat energy through a surface, is defined as

$$\mathbf{q} = -k_T \nabla T. \tag{16.5}$$

As derived in [21], the peridynamic heat flux can be expressed as

$$\mathbf{q} = -\frac{1}{2}\int_H \kappa_T \left[T'\left(\mathbf{x}',t\right) - T\left(\mathbf{x},t\right)\right] \frac{(\mathbf{x}'-\mathbf{x})}{|\mathbf{x}'-\mathbf{x}|} dV_{\mathbf{x}'}. \tag{16.6}$$

16.2.2 Moisture diffusion

If the potential, ψ, in Eq. (16.1) represents moisture concentration, C_M, the parameters in Eq. (16.1) are defined as $m_1 = 1$ and $m_2 = D_M$, with D_M corresponding to the moisture diffusion coefficient. The moisture concentration response function is denoted by $f_M\left(C'_M, C_M, \mathbf{x}', \mathbf{x}, t\right)$, and it enables the exchange of moisture between material points that are connected through hygro bonds. The PD bond constant, m, corresponds to the micromoisture diffusivity, κ_M, as

$$\kappa_M = \frac{2D_M}{A\delta^2} \text{ (1-D)}, \quad \kappa_M = \frac{6D_M}{\pi h\delta^3} \text{ (2-D)}, \quad \kappa_M = \frac{6D_M}{\pi\delta^4} \text{ (3-D)}. \tag{16.7}$$

Traditionally, moisture diffusion analysis is performed by using a thermal-moisture analogy. Since moisture concentration is not continuous along dissimilar interfaces, a new parameter called "wetness" is introduced to render it continuous. Wetness is the ratio of the moisture concentration to its value at the saturated state, and it is continuous along interfaces. Although this approach is commonly accepted, it is not always valid because the saturated moisture concentration is not always constant. PD removes this requirement because it is not necessary to impose such continuity conditions. Thus, it directly solves for the moisture concentration.

16.2.3 Electrical conduction

If the potential, ψ, in Eq. (16.1) represents electrical potential, Φ, the parameters in Eq. (16.1) are defined as $m_1 = c_E$ and $m_2 = k_E$, with k_E corresponding to the electrical conductivity and c_E to the electrical capacitance. The electrical response function is denoted by $f_E\left(\Phi', \Phi, \mathbf{x}', \mathbf{x}, t\right)$, and it enables the exchange of electrical current between material points that are connected through electrical bonds. The PD bond constant, m, corresponds to the microelectrical conductivity, κ_E, as

$$\kappa_E = \frac{2k_E}{A\delta^2} \text{ (1-D)}, \quad \kappa_E = \frac{6k_E}{\pi h\delta^3} \text{ (2-D)}, \quad \kappa_E = \frac{6k_E}{\pi\delta^4} \text{ (3-D)}. \tag{16.8}$$

The current density vector, $\mathbf{j}$, can be expressed in terms of electrical potential, Φ, as

$$\mathbf{j} = -k_E \nabla \Phi. \tag{16.9}$$

The current density vector in terms of the response function can be expressed as

$$\mathbf{j}(\mathbf{x},t) = -\frac{1}{2}\int_H f_E\left(\Phi',\Phi,\mathbf{x}',\mathbf{x},t\right)(\mathbf{x}'-\mathbf{x})dV_{\mathbf{x}'}. \tag{16.10}$$

16.3 Coupled field equations

Many different physical fields can have direct and significant effects on each other. If their degree of interactions is strong, their coupling needs to be included in the analysis. This section concerns the coupled field equations that govern thermomechanics, poroelasticity, hygrothermomechanics, thermoelectrics, and electromigration.

16.3.1 Thermomechanics

Similar to the derivation of classical thermomechanical equations ([15]), peridynamics for fully coupled thermomechanics is based on irreversible thermodynamics, i.e., the conservation of energy and the free energy density function. The first equation is the conservation of thermal energy (i.e., the heat transfer equation) with a contribution from deformational heating and cooling, and the second equation is the conservation of linear momentum (i.e., the equation of motion) with a thermoelastic constitutive relation.

16.3.1.1 Thermal diffusion with a structural coupling term

The bond-based peridynamic heat equation with structural coupling is derived as (see [20])

$$\rho c_v \dot{\Theta}(\mathbf{x},t) = \int_H \left(\kappa_T \frac{\Theta(\mathbf{x}',t)-\Theta(\mathbf{x},t)}{|\mathbf{x}'-\mathbf{x}|} - T_o\frac{c}{2}\alpha\dot{e}\right)dV' + q_T(\mathbf{x},t) \tag{16.11}$$

where $\Theta(\mathbf{x},t)$ is the temperature with respect to the reference temperature, T_0, and is defined as $\Theta(\mathbf{x},t) = T(\mathbf{x},t) - T_0$. The extension, e, is defined as

$$e = \left|(\mathbf{x}'+\mathbf{u}') - (\mathbf{x}+\mathbf{u})\right| - \left|\mathbf{x}'-\mathbf{x}\right| \tag{16.12}$$

and its time rate of change between the material points, $\dot{e}$, is obtained as

$$\dot{e} = \frac{(\mathbf{x}'+\mathbf{u}') - (\mathbf{x}+\mathbf{u})}{\left|(\mathbf{x}'+\mathbf{u}') - (\mathbf{x}+\mathbf{u})\right|}\cdot\left(\dot{\mathbf{u}}'-\dot{\mathbf{u}}\right) \tag{16.13}$$

where $\mathbf{u}$ and $\mathbf{u}'$ are the displacement vectors of material points $\mathbf{x}$ and $\mathbf{x}'$, respectively.

As derived by Silling and Askari in [24], the bond constant for an isotropic material can be expressed as

$$c = \frac{2E}{A\delta^2}\ \text{(1-D)}\,, \quad c = \frac{9E}{\pi h \delta^3}\ \text{(2-D)}, \quad c = \frac{12E}{\pi \delta^4}\ \text{(3-D)} \tag{16.14}$$

in which E is the Young's Modulus and α is the coefficient of thermal expansion.

16.3.1.2 Equation of motion with a thermal coupling term

The bond-based peridynamic deformation with thermal coupling is derived by Oterkus et al. (2014a) as

$$\rho \ddot{\mathbf{u}}(\mathbf{x},t) = \int_H c\left(s - \alpha \bar{\Theta}\right) \frac{(\mathbf{x}' + \mathbf{u}') - (\mathbf{x} + \mathbf{u})}{|(\mathbf{x}' + \mathbf{u}') - (\mathbf{x} + \mathbf{u})|} dV' + \mathbf{b}(\mathbf{x},t) \tag{16.15}$$

where $\mathbf{b}(\mathbf{x},t)$ is the body force density vector. The stretch between two material points is defined as

$$s = \frac{|(\mathbf{x}' + \mathbf{u}') - (\mathbf{x} + \mathbf{u})| - |\mathbf{x}' - \mathbf{x}|}{|\mathbf{x}' - \mathbf{x}|}. \tag{16.16}$$

The mean value of the change in temperatures, $\bar{\Theta}$, at material points $\mathbf{x}$ and $\mathbf{x}'$, is defined as

$$\bar{\Theta} = \frac{(T - T_0) + (T' - T_0)}{2} \tag{16.17}$$

in which T and T' represent the temperature at material points $\mathbf{x}$ and $\mathbf{x}'$ with T_0 being the reference temperature.

16.3.2 Porelasticity

Modeling of porelasticity involves at least three different coupled mechanisms: (i) mechanical deformation due to fluid pressure inside the porous medium, (ii) fluid flow inside the porous medium, and (iii) propagation of fractures. Hence, this process can be described by coupling the fluid flow in a porous medium and the deformation. The coupling can be achieved by including the effect of fluid pore pressure in the constitutive relation for the deformation field and time rate of dilatation for the flow field.

16.3.2.1 Mechanical deformation due to fluid pressure

The peridynamic equation of motion of a material point at $\mathbf{x}$ can be expressed as

$$\rho \ddot{\mathbf{u}}(\mathbf{x},t) = \int_H c\left(s - \alpha_B \gamma \bar{P}\right) \frac{(\mathbf{x}' + \mathbf{u}' - \mathbf{x} - \mathbf{u})}{|\mathbf{x}' + \mathbf{u}' - \mathbf{x} - \mathbf{u}|} dV_{\mathbf{x}'} + \mathbf{b}(\mathbf{x},t) \tag{16.18}$$

in which γ is the coefficient of fluid pore pressure and is defined as

$$\gamma = \frac{1}{E}\ \text{(1-D)}, \quad \gamma = \frac{(1-\nu)}{E}\ \text{(2-D)}, \quad \gamma = \frac{(1-2\nu)}{E}\ \text{(3-D)} \tag{16.19}$$

with ν representing the Poisson's ratio. The average fluid pore pressure, $\bar{P}$, at material points $\mathbf{x}$ and $\mathbf{x}'$, is defined as

$$\bar{P} = \frac{P(\mathbf{x}',t)+P(\mathbf{x},t)}{2}. \tag{16.20}$$

16.3.2.2 Fluid flow in porous medium

The governing equations for a single-phase fluid flow in a porous medium can be described by Darcy's Law as (see e.g. [7])

$$\frac{\partial(\rho\varphi)}{\partial t} = -\nabla.(\rho\mathbf{v})+q_f \tag{16.21}$$

where φ is the variation in the water content, ρ is the density of the fluid, and q_f is the mass of fluid produced per unit volume per unit time. The fluid velocity, $\mathbf{v}$, can be expressed as

$$\mathbf{v} = -\frac{k_P}{\mu}\nabla P \tag{16.22}$$

with

$$P = p_f - \rho g z \tag{16.23}$$

and p_f is the fluid pressure, k_P is the permeability of the porous medium, μ is the fluid viscosity, g is the magnitude of the gravitational acceleration, and z is the depth.

The variation in the water content can be defined as

$$\varphi = \alpha_B\theta + \frac{P}{Q_B}. \tag{16.24}$$

As given by Biot ([3]), Eq. (16.21) can be rewritten for constant material properties as

$$\alpha_B\frac{\partial\theta}{\partial t} + \frac{1}{Q_B}\frac{\partial P}{\partial t} = \frac{k_P}{\mu}\nabla^2 P + \frac{q_f}{\rho} \tag{16.25}$$

where Q_B, α_B, and θ represent the Biot modulus, Biot parameter, and the change in volume (dilatation), respectively.

The peridynamic form of Eq. (16.22) can be written as

$$\frac{1}{Q_B}\dot{P}(\mathbf{x},t) + \alpha_B\dot{\theta} = \int\limits_H \kappa_P \frac{P(\mathbf{x}',t)-P(\mathbf{x},t)}{|\mathbf{x}'-\mathbf{x}|}dV_{\mathbf{x}'} + \frac{q_f}{\rho} \tag{16.26}$$

in which κ_P can be expressed in terms of the permeability of the bulk material, k_P, and the fluid viscosity, μ, as

$$\kappa_P = \frac{2k_P}{\mu A\delta^2}\ \text{(1-D)}, \quad \kappa_P = \frac{6k_P}{\pi\mu h\delta^3}\ \text{(2-D)}, \quad \kappa_P = \frac{6k_P}{\pi\mu\delta^4}\ \text{(3-D)}. \tag{16.27}$$

The fluid velocity can be expressed as

$$\mathbf{v} = -\frac{1}{2}\int_H \kappa_P\left[P(\mathbf{x}',t)-P(\mathbf{x},t)\right]\frac{(\mathbf{x}'-\mathbf{x})}{|\mathbf{x}'-\mathbf{x}|}dV_{\mathbf{x}'}. \tag{16.28}$$

16.3.3 Electromigration

Electromigration concerns the coupling of deformation, heat transfer, electrical potential distribution and charge flow, and vacancy diffusion. The electrical potential and charge flow results in Joule heating, which results in a thermal field and, thus, thermal deformation. Material degradation occurs because of the presence of atomic flux divergence due to electromigration, thermomigration, and stress-migration ([12]).

The electrical field can have a direct effect on the thermal field due to Joule heating. The current density vector, Eq. (16.10), permits the determination of the Joule heating term as

$$q_E(\mathbf{x},t) = \frac{1}{k_E}\mathbf{j}(\mathbf{x},t)\cdot\mathbf{j}(\mathbf{x},t). \tag{16.29}$$

It requires the steady-state solution for the electrical conduction from

$$c_E\dot{\Phi}(\mathbf{x},t) = \int_H \kappa_E \frac{\Phi(\mathbf{x}',t)-\Phi(\mathbf{x},t)}{|\mathbf{x}'-\mathbf{x}|}dV_{\mathbf{x}'}. \tag{16.30}$$

The Joule heating term defines the volumetric heat generation $s(\mathbf{x},t) = q_E$ in Eq. (16.2); thus, the heat diffusion equation is modified as

$$\rho c_v\dot{T}(\mathbf{x},t) = \int_H f_T(T',T,\mathbf{x}',\mathbf{x},t)\,dV_{\mathbf{x}'} + q_E(\mathbf{x},t). \tag{16.31}$$

The thermal field arising from Joule heating can have a direct effect on the deformation field, and the deformation field is obtained by solving for Eq. (16.15).

The classical form of the atomic diffusion expression is given as

$$\nabla.(\mathbf{J}) + \frac{\partial C_A}{\partial t} = 0 \tag{16.32}$$

where C_A is the atomic concentration and $\mathbf{J}$ represents the total atomic flux due to electro-migration, $\mathbf{J}_{EM}$, thermo-migration, $\mathbf{J}_{TM}$, and stress-migration, $\mathbf{J}_{SM}$, as ([12])

$$\mathbf{J} = \mathbf{J}_{EM} + \mathbf{J}_{TM} + \mathbf{J}_{SM} \tag{16.33}$$

with

$$\mathbf{J}_{EM} = -\frac{C_A D_A}{k_b T}eZ^*\frac{1}{k_E}\mathbf{j} = -\frac{C_A D_A}{T}\left(\frac{eZ^*}{k_b}\right)(\nabla\Phi) \tag{16.34a}$$

$$\mathbf{J}_{TM} = \frac{C_A D_A}{T^2}\left(\frac{Q}{k_b}*\right)(\nabla T) \tag{16.34b}$$

$$\mathbf{J}_{SM} = \frac{C_A D_A}{T}\left(\frac{\Omega}{k_b}\right)(\nabla\sigma_H) \tag{16.34c}$$

in which e is the electron charge, Z^* is the effective charge of ions, k_b is the Boltzmann constant, Q^* is the specific heat of transport, Ω is the atomic volume, σ_H is the hydrostatic (mean) stress value, and D_A is the atomic diffusion coefficient. This coefficient varies with temperature as (see, e.g. [16])

$$D_A = D_o e^{-Q/k_b T} \tag{16.35}$$

where D_o represents the diffusion coefficient at infinite temperature and Q the activation energy. Also, the atomic concentration of vacancies is assumed to be a function of temperature only as ([16])

$$C_A = C_0 e^{-E_v/k_b T} \tag{16.36}$$

where E_v is the energy for vacancy formation, and C_0 is the vacancy concentration at the melting temperature of the material.

The peridynamic representation of the atomic flux due to electromigration, $\mathbf{J}_{EM}$, thermomigration, $\mathbf{J}_{TM}$, and stress-migration, $\mathbf{J}_{SM}$, can be written as ([19])

$$\mathbf{J}_{EM} = \frac{1}{2}\int_H \frac{\bar{C}_A \bar{D}_A}{\bar{T}} f_{EM}\left(\Phi', \Phi, \mathbf{x}', \mathbf{x}, t\right)(\mathbf{x}' - \mathbf{x}) dV_{\mathbf{x}'} \tag{16.37a}$$

$$\mathbf{J}_{TM} = \frac{1}{2}\int_H \frac{\bar{C}_A \bar{D}_A}{\bar{T}^2} f_{TM}\left(T', T, \mathbf{x}', \mathbf{x}, t\right)(\mathbf{x}' - \mathbf{x}) dV_{\mathbf{x}'} \tag{16.37b}$$

$$\mathbf{J}_{SM} = \frac{1}{2}\int_H \frac{\bar{C}_A \bar{D}_A}{\bar{T}} f_{SM}\left(\sigma'_H, \sigma_H, \mathbf{x}', \mathbf{x}, t\right)(\mathbf{x}' - \mathbf{x}) dV_{\mathbf{x}'}. \tag{16.37c}$$

In Eq. (33a–c), the parameters $\bar{T}$, $\bar{C}_A$, and $\bar{D}_A$ are defined as

$$\bar{T} = \frac{T(\mathbf{x}', t) + T(\mathbf{x}, t)}{2} \tag{16.38a}$$

$$\bar{C}_A = \frac{C_A(\mathbf{x}', t) + C_A(\mathbf{x}, t)}{2} \tag{16.38b}$$

$$\bar{D}_A = \frac{D_A(\mathbf{x}', t) + D_A(\mathbf{x}, t)}{2}. \tag{16.38c}$$

The response functions are defined as

$$f_{EM}\left(\Phi', \Phi, \mathbf{x}', \mathbf{x}, t\right) = \kappa_{EM} \frac{\Phi'(\mathbf{x}', t) - \Phi(\mathbf{x}, t)}{|\mathbf{x}' - \mathbf{x}|} \tag{16.39a}$$

$$f_{TM}\left(T', T, \mathbf{x}', \mathbf{x}, t\right) = \kappa_{TM} \frac{T'(\mathbf{x}', t) - T(\mathbf{x}, t)}{|\mathbf{x}' - \mathbf{x}|} \tag{16.39b}$$

$$f_{SM}\left(\sigma'_H, \sigma_H, \mathbf{x}', \mathbf{x}, t\right) = \kappa_{SM} \frac{\sigma'_H(\mathbf{x}', t) - \sigma_H(\mathbf{x}, t)}{|\mathbf{x}' - \mathbf{x}|} \tag{16.39c}$$

with

$$\kappa_{EM} = \frac{2}{A\delta^2}\left(\frac{eZ^*}{k_b}\right),\ \kappa_{TM} = \frac{2}{A\delta^2}\left(\frac{Q}{k_b}^*\right),\ \kappa_{SM} = \frac{2}{A\delta^2}\left(\frac{\Omega}{k_b}\right)\text{(1-D)} \tag{16.40a}$$

$$\kappa_{EM} = \frac{6}{\pi h \delta^3}\left(\frac{eZ^*}{k_b}\right), \ \kappa_{TM} = \frac{6}{\pi h \delta^3}\left(\frac{Q_*}{k_b}\right), \ \kappa_{SM} = \frac{6}{\pi h \delta^3}\left(\frac{\Omega}{k_b}\right) \text{ (2-D)} \quad (16.40b)$$

$$\kappa_{EM} = \frac{6}{\pi \delta^4}\left(\frac{eZ^*}{k_b}\right), \ \kappa_{TM} = \frac{6}{\pi \delta^4}\left(\frac{Q_*}{k_b}\right), \ \kappa_{SM} = \frac{6}{\pi \delta^4}\left(\frac{\Omega}{k_b}\right) \text{ (3-D).} \quad (16.40c)$$

The hydrostatic stress, σ_H, can be calculated using the stress field, $\text{œ}(\mathbf{x},t)$, obtained from Zimmerman [27]:

$$\text{œ}(\mathbf{x},t) = \frac{1}{2}\int_H cs \frac{(\mathbf{x}'+\mathbf{u}')-(\mathbf{x}+\mathbf{u})}{|(\mathbf{x}'+\mathbf{u}')-(\mathbf{x}+\mathbf{u})|} \otimes \left((\mathbf{x}'+\mathbf{u}')-(\mathbf{x}+\mathbf{u})\right) dV_{\mathbf{x}'}. \quad (16.41)$$

As a result of vacancy diffusion, the cross-sectional area, A, of the metal strip changes as a function of time. Rather than solving for the vacancy concentration, Gerstle et al. ([8]) assumed the rate of change of area as

$$\frac{\partial A}{\partial t} = -\Omega A \nabla \cdot \mathbf{J}. \quad (16.42)$$

The PD expression for the change in area can be rewritten as

$$\begin{aligned}\frac{\partial A}{\partial t} &= \frac{-\Omega A \kappa_J}{2} \int_H \left(\left(J_x'(\mathbf{x}',t) - J_x(\mathbf{x},t)\right) \frac{x'-x}{|\mathbf{x}'-\mathbf{x}|} \right. \\ &\left. + \left(J_y'(\mathbf{x}',t) - J_y(\mathbf{x},t)\right) \frac{y'-y}{|\mathbf{x}'-\mathbf{x}|} \right) dV_{\mathbf{x}'} \end{aligned} \quad (16.43a)$$

with

$$\kappa_J = \frac{2}{A\delta^2} \text{ (1-D)}, \quad \kappa_J = \frac{6}{\pi h \delta^3} \text{ (2-D)}, \quad \kappa_J = \frac{6}{\pi \delta^4} \text{ (3-D).} \quad (16.43b)$$

16.3.4 Hygrothermomechanics

Both the changes in temperature and moisture concentration can result in a change in deformation due to coupling between these fields. Therefore, the modeling of hygrothermomechanics requires, first, the solution for moisture and thermal diffusion, and then the equation of motion for the displacement field. The effects of temperature and moisture on deformation can be invoked into the peridynamic equation of motion as

$$\rho \ddot{\mathbf{u}}(\mathbf{x},t) = \int_H c\left(s - \alpha\bar{\Theta} - \beta\bar{C}_M\right) \frac{(\mathbf{x}'+\mathbf{u}')-(\mathbf{x}+\mathbf{u})}{|(\mathbf{x}'+\mathbf{u}')-(\mathbf{x}+\mathbf{u})|} dV' + \mathbf{b}(\mathbf{x},t) \quad (16.44)$$

where α is the thermal expansion coefficient, β is the coefficient of hygroscopic swelling, and $\bar{\Theta}$ and $\bar{C}_M$ represent the mean values of temperature difference and moisture concentration difference, respectively, between material points $\mathbf{x}$ and $\mathbf{x}'$.

16.4 Numerical solution to peridynamic field equations

The solution to the PD field equations requires time and spatial integrations while considering constraints and/or loading conditions, as well as initial conditions. The spatial integration is performed by using a Gaussian integration (meshless) scheme, due to its simplicity, and time integration by using backward and forward difference explicit integration schemes. The domain is divided into a uniform grid, with integration or collocation (material) points associated with specific volumes. Associated with a particular material point, numerical implementation of spatial integration involves the summation of the volumes of material points within its horizon. However, the volume of each material point may not be embedded in the horizon in its entirety, i.e., the material points located near the surface of the horizon may have truncated volumes. As a result, the volume integration over the horizon may be incorrect if the entire volume of each material point is included in the numerical implementation. Therefore, a volume correction factor is necessary to correct for the extra volume, as explained in, for example, reference [14]. If an explicit integration scheme is adopted, a stability criterion on the value of the incremental time step is necessary to ensure convergence. Details of the time integration scheme and stability criterion are given in [14].

The steady-state solution to the PD field equation can be achieved through different techniques. In this study, however, an adaptive dynamic relaxation method is employed as described in detail in [14].

16.4.1 Correction of PD material parameters

The PD material parameters, κ_T ,κ_M, κ_E ,κ_P, κ_{EM}, κ_{TM}, κ_{SM}, and c, are determined by computing either the potential or strain energy density of a material point whose horizon is completely embedded in the material. The values of these parameters depend on the domain of integration defined by the horizon. Therefore, their values require correction if the material point is close to free surfaces or material interfaces (Fig. 16.1). Since the presence of free surfaces is problem dependent, it is impractical to resolve this issue analytically. The correction of the material parameters is achieved by numerically integrating the potential and strain energy density at each material point inside the body for simple field variation and comparing them to their counterparts obtained from classical solutions. The details of this approach are given by [14].

The correction at the interface is achieved by using equivalent PD material parameters ([20]). As shown in Fig. 16.2, the material point $\mathbf{x}_{(i)}$ may interact with material points $\mathbf{x}_{(j)}$ and $\mathbf{x}_{(m)}$. Material points $\mathbf{x}_{(i)}$ and $\mathbf{x}_{(j)}$ are embedded in material 1, and $\mathbf{x}_{(m)}$ is embedded in material 2. The PD material parameter between points $\mathbf{x}_{(i)}$ and $\mathbf{x}_{(j)}$ is $c_{(i)(j)}$, and it differs from $c_{(i)(m)}$ between material points $\mathbf{x}_{(i)}$ and $\mathbf{x}_{(m)}$. Because the material points $\mathbf{x}_{(i)}$ and $\mathbf{x}_{(m)}$ are embedded in two different materials, their material parameter, $c_{(i)(m)}$, can be expressed in terms of an equivalent material constant as

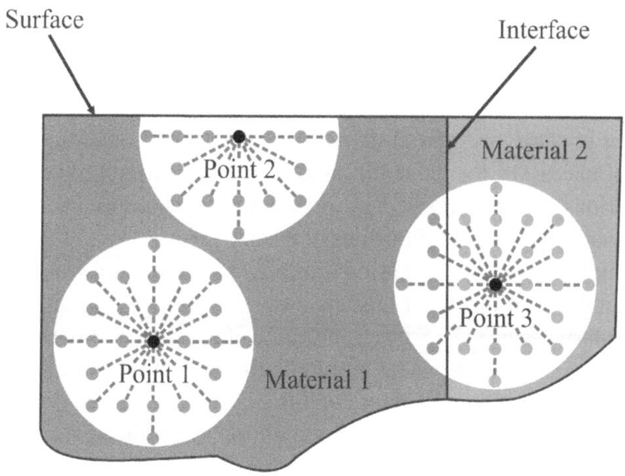

Figure 16.1: Surface effects in the domain of interest.

$c_{(i)(m)} = (\ell_1 + \ell_2)/(\ell_1/c_1 + \ell_2/c_2)$, with ℓ_1 representing the segment of the distance between material points $\mathbf{x}_{(i)}$ and $\mathbf{x}_{(m)}$ in material 1 whose material parameter is c_1, and ℓ_2 representing the segment in material 2 whose material parameter is c_2.

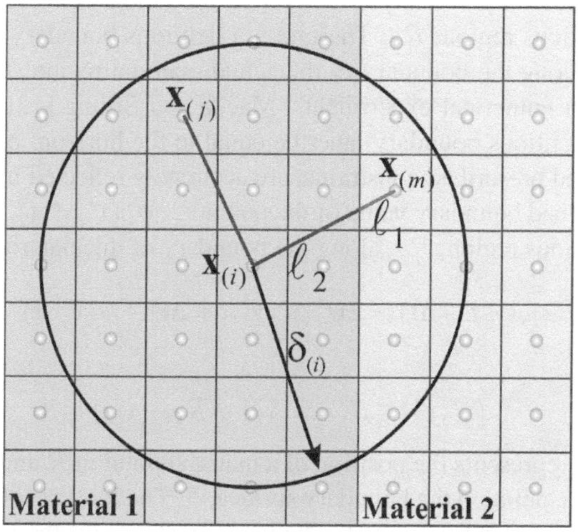

Figure 16.2: Interaction of material points across the interface.

16.4.2 Boundary conditions

Unlike the local theory, the boundary conditions are imposed through a nonzero volume of fictitious boundary layers. This necessity arises because the PD field equations do not contain any spatial derivatives; therefore, constraint conditions are, in general, not necessary for the solution of an integro-differential equation. However, such conditions can be imposed by prescribing constraints on the potential or its spatial derivatives (flux) in a fictitious boundary layer.

The essential boundary condition is imposed by assigning constraints to the material points in the fictitious layer in such a way that the condition is satisfied explicitly on the boundary surface. Therefore, the value of the potential in the fictitious layer is approximated based on the linear extrapolation of the values in the real domain and the specified value of the boundary condition.

Similarly, the flux-type boundary condition is enforced by approximating the value of the potential in the fictitious region so that the variation of the potential both in the real domain and the fictitious layer recovers the applied flux value on the surface. However, in the case of zero flux, it is not necessary to employ a fictitious layer; it can be satisfied in a natural way without enforcing any specific constraints.

16.4.2.1 Essential boundary conditions

The essential boundary condition concerns the imposition of a prescribed value of the potential, ψ^*, or displacement components, U^* and V^*, along the boundary of the domain in the case of 2-D analysis. This type of boundary condition can be achieved through a fictitious region, $\mathcal{R}_f$. Therefore, a fictitious boundary layer with depth δ is introduced along the boundary of the actual material region, $\mathcal{R}$, as shown in Fig. 16.3. Based on numerical experiments, Macek and Silling [13] suggested that the extent of the fictitious boundary layer be equal to the horizon, δ, in order to ensure that the imposed prescribed constraints are accurately reflected in the real domain.

The prescribed boundary value of the potential, $\psi^*(x^*, y^*, t)$, is imposed through a layer of fictitious region, $\mathcal{R}_f$, along the boundary of the material surface, $\mathcal{S}$, as

$$\psi_f(x_f, y_f, t+\Delta t) = 2\psi^*(x^*, y^*, t+\Delta t) - \psi(x, y, t) \tag{16.45a}$$

with

$$(x^*, y^*) \in \mathcal{S}, \ \left(x_f, y_f\right) \in \mathcal{R}_f, \ (x, y) \in \mathcal{R} \tag{16.45b}$$

in which (x, y) represents the position of a material point in $\mathcal{R}$ and (x^*, y^*) represents the location of a point on the boundary surface, $\mathcal{S}$. The location of the image material point in $\mathcal{R}_f$ is denoted by $\left(x_f, y_f\right)$. The implementation of the prescribed boundary value of the potential is depicted in Fig. 16.3. In the case of $\psi^*(x^*, y^*, t) = 0$, this condition becomes

$$\psi_f(x_f, y_f, t+\Delta t) = -\psi(x, y, t) \tag{16.46}$$

The applied displacement constraints, U^* and V^*, in the x- and y-directions can be enforced as

$$u_f(x_f, y_f, t+\Delta t) = 2U^*(x^*, y^*, t+\Delta t) - u(x, y, t) \tag{16.47a}$$

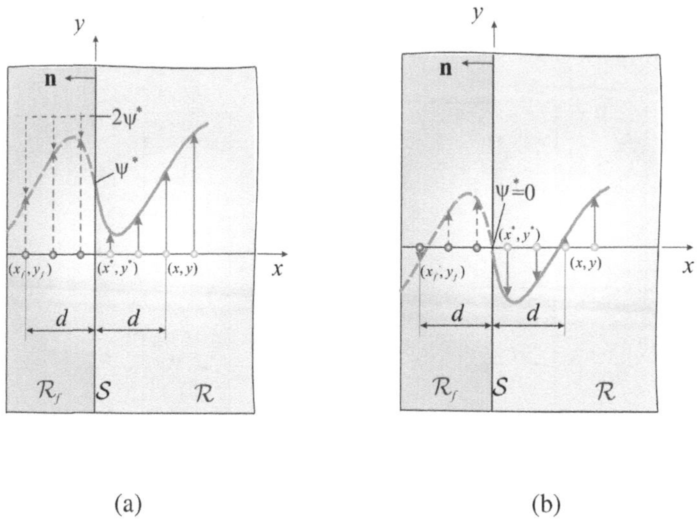

Figure 16.3: Imposing essential boundary conditions on potentials: (a) nonzero potential $\psi(x^*, y^*, t) = \psi^*$ and (b) zero potential $\psi(x^*, y^*, t) = \psi^* = 0$.

$$v_f(x_f, y_f, t+\Delta t) = 2V^*(x^*, y^*, t+\Delta t) - v(x, y, t) \tag{16.47b}$$

which represent the clamped boundary conditions for the prescribed values of $U^* = V^* = 0$ (Fig. 16.4).

If only the displacement constraint U^* is applied and there is no displacement constraint in the other direction, the boundary conditions are imposed as

$$u_f(x_f, y_f, t+\Delta t) = 2U^*(x^*, y^*, t+\Delta t) - u(x, y, t), \tag{16.48a}$$

$$v_f(x_f, y_f, t+\Delta t) = v(x, y, t) \tag{16.48b}$$

which represents the roller support boundary conditions for the prescribed value, $U^* = 0$ (Fig. 16.5).

16.4.2.2 Natural boundary conditions

Similar to the essential boundary conditions, the natural (potential gradient) boundary conditions are imposed through a fictitious region, $\mathcal{R}_f$. The prescribed gradient on the boundary, $q^*(x^*, y^*, t)$, is imposed as (Fig. 16.6a)

$$\Theta_f(x_f, y_f, t+\Delta t) = q^*\left(x^*, y^*, t+\Delta t\right)(x_f - x) + \Theta(x, y, t). \tag{16.49}$$

For a zero gradient condition, i.e., $q^* = 0$, the expression reduces to

$$\Theta_f(x_f, y_f, t+\Delta t) = \Theta(x, y, t). \tag{16.50}$$

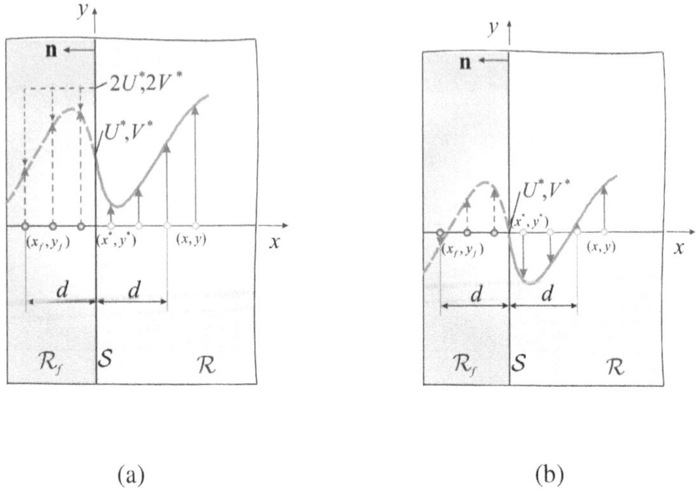

Figure 16.4: Imposing boundary conditions on displacement components: (a) constant displacement components $u(x^*, y^*, t) = U^*$ and $v(x^*, y^*, t) = V^*$ and (b) zero displacement components $u(x^*, y^*, t) = U^* = 0$ and $v(x^*, y^*, t) = V^* = 0$.

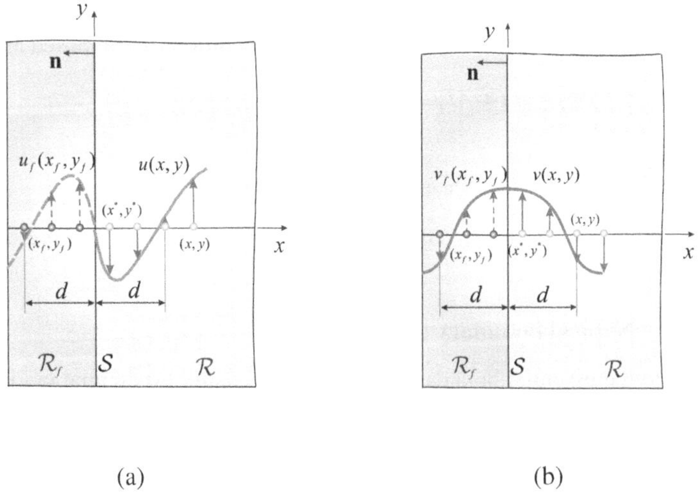

Figure 16.5: Imposing boundary conditions on displacement components: (a) fixed in the normal direction, $U^* = 0$ and (b) free to move in the tangent direction.

Thus, the zero gradient can be achieved by assigning the mirror image of the potential near the boundary in the actual domain to the material points in the fictitious region, as shown in Fig. 16.5b.

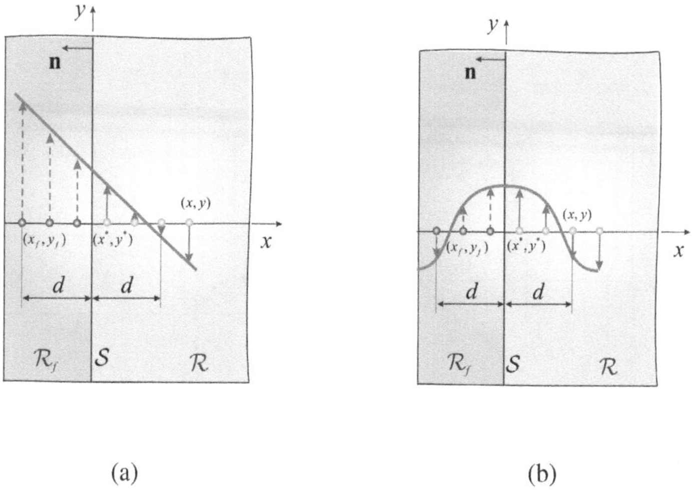

Figure 16.6: Material point and its image in the fictitious region for imposing (a) $q^* \neq 0$ and (b) $q^* = 0$.

On the boundary with a unit normal $\mathbf{n} = \mathbf{e}_x$, the applied stresses, σ_0 and τ_0, normal and tangent to the boundary can be enforced by assigning the following displacement values in the fictitious region:

$$u_f(x_f, y_f, t+\Delta t) = \left[\frac{\sigma_o\left(1-\nu^2\right)}{E} - \nu\frac{v(x,y^+,t)-v(x,y^-,t)}{y^+-y^-}\right]\left(x_f - x\right) + u(x,y,t) \tag{16.51a}$$

$$v_f(x_f, y_f, t+\Delta t) = \left[\frac{2\left(1+\nu\right)\tau_o}{E} - \frac{u(x,y^+,t)-u(x,y^-,t)}{y^+-y^-}\right]\left(x_f - x\right) + v(x,y,t) \tag{16.51b}$$

in which (x,y) are the coordinates of a material point in the real boundary layer, $\mathcal{R}$. As shown in Fig. 16.6a, the locations of material points above and below this point are denoted by (x,y^+) and (x,y^-), respectively.

Similarly, on the boundary with a unit normal $\mathbf{n} = \mathbf{e}_y$, the applied stresses, σ_0 and τ_0, normal and tangent to the boundary can be enforced in the fictitious region as

$$u_f(x_f, y_f, t+\Delta t) = \left[\frac{2\left(1+\nu\right)\tau_o}{E} - \frac{v(x^+,y,t)-v(x^-,y,t)}{x^+-x^-}\right]\left(y_f - y\right) + u(x,y,t) \tag{16.52a}$$

$$v_f(x_f,y_f,t+\Delta t) = \left[\frac{\sigma_o\left(1-v^2\right)}{E} - v\frac{u(x^+,y,t)-u(x^-,y,t)}{x^+-x^-}\right](y_f-y)+v(x,y,t)$$
(16.52b)

As shown in Fig. 16.7b, the locations of the material points on the left and on the right of this point are denoted by (x^-,y) and (x^+,y), respectively.

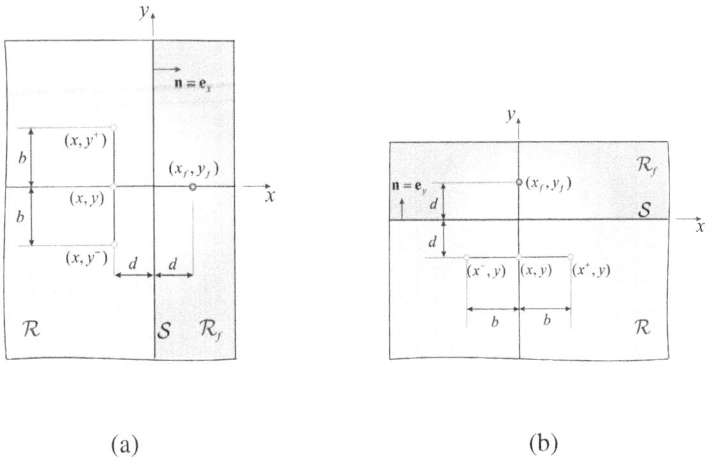

Figure 16.7: Fictitious region along the boundary with a unit normal (a) $\mathbf{n} = \mathbf{e}_x$ and (b) $\mathbf{n} = \mathbf{e}_y$.

If both normal and tangential stress components are zero along the boundary, a fictitious region is not necessary as they are satisfied in a natural way.

The validity and accuracy of such implementation for imposing the boundary conditions are demonstrated by considering a bar subjected to constant temperature at both ends, a plate under uniform stretch at both ends, and a clamped plate under uniform stretch. Throughout this study, the comparisons are performed against finite element analysis (FEA) by using ANSYS, a commercially available program.

16.4.2.3 Example 1

The bar has a length of $L = 0.01$ m and a cross-sectional area of $A = 1\times10^{-6}\text{m}^2$ with material properties $\alpha = k/\rho c_v = 4.5\times10^{-8}\text{m}^2/\text{s}$. It has an initial temperature field of $\Theta(x,t=0) = 0\,°\text{C}$ and its ends are subjected to a constant temperature of $\Theta(x=0,t) = \Theta(x=L,t) = 1\,°\text{C}$. These boundary conditions are enforced as

$$\Theta_f(x_f,t+\Delta t) = 2\Theta^*(x^*=0,t+\Delta t) - \Theta(x,t) \quad \text{with } x_f \in \mathcal{R}_{f(x=0)} \tag{16.53a}$$

$$\Theta_f(x_f,t+\Delta t) = 2\Theta^*(x^*=L,t+\Delta t) - \Theta(x,t) \quad \text{with } x_f \in \mathcal{R}_{f(x=L)}. \tag{16.53b}$$

In the PD model, the uniform spacing between material points is $\Delta = 1 \times 10^{-4}$ m, resulting in 100 material points, and the horizon is specified as $\delta = 3.015\Delta$.

The temperature variation along the bar is shown in Fig. 16.8 at different times, and it is compared with the FEA predictions at the final time step. As evident from Fig. 16.9, the implementation of the boundary conditions is very accurate. The change of temperature at the center of the bar as a function of time and its comparison with the FEA prediction are shown in Fig. 16.10.

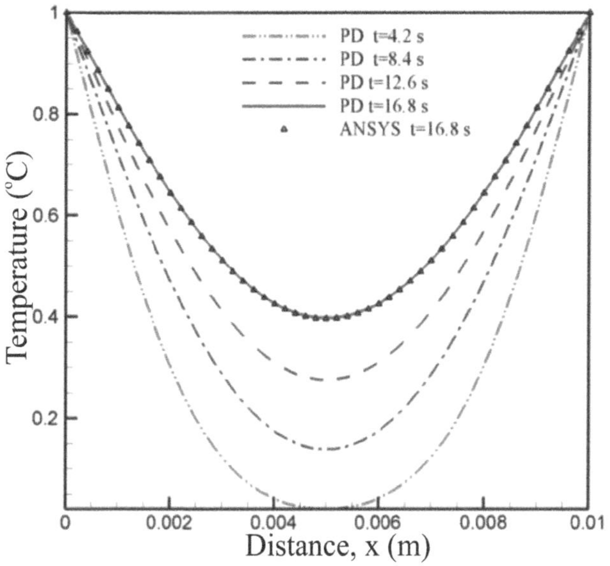

Figure 16.8: PD temperature variation along the bar at different times and its comparison with FEA solutions at the final time step.

16.4.2.4 Example 2

As shown in Fig. 16.11, a rectangular plate is subjected to displacement constraints of $u(x=0,y,t) = U^* = -0.0005$ m and $u(x=L,y,t) = U^* = 0.0005$ m. It is free of tractions. The plate has a length of $L = 1$ m, width of $H = 0.5$ m, and thickness of $h = 0.01$ m. Its Young's modulus and Poisson's ratio are specified as $E = 200 \times 10^9$ Pa and $\nu = 1/3$, respectively, with mass density, $\rho = 7850\,\text{kg/m}^3$. The boundary conditions are enforced as

$$\begin{aligned} u_f(x_f, y_f, t+\Delta t) &= 2U^*(x^* = 0, y^*, t+\Delta t) - u(x,y,t) \\ v_f(x_f, y_f, t+\Delta t) &= v(x,y,t) \quad \text{with } x_f \in \mathcal{R}_{f(x=0)}. \end{aligned} \tag{16.54a}$$

$$\begin{aligned} u_f(x_f, y_f, t+\Delta t) &= 2U^*(x^* = L, y^*, t+\Delta t) - u(x,y,t) \\ v_f(x_f, y_f, t+\Delta t) &= v(x,y,t) \quad \text{with } x_f \in \mathcal{R}_{f(x=L)}. \end{aligned} \tag{16.54b}$$

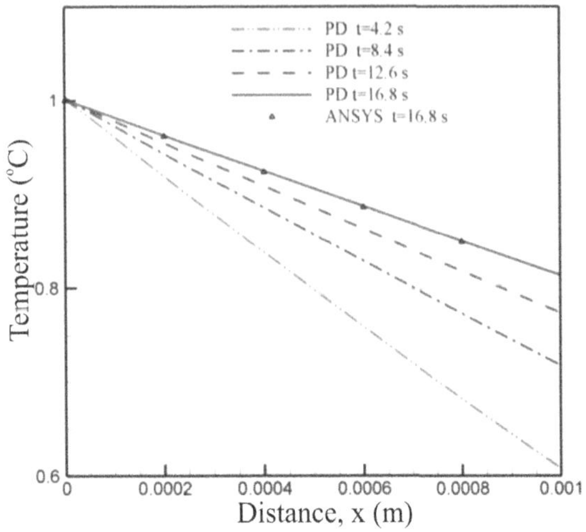

Figure 16.9: PD temperature variation close to the left end of the bar at different times and its comparison with FEA at the final time step.

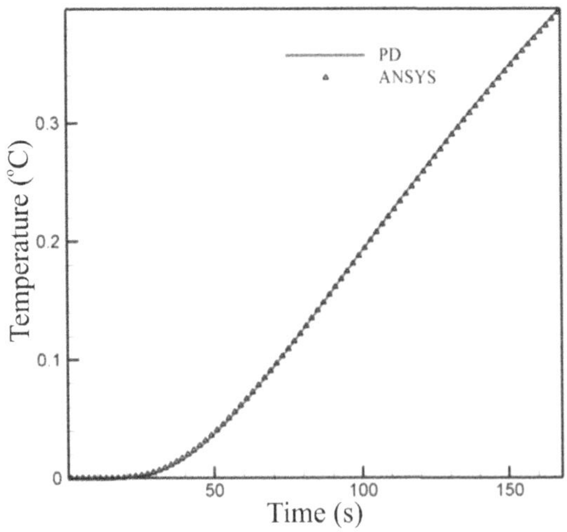

Figure 16.10: PD temperature evolution at the center of the bar as a function of time and its comparison with FEA solution.

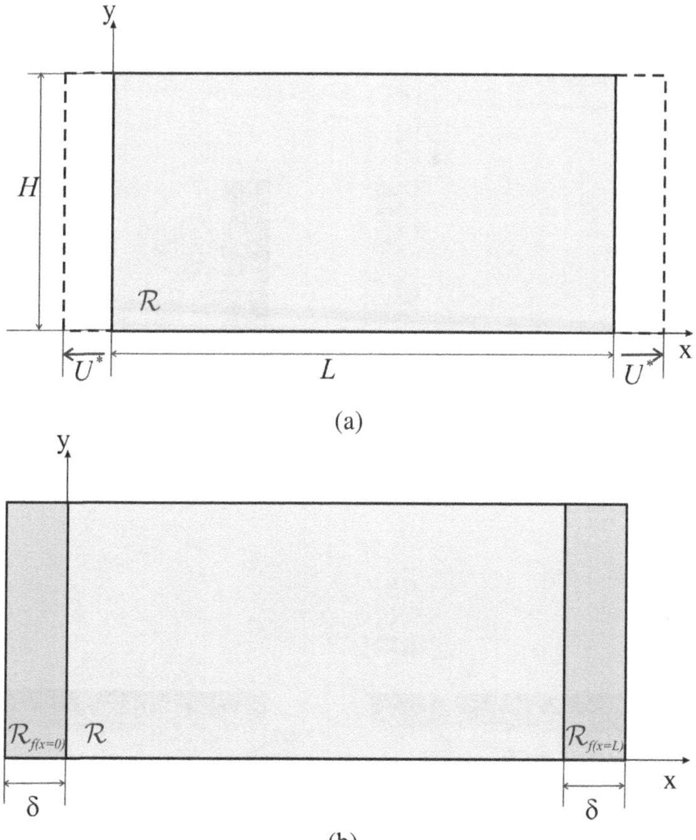

Figure 16.11: Rectangular plate under uniform stretch: (a) geometry and loading and (b) PD model with fictitious regions.

In the PD model, the uniform spacing between material points is $\Delta = 1 \times 10^{-2}$ m, resulting in 100 and 50 material points in the x- and in the y- directions, respectively. The horizon is specified as $\delta = 3.015\Delta$.

The steady-state PD displacement predictions are compared against FEA solutions. As shown in Figs. 16.12 and 16.13, the results exhibit remarkable agreement for both displacement components, and validate the implementation for enforcing displacement boundary conditions.

16.4.2.5 Example 3

The rectangular plate considered previously is subjected to uniform normal stress of $\sigma_0 = 100 \times 10^6$ Pa on the right side and fixed on the left side, $u(x = 0, y, t) = U^* = 0$, and $v(x = 0, y, t) = V^* = 0$, as shown in Fig. 16.14.

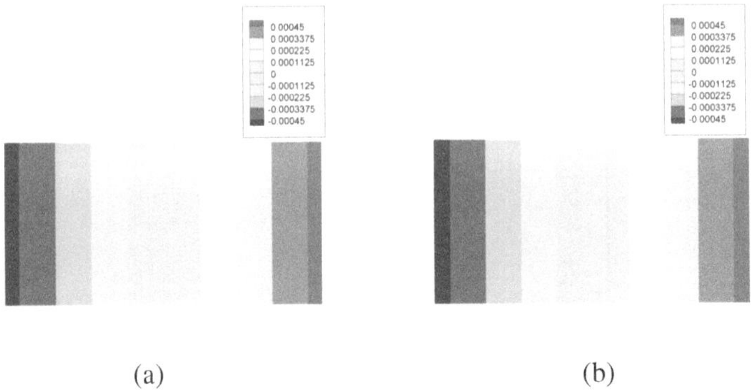

Figure 16.12: Horizontal displacement variation: (a) PD solution and (b) FEA solution.

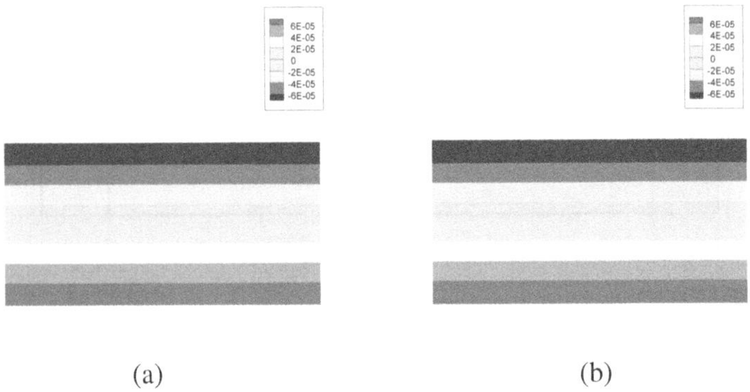

Figure 16.13: Vertical displacement variation: (a) PD solution and (b) FEA solution.

These boundary conditions along $x = 0$ with $x_f \in \mathcal{R}_{f(x=0)}$ are enforced as

$$\begin{aligned} u_f(x_f, y_f, t+\Delta t) &= -u(x,y,t) \\ v_f(x_f, y_f, t+\Delta t) &= -v(x,y,t). \end{aligned} \tag{16.55a}$$

Along $x = L$ with $x_f \in \mathcal{R}_{f(x=L)}$, the applied nonzero normal and zero shear stresses are enforced as

$$\begin{aligned} u_f(x_f, y_f, t+\Delta t) &= \left[\frac{\sigma_o(1-\nu^2)}{E} - \nu \frac{v(x,y^+,t) - v(x,y^-,t)}{y^+ - y^-}\right](x_f - x) + u(x,y,t) \\ v_f(x_f, y_f, t+\Delta t) &= \left[-\frac{u(x,y^+,t) - u(x,y^-,t)}{y^+ - y^-}\right](x_f - x) + v(x,y,t). \end{aligned} \tag{16.55b}$$

The steady-state PD displacement predictions are compared against FEA solutions. As shown in Figs. 16.15 and 16.16, the results exhibit remarkable agreement

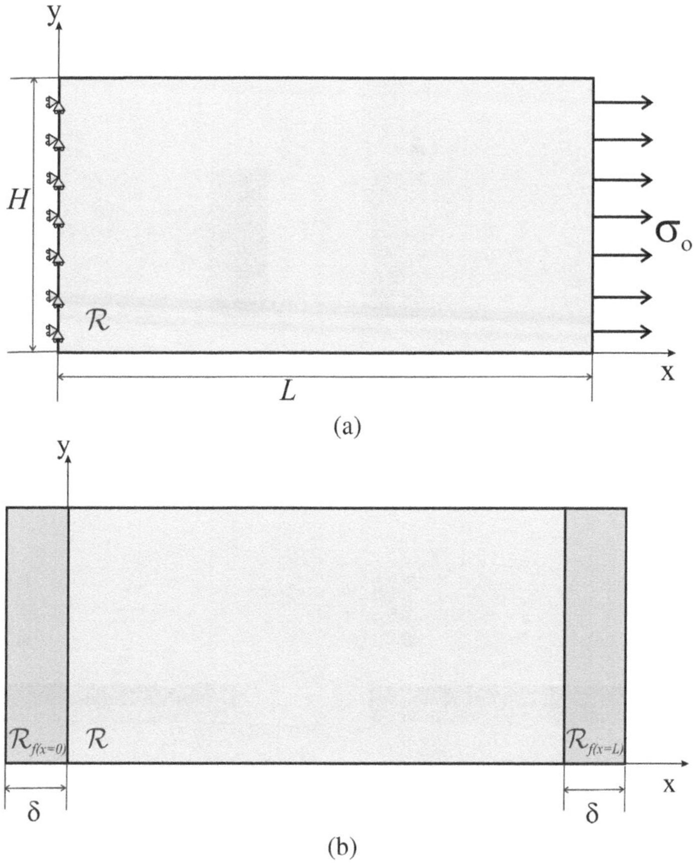

Figure 16.14: Fixed rectangular plate under tension: (a) geometry and loading and (b) PD model with fictitious regions.

for both displacement components, and validate the implementation for enforcing displacement constraints and traction boundary conditions.

16.5 Applications

16.5.1 Coupled nonuniform heating and deformation

We are now concerned with the deformation of a rectangular plate with a hole due to nonuniform heating. The geometry of the plate is given in Fig. 16.17a. Its length and width are 15 cm and 5 cm, respectively, with the hole radius being 1 cm. The plate has a density $\rho = 7850 \text{ kg/m}^3$, thermal conductivity $k = 20$ W/m °K, and spe-

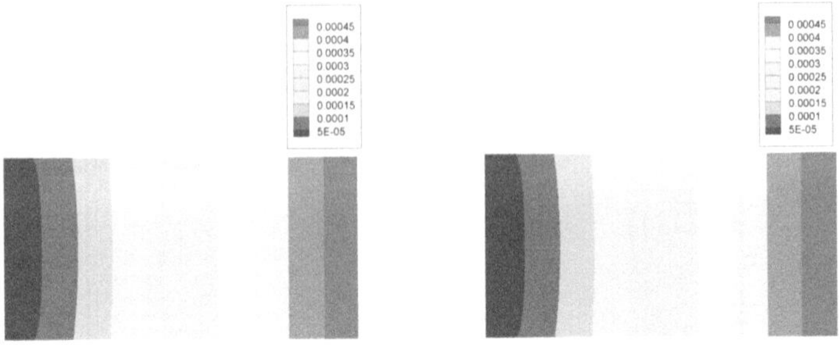

(a) (b)

Figure 16.15: Horizontal displacement variation: (a) PD solution and (b) FEA solution.

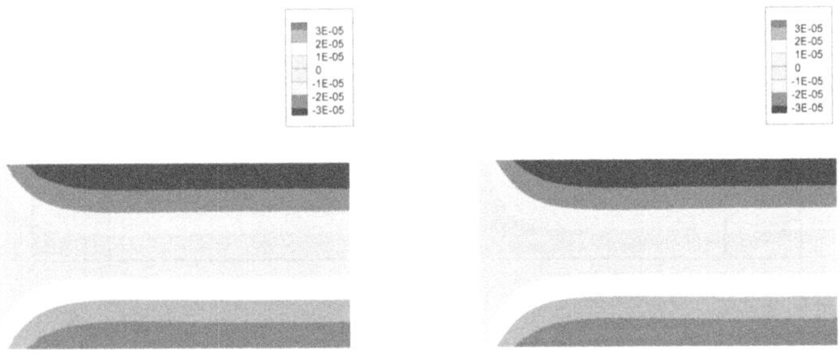

(a) (b)

Figure 16.16: Vertical displacement variation: (a) PD solution and (b) FEA solution.

cific heat $c_v = 500$ J/kg°K. Its Young's modulus and Poisson's ratio are specified as $E = 200 \times 10^9$ Pa and $\nu = 1/3$, respectively, with a thermal expansion coefficient of $\alpha = 16 \times 10^{-6}$ 1/°K. Its initial temperature is 20°C. The left and right boundaries are subjected to constant temperatures of $\Theta^*(x = -L/2, y, t) = 20$°C and $\Theta^*(x = +L/2, y, t) = 500$°C, respectively. The remaining external and internal boundaries are insulated. Also, the plate is subjected to zero displacement constraints of $u(x = -L/2, y, t) = u(x = L/2, y, t) = 0$, while being free of tractions. The time integration interval for heat diffusion is $\Delta t = 1 \times 10^{-2} s$, for a duration of $t = 1 \times 10^3 s$. The plate is discretized with a uniform grid spacing of $\Delta x = \Delta y = 5 \times 10^{-4}$ m, resulting in 300 and 100 material points in the x- and in the y-directions, respectively. The horizon is specified as $\delta = 3.015\Delta$.

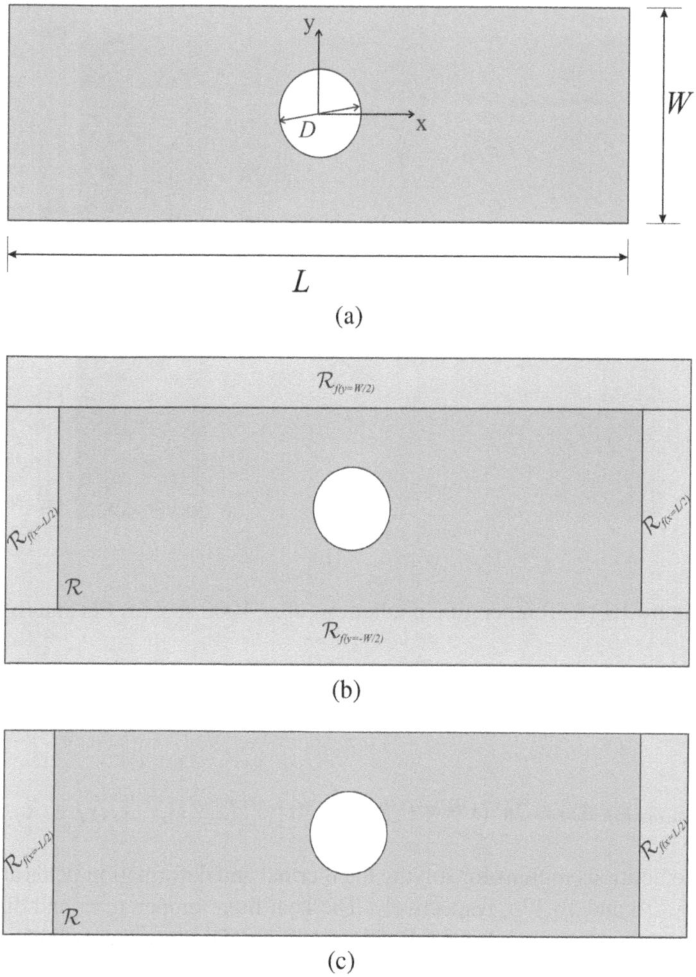

Figure 16.17: Description of (a) plate geometry, (b) fictitious regions for the thermal field, and (c) fictitious regions for the displacement field.

The boundary conditions are imposed as

$$\Theta(x_f, y_f, t+\Delta t) = 2\Theta^*(x=-L/2, y, t+\Delta t) - \Theta(x, y, t), \quad x_f, y_f \in \mathcal{R}_{f(x=-L/2)} \tag{16.56a}$$

$$\Theta(x_f, y_f, t+\Delta t) = 2\Theta^*(x=L/2, y, t+\Delta t) - \Theta(x, y, t), \quad x_f, y_f \in \mathcal{R}_{f(x=L/2)} \tag{16.56b}$$

$$\Theta(x_f, y_f, t+\Delta t) = \Theta^*(x, y=-W/2, t+\Delta t), \quad x_f, y_f \in \mathcal{R}_{f(y=-W/2)} \tag{16.56c}$$

(a)

(b)

Figure 16.18: Temperature predictions after 1000 sec: (a) PD and (b) FEA.

$$\Theta(x_f, y_f, t+\Delta t) = \Theta^*(x, y = W/2, t+\Delta t), \quad x_f, y_f \in \mathcal{R}_{f(y=W/2)} \tag{16.56d}$$

$$u(x_f, y_f, t+\Delta t) = 2u^*(x = \mp L/2, y, t+\Delta t) - u(x, y, t), \quad x_f, y_f \in \mathcal{R}_{f(x=\pm L/2)}. \tag{16.56e}$$

The fictitious regions for solving the thermal and deformation fields are shown in Fig. 16.17b and 16.17c, respectively. The heat flux, temperature, and displacement constraints are imposed through fictitious regions. The zero flux boundary condition around the hole is satisfied in a natural way without imposing any explicit conditions. Similarly, the stress-free boundary conditions are also satisfied in a natural way.

The comparison of PD and FEA predictions for the steady-state temperature field is shown in Fig. 16.18. The horizontal and vertical displacement variations for the corresponding temperature profile and their comparison to FEA predictions are shown in Figs. 16.19 and 16.20. The comparison indicates remarkable agreement.

16.5.2 Coupled nonuniform moisture and deformation in a square plate

We now consider the deformation of a square plate arising from nonuniform moisture concentration. The plate is made of a molding compound and copper, as shown in Fig. 16.21a.

Its length and width are $L = W = 1.2 \times 10^{-3}$m, with thickness $h = 1 \times 10^{-5}$m. The copper region is located at the center with length $a = 0.4 \times 10^{-3}$m. The molding

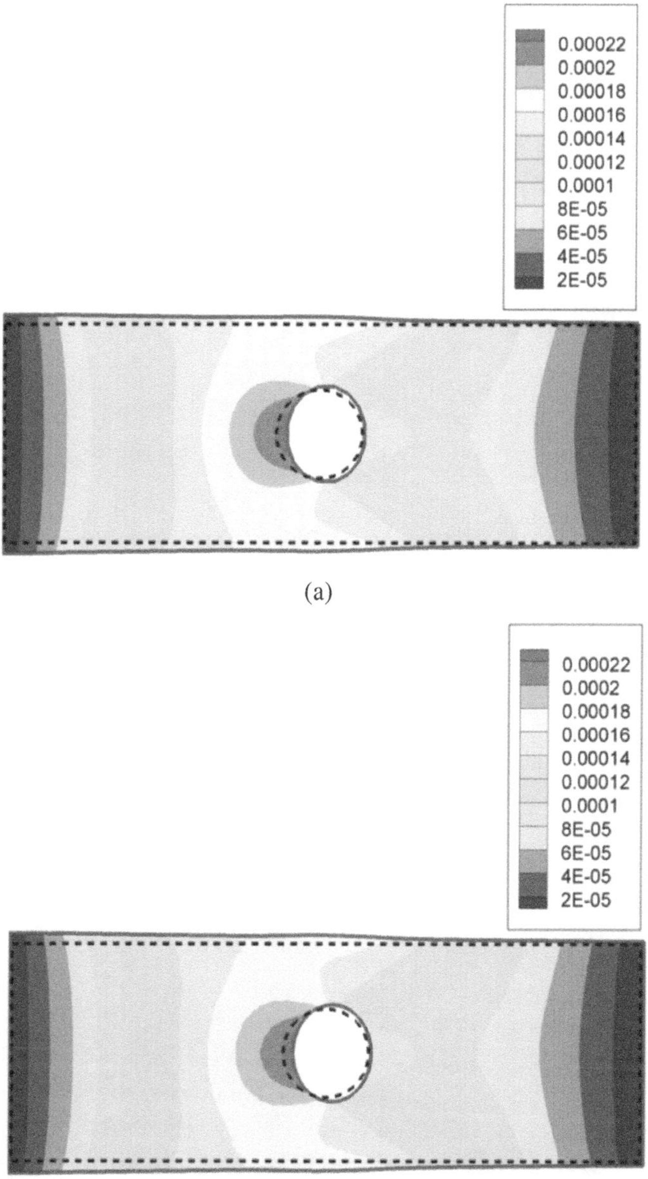

(b)

Figure 16.19: Horizontal displacement variation and deformed configuration (magnified by 10): (a) PD and (b) FEA.

compound has a density of $\rho_1 = 1180$ kg/m^3 and diffusivity of $D_1 = 2.66 \times 10^{-9}$ m^2/hour. Its Young's modulus and Poisson's ratio are specified as $E_1 = 15 \times 10^9$ Pa

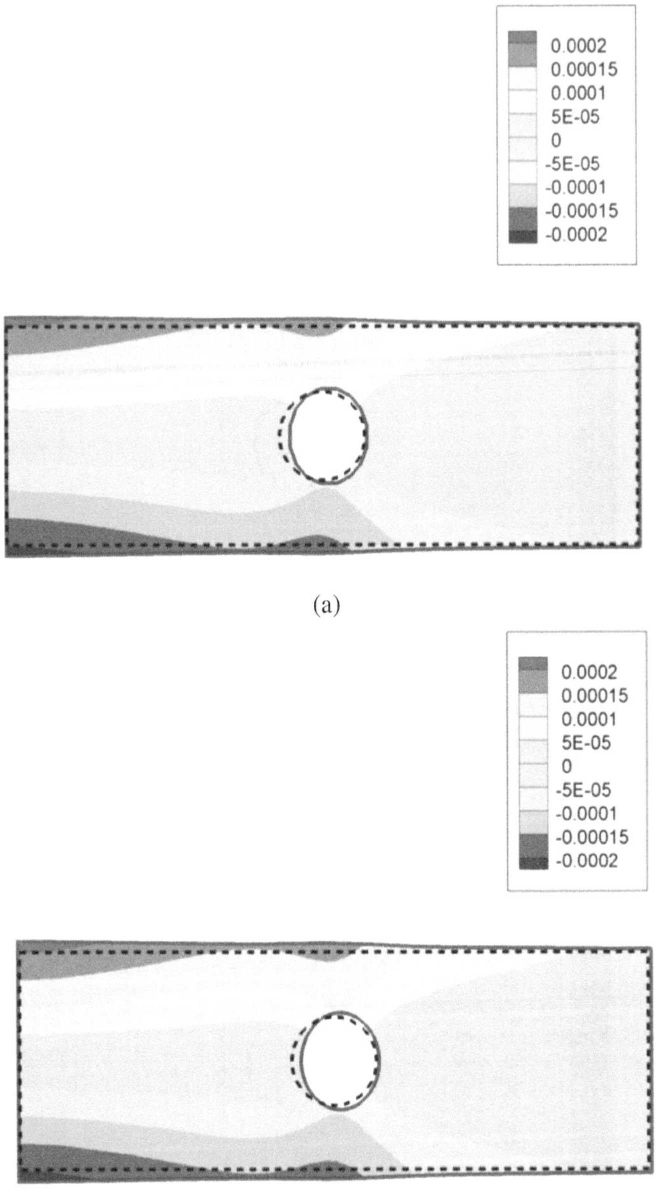

(b)

Figure 16.20: Vertical displacement variation and deformed configuration (magnified by 10): (a) PD and (b) FEA.

and $\nu_1 = 1/3$, respectively, with a moisture expansion coefficient of $\beta_1 = 22.2 \times 10^{-5} \mathrm{m}^3/\mathrm{kg}$. Its saturated moisture concentration is $C_{sat_1} = 7.06\,\mathrm{kg/m}^3$. The copper

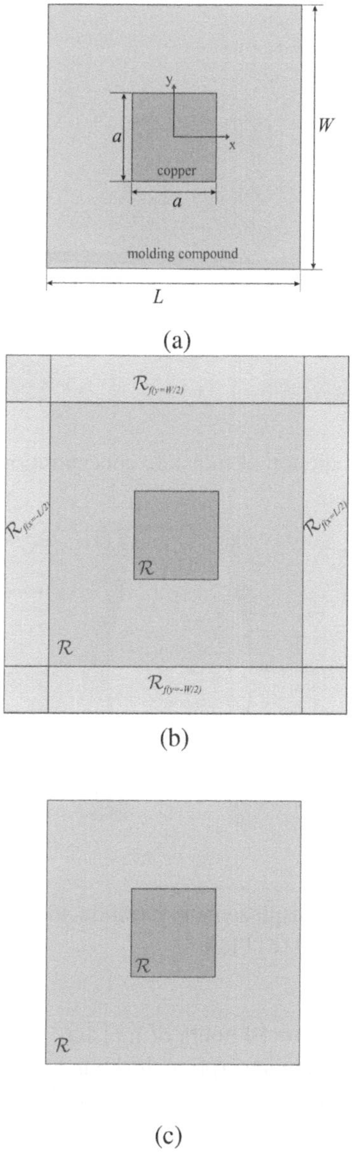

Figure 16.21: Description of (a) plate geometry, (b) fictitious regions for the concentration field, and (c) no fictitious region for the displacement field.

has a density of $\rho_2 = 8940$ kg/m^3, with Young's modulus and Poisson's ratio specified as $E_2 = 129 \times 10^9$ Pa and $\nu_2 = 1/3$, respectively. It has a zero moisture expansion coefficient, saturated moisture concentration, and diffusivity.

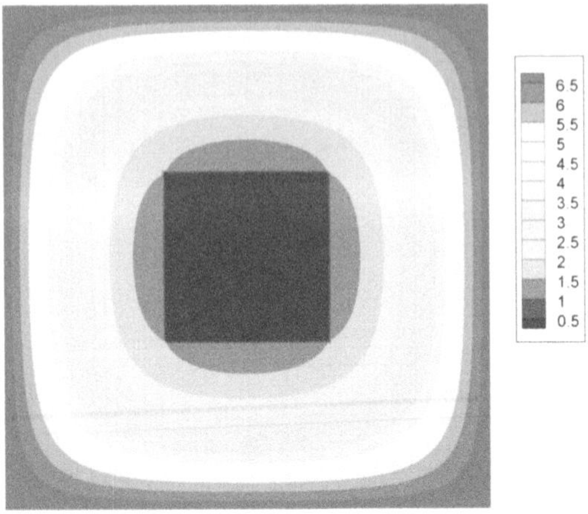

Figure 16.22: PD prediction of moisture concentration after 10 h of absorbtion.

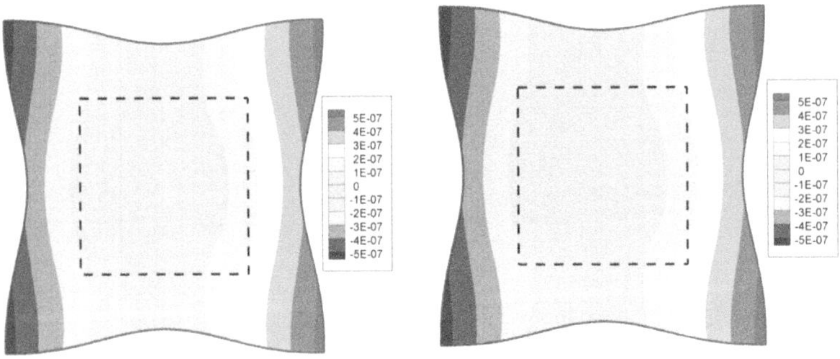

(a) (b)

Figure 16.23: Horizontal displacement variation and deformed configuration (magnified by 1000): (a) PD and (b) FEA.

The plate is subjected to 10 hours of moisture absorption. The initial moisture concentration is zero; during absorption the boundaries of the plate are exposed to a constant concentration of $C^*(x = \mp L/2, y = \mp W/2, t) = 7.06\,\text{kg/m}^3$. Also, the plate is free of any boundary constraints. Time integration interval is $\Delta t = 1 \times 10^{-3}$ h for a duration of $t = 10$ h. The plate is discretized with a uniform grid spacing of $\Delta x = \Delta y = 1 \times 10^{-5}$ m, resulting in 120 material points in the x- and in the y-directions. The horizon is specified as $\delta = 3.015\Delta$.

The boundary conditions are imposed as

$$C(x_f, y_f, t+\Delta t) = 2C^*(x = \mp L/2, y, t+\Delta t) - C(x, y, t), \quad x_f, y_f \in \mathcal{R}_{f(x=\pm L/2)} \tag{16.57a}$$

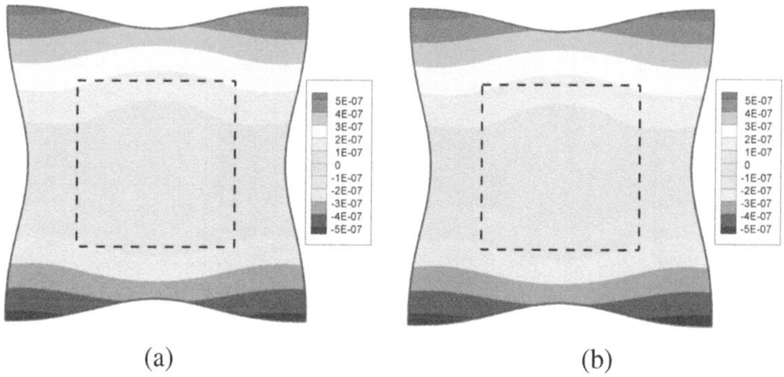

Figure 16.24: Vertical displacement variation and deformed configuration (magnified by 1000): (a) PD and (b) FEA.

$$C(x_f, y_f, t+\Delta t) = 2C^*(x, y = \pm W/2, t+\Delta t) - C(x, y, t), \quad x_f, y_f \in \mathcal{R}_{f(y=\pm W/2)}. \tag{16.57b}$$

The fictitious regions for solving the concentration and deformation fields are shown in Fig. 16.21b and 16.21c, respectively. The moisture concentration is imposed through fictitious regions. The stress-free boundary conditions are satisfied in a natural way. The concentration distribution in the plate after 10 h of absorption is shown in Fig. 16.22. The corresponding horizontal and vertical displacement variations and their comparison to FEA predictions are shown in Figs. 16.23 and 16.24. The comparison indicates remarkable agreement.

16.5.3 Coupled fluid pore pressure and deformation

We now look at the deformation of a square domain arising from nonuniform fluid pressure. The length and width of the domain are 10 m. Its density and the elastic modulus are specified as $\rho = 2000\,\text{kg/m}^3$ and $E = 9\,\text{GPa}$, respectively. The Biot parameter is specified as $\alpha_B = 1$, and the Biot modulus, Q_B, is assumed to approach infinity in order to simulate steady flow. The medium permeability and fluid viscosity are defined as $k = 10^{-9}$ m^2 and $\mu = 10^{-3}$ kg/(m·s), respectively. As a demonstration case, the effect of fluid flow on the deformation is considered, but not the effect of mechanical deformation on the fluid flow. All boundaries except for the top surface are constrained in the normal direction to the surface. These boundary conditions are imposed as

$$u_f(x_f, y_f, t+\Delta t) = 2U^*(x = \pm L/2, y, t+\Delta t) - u(x, y, t) \quad x_f, y_f \in \mathcal{R}_{f(x=\pm L/2)} \tag{16.58a}$$

$$v_f(x_f, y_f, t+\Delta t) = v(x, y, t) \tag{16.58b}$$

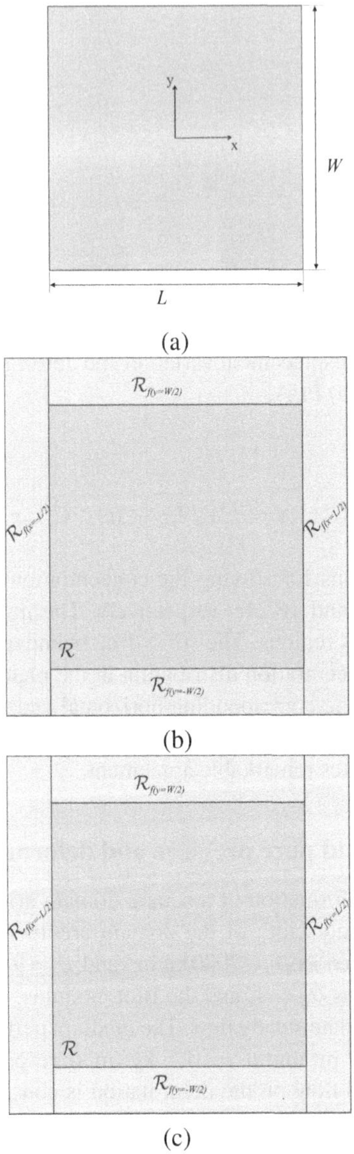

Figure 16.25: Description of (a) plate geometry, (b) fictitious regions for the fluid pressure, and (c) fictitious regions for the displacement field.

$$v_f(x_f, y_f, t+\Delta t) = 2V^*(x, y=-W/2, t+\Delta t) - v(x,y,t), \quad x_f, y_f \in \mathcal{R}_{f(y=-W/2)} \tag{16.58c}$$

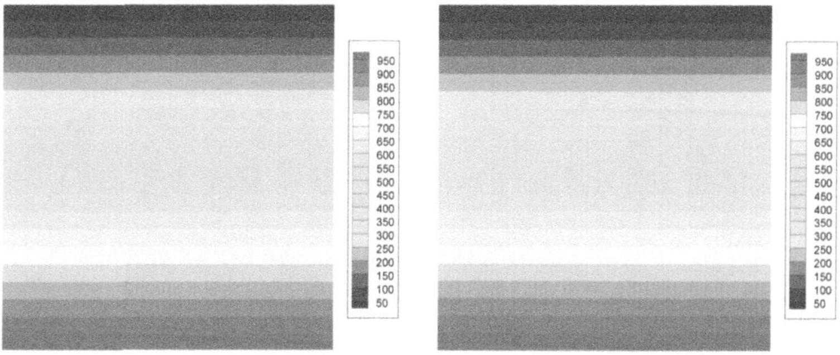

(a) (b)

Figure 16.26: Fluid pore pressure predictions (Pa): (a) PD and (b) FEA.

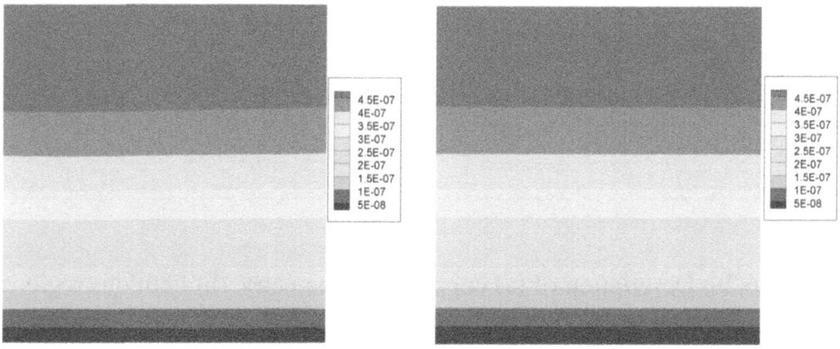

(a) (b)

Figure 16.27: Vertical displacement variation (m): (a) PD and (b) FEA.

$$u_f(x_f, y_f, t+\Delta t) = u(x, y, t) \tag{16.58d}$$

in which $U^* = V^* = 0$.

The fluid pressure at the top surface is assumed to be zero and is implemented as

$$P(x_f, y_f, t+\Delta t) = -P(x, y, t), \quad \text{with } x_f, y_f \in \mathcal{R}_{f(y=W/2)}. \tag{16.59}$$

The left and right edges are specified as impermeable. Furthermore, a fluid flux of $-(\kappa/\mu)\partial P(x=-W/2, y, t)\partial y = 10^{-4}$m/s is applied at the bottom surface which is imposed as

$$P(x_f, y_f, t+\Delta t) = \frac{\partial P}{\partial y}(x=-W/2, y, t+\Delta t)(y_f - y) + P(x, y, t),$$
$$x_f, y_f \in \mathcal{R}_{f(y=-W/2)}. \tag{16.60}$$

On the other hand, no-flux boundary conditions on the left and right are imposed as

$$P(x_f, y_f, t+\Delta t) = P(x, y, t), \quad x_f, y_f \in \mathcal{R}_{f(x=-L/2)} \tag{16.61a}$$

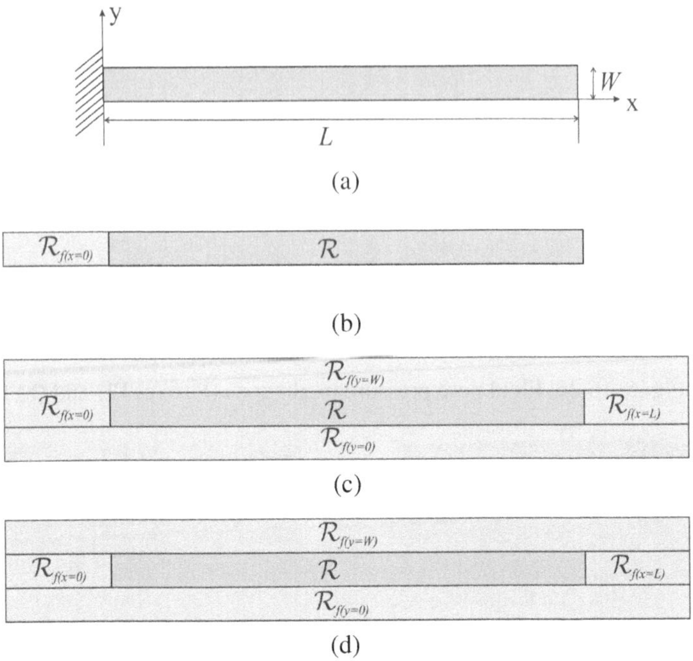

Figure 16.28: Description of (a) copper strip geometry, (b) fictitious regions for the displacement field, (c) fictitious regions for the electrical potential, and (d) fictitious regions for the temperature field.

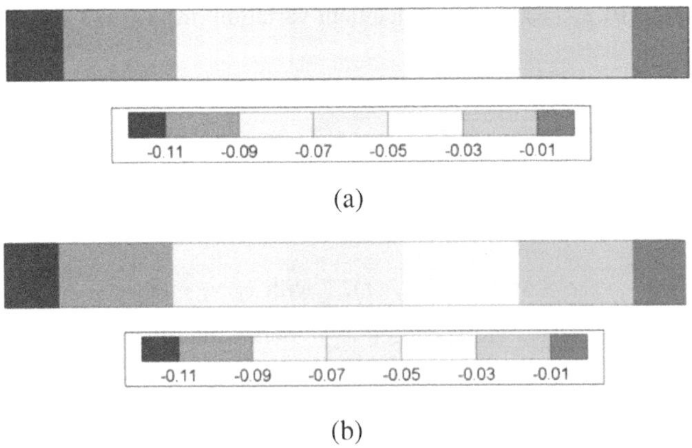

Figure 16.29: Comparison of voltage distribution: (a) ANSYS solution and (b) PD solution at $t = 5 \times 10^{-5}$s.

$$P(x_f, y_f, t + \Delta t) = P(x, y, t), \quad x_f, y_f \in \mathcal{R}_{f(x=-L/2)} \tag{16.61b}$$

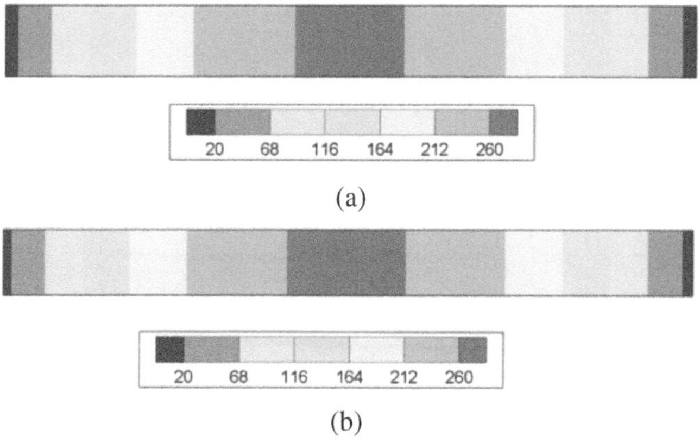

Figure 16.30: Comparison of temperature distribution(°C): (a)ANSYS solution and (b) PD solution at $t = 5 \times 10^{-5}$s.

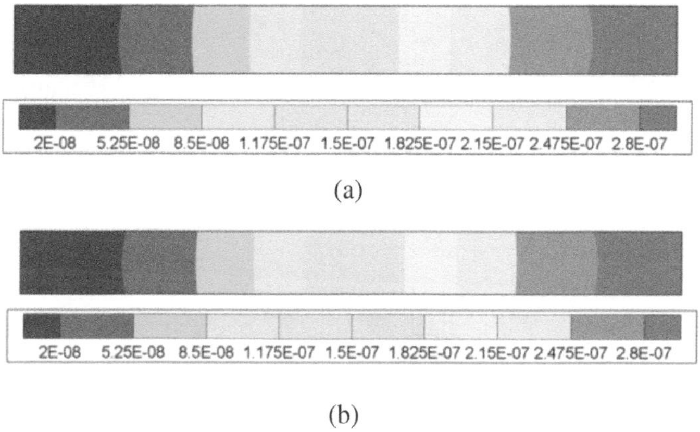

Figure 16.31: Comparison of axial displacement (m): (a) ANSYS solution and (b) PD solution at $t = 5 \times 10^{-5}$s.

The plate is discretized with a uniform grid spacing of $\Delta x = \Delta y = 1 \times 10^{-5}$ m, resulting in 120 material points in the x- and in the y-directions. The horizon is specified as $\delta = 3.015\Delta$.

The fictitious regions for solving the fluid pressure and deformation fields are shown in Fig. 16.25b and 16.25c, respectively. The moisture concentration is imposed through fictitious regions. The stress-free boundary conditions are satisfied in a natural way.

The variation of the fluid pore pressure and the vertical displacements inside the domain are shown in Figs. 16.26 and 16.27. The finite element predictions with ANSYS are in excellent agreement with the peridynamic predictions for both fluid pore pressure and vertical displacement.

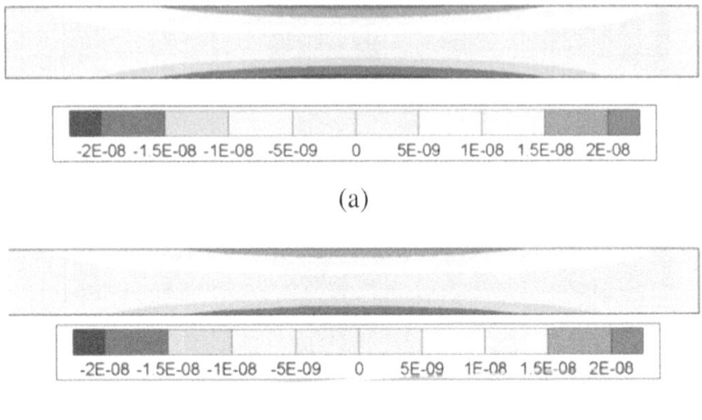

Figure 16.32: Comparison of transverse displacement (m): (a) ANSYS solution and (b) PD solution at $t = 5 \times 10^{-5}$s.

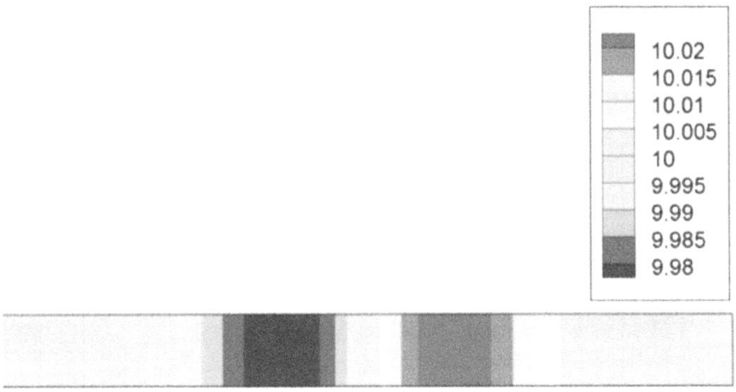

Figure 16.33: Variation of cross-sectional area (μm^2) due to electromigration at $t = 1\,h$.

16.5.4 Coupled electrical, temperature, deformation, and vacancy diffusion

We now consider the two-dimensional analysis of the copper strip considered previously by Gerstle et al. [8] as a one-dimensional analysis. As shown in Fig. 16.28, the geometry of the metal strip is defined by $L = 100\,\mu$m, $W = 10\,\mu$m, and thickness, $h = 1\,\mu$m.

The material properties are specified as Young's modulus $E = 120 \times 10^9\,\text{N/m}^2$, Poisson's ratio $\nu = 0.34$, the coefficient of thermal expansion $\alpha = 16.5 \times 10^{-6}\,1/^\circ$K, the mass density $\rho = 8960\,\text{kg/m}^3$, the thermal conductivity $k_T = 401$ W/(m °K, the heat capacity, $c_v = 385\,\text{J/kg}^\circ$K, the electrical conductivity $k_E = 59.6 \times 10^6$ 1/(Ohm·m), the effective charge of ions $Z^* = 1.5$, the atomic

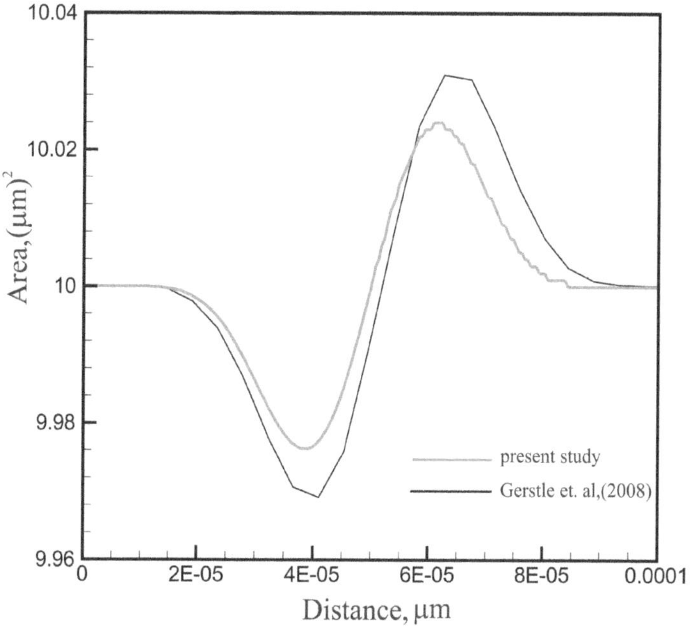

Figure 16.34: Comparison of area change due to electromigration at $t = 1\,\text{h}$.

mass $m_A = 63.546 \times 10^{-3}$kg/mol, the electron charge $e = 1.602 \times 10^{-19}$ Coulomb, the Boltzmann constant $k_b = 1.3807 \times 10^{-23}\, J/^{\circ}K$, the energy for vacancy movement $Q = 3.4877 \times 10^{-23}$ J/vacancy, the energy for vacancy formation $E_v = 1.602 \times 10^{-19}$ J/vacancy, diffusion at infinite temperature $D_o = 6.9 \times 10^{-5}\,\text{m}^2/s$, the vacancy concentration at the melting temperature $(T_m)C_m = 0.01$ vacancies/atom, and the atomic volume $\Omega = 1.182 \times 10^{-29}\,\text{m}^3$.

The strip is discretized with a uniform grid spacing of $\Delta x = \Delta y = 1 \times 10^{-6}\,\text{m}$, resulting in 100 and 10 material points in the x- and in the y-directions, respectively. The horizon is specified as $\delta = 3.015\Delta$.

The initial conditions in the copper strip are specified as

$$u(x,y,t=0) = 0 \tag{16.62a}$$

$$v(x,y,t=0) = 0 \tag{16.62b}$$

$$T(x,y,t=0) = 0\,^{o}\text{C} \tag{16.62c}$$

$$\Phi(x,y,t=0) = 0\,\text{volt}. \tag{16.62d}$$

The difference in electrical potential between the ends of the copper strip is specified by

$$\Phi(x=0,y,t) = -0.12\,\text{volt} \tag{16.63a}$$

$$\Phi(x = L, y, t) = 0 \text{ volt.} \tag{16.63b}$$

Due to the geometry of the copper strip, the current density in the vertical direction is also specified as

$$j_y(x, y = 0, t) = 0 \tag{16.64a}$$

$$j_y(x, y = W, t) = 0. \tag{16.64b}$$

The copper strip is fixed on the left, and all boundaries are assumed to be stress free. Also, it is insulated on the top and bottom surfaces, but subjected to zero temperature at the ends. These boundary conditions are

$$u(x = 0, y, t) = 0 \tag{16.65a}$$

$$v(x = 0, y, t) = 0 \tag{16.65b}$$

$$\sigma_{xx}(x = L, y, t) = 0 \tag{16.65c}$$

$$\sigma_{yx}(x = L, y, t) = 0 \tag{16.65d}$$

$$\sigma_{yy}(x, y = 0, t) = \sigma_{xy}(x, y = 0, t) = 0 \tag{16.65e}$$

$$\sigma_{yy}(x, y = W, t) = \sigma_{xy}(x, y = W, t) = 0 \tag{16.65f}$$

$$T(x = 0, y, t) = 0\,^{o}\text{C} \tag{16.65g}$$

$$T(x = L, y) = 0\,^{o}\text{C} \tag{16.65h}$$

$$T_{,y}(x, y = 0, t) = 0 \tag{16.65i}$$

$$T_{,y}(x, y = W, t) = 0. \tag{16.65j}$$

The boundary conditions for the PD model are enforced as

$$u_f(x_f, y_f, t + \Delta t) = -u(x, y, t),\ x_f, y_f \in \mathcal{R}_{f(x=0)} \tag{16.66a}$$

$$v_f(x_f, y_f, t + \Delta t) = -v(x, y, t),\ x_f, y_f \in \mathcal{R}_{f(x=0)} \tag{16.66b}$$

$$T(x_f, y_f, t + \Delta t) = -T(x, y, t),\ x_f, y_f \in \mathcal{R}_{f(x=0)} \tag{16.66c}$$

$$T(x_f, y_f, t+\Delta t) = -T(x, y, t),\ x_f, y_f \in \mathcal{R}_{f(x=L)} \tag{16.66d}$$

$$T(x_f, y_f, t+\Delta t) = T(x, y, t),\ x_f, y_f \in \mathcal{R}_{f(y=0)} \tag{16.66e}$$

$$T(x_f, y_f, t+\Delta t) = T(x, y, t),\ x_f, y_f \in \mathcal{R}_{f(y=W)} \tag{16.66f}$$

$$\Phi(x_f, y_f, t+\Delta t) = 2\Phi(x=0, y, t) - \Phi(x, y, t), \quad x_f, y_f \in \mathcal{R}_{f(x=0)} \tag{16.66g}$$

$$\Phi(x_f, y_f, t+\Delta t) = 2\Phi(x=L, y, t) - \Phi(x, y, t),\ x_f, y_f \in \mathcal{R}_{f(x=L)} \tag{16.66h}$$

$$\Phi(x_f, y_f, t+\Delta t) = \Phi(x, y, t), \quad x_f, y_f \in \mathcal{R}_{f(y=0)} \tag{16.66i}$$

$$\Phi(x_f, y_f, t+\Delta t) = \Phi(x, y, t),\ x_f, y_f \in \mathcal{R}_{f(y=W)}. \tag{16.66j}$$

The PD model is first verified for the electrical, thermal, and mechanical fields by comparing against FEA results. The temperature, electrical potential, and displacement fields are predicted at $t = 5 \times 10^{-5}$s. As the metal strip reaches a steady-state condition, the electrical potential approaches a linear variation, as shown in Fig. 16.29. Due to the linear potential variation, the joule heating becomes uniform, which gives rise to a symmetric parabolic temperature profile in the thermal fields, as shown in Fig. 16.30. As observed in Figs. 16.29 and 16.30, both the thermal response and the electrical response predicted by the PD formulation are in close agreement with the ANSYS predictions. The PD predictions for mechanical response caused by the electrical joule heating are shown in Figs. 16.31 and 16.32, and they are also in close agreement with the ANSYS predictions.

After verifying the three coupled fields, the peridynamic model is extended to include atomic diffusion. The verification is achieved by comparing the change in cross-sectional area due to atomic diffusion at $t = 1$ h.

Due to the applied voltage difference, the electrons move from left to the right. This electron movement drives the atoms to the right and vacancies to the left. Thus, the thickness reduces on the left side but increases on the right side. Since the metal strip is stress free in all three sides, the area change due to stress migration is ignorable. The total variation of the cross-sectional area is given in Fig. 16.33. The change in cross-sectional area is compared against the previous one-dimensional result from [8] in Fig. 16.34.

16.6 Remarks

This study presents the PD field equations for thermal diffusion, moisture diffusion, electric potential distribution and vacancy diffusion in either an uncoupled or a coupled manner. Inherent with the nonlocal formulations, the imposition of the boundary conditions requires volume constraints. This study also describes the implementation of the essential and natural boundary conditions, and demonstrates the accuracy of their implementation. Solutions to specific coupled field problems concern thermo-mechanics, poroelasticity, and electromigration, and their comparisons with the finite element predictions establish the validity of the PD field equations for coupled analysis.

References

[1] A. Agwai. A peridynamic approach for coupled fields. Ph.D. dissertation, University of Arizona, 2011.

[2] R. Ardito and C. Comi. Nonlocal thermoelastic damping in microelectromechanical resonators. *Journal of Engineering Mechanics–ASCE*, 135: 214–220, 2009.

[3] M.A. Biot. General theory of three-dimensional consolidation. *Journal of Applied Physics*, 12: 155–164, 1941.

[4] F. Bobaru and M. Duangpanya. The peridynamic formulation for transient heat conduction. *International Journal of Heat and Mass Transfer*, 53: 4047–4059, 2010.

[5] F. Bobaru and M. Duangpanya. A peridynamic formulation for transient heat conduction in bodies with evolving discontinuities. *Journal of Computational Physics*, 231: 2764–2785, 2012.

[6] G. Chen. Ballistic-diffusive equations for transient heat conduction from nano to macroscales. *Journal of Heat Transfer*, 124: 320–328, 2002.

[7] Z. Chen, G. Huan, and Y. Ma. *Computational methods for multiphase flows in porous media. Computational Science & Engineering*. Society for Industrial and Applied Mathematics (SIAM), Philadelphia, PA, 2006.

[8] W. Gerstle, S.A. Silling, Read, V. Tewary, and R. Lehoucq. Peridynamic simulation of electromigration. *CMC-Computers, Materials & Continua*, 8: 75–92, 2008.

[9] A. Katiyar, J. Foster, H. Ouchi, M. Sharma. A peridynamic formulation of pressure driven convective fluid transport in porous media. *Journal of Computational Physics*, 261: 209–229, 2014.

[10] B. Kilic and E. Madenci. Peridynamic theory for thermomechanical analysis. *IEEE Transactions on Advanced Packaging*, 33: 97–105, 2010.

[11] D.L. Koch and J.F. Brady. A nonlocal description of advective-diffusion with application to dispersion in porous media. *Journal of Fluid Mechanics*, 180: 387–403, 1987.

[12] Y. Liu, L. Liang, S. Irving, and T. Luk. 3D Modeling of electromigration combined with thermal-mechanical effect for IC device and package. *Microelectronics Reliability*, 48: 811–824, 2008

[13] R.W. Macek and S.A. Silling. Peridynamics via finite element analysis. *Finite Elements in Analysis and Design*, 43: 1169–1178, 2007.

[14] E. Madenci and E. Oterkus. *Peridynamic Theory and Its Applications*, Springer, Boston, MA, 2014.

[15] J.L. Nowinski. *Theory of Thermoelasticity with Applications*. Sijthoff & Noordhoff International Publishers, 1978.

[16] M.A. Omar. *Elementary Solid State Physics*. Addison-Wesley, London, 1975.

[17] S. Oterkus and E. Madenci. Crack Growth Prediction in Fully-Coupled Thermal and Deformation Fields Using Peridynamic Theory. 54th AIAA/ASME/ASCE/AHS/ASC Structures, Structural Dynamics and Materials Conference, Boston, Massachusetts, Paper No. AIAA 2013-1477, 2013.

[18] S. Oterkus and E. Madenci. Peridynamics for Fully Coupled Thermomechanical Analysis of Fiber Reinforced Laminates. 55th AIAA/ASME/ASCE/AHS/ASC Structures, Structural Dynamics and Materials Conference, National Harbor, Maryland, Paper No. AIAA 2014-0694, 2014.

[19] S. Oterkus, J. Fox, and E. Madenci. Simulation of Electromigration through Peridynamics. 63rd Electronic Components & Technology Conference, Las Vegas, NV, pp. 1488-1493, 2013.

[20] S. Oterkus, E. Madenci and A. Agwai. Fully coupled peridynamic thermomechanics. *Journal of the Mechanics and Physics of Solids*, 64: 1–23, 2014.

[21] S.Oterkus, E. Madenci and A. Agwai. Peridynamic thermal diffusion. *Journal of Computational Physics*, 265: 71-96, 2014.

[22] S. Oterkus, E. Madenci, E. Oterkus, Y. Hwang, J. Bae, and S. Han. Hygro-Thermo-Mechanical Analysis and Failure Prediction in Electronic Packages by Using Peridynamics. 64th Electronic Components & Technology Conference, Lake Buena Vista, FL, USA, 2014.

[23] M. Sen and E. Ramos. A spatially nonlocal model for flow in porous media. *Transport in Porous Media*, 92: 29–39, 2012.

[24] S. A. Silling and E. Askari. A meshfree method based on the peridynamic model of solid mechanics. *Computers and Structures*, 83: 1526–1535, 2005.

[25] S.L. Sobolev. Equations of transfer in nonlocal media. *International Journal of Heat and Mass Transfer*, 37: 2175–2182, 1994.

[26] S.L. Sobolev. Nonlocal diffusion models: Application to rapid solidification of binary mixtures. *International Journal of Heat and Mass Transfer*, 71: 295–302, 2014.

[27] M. Zimmerman. A continuum theory with long-range forces for solids, Ph.D. dissertation, Department of Mechanical Engineering, MIT, 2013.

Index

A

Absorbing boundary conditions with verification, 201–225
cone and center crack problems, verification of, 213–219
bond-based verification of center crack, 218–219
dimensional analysis of Hertzian cone crack development in brittle elastic solids, 213–216
glass, cone crack development in, 214
state-based verification of cone crack, 216–218
infinite medium, simulation of, 202
perfectly matched layer, 201
PML for state-based peridynamics, 202–213
auxiliary field formulation and PML application, 204–210
"bilinear transformation," 208
"convolution quadrature," 208
deformation gardient, 203
"filter design by approximation of derivatives," 208
Gaussian kernel, 211
Hooke's law, 203
Laplace domain, 204, 207
numerical examples, 210–213
temporal discretization, 207
two-dimensional, state-based peridynamics review, 202–204
verification of axisymmetric indentation problem, 219–222
analytical verification, 221–222
formulation, 220–221
Laplace domain, 221
Action-reaction, law of, 181
Activation-controlled corrosion, 452, 459, 473
ANSYS, 506, 523
Asymptotically compatible schemes, 89
Atomic bond rupture, 246
Atomistic-to-continuum (AtC) coupling methods, 409
Atomistic models, links between peridynamic and, 179–199
coarse-graining methods, 180
computational speedup through upscaling, 194–196
continuous upscaling, 180
MARTINI force field, 180
meshfree discretization of peridynamic models, 182–184
equation of motion, 183
quadrature weight, 183
molecular dynamics, 181–182
continuum formulation of molecular dynamics, 179
cutoff radius, 181
law of action-reaction, 181
Newton's third law, 181
multibody interactions, 180
upscaling molecular dynamics to peridynamics, 184–194
higher-order gradient partial differential equation, 185, 187
inner expansion technique, 185
isotropic deformation, 186, 190
linear peridynamic solid model, 191
one-dimensional nonlocal linear springs model, 184–189
"safety factor," 188

three-dimensional embedded-atom model, 189–194
upscaling process, 189
wave propagation, simulations of, 188, 192

B

"Bilinear transformation," 208
Biot modulus, 496
Blending methods, 409
Bond-based verification of center crack, 218–219
Bond-breakage criterion, 158
Butler–Volmer equation, 443

C

Cauchy local theory, heterogeneous material system, 12
Cauchy stress tensor, 38, 411
CCM, *see* Classical continuum mechanics
CDD model, *see* Concentration-dependent damage model
Cellular Automata (CA) techniques, 440
CFL condition, *see* Courant–Friedrichs–Levy condition
Classical cohesive models, relations between peridynamic and, 317–338
"bond stretch" criterion, 318
failure process, 319
mixed-mode cohesive law, PD-based, 333–337
normal cohesive law, PD-based, 319–329
bonds have exceeded critical stretch (case), 323–325
cubic particle arrangement, 326
no bonds have reached critical stretch (case), 322
normal traction, 321
numerical approximation, 325–329
tangential cohesive law, PD-based, 329–333
bonds have exceeded the critical stretch (case), 331–333
failure process, 331
no bonds have reached critical stretch (case), 331
shear traction, 330
"steeper" failure process, 333
"upper dome" of the horizon, 333
Classical continuum mechanics (CCM), 408, 431
Coarse graining, 10
Cohesive zone modeling (CZM), 317, *see also* Classical cohesive models, relations between peridynamic and
Commutative diagram, 89
Compressive-strain dependence (CSD), 386
Concentration-dependent damage (CDD) model, 459–465
Constitutive modeling in peridynamics, 143–177
correspondence models, 144, 163–166
arbitrary strain-energy functionality, 163
non-ordinary correspondence models for solid mechanics, 164–165
ordinary correspondence models for solid mechanics, 165–166
Piola–Kirchhoff stress, 164
zero-energy deformation modes, 165
early research, 144
finite deformations, 159–163
identity scalar-state, 160
invariants of peridynamic scalar-states, 162–163

peridynamic Lamé constants, 160
Seth–Hill strain, 162
isotropic elastic model, 145
kinematics, momentum conservation, and terminology, 146–149
calculus-of-variations, 146
deformation vector-state, 147
dummy variables-of-integration, 148
non-ordinary materials, 149
ordinary materials, 149
linear peridynamic isotropic solid, 150–159
"bond-based" theories as a special case, 156–157
bond-breakage criterion, 158
bond-strain, 157
deformation scalar-state, 150
deviatoric extension scalar-state, 151
force scalar-state, 150
Fréchet differentiation, 150
Generalized Hooke's law, 153
improved prototype microelastic brittle model, 157
micropotential, 157
other elasticity theories, 159
plane elasticity, 152–156
plane strain, 156
plane stress, 153–155
prototype microelastic brittle material model, 157
role of influence function, 158–159
strength of locality, 158
unit-state, 150
non-ordinary materials, 144, 149, 164
non-ordinary models, 169–171
bond-multiple models, 171
bond-pair model, 169
Euler–Bernoulli beam strain-energy density, 170
micropolar peridynamic material, 171
non-ordinary beam model, 169–171
non-ordinary plate/shell model, 171
other non-ordinary models, 171
peridynamic states, mathematics of, 144
plasticity, 166–168
discussion, 168
Drucker–Prager plasticity model, 168
force scalar-state, 166
isotropic force scalar-state 167
loading/unloading and consistency, 167–168
plastic deviatoric extension scalar-state, 166
yield surface and flow rule, 167
smooth fields, 145
Timeshenko beam theory, 172
Continuum difference, 97
"Convolution quadrature," 208
Copper strip, 522, 524
Correspondence models, 144, 163–166
arbitrary strain-energy functionality, 163
non-ordinary correspondence models for solid mechanics, 164–165
ordinary correspondence models for solid mechanics, 165–166
Piola–Kirchhoff stress, 164
zero-energy deformation modes, 165
Corrosion damage, peridynamic model for, 437–487
Cellular Automata techniques, 440
convergence study for 1D diffusion-controlled corrosion, 481–482
convergence study for 2D activation-controlled corrosion with CDD model, 482–483

corrosion damage and concentration-dependent damage model, 459–465
damage evolution, 462–464
orientation-dependent damage evolution, 461
saturated concentration, 464–465
effective diffusion, 441, 444
electrochemical kinetics, 443–446
Butler–Volmer equation, 443
electrical double layer, 445
Faraday constant, 443
Tafel law, 443
energy dispersive X-ray spectroscopy mapping, 441
formulation and results of 2D and 3D pitting corrosion, 465–476
activation-controlled corrosion, 470, 473
calculation kernel, 466
CDD model for pitting corrosion (example), 469–472
coupled corrosion/damage model for pitting corrosion (example), 472–474
damage index, 472
diffusivity affects corrosion rate, 474
micro-diffusivity, 467
PD formulation of 2D and 3D pitting corrosion, 466–469
pitting corrosion with CDD+DDC model in 3D, 475–476
peridynamic formulation for 1D pitting corrosion, 449–452
activation-controlled corrosion, 452
diffusion bonds, 449
horizon factor, 450
metal concentration, evolution of, 452
micro-diffusivity, 451
"triangular profile," 451
"pit surface," 442
pitting corrosion in heterogeneous materials (examples), 476–480
constituent particles, 476
heterogeneity, 476
pitting corrosion in layer structures, 476–478
pitting corrosion in material with inclusions (example), 479–480
problem formulation of 1D pitting corrosion, 446–449
diffusion-controlled corrosion, 448
pitting depth, 447
Stefan condition, 448
results and discussion of 1D pitting corrosion, 453–459
activation-controlled, diffusion-controlled, and IR-controlled corrosion, 456–459
overpotential/current density (VI) relationship, 459
pit corrosion depth, 453–456
quantitative analysis, 455
single data-point calibration, 458
zero-overpotential diffusivity, 457
Coupled field equations, peridynamics for, 489–530
applications, 511–527
coupled electrical, temperature, deformation, and vacancy diffusion, 524–527
coupled fluid pore pressure and deformation, 519–523
coupled nonuniform heating and deformation, 511–514
coupled nonuniform moisture and deformation in square plate, 514–519
fluid pore pressure, 523
zero flux boundary condition, 514
Boltzmann equation, 490
coupled field equations, 494–499

atomic flux, 498
Biot modulus, 496
electromigration, 497–499
equation of motion with thermal coupling term, 495
fluid flow in porous medium, 496
heat transfer equation, 494
hydrostatic stress, 499
hygrothermomechanics, 499
mechanical deformation due to fluid pressure, 495–496
porelasticity, 495–496
thermal diffusion with structural coupling term, 494–495
thermomechanics, 494–495
diffusion equation, 492–494
electrical conduction, 493–494
moisture diffusion, 493
thermal diffusion, 492–493
"wetness," 493
Fourier's law, 490
nano-electromechanical systems, 491
nanoscale microelectromechanical systems, 490–491
numerical solution to peridynamic field equations, 500–511
boundary conditions, 502–511
correction of PD material parameters, 500–501
displacement constraints, 502
essential boundary conditions, 502–503
natural boundary conditions, 503–506
volume integration over the horizon, 500
zero gradient, 505
Coupling local and nonlocal models, 407–435
Arlequin method, 410–415
description of coupling model, 410–413
Dirichlet boundary, 410
displacement solutions, 413
Lagrange multiplier field, 414
Neumann boundary, 410
numerical example, 413–415
weight functions, 412
atomistic-to-continuum coupling methods, 409
blending methods, 409
classical continuum mechanics, 408
energy-based blending schemes, 409, 410–423
Arlequin method, 410–415
morphing method, 415–423
force-based blending schemes, 409, 423–430
blending function, 425, 429
bridge region, 426
convergence of peridynamic models to classical models, 424–425
derivation of force-based blending schemes, 425–427
domain decomposition geometry, 430
Navier–Cauchy equation of classical elasticity, 424, 426
numerical example, 427–430
patch-test consistent model, 427
Taylor expansion, 425
morphing function, 409
morphing method, 415–423
benchmark simulations, 419–420
bond-based PD model, 415
constraints based on local strain energy density, 418–419
cracked block, 422
defining the morphing, 418
description, 417–419
energy equivalence, 416
morphing between local and nonlocal models, 417–418
neighborhood, 415
numerical examples, 419–423
one-dimensional analysis of ghost forces, 419
overview, 415–417

PD horizon, 415
traction test, 420
Courant–Friedrichs–Levy (CFL) condition, 54, 128
Crack branching, *see* Dynamic brittle fracture, crack branching in
Critical bond breakage strain, 45
Critical stretch bond failure law, 132
Critical time step, 127
CSD, *see* Compressive-strain dependence
Cubit mesh generation code, 132
CZM, *see* Cohesive zone modeling

D

Damage-dependence (DD), 386
Damage-Dependent Corrosion (DDC) model, 471
Damage-dependent peridynamic model, 259
Deformation scalar-state, 150
Degree of freedom (DOF), 10
alternate, 11
elimination of, 12
Deviatoric extension state, 37, 151, 167
Diffusion-controlled corrosion, 448
Dirac delta functions
integral equations, 26
material modeling, 40
nonlocal calculus of variations, 70
origins of nonlocality in a model, 15
Discretization schemes, *see* Local limits and asymptotically compatible discretizations
DOF, *see* Degree of freedom
Drucker–Prager plasticity model, 168
Dynamic brittle fracture, crack branching in, 245–316
angle of branching, 249
atomic bond rupture, 246
bond-based peridynamic model, 256–259
damage-dependent peridynamic model, 259
ghost nodes, 257
horizon region, 256
linear microelastic material, 257
relative elongation of the bond, 257
"skin effect," 257
branching of a crack, 247
crack branching, literature on, 249–256
atomistic modeling, dynamic brittle fracture results based on, 253
cohesive zone FEM models, 252
computations of dynamic brittle fracture based on FEM, 251–252
crack surface, tracking of, 252
displacement field, 255
dynamic crack curving criterion, 250
"element-erosion" method, 252
failure criterion, 255
"forbidden velocities," 253
"infinite specimens," 253
"intermittency" phenomenon, 253
lattice dynamics models, 253
laws of propagation, 252
"particle cracking model," 254
particle and lattice-based methods, dynamic brittle fracture based on, 253–254
phase-field models in dynamic fracture, 254–255
Rayleigh wave speed, 250
reproducing kernel particle method, 254
results on dynamic brittle fracture from peridynamic models, 255–256
"river lines" nucleation, 251
theoretical models and experimental results, 249–251
Voronoi structure, 254
crack path instabilities, 248, 259

discussion of crack branching results, 291–295
crack branching angles, 294
range of speeds at branching, 291
Rayleigh wave speeds, 291
importance of nonlocal modeling in crack branching, 304–307
arbitrary directions, generating damage in, 306
nearest-neighbor-only interactions, 305
origin of instabilities, 307
Linear Fracture Mechanics, 247
microcracks, coalescence of, 247
peridynamic results for dynamic fracture and crack branching, 262–291
homalite, crack branching in, 275–285
loading conditions, 263
quasi-statically growing cracks, 264
sample geometry, influence of, 285–291
soda-lime glass, crack branching in, 265–275
stress waves, 263
quadrature scheme, 260–262
conical micromodulus, algorithm for, 262
discretization grids, 260
Guass quadrature method, 260
points outside the horizon, 261
why cracks branch, 296–303
damage index, 298
dynamic crack propagation, 296
macroscopic mechanism of dynamic brittle branching, 300
migration of damage and branching, 296
pile up of stress waves, 296
Rayleigh waves, 298
strain energy, 299
velocity vector plot, 297
Dynamic brittle fracture as upscaling of unstable mesoscopic dynamics, 229–244
bounds on kinetic and potential energy for mesoscopic evolution (theorem), 233–234
continuous mesoscopic evolution in mean square norm (theorem), 234
"Dirichlet" boundary conditions, 232
dynamic instability and fracture initiation, 238
energy balance (theorem), 234
fracture nucleation condition, 239
free crack propagation in small horizon-macroscopic limit, 241–242
hypotheses, 241
neutrally stable neighborhoods, 241
wave equation, 242
inflection point, 231
Linear Elastic Fracture Mechanics, 232
localization of dynamic instability in small horizon-macroscopic limit, 240–241
neighborhood, 240
nucleation sites, 241
unstable subset, 240
macroscopic evolution of brittle fracture as small horizon limit of mesoscopic dynamics, 234–238
bottom-up approach, 235
convergence of mesoscopic dynamics to macroscopic dynamics (theorem), 236–237
energy inequality (theorem) 237–238
Hausdorff measure, 237
jump discontinuity, 238

macroscopic dynamics has bounded LEFM energy (theorem), 237
mesoscopic model, 230
Mumford–Shah functional, 243
theorem, 233, 234

E

EAM, *see* Embedded-atom model
EDS mapping, *see* Energy dispersive X-ray spectroscopy mapping
Effective diffusion (corrosion), 441, 444
E-glass, 340
Einstein summation convention for repeated indices, 424
Electrical double layer (EDL), 445
Electromigration, 497–499
"Element-erosion" method, 252
Embedded-atom model (EAM), 189
Emu numerical method, 54–56
claim, 56
Courant–Friedrichs–Levy condition, 54
material "calibrated in place," 55
micromodulus function, 55
Energy-based blending schemes, 409, 410–423
Arlequin method, 410–415
description of coupling model, 410–413
Dirichlet boundary, 410
displacement solutions, 413
Lagrange multiplier field, 414
Neumann boundary, 410
numerical example, 413–415
weight functions, 412
morphing method, 415–423
benchmark simulations, 419–420
bond-based PD model, 415
constraints based on local strain energy density, 418–419
cracked block, 422
defining the morphing, 418
description, 417–419
energy equivalence, 416
morphing between local and nonlocal models, 417–418
neighborhood, 415
numerical examples, 419–423
one-dimensional analysis of ghost forces, 419
overview, 415–417
PD horizon, 415
traction test, 420
Energy dispersive X-ray spectroscopy (EDS) mapping, 441
Euler–Bernoulli beam strain-energy density, 170
Euler–Lagrange equation, 77
Explicit time integration, 125

F

FEA, *see* Finite element analysis
FEM, *see* Finite element method
Fiber-reinforced composites, peridynamic modeling of, 339–377
critical stretch values for bond constants, 374–375
critical stretch values for bond constants, critical shear angle, 375
interlayer bond, 374
E-glass, 340
numerical results, 349–354
lamina under uniaxial tension and uniform temperature change, 349–350
lamina with a pre-existing central crack under tension, 351–353
laminates under uniaxial tension, 351
laminates with pre-existing central crack under tension, 353
splitting mode, 353
"Z" type failure mode, 353

PD interlayer and shear bond constants of laminate, 368–374
deformed bond length, 373
ply thickness, 371
shear angles, 370
total strain energy density, 369
PD material constants of lamina, 354–363
biaxial stretch, 359–363
engineering constants, 361
mechanical stretch, 358
simple shear, 355–356
uniaxial stretch in fiber direction, 356–358
uniaxial stretch in transverse direction, 358–359
peridynamic analysis of lamina, 341–347
engineering material constants, 342
failure parameter, 346
force-stretch relation, 345
material point, 342
stiffness matrix, 343
strain energy density, 343, 344
surface correction, 346
uniaxial tension test cases, 345
peridynamic analysis of laminate, 347–349
bond constants, 347
critical stretch value, 348
interlayer bonds, 347
shear bonds, 348
surface correction factors for composite lamina, 363–368
applied constant displacement gradient, 363
correction factors, 364
decomposition, 365, 366, 367
material point, strain energy density of, 363
uniaxial strain loading, 363
Fictitious nodes, 257
"Filter design by approximation of derivatives," 208
Finite element analysis (FEA), 506
Finite element method (FEM), 251–252
basis of existing models, 252
elastic response of multilayered system and, 391
models, cohesive zone, 252
Flux, 30
"Forbidden velocities," 253
Force-based blending schemes, 409, 423–430
convergence of peridynamic models to classical models, 424–425
Navier–Cauchy equation of classical elasticity, 424, 426
Taylor expansion, 425
derivation, 425–427
blending function, 425, 429
bridge region, 426
patch-test consistent model, 427
domain decomposition geometry, 430
numerical example, 427–430
Fourier's law, 490
Fragmentation, *see* Impact and fragmentation, peridynamic modeling of
Fréchet derivatives
constitutive modeling, 148, 152, 164
damage and fracture, 50
material modeling, 34, 36
Fubini theorem, 69

G

Gauss quadrature scheme, 110
Generalized Hooke's law, 153
Ghost nodes, 257
Glass
cone crack development in, 214
E-, 340
multilayered, impact on, 388–393

comparison between FEM and peridynamics for elastic response, 391–393
material properties, 390
model description, 389–391
polycarbonate layer, 389
quasi-brittle material, glass behavior as, 391
"reduced" system, 391
rigid projectile, 390
strain-rate dependence, 391
verification test, 392
zero-vertical displacements, 389
seven-layer, computational results for damage progression in, 393–401
bending deformation, 394
circumferential cracks, 399
cross section, 394–396
damage index, 393
first layer, 396–398
"floret"-type crack, 397
fourth layer, 399–400
quadrant cracks, 399
radial cracks, 398, 400
second layer, 398
seventh layer, 400–401
spall-type fractures, 395
soda-lime, crack branching in, 265–275
crack velocity, calculation of, 265, 267
damage index, 265
electromagnetic loading of crack surfaces, 269
fish-bone like pattern of branchings, 273
Rayleigh wave propagation, 271
stress on boundaries, 265–268
stress on pre-crack surfaces, 269–272
stress waves, reinforcement of, 270
velocity boundary conditions, 272–275
Green's identity (nonlocal), 72, 96

H

Higher-order gradient (HOG) partial differential equation, 185, 187
Homalite, crack branching in, 275–285
damage maps, 281
dynamic loading, 283
location of branching, 280
paradox, 279
process zones, 284
stress on boundaries, 276–278
stress on pre-crack surfaces, 278–282
velocity boundary conditions, 282–285
Horizon
factor (1D pitting corrosion), 450
parameter, 70
setting, 40–41
"upper dome" of, 333
volume integration over, 500
Hygrothermomechanics, 499

I

Identity scalar-state, 160
Impact and fragmentation, peridynamic modeling of, 379–404
convergence studies and damage models, 382–388
compressive-strain dependence, 386
convergence study for impact on glass plate, 384–388
critical bond strain dependence on compressive strains along other directions, 383
damage-dependence, 386
damage-dependent critical bond strain, 382–383
damage index, 383
damage patterns, 385
grid symmetry, 388
surface correction, 386

surface effect in impact problems, 383–384
glass laminates (multi-layer), penetration resistance of, 381
multilayered glass system, impact on, 388–393
comparison between FEM and peridynamics for elastic response, 391–393
material properties, 390
model description, 389–391
polycarbonate layer, 389
quasi-brittle material, glass behavior as, 391
"reduced" system, 391
rigid projectile, 390
strain-rate dependence, 391
verification test, 392
zero-vertical displacements, 389
seven-layer glass system, computational results for damage progression in (damage evolution), 393–401
bending deformation, 394
circumferential cracks, 399
cross section, 394–396
damage index, 393
first layer, 396–398
"floret"-type crack, 397
fourth layer, 399–400
quadrant cracks, 399
radial cracks, 398, 400
second layer, 398
seventh layer, 400–401
spall-type fractures, 395
Implicit time integration, 125
Influence function, 36
"Intermittency" phenomenon, 253
Introduction to peridynamics, 25–60
boundaries and interfaces, treatment of, 51–54
bond-based materials, 51–53
common software bug, 52
end-to-end partial bond, 53
interfacial crack growth, 53
state-based materials, 53–54
damage and fracture, 45–51
autonomous crack growth, 45
bond-based models, damage in, 45–47
correspondence material models, damage in, 49–50
critical bond breakage strain, 45
jump discontinuity, 51
net damage field, 45
nucleation strain, 50–51
ordinary state-based material models, damage in, 47–48
prototype microelastic brittle material, 45
Emu numerical method, 54–56
claim, 56
Courant–Friedrichs–Levy condition, 54
material "calibrated in place," 55
micromodulus function, 55
equilibrium in terms of integral equations, 26–28
body force density field, 26
family, 27
horizon, 27
pairwise bond force density function, 26
material modeling, 28–41
bond-based materials, 28–30
bond extension, 28
Cauchy stress tensor, 38
correspondence materials, 37–39
definition of, 33
deformation state, 33
deviatoric extension state, 37
discrete particles as peridynamic bodies, 39–40
flux, 30
influence function, 36
linearized peridynamics, 41
linear peridynamic solid model, 37
micromodulus, 28
micropotential, 29

nonlocal calculus, 32
non-polarity, 35
ordinary state-based materials, 35–37
peridynamic states, 32–35
prototype microelastic, 30
relation between bond densities and flux, 30–32
setting the horizon, 40–41
spring constant, 30
theory of elasticity, 29
zero energy nodes, 39
plasticity, 42–44
bond-based microplastic material, 42–43
LPS material with plasticity, 43–44
IR (current resistance)-controlled corrosion, 459
Isotropic force scalar-state 167

J

Jump discontinuity, 51, 238, 453

K

Kalthoff–Winkler test, 255
Kirchhoff–Love plate theory, 171

L

Lattice dynamics models, 253
Lebesgue function space, 66
Linear elastic fracture mechanics (LEFM), 247
bounded energy, 237
breaking down of, 247
initial data, 233, 236
origins of nonlocality, 6, 7
upscaling of unstable mesoscopic dynamics, 232
Linear microelastic material, 257
Linear peridynamic isotropic solid, 150–159
Linear peridynamic solid (LPS), 116
breaking down of, 38
equation of motion, 192
material with plasticity, 43
model, 37, 43, 191
time integration, 128
Local limits and asymptotically compatible discretizations, 87–108
asymptotically compatible schemes for peridynamics, 89, 100–103
asymptotically compact embedding property, 101
best approximation and asymptotic compatibility, 101–102
conforming Galerkin discretization, 101
definition of, 100
key properties and assumptions, 100–101
recap of AC schemes via examples, 102–103
benchmark test, 87
commutative diagram, 89
consistency issue, 87
discretization schemes and discrete local limits, 97–99
continuum difference, 97
over-estimated constant, 98
quadrature based finite difference discretization, 97
Riemann sum quadrature, 98
linearized peridynamics, 88
local PDE limits of linear peridynamic models, 90–97
consistency via Taylor expansion, 92–93
convergence to local limit, 91–92
convergence of weak solutions, 95–96
Fourier analysis and dispersion relation, 93–94
Fourier analysis, point spectrum, stiffness matrix conditioning, 94–95

local limit, 91
more general systems and nonlocal constraints, 96–97
one-dimensional model, 90
patch test, 90, 91
questions, 88
Smoothed Particle Hydrodynamics, 103
LPS, *see* Linear peridynamic solid

M

MARTINI force field, 180
Material modeling, 28–41
bond-based materials, 28–30
bond extension, 28
micromodulus, 28
micropotential, 29
prototype microelastic, 30
spring constant, 30
theory of elasticity, 29
correspondence materials, 37–39
Cauchy stress tensor, 38
zero energy nodes, 39
discrete particles as peridynamic bodies, 39–40
linearized peridynamics, 41
ordinary state-based materials, 35–37
deviatoric extension state, 37
influence function, 36
linear peridynamic solid model, 37
peridynamic states, 32–35
definition of, 33
deformation state, 33
non-polarity, 35
relation between bond densities and flux, 30–32
flux, 30
nonlocal calculus, 32
setting the horizon, 40–41
Microelectromechanical systems (MEMS), 490–491
Micropolar peridynamic material, 171
Mixed-mode cohesive law, PD-based, 333–337
Molecular dynamics (MD), 181–182
continuum formulation of, 179
cutoff radius, 181
law of action-reaction, 181
Newton's third law, 181
Mumford–Shah functional, 243

N

Nano-electromechanical systems (NEMS), 491
Nearest-neighbor-only interactions (NNOI), 305
Neighbor list, 124
Nernst–Planck equation, 445
Net damage field (bond-based models), 45
Neumann boundary (coupling model), 410
Neutrally stable neighborhoods, 241
NNOI, *see* Nearest-neighbor-only interactions
Nodal volume, 111
Nonlocal calculus of variations and well-posedness of peridynamics, 63–85
linearized peridynamics, 65
mathematical questions, 63–64, 81
nonlocal balance laws and nonlocal vector calculus, 67–72
formulating nonlocal models via nonlocal operator, 70
Fubini theorem, 69
Green's identities, 72
horizon parameter, 70
nonlocal balance laws, 68–69
nonlocal flux, 68
nonlocal Laplacian, 70
nonlocal operators, 69–70
peridynamic Navier operator, 70–71
Stokes theorem, nonlocal extensions to, 70

weighted nonlocal operators, 71
nonlocal calculus of variations (further discussions), 79–81
energy functional, 79
Navier operator, 80
time-dependent peridynamic models, 80
nonlocal calculus of variations (illustration), 72–79
assumptions on nonlocal interaction kernel, 75
bilinear form and weak form, 77
energy functional, 74–75
energy space, 76
Euler–Lagrange equation, 77
nonlocal boundary conditions and nonlocal constraints, 73–74
nonlocal constrained value problem, 74
nonlocal Poincare's inequality, 78
solution space, 75–76
steady-state problem, 72–73
variational principle, 77
well-posedness of the variational problem, 78–79
peridynamic horizon, 65
well-posedness results, review of, 66–67
Lebesgue function space, 66
peridynamic Navier equation, 67
Non-ordinary materials, 144, 149, 164

O

Orientation-dependent damage evolution, 461
Overpotential/current density (VI) relationship, 459

P

Partial differential equations (PDEs), 4
creative aspects of applying, 4
limits of linear peridynamic models, 90–97
reliance of standard theory on, 5
"Particle cracking model," 254
Patch-test consistent model, 427
Perfectly matched layer (PML) (state-based peridynamics), 202–213
auxiliary field formulation and PML application, 204–210
"bilinear transformation," 208
"convolution quadrature," 208
deformation gardient, 203
"filter design by approximation of derivatives," 208
Gaussian kernel, 211
Hooke's law, 203
Laplace domain, 204, 207
numerical examples, 210–213
temporal discretization, 207
two-dimensional, state-based peridynamics review, 202–204
Peridigm code, 112, 124
Piola–Kirchhoff stress, 164
Plastic deviatoric extension scalar state, 166
Plasticity, 166–168
discussion, 168
Drucker–Prager plasticity model, 168
force scalar-state, 166
isotropic force scalar-state 167
loading/unloading and consistency, 167–168
plastic deviatoric extension scalar-state, 166
yield surface and flow rule, 167
Poincare's inequality (nonlocal), 78
Porelasticity, 495–496
Prototype microelastic (PM), 30
Prototype microelastic brittle (PMB) material
bond-based models and, 45
critical strain for bond failure, 382
model, 128, 157

nucleation strain, 50
spring constant, 52
Purpose of peridynamics, 3–23
mixed blessing of locality, 3–6
"action at a distance," 5
assumptions, 4
contact forces in a continuum, 5
internal forces, 6
kinetic relation, 6
micropolar Cosserat continuum, 3
repairing or adding onto theory, 5
mixed blessing of nonlocality, 20–21
difficulties in modeling problems, 20
multiscale methods, 21
nonlocality at the macroscale, 18–20
open hole tension test, 20
stress decay, 18
origins of nonlocality in a model, 6–18
bond force densities, 17
coarsening a fine-scale material system, 10–12
concrete, 12
Dirac delta function, 15, 16
force balance, 14
length scale, 17
linear elastic fracture mechanics, 6
long-range forces, 7–10
micromodulus, 16
modeling decision, 17
physical interaction, 17
ratio of stress to strain, 12
self-assembly of structures, 9
smoothing of a heterogeneous material system, 12–18
stress intensity factor, 6
uniaxial tension test, 14

Q

Quadrant cracks, 399
Quadrature based finite difference discretization, 97

R

Rayleigh wave
propagation, 271
speed, 247, 250
Recursive Coordinate Bisection (RCB) algorithm, 124
Reproducing kernel particle method (RKPM), 254
Riemann sum quadrature, 98
"Ripple cracks," 399
"River lines" nucleation, 251

S

Sandia Red Sky supercomputer, 193
SC, *see* Surface correction
Short-range force model, 121
SIF, *see* Stress intensity factor
Smoothed Particle Hydrodynamics (SPH), 103
Sobolev spaces, 100
Soda-lime glass, crack branching in, 265–275
crack velocity, calculation of, 265, 267
damage index, 265
electromagnetic loading of crack surfaces, 269
fish-bone like pattern of branchings, 273
Rayleigh wave propagation, 271
stress on boundaries, 265–268
stress on pre-crack surfaces, 269–272
stress waves, reinforcement of, 270
velocity boundary conditions, 272–275
Software implementation, roadmap for, 109–139
bond damage and failure, 115
example simulations, 131–134

critical stretch bond failure law, 132
Cubit mesh generation code, 132
fragmentation of brittle disk resulting from impact, 131–133
quasi-static simulation of tensile test, 133–134
Gauss quadrature scheme, 110
goal of peridynamic simulation, 110
internal force density, evaluation of, 112–114
algorithm, 113, 114
constitutive model routines, 112
linear peridynamic solid constitutive model, 113
tracking nodal positions, 112
meshfree discretizations for peridynamics, 122–123
mesh generation tools, 122
nodal volumes, 122
options, 122
prototype microelastic brittle material, 123
meshless approach, 112
modeling contact, 119–122
algorithm, 120
Peridigm code, 119
self-contact, 121
short-range force model, 121
nodal volume, 111
nomenclature, 111
Peridigm code, 112, 117
progression of simulation, 111
proximity search for identification of pairwise interactions, 124–125
neighbor list, 124
Peridigm code, 124
Recursive Coordinate Bisection algorithm, 124
serial proximity search, 125
stand-alone software, 111
tangent stiffness matrix, 115–118
definition of, 116
finite difference schemes, 117
linear peridynamic constitutive models, 116
modulus state, 116
Peridigm code, 117
time integration, 125–131
critical time step, 127
explicit time integration, 125
implicit time integration for quasi-statics, 129–131
linear peridynamic solid, 128
maximum stable time step, estimation of, 128–129
nodal accelerations, 127
prototype microelastic brittle material, 128
reaction forces, 130
time-stepping process, 127
transient dynamics, explicit time integration for, 126–128
SPH, *see* Smoothed Particle Hydrodynamics
Spring constant, 30
Stress intensity factor (SIF), 6
Surface correction (SC), 386

T

Tafel law, 443
Tangential cohesive law, PD-based, 329–333
bonds have exceeded the critical stretch (case), 331–333
failure process, 331
no bonds have reached critical stretch (case), 331
shear traction, 330
"steeper" failure process, 333
"upper dome" of the horizon, 333
Tangent stiffness matrix (software implementation), 115–118
Timeshenko beam theory, 172

U

Unit-state, 150
Upscaling molecular dynamics to peridynamics, 184–194
  higher-order gradient partial differential equation, 185, 187
  inner expansion technique, 185
  isotropic deformation, 186, 190
  linear peridynamic solid model, 191
  one-dimensional nonlocal linear springs model, 184–189
  "safety factor," 188
  three-dimensional embedded-atom model, 189–194
  upscaling process, 189
  wave propagation, simulations of, 188, 192
Upscaling of unstable mesoscopic dynamics, dynamic brittle fracture as, 229–244
  bounds on kinetic and potential energy for mesoscopic evolution (theorem), 233–234
  continuous mesoscopic evolution in mean square norm (theorem), 234
  "Dirichlet" boundary conditions, 232
  dynamic instability and fracture initiation, 238
  energy balance (theorem), 234
  fracture nucleation condition, 239
  free crack propagation in small horizon-macroscopic limit, 241–242
    hypotheses, 241
    neutrally stable neighborhoods, 241
    wave equation, 242
  inflection point, 231
  Linear Elastic Fracture Mechanics, 232
  localization of dynamic instability in small horizon-macroscopic limit, 240–241
    neighborhood, 240
    nucleation sites, 241
    unstable subset, 240
  macroscopic evolution of brittle fracture as small horizon limit of mesoscopic dynamics, 234–238
    bottom-up approach, 235
    convergence of mesoscopic dynamics to macroscopic dynamics (theorem), 236–237
    energy inequality (theorem) 237–238
    Hausdorff measure, 237
    jump discontinuity, 238
    macroscopic dynamics has bounded LEFM energy (theorem), 237
  mesoscopic model, 230
  Mumford–Shah functional, 243
  theorem, 233, 234

V

Variations, nonlocal calculus of, *see* Nonlocal calculus of variations and well-posedness of peridynamics
Voronoi structure, 254

W

Wave propagation, simulations of, 188, 192
Well-posedness, *see* Nonlocal calculus of variations and well-posedness of peridynamics

Y

Young's modulus
  bond-based peridynamic model, 258
  center crack, bond-based verification of, 218

copper strip, 524
coupling model, 413
crack branching, 262
definition of, 12
force-based blending schemes, 425
homalite, 275
material modeling, 29
morphing method, 419
soda-lime glass, 265, 384
state-based peridynamics, 202
tensile bar, 133, 134

Z

Zero-energy deformation modes, 165
Zero-overpotential diffusivity (corrosion), 457
Zoltan software package, 124
"Z" type failure mode (laminates), 353

For Product Safety Concerns and Information please contact our EU representative GPSR@taylorandfrancis.com Taylor & Francis Verlag GmbH, Kaufingerstraße 24, 80331 München, Germany